Proceedings

D1806708

Ein stetig steigender Fundus an Informationen ist heute notwendig, um die immer komplexer werdende Technik heutiger Kraftfahrzeuge zu verstehen. Funktionen, Arbeitsweise, Komponenten und Systeme entwickeln sich rasant. In immer schnelleren Zyklen verbreitet sich aktuelles Wissen gerade in Konferenzen, Tagungen und Symposien in die Fachwelt. Den raschen Zugriff auf diese Informationen bietet diese Reihe Proceedings, die sich zur Aufgabe gestellt hat, das zum Verständnis topaktueller Technik rund um das Automobil erforderliche spezielle Wissen in der Systematik aus Konferenzen und Tagungen zusammen zu stellen und als Buch in Springer.com wie auch elektronisch in SpringerLink und Springer für Professionals bereit zu stellen.

Die Reihe wendet sich an Fahrzeug- und Motoreningenieure sowie Studierende, die aktuelles Fachwissen im Zusammenhang mit Fragestellungen ihres Arbeitsfeldes suchen. Professoren und Dozenten an Universitäten und Hochschulen mit Schwerpunkt Kraftfahrzeug- und Motorentechnik finden hier die Zusammenstellung von Veranstaltungen, die sie selber nicht besuchen konnten. Gutachtern, Forschern und Entwicklungsingenieuren in der Automobil- und Zulieferindustrie sowie Dienstleistern können die Proceedings wertvolle Antworten auf topaktuelle Fragen geben.

Michael Bargende · Hans-Christian Reuss ·
Jochen Wiedemann
Herausgeber

16. Internationales Stuttgarter Symposium

Automobil- und Motorentechnik

Band 1

 Springer Vieweg

Herausgeber

Prof. Dr.-Ing. Michael Bargende
Prof. Dr.-Ing. Hans-Christian Reuss
Prof. Dr.-Ing. Jochen Wiedemann
Forschungsinstitut für Kraftfahrwesen und Fahrzeugmotoren Stuttgart -FKFS-
Stuttgart, Deutschland

ISSN 2198-7432   ISSN 2198-7440 (electronic)
Proceedings
ISBN 978-3-658-13254-5   ISBN 978-3-658-13255-2 (eBook)
DOI 10.1007/978-3-658-13255-2

Die Deutsche Nationalbibliothek verzeichnet diese Publikation in der Deutschen Natio-
nalbibliografie; detaillierte bibliografische Daten sind im Internet über http://dnb.d-nb.de
abrufbar.

Umschlagbild: © [M] Peugeot

Gedruckt auf säurefreiem und chlorfrei gebleichtem Papier.

Springer Vieweg ist Teil von Springer Nature
Die eingetragene Gesellschaft ist Springer Fachmedien Wiesbaden GmbH

WELCOME IN STUTTGART

What is it that has elevated the Stuttgart International Symposium – hosted by the Research Institute of Automotive Engineering and Vehicle Engines Stuttgart (FKFS) – to a real can't-miss event for years now? Is it the broad range of current and important topics that line the agenda? The expert presentations guiding the way for the future? Or is it because the event takes place in Stuttgart, the heart of the automotive state, Baden-Wuerttemberg, and is attended by prominent participants from both the business and research sectors? The answer is: All of the above!

For the 16th installment of the Conference, I am glad to be assuming patronage for the event and welcome all the participants and speakers.

This year's focus topic, "Global products versus non-global requirements", addresses the challenges facing both the economy in general and the automobile industry in particular: What are the consequences resulting from the contrast between global sales and adaptation to country-specific norms? How can these be overcome?

The Transatlantic Trade and Investment Partnership (TTIP) that is currently being negotiated between political decision makers in the US and the EU is intended to resolve exactly these issues, such as this conflict of interests in the automotive sector. One of the key intentions of this agreement is to standardize the respective national regulations and requirements for products sold globally on both sides of the Atlantic – such as: side mirrors, blinkers and crash tests – and introduce a unified, high standard of safety and security. This will lead to the removal of expensive duplicate standards, which, despite nearly identical safety norms, have up to now required redundant development and certification processes. Consumers ultimately stand to benefit once this burden is lifted. Such a development will likewise be significant for the automotive industry in Baden-Wuerttemberg – and for the associated machinery, metal and electronic industries – in that Germany, as an export country, will be able to set the standards according to which our worldwide trading partners will have to adapt – not the other way around! Our high level of precision, reliability, quality and safety together present a strong sales argument. As such, these high standards should not be weakened in the course of the TTIP negotiations. The Government of Baden-Wuerttemberg fundamentally supports the TTIP agreement, but also insists that the high safety standards associated with products "made in Baden-Wuerttemberg" are not lowered in any way.

I would like to thank all those who have been involved in organizing and executing this event. I hope that all the attendees will benefit from the exciting discussions, interesting presentations and also gain new ideas for their own work.

Winfried Kretschmann
Prime Minister of the State of Baden-Wuerttemberg

A WARM WELCOME

Demands on the automotive industry in terms of research and development are constantly in flux. Producers and suppliers are forced to create global solutions while also considering individual customer needs as well as the legislative requirements in each market. Even emissions regulations are anything but unified globally. Starting in September 2017, Europe plans to implement a measurement termed "real-driving emissions" (RDE). Evaluating pollutant emissions will be done on the street rather than at a testing station, with far-reaching consequences for engine development. Many regions around the globe are, at the same time, calling for localized zones with emissions-free traffic. This all overlaps with the ongoing process of reducing CO_2 limits for vehicle fleets, with all world regions having defined various steps to achieve reductions in this area. This is topped by rising demands in terms of the comfort and emotionality of cars. How will the automotive industry respond to the growing conflict between increasing globalization and maximized global product marketing, on the one hand, and diverse, particular regional requirements for vehicles, on the other? What are the technical impacts of this? Industry and research experts will report on and discuss these issues and many more at the

16th Stuttgart International Symposium for "Automotive and Engine Technology" on 15 – 16 March 2016.

In six parallel sessions with over one hundred presentations, leading experts will address the current state of technological development, their most recent research findings, and concepts for the future. The program spans the entire process of vehicle creation, from research and development through to production. This year, we have also been able to invite excellent keynote speakers and participants for the subsequent podium discussion. This will be supplemented by numerous opportunities for exchanging ideas, be it in relation to technical discussions, at the accompanying trade exhibition, or in the social context of breaks and the evening event.

We look forward to seeing you in Stuttgart, the birthplace of the automobile!

Prof. Dr.-Ing. Michael Bargende
Prof. Dr.-Ing. Hans-Christian Reuss
Prof. Dr.-Ing. Jochen Wiedemann

INDEX – Volume 1

SECTION 2

ELECTRIC POWERTRAIN
Chairperson: Prof. Dr. Nejila Parspour

TIRES
Chairperson: Prof. Dr. Thomas Vietor

SOFTWARE AND DEVELOPMENT
Chairperson: Prof. Dr. Tobias Flämig-Vetter

ARENA2036
Chairperson: Peter Froeschle

STEERING
Chairperson: Prof. Dr. Clemens Gühmann

ERGONOMICS IN AUTOMOTIVE PRODUCTION
Chairperson: Urban Daub

SPEAKERS, CHAIRPERSONS – Volume 1

Adam Babik
Robert Bosch GmbH

Prof. Carla Bailo
The Ohio State University

Prof. Dr. Michael Bargende
FKFS/IVK, Universität Stuttgart

Katharina Bause
IPEK/Karlsruher Institut für Technologie (KIT)

Dr. Bernd-Dietmar Becker
FARO

Lukas Behr
Robert Bosch Battery Systems GmbH

Prof. Dr. Christian Beidl
TU Darmstadt

Florian Blab
Fraunhofer IPA

Prof. Dr. Stefan Böttinger
Universität Hohenheim

Tillmann Braun
Daimler AG

Dr. Bernhard Budaker
Fraunhofer IPA

Rene Budich
HTW Dresden

Clemens Buschhoff
Fraunhofer IPT

Jing Cheng
Universität Stuttgart

Urban Daub
Fraunhofer IPA

Ronnie Dessort
TESIS DYNAware GmbH

Prof. Dr. Klaus Dietmayer
Universität Ulm

Prof. Dr. Lutz Eckstein
RWTH Aachen University

Prof. Dr. Helmut Eichlseder
TU Graz

Prof. Dr. Peter Eilts
TU Braunschweig

Tobias Engelhardt
Dr. Ing. h.c. F. Porsche AG

Stefan Epple
IKT, Universität Stuttgart

Marius Feilhauer
ETAS GmbH

Dr. Holger Fink
Robert Bosch Battery Systems GmbH and Lithium Energy and Power GmbH

Prof. Dr. Tobias Flämig-Vetter
Duale Hochschule BW Stuttgart

Dr. Günter Fraidl
AVL LIST GMBH

Arthur Frick
Daimler AG

Prof. Dr. Horst E. Friedrich
DLR Institut für Fahrzeugkonzepte

Peter Froeschle
ARENA2036 e.V.

Prof. Agostino Gambarotta
University of Parma

Prof. Dr. Frank Gauterin
Karlsruher Institut für Technologie (KIT)

Prof. Dr. Bernhard Geringer
TU Wien

Jan Gerstenberg
Bosch Engineering GmbH

Vivan Govender
Daimler AG

Prof. Dr. Clemens Gühmann
TU Berlin

Andreas Haag
Promotionskolleg Hybrid

Andreas Hackl
TU Graz

Prof. Dr. Horst Harndorf
Universität Rostock

Frank Hermsdorf
Technische Universität Dresden

Prof. Dr. Dr. Gerhard Hettich
EAST Consulting

Tilmann Hilbert
J. Schmalz GmbH

Detlef Hoffmann
SGS Germany GmbH

Prof. em. Dr. Günter Hohenberg
IVD Prof. Hohenberg GmbH

Stefan Jetter
Daimler AG

Dr. Stefan Junker
Robert Bosch GmbH

Benjamin Kaal
FKFS

Dr. Stefan Kampmann
Robert Bosch GmbH

Dan Keilhoff
Universität Stuttgart

Mahir Tim Keskin
IVK, Universität Stuttgart

Matthias Klingbeil
Dr. Ing. h.c. F. Porsche AG

Prof. Dr. Thomas Koch
Karlsruher Institut für Technologie (KIT)

Dr. Karl Kollmann

Silke Krebs
Staatsministerium Baden-Württemberg

Prof. Dr. Karl-Ludwig Krieger
Universität Bremen

Prof. Dr. Ferit Küçükay
TU Braunschweig

Thorsten Lajewski
Daimler AG

Thomas Landwehr
IVK, Universität Stuttgart

Christian Lersch
Velamed GmbH

Yujun Liao
Empa – Swiss Federal Laboratories for
Materials Science and Technology

Mike Liebers
TU Dresden

Dr. René Linssen
Daimler AG

Prof. Dr. Thomas Maier
Universität Stuttgart

Gian Mauro Mancia
Ce.S.I. Centro Studi Industriali Srl

Zdenek Mestenhauser
MTS Systems Corporation

Prof. Dr. Peter Middendorf
IFB, Universität Stuttgart

Dr. Nebojsa Milovanovic
MAHLE Powertrain

Marco Münster
DLR Institut für Fahrzeugkonzepte

Dirk Naber
Robert Bosch GmbH

Walter Nagler
ZF Friedrichshafen AG

Dr. Hiroshi Nakamura
HORIBA Ltd.

Dr. Harald Naunheimer
ZF Friedrichhafen AG

Dirk Neumann
IAV GmbH

Prof. Karl-Ernst Noreikat
NorCon

Markus Orner
FKFS

Helena Ortwein
Rennteam Uni Stuttgart e.V.

Prof. Dr. Nejila Parspour
Universität Stuttgart

Dr. Jörg Paschedag
ITK Engineering AG

Raphael Pfeil
FKFS

Vincent Raimbault
MANN+HUMMEL France SAS

Prof. Dr. Dr. Wolfram Ressel
Universität Stuttgart

Prof. Dr. Hans-Christian Reuss
FKFS/IVK, Universität Stuttgart

Christian Riese
Robert Bosch GmbH

Prof. Dr. Hermann Rottengruber
OvGU Magdeburg

Prof. Dr. Eric Sax
Karlsruher Institut für Technologie (KIT)

Markus Schäfer
Daimler AG

Tanja Schembera-Kneifel
AUDI AG

Dr. Axel Schloßer
FEV GmbH

Helge Schmidt
TÜV NORD Mobilität GmbH & Co. KG

Patrick Schmidt
LBST – Ludwig-Bölkow-Systemtechnik GmbH

Dr. Marco Schneider
Fraunhofer IPA

Dr. Simon Schneider
MAHLE International GmbH

Ulrich Schulmeister
Robert Bosch GmbH

Manuel Schuster
Fraunhofer IPA

Dr. Thomas Schütz
BMW AG

Dr. Michael Steiner
Dr. Ing. h.c. F. Porsche AG

Prof. Dr. Thomas Vietor
TU Braunschweig

Matthias Vogt
bridgingIT GmbH

Prof. Dr. Georg Wachtmeister
TU München

Prof. Dr. Karl-Heinz Wehking
IFT, Universität Stuttgart

Ulrike Weinrich
FKFS

Prof. Dr. Jochen Wiedemann
FKFS/IVK, Universität Stuttgart

Johannes Winterhagen
Redaktionsbüro delta eta

Prof. Dr. Oliver Zirn
Hochschule Esslingen

WLTP – On the increased importance of aerodynamics and impact on development procedures

Thomas Schütz, BMW Group

Abstract

For some years a new test procedure has been developed among the UN for the determination of pollutant and CO_2 emissions and fuel consumption, which should represent the average customer behaviour. Beside the EU various other countries are involved (e.g., India, Japan). The test procedure WLTP and the underlying driving cycle WLTC was developed with the help of worldwide accumulated driving data and covers driving situations from city up to highway traffic.

Beside the driving cycle also the consideration of optional extra equipment for the CO_2 and consumption determination is a reason for the increasing work load on aerodynamics. Compared to today the influence of the aerodynamic drag on the whole consumption will increase with the introduction of WLTP. In addition, this generally increases the development and test expenditures of the OEMs – not only in aerodynamic development. Decisive for the future expenditure increase in the aerodynamics development due to the type approval procedure of all vehicle variations will be whether these are measured in coast-down tests like today, or whether suitable CO_2 determination methods will be used which enclose wind tunnel measurements and CFD.

1 Introduction

In the late 1960s the first exhaust gas norm was decided in Germany, which led to the European test procedure written in the directive 70/220 EWG in March 1970 [3]. Among other things, the first driving cycle regulation was already included here. This is a velocity-time function the vehicle must follow within the scope of the type approval. This directive was steadily developed during the subsequent years. In 1992 the new European driving cycle (NEDC) was defined and since 1997 the fuel consumption must be calculated from the amount of exhaust gas, which is determined in the driving cycle of the exhaust gas norm. Both are still valid today. The driving cycle lasts a total of 1180 seconds, so just 20 minutes. The city cycle lasts two thirds of this time and the extra-urban cycle one third. In June 2007 the abolition of the directive 70/220/EWG was decided for January 2013, and it was recommended to critically check the driving cycle [4].

The fuel consumption of passenger cars is usually part of the published technical data with every vehicle. In the past, it as been observed that the consumption values in the manufacturer's data deviates considerably from the customers experience. With larger deviations, increasing dissatisfaction will be the result. Due to this the type approval authorities became attentive. Subsequently three aspects were elaborated within the scope of the type approval procedure, which lead to this difference:

2

- Low velocities during the driving cycle
- Almost no unsteady acceleration and
- No consideration of optional extra equipment

Especially the last point leads to the fact that, on the one hand, every customer vehicle differs from the other and, on the other hand, almost every delivered vehicle differs from the so-called report model. However, with the help of the report model the CO_2 emissions are determined. This is often an unattractive and minimally equipped or even not buildable vehicle. Here, inconsistent aerodynamics- and weight-optimum equipment are tied together to be able to communicate the best possible consumption value. However, considering the example of light weight and aerodynamically optimized rims it becomes clear that this doesn´t necessarily always agree.

As a result of this a study was published in 2013 by T&E, which shows clearly how the car manufacturers have (miss)used this flexibility [6], which leads to a consumption difference of 10 - 20%. Some examples are:

- No reload of the vehicle battery during the cycle
- Taped exterior gaps
- Change of track and camber angle settings of the wheels
- Increased tyre pressure
- Use of the minimum vehicle weight
- Deduction of 4 % tolerance on the measured value
- Avoidance of trailing brakes
- Adaptation of the engine control

After that a new type approval method was discussed and developed at UN level which should counteract against the mentioned shortcomings from a customers point of view. The final version was written down in the end of 2015.

2 European Type Approval Today

Today's type approval runs according to the regulations in the EU directive 70/220/EWG and, amongst other things, contains the definition of the new European driving cycle (NEDC). The procedure is described in Figure 1. By the way, it will be the same after the introduction of WLTP.

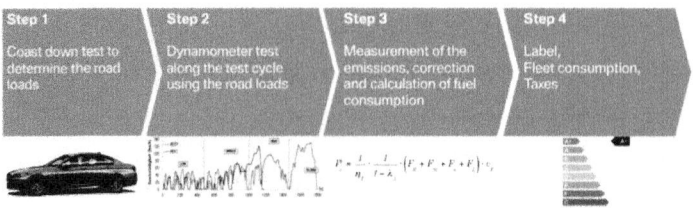

Figure 1: Type Approval Procedure

In the first step a coast down test is carried out with the examining vehicle to be able to calculate the road loads from the speed time behaviour. In the second step the vehicle is put on a dynamometer test bench and follows the driving cycle. The dynamometer acts as a power brake and loads the vehicle with the power of the road loads according to the velocities in the driving cycle. The emissions contained in the exhaust gas are measured and the fuel consumption can be calculated (step 3). The last step uses this data to determine the CO_2 label, the tax amounts and the fleet consumption.

For some years now there have been legal limits in terms of the fleet consumption for almost every market which can be more or less fulfilled by the vehicle manufacturers. During the next years these legal limits will be tightened, step by step. This development is shown in Figure 2. For comparison: In 2013 the BMW fleet consumption amounted to 134.4 g/km CO_2 [5].

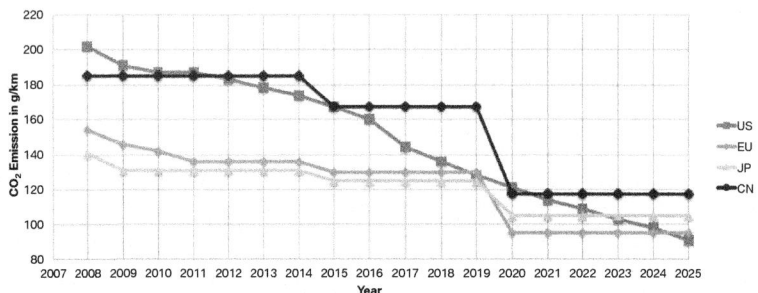

Figure 2: Legal CO_2 limits in different markets

Probably from 2018 onwards the EU directive 70/220/EWG will be replaced with the new type approval procedure WLTP. The legal targets will remain untouched. However, according to today's status (December, 2015) a specific correction for each manufacturer should occur which takes into account the fact that in the WLTP an identical vehicle fleet generates higher consumption values than in 70/220/EWG. Figure 3 shows an overview to today's worldwide regulations, valid in the different markets. Not all markets will use the WLTP according to today's status (December, 2015). The EU introduces the WLTP very probably in 2018 and Japan probably in 2020. In China the introduction is uncertain. The USA won't introduce WLTP for the moment.

The original political attempt was to evaluate the worst vehicle variation of a type only within the scope of the WLTP. With this approach the manufacturers wouldn't in future have been able to fulfil the demanding CO_2 targets. Therefore, the automotive industry also achieved the permission to publish a vehicle specific CO_2 value. Beside the new driving cycle this is the second major change in WLTP compared to 70/220/EWG.

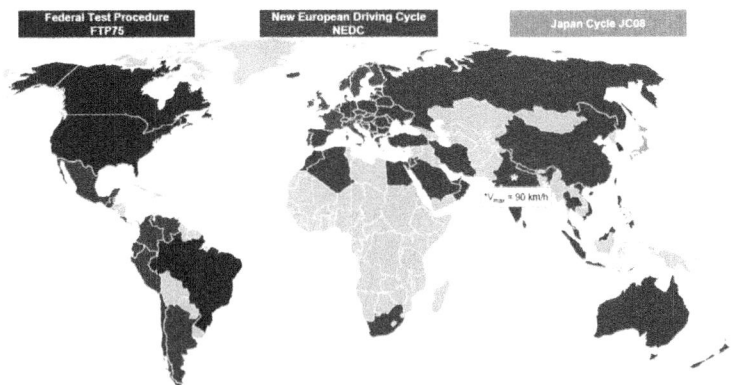

Figure 3: Driving cycles worldwide

3 Influence of Aerodynamics on CO_2 Emissions

The impact of aerodynamics on the power requirement of automobiles has often been discussed, for example in [8] und [9]. The non-engine road loads on a level track are air drag F_D, rolling resistance F_R and acceleration resistance F_a. Using the drag coefficient c_D, the frontal area A_x, the air density ρ_a, the speed v_V, rolling resistance coefficient f_r, the vehicle mass m_V, the lift force F_L, the translational acceleration a and the mass factor e to account for rotating masses we get

$$F_D = c_D \cdot A_x \cdot \frac{\rho_a}{2} v_v^2$$

(Gl. 1)

$$F_R = \sum_i f_{R,i} \cdot F_{N,i} = f_R \cdot \left(m_V \cdot g - F_L \right)$$

(Gl. 2)

$$F_a = m_V \cdot a + \sum_i J_i \cdot \dot{\omega}_i = e \cdot m_V \cdot a$$

.

(Gl. 3)

By multiplying with the driving speed the road load power on level ground is:

$$P_{RL} = \left(F_R + F_a + F_D \right) \cdot v_V$$

.

(Gl. 4)

Obviously beside the already mentioned vehicle parameters the velocity is important. In the case of the type approval, the velocity is specified within the driving cycle. Figure 4 shows the driving cycles of 70/220/EWG and the WLTP in comparison.

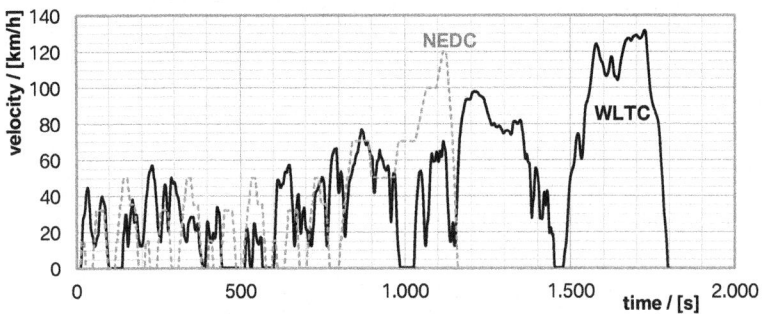

Figure 4: New European Driving Cycle (NEDC) and Worldwide harmonized Light Vehicles Test Cycle (WLTC)

The most important characteristics can be taken from Chart 1. In comparison to NEDC (described above) the WLTC was developed with the help of worldwide accumulated driving data. It lasts 30 minutes instead of 20 minutes, shows only 13% of stop time (formerly 25%), 1.5 m/s² maximum acceleration (1.0 m/s² in the NEDC) and higher average (46.5 km/h instead of 34 km/h) and maximum speeds (131 km/h instead of 120 km/h). All of this is clearly closer to the real customer driving behaviour.

Chart 1: Important Properties of both Cycles NEDC and WLTC

	WLTC	NEDC
Cycle time	30 min	20 min
Idle	13 %	25 %
Cycle length	23,3 km	11 km
$v_{average}$	46.5 km/h	34 km/h
v_{max}	131 km/h	120 km/h
a_{max}	1.5 m/s²	1 m/s²

The road load power (Gl. 4) can be calculated using the driving cycles shown in Figure 4 in order to indicate the importance of the aerodynamic drag for the CO_2 emissions. The vehicle properties shown in Chart 2 shall be applied. Lift force, rolling resistance and mass factor remain unchanged over all configurations.

Chart 2: Vehicle properties used to compute the aerodynamic impact on CO_2 emissions

	Micro Car	Compact Car	Luxury Class
Air Drag c_D	0.33	0.31	0.24
Frontal Area A_x	1.8 m²	2.0 m²	2.4 m²
Vehicle Mass m_F	900 kg	1200 kg	1700 kg

Figure 5 shows the results. For NEDC the aerodynamic contribution to the complete energy cycle is 38.0% for the micro car, 32.5 % for the compact car and 24.0 % for the luxury car. Applying WLTC these values increase due to the higher velocities up to 42.2 %, 36.4 % and 27.3 %. The proportion of acceleration also increases because of the increased unsteadiness and the influence of the rolling resistance is reduced. Now it is assumed for WLTP, that these vehicles shall win back 50% of the acceleration energy by recuperation, so that the impact due to aerodynamics increases further, up to 50.7%, 44.6% and 34.5%. The proportion of rolling resistance also rises and the acceleration proportion decreases. In terms of WLTP one can derive from this that the aerodynamics will have a larger influence compared to the other road loads. If electrified vehicles are considered, this influence will further increase.

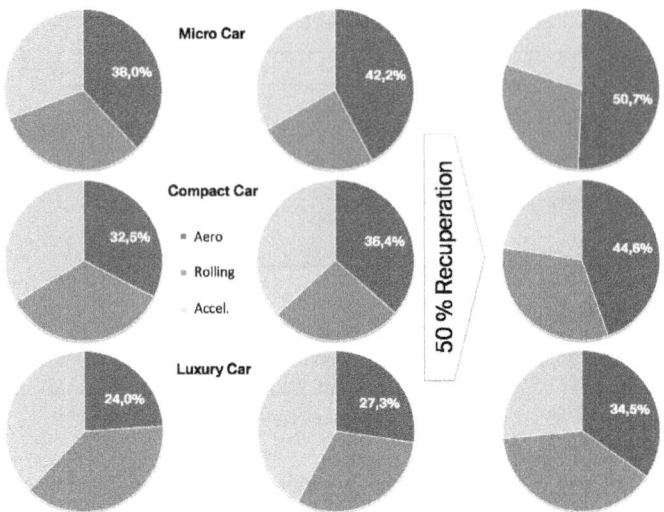

Figure 5: Aerodynamic proportion of total road load using NEDC (left), WLTC (middle) and WLTC and 50 % recuperation (right) for three vehicle types

The values shown in Figure 6 result if the average cycle energy per second is calculated for these three example vehicles in both cycles. Because of the higher velocities and the increased unsteadiness the absolute values lie higher in the case of WLTC compared to NEDC. In terms of absolute fuel consumption an increased importance of the aerodynamic drag also arises.

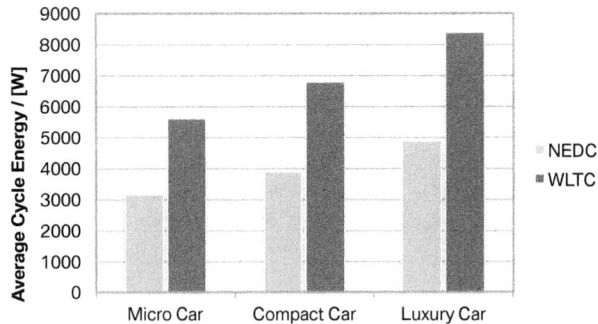

Figure 6: Average cycle energy of three vehicle types using NEDC and WLTC

To answer the question on the expenditure increase caused by WLTP, due to the consideration of optional extra equipment, a look on the web pages of the OEMs can help. With the example of the BMW web page and the current 3 series (year 2015) one can see how many optional extra equipment items are available. However, from an aerodynamic point of view it is not enough to evaluate the items separately as they may influence each other. Rather in the worst case, all conceivable configurations of optional extra equipment must be considered. In this manner Chart 3 derives 1.890 aerodynamically relevant permutations. Among these about 500 are buildable.

Chart 3: Derivation of aerodynamically relevant permutations of optional extra equipment, [2]

Equipment (w/o. colours)	Number	Aerodynamically relevant
Drive Trains	41	5 (Road-load family, cooling packages)
Rims	21	21
Tyres	3	3
Further Extra Equipment	$88 - 14^1 = 74$	6
Permutations	$41 \cdot 21 \cdot 3 \cdot 74 = \mathbf{191,142}$	$5 \cdot 21 \cdot 3 \cdot 6 = \mathbf{1,890}$
Final Configurations	**Ca. 40,000**	**Ca. 500**

As an example of optional extra equipment with a particularly large effect on aerodynamics, some of the rims offered for a mid class vehicle should be discussed. Figure 7 shows the bandwidth of wheel dimensions of 16 to 19 inches for which a drag value difference can be observed of more than 30 counts. A CO_2 range of about 3 g/km CO_2 follows for an evaluation using the WLTP. The reason for the worsening of the drag is on the one hand the larger width of the tyres with rising diameter and also the more openness and sportier design of the rims, [8, [10].

---

[1] Without any relevance for the road loads

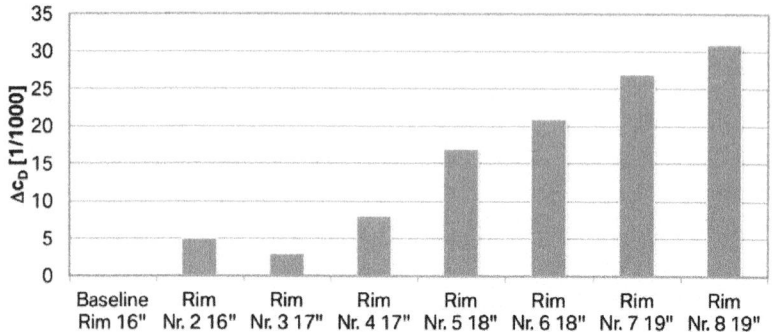

Figure 7: Wheels – Example for the influence of extra equipment on the aerodynamic drag

4 Aerodynamic Aspects within the new Regulations

The regularities within the scope of the WLTP are written down in the general technical requirements (GTR) US/ECE/WP29. They contain all necessary defaults that define how the type approval must proceed. In particular the driving cycle is defined here. Chart 4 shows the contents of the 1st arrangement level. Below the subjects especially relevant for aerodynamics (besides the cycle) are discussed.

In order to reduce the necessary number of the vehicle assessments within the scope of the type approval, certain vehicle families were defined in the GTR. The non-engine road loads are almost independent of the engine. Hence, the road loads of an equipment variation can be determined once for all offered engines. All these vehicles of different motorizations establish a road load family. It must be said that also the car body shape (estate car, coupé, convertible) is treated like an optional equipment within a road load family. The CO_2 emissions depend on the engine properties and the non-engine road loads. The vehicles of the road load family are separated according to engine types and constitute a CO_2 family like shown in Figure 8.

Chart 4: Structure of the WLTP GTR, [7]

Main part	Definitions	Vehicle families, etc.
Annex 1	Cycles	"WLTC", Downscaling
Annex 2	Shift points	Procedure to manual transmission
Annex 3	Fuels	Reference fuels
Annex 4	Road loads	Road load determination, dynamometer setting
Annex 5	Test bench	Test equipment and calibration
Annex 6	Test procedure	Test procedures and test conditions
Annex 7	Calculations	Emissions, particles, combined approach, …
Annex 8	Electrified Vehicles	Pure electric, hybrid electric, fuel cell hybrid vehicles
Annex 9	Equivalency	Determination of method equivalency

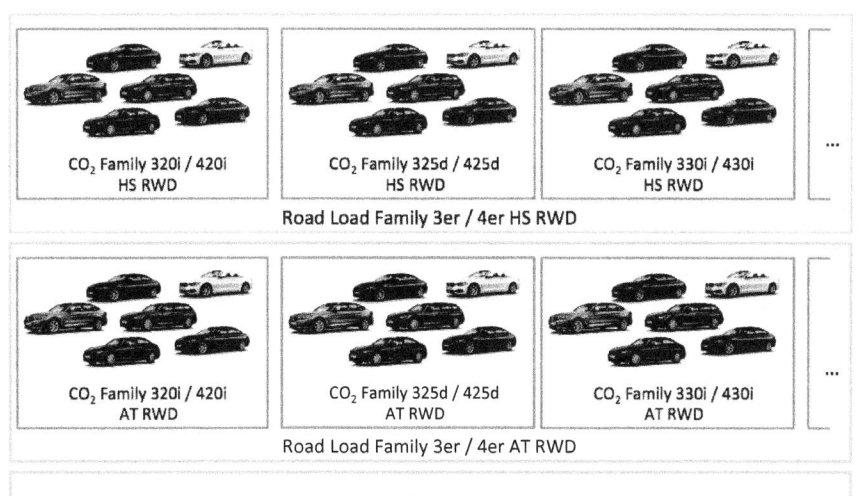

Figure 8: Definition of CO_2 and road load families

The advantage of the family definition arises from the reduction of the necessary coast down tests. On the dynamometer test bench the driving cycles including all road loads of the members of the family can be achieved with one single physical vehicle of the respective CO_2 family. This clearly reduces the required number of physical vehicles. Afterwards, within a coast down procedure all variations independent of the engine must be considered, namely for one single motorization within the road load family.

During the development of the GTR, the car manufacturers have worked towards the fact that when required the coast down test can be substituted by suitable procedures to determine the road loads. Reasons for this are:

– Coast down tests are costly and expensive

– The variation variety within the scope of the WLTP requires a lot of physical vehicles

– The availability of coast down tracks is limited.

The first method is called "combined approach" and is designed to calculate the road loads for the heads of a family, in other words to calculate the best and the worst case vehicle, from wind tunnel (air drag), flat belt or dynamometer test bench and balance. In the GTR this method is called wind tunnel method. These two vehicles define the road load family. For this method the wind tunnel criteria is held strict because absolute drag values must be determined properly. The wind tunnel value should correspond to the street value and yield a reproduction of $c_D A_x = 0,015$. Wind tunnel corrections are allowed.

Then the second method allows the road load difference of any vehicle configuration to be determined between the heads, either by wind tunnel, dynamometer and balance measurement or by any other appropriate procedures (for example CFD simulation in terms of air drag). The wind tunnel criteria for this method is less demanding, because merely the correct measurement of the drag differences of different optional equipment is necessary, cf. Figure 9.

Movable aerodynamic body parts on the test vehicles shall operate during road load determination as intended during the on-road drive. Every vehicle system that dynamically modifies the vehicle's aerodynamic drag (e.g. vehicle height control) shall be considered to be a movable aerodynamic body part. Appropriate requirements shall be added if future vehicles are equipped with movable aerodynamic items of optional equipment whose influence on aerodynamic drag justifies the need for further requirements.

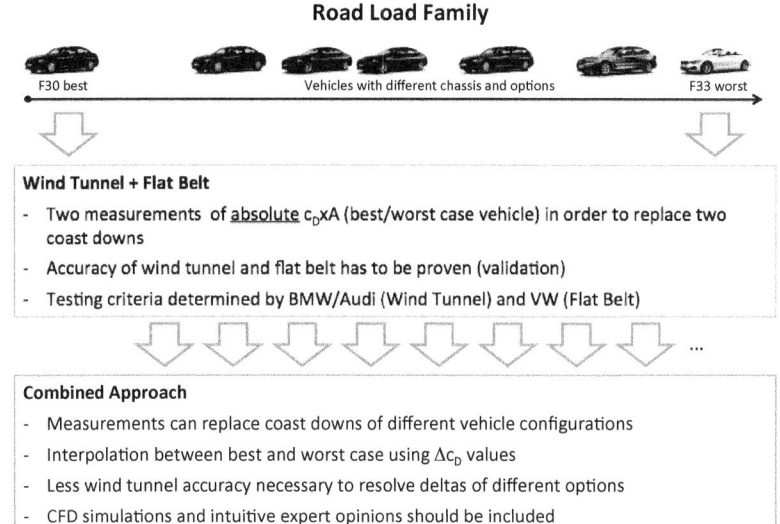

Figure 9: Coast down substituting methods within the WLTP GTR

For all wind tunnel tests moving ground is presumed between the left and right track and underneath the vehicle over the complete vehicle length. Moreover, the wheels must turn. The remaining wind tunnel criteria for both methods are shown in Chart 5.

Among them the pressure difference in front of and behind the vehicle and the nozzle blockage (ratio between nozzle cross section and vehicle frontal area) are especially demanding criteria. As shown in Figure 10 most European wind tunnels lie within the demanded pressure range of $\Delta c_p = 0,02$. Also with regard to the test section length all relevant wind tunnels in Europe fulfil the demand of at least 6 m. With usual vehicle frontal areas between 1.8 m² and 3 m² no exclusion reasons arise for most wind tunnels, as can be seen in Chart 6.

Chart 5: Wind tunnel criteria for both methods

Criterion	Combined Approach GTR WLTP Annex 4-3.2	Wind tunnel method GTR WLTP Annex 4-6
Wind speed	>140 ± 2 km/h	
Air temperature	±3 °C	
Degree of turbulence at nozzle exit	< 1 %	
Nozzle blockage	< 35 %	< 25 %
Wheel velocity rel. to v_{Wind}	±3 km/h	
Belt velocity rel. to v_{Wind}	±3 km/h	
Horizontal jet deflection at nozzle exit	< 1°	
Total pressure deviation at nozzle exit (dimensionless)	< 0.02	< 0.01
Pressure difference 3 m in front and behind turntable centre (dimensionless)	< 0.02	
Boundary layer thickness (δ_{99}) at turntable centre	< 30 mm	
Restraint blockage rel. to vehicle frontal area	< 10 %	< 3 %
Accuracy of the balance (resolution and repeatability)	< 5 N, ± 3 N	
Position of the vehicle along y = 0		< ±10 mm
WRU surface area rel. to tire patch area		> +20 %
Yaw angle deviation		< 0.1°
Measuring time		> 60 s; > 5 Hz

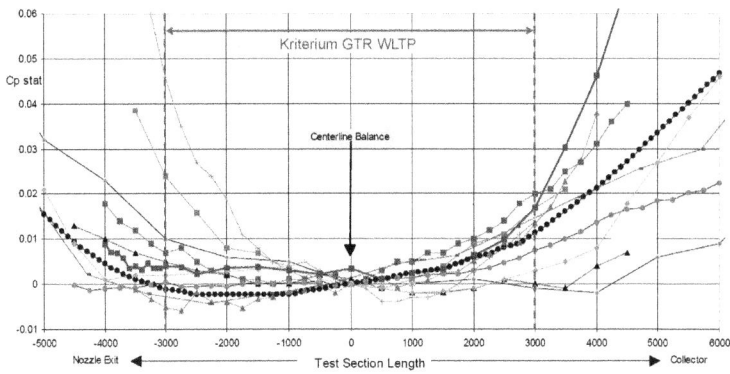

Figure 10: Pressure distribution over test section length of some European Wind Tunnels, [1]

Chart 6: Nozzle cross section and test section length of some European Wind Tunnels

Wind Tunnel	Nozzle cross section [m²]	Test section length [m]
DNW LLF 8x6	48,0	20,0
MIRA	34,9	15,2
Volkswagen	37,5	10,0
FIAT	30,5	10,5
FORD	20,0	9,7
IVK	22,5	9,9
BMW	25,0	22,0
Audi	11,0	9,5
Pininfarina	11,0	8,0
Volvo	27,0	15,8
IAT S10	15,0	10,2

15

5   Outlook

Now after the completion of work on the WLTP regulation it will become clear how the vehicle manufacturers will react to the changed conditions. This concerns the following points:

- Which strategy will be chosen in order to publish aerodynamic data? The manufacturers can use overall the worst value among all variations of a type, evaluate some single sensible intermediate steps or give individual data for every vehicle configuration.

- The latter possibility means a considerable increase in the development and type approval expenditure. This would increase the demand for OEM and supplier wind tunnels as well as clearly larger mainframe capacities for CFD.

- The combined approach within the GTR leaves open the opportunity for alternative air drag determination methods. The question is which possibilities the manufacturers will adapt.

- How will the vehicles of the future look, that have to be approved within the WLTP and fulfil the stricter CO_2 targets? Probably a noticeable reduction of the drag values will arise during the coming years across all vehicle classes. This was already a result of the oil crisis during 1970s. Vehicles with a drag coefficient of $c_D = 0.20$ and lower will be represented in the market.

Not least, the increased type approval criteria will also push aerodynamic research of the OEMs and at the universities because both the quality of today's developing methods as well as today's air drag optimization measures are noticeably at their limits.

6   References

1. EADE – European Aerodynamics Data Exchange. EADE Correlation Test 2010 – Final Report.

2. http://www.bmw.de; am 20. Januar 2016.

3. http://eur-lex.europa.eu/legal-content/de/ALL/?uri=CELEX:31970L0220; am 20. Januar 2016.

4. http://eur-lex.europa.eu/legal-content/de/ALL/?uri=CELEX:32007R0715; am 20. Januar 2016.

5. http://www.transportenvironment.org/publications/how-clean-are-europe%E2%80%99s-cars-2014-%E2%80%93-part-1; am 20. Januar 2016.

6. http://www.transportenvironment.org/publications/mind-gap-why-official-car-fuel-economy-figures-don%E2%80%99t-match-reality; am 20. Januar 2016.

7. https://www2.unece.org/wiki/pages/viewpage.action?pageId=2523179; am 20. Januar 2016.

8. Schütz, T. (Hrsg.).: Hucho – Aerodynamik des Automobils. 6. Auflage, Springer-Vieweg Verlag, 2013. ISBN 978-3-8348-1919-2.

9. Schütz, T.: Verbesserte Aerodynamik des Audi Q3 und Q5. Haus der Technik, 10. Tagung: Fahrzeug-Aerodynamik „Neue Chancen und Perspektiven für die Kraftfahrzeugaerodynamik durch CO2-Gesetzgebung und Energiewende", München, 2012.

10. Wittmeier, F.; Kuthada, T.: The Influence of Wheel and Tire Aerodynamics in WLTP. In: Pfeiffer, P. E. (Ed.): 6th International Munich Chassis Symposium 2015. Proceedings. Springer Verlag, 2015. ISBN 978-3-658-09711-0.

Test bench technologies for improving WLTP measurement results

Zdenek Mestenhauser, V. Senft

MTS Systems Corporation

This manuscript is not available according to publishing restriction. Thank you for your understanding.

On- and off-cycle energy efficiency technologies of a global supplier

Ulrich Schulmeister
Robert Bosch GmbH

Co-Authors:

Axel Lang, Martin Johannaber, Daniel Rieker
Robert Bosch GmbH

Florian Götz, Matthias Rauscher, Frank Schürg
Bosch Engineering GmbH

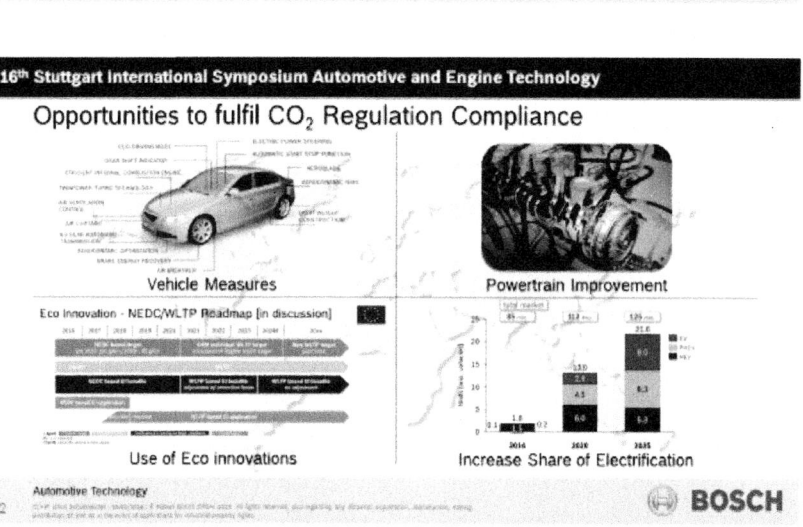

1 Vehicle Measures for Energy Saving

Vehicle Measures for Energy Saving

ECO DRIVING MODE

GEAR SHIFT INDICATOR

EFFICIENT INTERNAL COMBUSTION ENGINE

TWINPOWER TURBO TECHNOLOGY

AIR VENTILATION CONTROL

AIR CURTAIN

8-9 GEAR AUTOMATIC TRANSMISSION

AERODYNAMIC OPTIMIZATION

BRAKE ENERGY RECOVERY

AIR BREATHER

ELECTRIC POWER STEERING

AUTOMATIC START STOP FUNCTION

AEROBLADE

AERODYNAMIC RIMS

LOW ROLL RESISTANCE TIRES

LIGHT WEIGHT CONSTRUCTION

Automotive Technology

BOSCH

23

2 Powertrain Improvement

16th Stuttgart International Symposium Automotive and Engine Technology

Platform model for on- and off-cycle vehicle simulation

→ Real world energy efficiency optimization requires cross-domain simulation

→ Bosch platform environment integrates expert sub-models from all mobility sector business units

→ Toolchain is available for customer individual simulation services

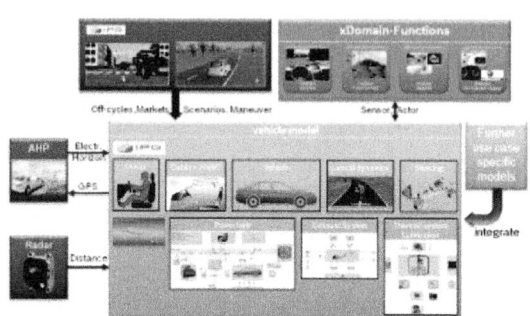

Automotive Technology

4

BOSCH

16th Stuttgart International Symposium Automotive and Engine Technology

Effect of Engine Technologies in NEDC and WLTC

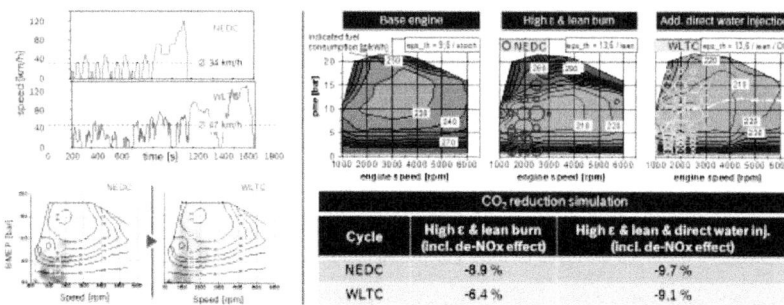

Cycle	High ε & lean burn (incl. de-NOx effect)	High ε & lean & direct water inj. (incl. de-NOx effect)
NEDC	-8.9 %	-9.7 %
WLTC	-6.4 %	-9.1 %

Engine efficiency improvement at medium and high loads beneficial for WLTC

Automotive Technology   Source: Bosch Corporate Research

5

BOSCH

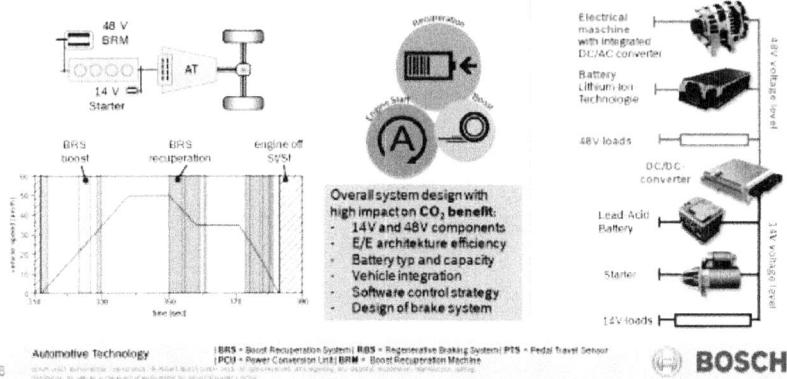

CO$_2$ potential of 48V system & corresponding brake system

Overall system design with high impact on CO$_2$ benefit:
- 14V and 48V components
- E/E architekture efficiency
- Battery typ and capacity
- Vehicle integration
- Software control strategy
- Design of brake system

Automotive Technology | BRS = Boost Recuperation System| RBS = Regenerative Braking System| PTS = Pedal Travel Sensor | PCU = Power Conversion Unit| BRM = Boost Recuperation Machine

BOSCH

Detailed Simulation Results BRS+RBS w/o, w/ PTS

Boundary conditions for simulation
- BRM with 12 kW maximal power
- Gear ratio BRM to ICE. 2.7.1
- 48V Li-Ion battery
- 3 kW DC/DC converter

Vehicle results.
- benefit depending on RBS strategy w/o or w/ PTS

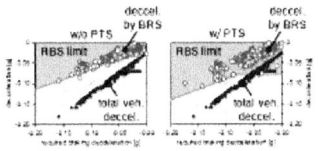

- BRS recuperation not only limited by RBS but also by system limits (BRM, battery, ... etc)

Reference: Large SUV

Automotive Technology | BRS = Boost Recuperation System| RBS = Regenerative Braking System| PTS = Pedal Travel Sensor | PCU = Power Conversion Unit| BRM = Boost Recuperation Machine

BOSCH

3 Eco Innovations

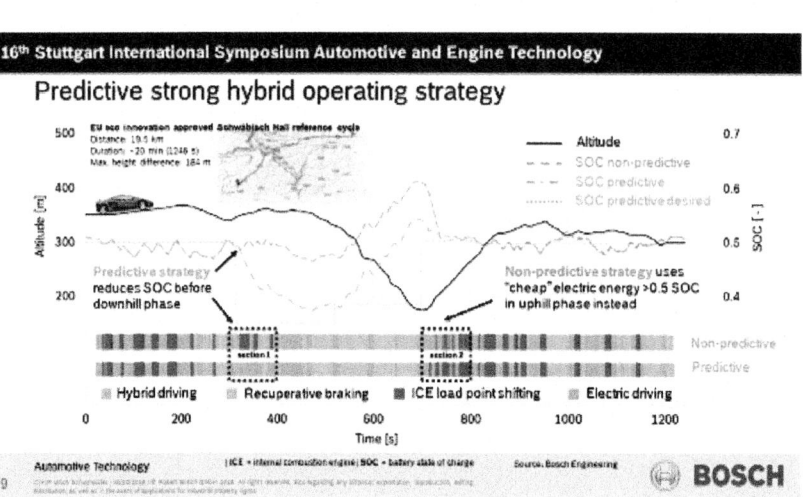

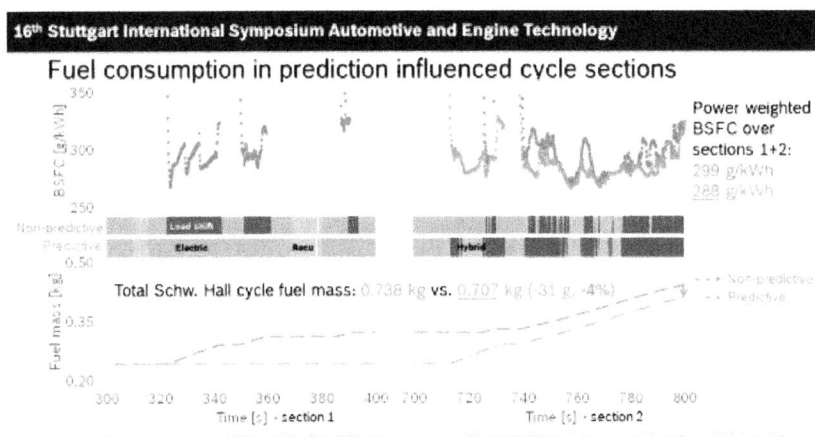

16th Stuttgart International Symposium Automotive and Engine Technology

Estimation of Real-Life CO$_2$ from Passenger Cars in the EU

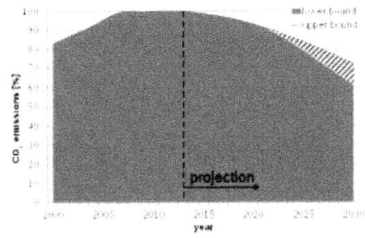

- Premise that future fleet targets will be met
- Peak of total driven distance in 2020, peak car between 2020 and 2030
- Average age of vehicle population 8.5 years
- Upper and lower bounds: Different target values for 2030 and differences between real life and test procedure

→ Effect of CO$_2$ fleet targets in Europe:
- Bring significant CO$_2$ reduction, but
- With high delay (average car age > 8a)

→ Delta between real-life and certified CO$_2$:
- Important parameter in projections
- WLTP will reduce, but can not fully close gap (different driving style, slope, extreme temperatures)

→ Off-cycle CO$_2$ technologies:
- Bring real benefit for car drivers
- Reduce on-the-road CO$_2$ emissions
- Need promotion in todays cycle focused R&D → eco innovations are the right way

Eco Innovations are an important means for real-life CO$_2$ reduction

Automotive Technology

 BOSCH

27

Slide 12

Eco Innovation - Status

1st January 2016

Technology	Applicant	
	Supplier	OEM
LED lights	Automotive Lighting	Audi, Daimler, Toyota[1], Mazda[1], Honda[1]
Efficient alternators[2]	BOSCH, Valeo, Denso, Melco	-
Engine encapsulation	-	Daimler
Predictive Hybrid operating stgy.	BOSCH	-
Solar roof	Webasto, Asola, a2-solar[1]	-
Coasting	-	Porsche
Enthalpy storage tank	Mahle Behr[1]	-
Adv. Multi Air Technology	-	Fiat Chrysler[1]

**Up to now, several Eco Innovations (NEDC based) approved.
Measure Eco Innovation will also exist in WLTP regime...NEDC/WLTP transition in discussion.**

| LED = Light emitting diode | OEM= original equipment manufacturer |[1) shaded = application in approval process |[2) = several different technologies, some applications still in approval process
| Source: EU KOM Website http://ec.europa.eu/clima/policies/transport/vehicles/cars/documentation_en.htm|

Automotive Technology   Source: Bosch Diesel Systems, Gasoline Systems

12

BOSCH

Slide 13

Eco Innovation - Effect of WLTP testing conditions

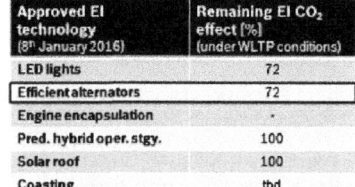

Approved EI technology (8th January 2016)	Remaining EI CO_2 effect [%] (under WLTP conditions)
LED lights	72
Efficient alternators	72
Engine encapsulation	-
Pred. hybrid oper. stgy.	100
Solar roof	100
Coasting	tbd

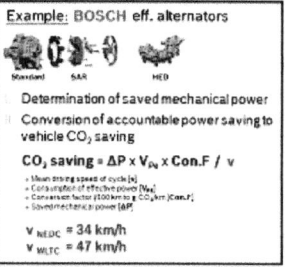

Example: BOSCH eff. alternators

Standard   SAR   HED

- Determination of saved mechanical power
- Conversion of accountable power saving to vehicle CO_2 saving

$$CO_2 \text{ saving} = \Delta P \times V_{FN} \times Con.F / v$$

- Mean driving speed of cycle [v]
- Consumption of effective power [V_FN]
- Conversion factor (100 km to g CO_2/km) [Con.F]
- Saved mechanical power [ΔP]

$V_{NEDC} = 34$ km/h
$V_{WLTC} = 47$ km/h

**Some EI technologies show reduced CO_2 effect under WLTP conditions.
Correlation needed to ensure "comparable stringency".**

| EI = Eco Innovation | HED = high efficiency diode | LED = Light emitting diode | SAR = Synchronous Active Rectification
| Source: Bosch internal estimation/ analysis

Automotive Technology   Source: Bosch Diesel Systems, Gasoline Systems

13

BOSCH

4   Increase Share of Electrification

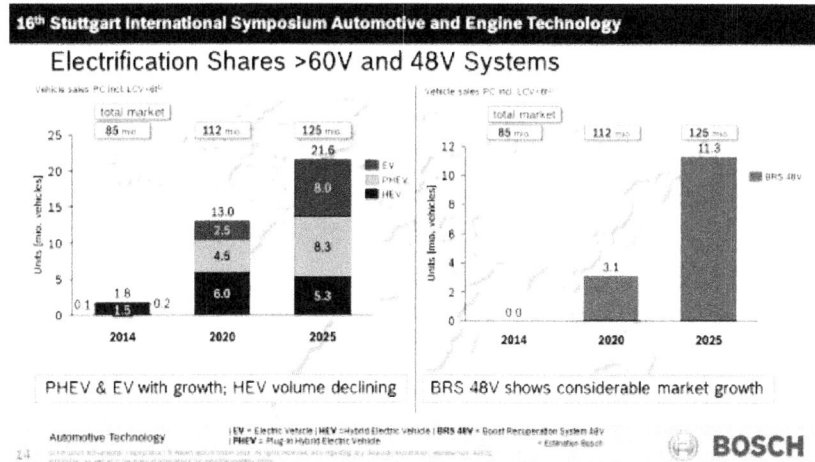

The novel SCR and PNA exhaust gas after treatment systems for diesel passenger cars

Nebojsa Milovanovic, Shant Hamalian, MAHLE Powertrain UK

Magnus Lewander, Kenneth Larsen, Haldor Topsoe, Denmark

1 Abstract

The future emissions legislation for diesel passenger cars is likely to include more dynamic test cycles than we have today, such as the WLTP and RDE cycles in the EU and challenging SULEV legislations in the USA. In order to meet these emissions legislation more complex exhaust gas after treatment systems are needed.

The aim of this paper is to describe a novel exhaust gas after treatment system that consists of a passive NOx adsorber (PNA) combined with the uf-SCR (Underfloor Selective Catalytic Reduction) or SCRonDPF (Selective Catalytic Reduction on Diesel Particulate Filter). The novel PNA stores NOx at low temperatures and self-releases it at high temperatures without the need for a rich engine operation purge.

The experimental results from a D segment vehicle using different PNA and SCR configurations are presented and the potentials and limitations of each configuration are discussed. Furthermore the trade-off between fuel consumption and NOx emissions are presented.

2 Introduction

The future emissions legislation for diesel passenger cars is likely to include different test cycles than we have today, such as the WLTC and RDE in the EU and very challenging SULEV legislations in the USA. In parallel with these more stringent emissions legislation many countries are implementing challenging vehicle CO_2 (fuel consumption) reduction targets. For example, the EU will make changes from 2021, China 2025 and the Green House Gas GHG targets in the US took effect from 2014. The calibration strategies across all vehicle applications (passenger cars, light commercial vehicles, on and off road heavy duty vehicles) need to provide the best possible fuel consumption whilst keeping the exhaust gas emissions within legislative limits across the entire engine operating map for all tested cycles.

With the introduction of new testing cycles such as WLTC or RDE, the engine and vehicle behaviour is quite different compared to the current New European Driving Cycle (NEDC). The WLTC is longer and more transient than the NEDC resulting in vehicle operation that reaches higher speeds and accelerations. For the RDE, the testing will have to be carried on public roads in real driving conditions using Portable Emission Measurement Systems (PEMS). The testing is composed of urban, rural and motorway roads and should last 90-120min with CO_2, NOx and particulate number emissions adhering to a Compliance Factor (CF) that is a multiple of the legislative levels.

In order to meet these emissions legislation and challenging CO_2 targets, more complex Exhaust Gas Aftertreatment System (EGATS) and corresponding calibration strategies are needed. The EGATS, depending on the vehicle weight, usually consists of a Diesel Oxidation Catalyst (DOC), Diesel Particular Filter (DPF) and NOx control technologies: NOx Adsorber Catalyst (NAC) and Selective Catalyst Reduction (SCR). The NAC is mainly used for lighter vehicles (<1600kg) while the SCR is more applicable for heavier vehicles.

For both NOx control technologies it is important to have a dedicated calibration strategy that provides the highest NOx reduction efficiency during real life driving conditions and day to day usage. The overall vehicle calibration includes engine, EGATS and transmission calibrations that provide the desired vehicle performance, efficiency and drivability, while satisfying the legal exhaust emissions requirements. The usual trade-off in vehicle calibration is between the engine efficiency and emissions (e.g. fuel consumption vs engine out NOx), where the fuel consumption is sacrificed in order to reduce NOx emissions. With the use of EGATS however, it is possible to calibrate the engine and EGATS individually for the optimum fuel consumption and lowest NOx emissions. To achieve this calibration strategy has to be selected to match the EGATS capability.

The aim of this paper is to evaluate different EGTAS configurations and calibration changes for a passenger diesel car and to understand their effects on the tailpipe emissions and fuel consumption. The testing was performed on a D segment vehicle (>1600kg) equipped with a novel EGATS consisting of a Diesel Oxidation Catalyst (DOC), Selective Catalytic Reduction on Diesel Particulate Filter (SCRonDPF) and small underfloor SCR (ufSCR) capable of satisfying RDE and SULEV in different configurations. The obtained results are presented and discussed.

3   Experimental Set up

3.1   Vehicle specification

The vehicle used was D class, EU6 with a curb weight of 1670 kg. The production ECU was replaced with a MAHLE Flexible ECU (MFE) which provides full control of the engine calibration and Adblue dosing strategy. The vehicle main specifications are shown in Table 1.

Table 1. Vehicle specifications

	Specifications
Vehicle class/Emission level	D/EU6
Engine	2L , 4 cylinder inline
Power/Torque (kW/Nm)	103/430
Gears	6 speed manual

3.2 EGATS Selection and specification

The EGATS, supplied by Haldor Topsoe, consisted of a DOC, SCRonDPF and uf-SCR [1, 2]. The first layout with a single large under floor SCR catalyst, shown in Figure 1, is the most common layout already in production on several EU6 applications [3]. This layout has significant distance between the urea injection point and the SCR catalyst (300 mm in our case), providing very good mixing and uniformity index (UI).

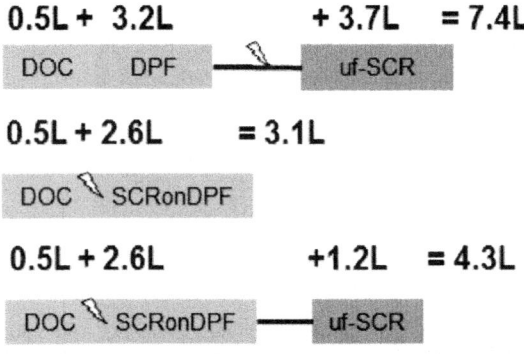

Figure 1. The EGTAS layouts

However, the SCR light–off is delayed due to being located far from the turbine outlet (>1m) and the presence of high thermal inertia components upstream (DOC, coated DPF, and the intermediate pipe). The second layout, closed coupled SCRonDPF, is closer to the engine (i.e. mounted at the turbocharger exit) and therefore benefits from a higher exhaust gas temperature and faster light-off. The drawbacks are a limited volume and length available for urea mixing due to the packaging constraints. The Adblue doser and supply lines also have to be redesigned for a hotter environment [4, 5].The third layout includes SCRonDPF and additional SCR volume at the underfloor location. This layout has all the advantages of the SCRonDPF alone layout, with additional benefit due to the extra catalyst volume, giving the potential to provide a high NOx efficiency at higher engine loads. For the SCRonDPF and SCRonDPF + ufSCR layouts the Adblue doser used was a water cooled Delphi B3.2 prototype [4]. The AdBlue spray shape was tailor made for this particular application and EGATS layouts.

3.3   Tested cycles

The three EGATS layouts were tested over four different cycles: NEDC, WLTC, EPA Federal Test Procedure (FTP-75) and a MAHLE developed High Speed Cycle (HSC). Table 2 summarises the cycles length, duration, maximum and average speed observed during the cycle.

Table 2. Tested cycles summary

	NEDC	**WLTC**	**HSC**
Length (km)	11.8	23.3	36.6
Duration (s)	1180	1800	1800
Average Speed (km/h)	33.3	46.5	73.1
Maximum Speed (km/h)	120	130	115

Results with the different EGATS Layouts on tested drive cycles

The WLTC is comprised of four phases (Figure 2): low speed (city driving), medium speed (suburban driving), high speed (A road driving) and extra high speed (motorway driving).

When testing under WLTC, the SCRonDPF, during the Low speed part, has and average temperature around 180°C (Figure 2). During the medium speed section the temperature rises to 250°C. And for the high-speed and extra-high speed sections the temperature is around 300°C which is sufficient for high NOx conversion efficiency. The uf-SCR is always around 50°C cooler bringing the average temperature of each phase to 130°C, 200°C and 250°C.

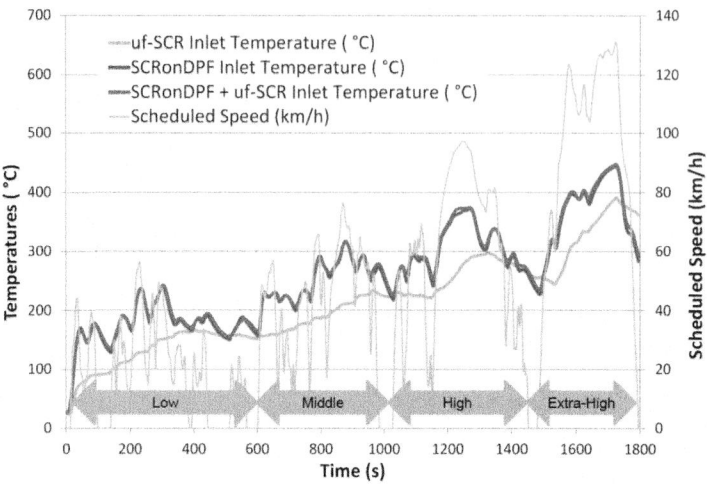

Figure 2. WLTP Temperature Profile for the tested EGTAS layouts

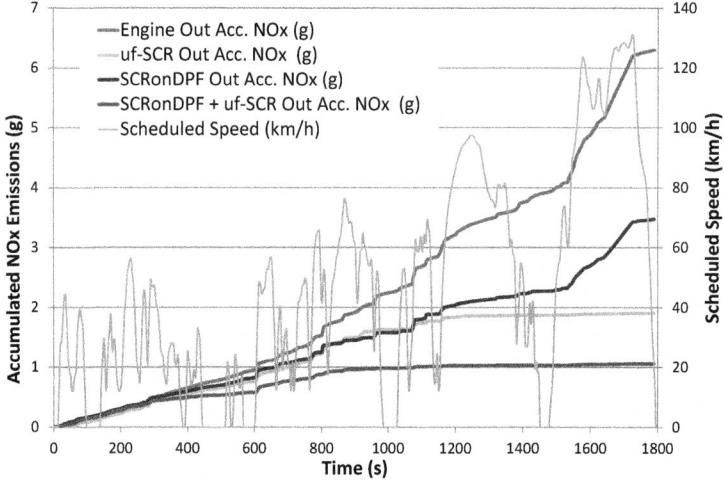

Figure 3. WLTC NOx Emissions for the tested EGATS layouts

During the low and medium phases, the SCRonDPF and uf-SCR layouts have comparable NOx emissions results (Figure 3) even though there is a difference in SCR volume

(2.6L vs 3.7L). This is mainly due to the higher temperature of SCRonDPF which enables it to provide comparable NOx reduction efficiency with a smaller SCR volume. However, during the high speed and extra-high speed sections, the lower volume of the SCRonDPF could not cope with the entire engine out NOx as it became space velocity limited. The SCRonDPF with the small uf-SCR showed the best results by combining the benefits of being sufficiently close to the engine and having an additional SCR volume to cope with the higher engine load NOx.

In the HSC (Figure 4) the SCRonDPF layout was not able to reduce the NOx emitted in the high speed section due to having the lowest SCR volume. The other two layouts uf-SCR and SCRonDPF+uf-SCR performed well.

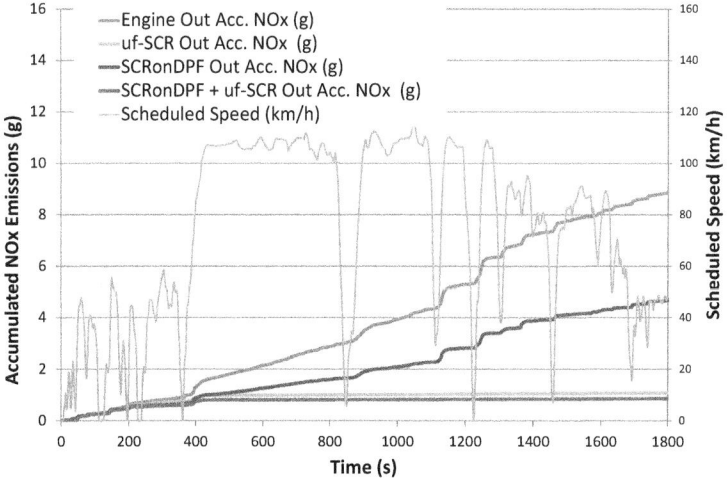

Figure 4. HSC NOx Emissions

Figure 5 summarises the NOx emissions results from testing the three EGTAS over the cycles. The SCRonDPF with the small uf-SCR layout gave the best results on all tested cycles and was chosen for the fuel consumption optimisation study.

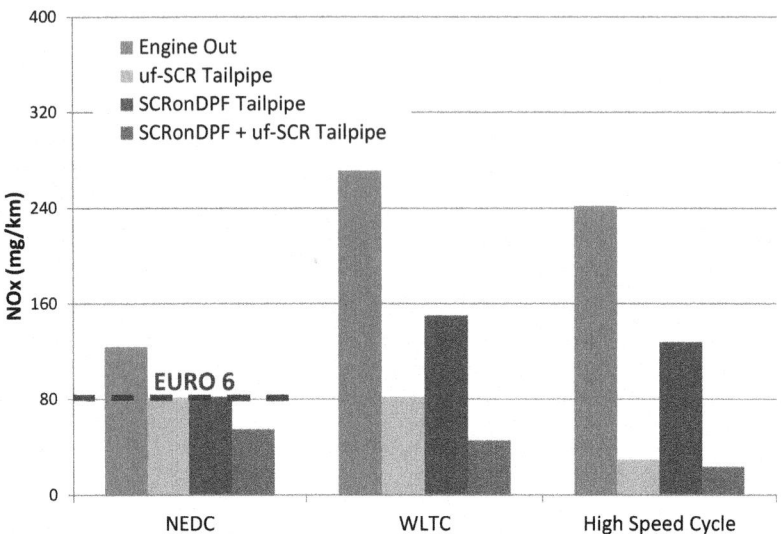

Figure 5. Bar Chart Summary of NOx Emissions over the tested cycles

4 Engine calibration optimisation

The aim of the calibration optimisation was to provide the lowest fuel consumption and the highest EGATS efficiency across the entire operating map, by optimising the engine calibration and Adblue dosing strategies. The EGATS calibration dosing strategy was based on NOx measurement using a direct input from a NOx sensor position at the engine out location. The EGATS was kept unchanged and fixed Ammonia to NOx Ratio (ANR) was used to explore the effects of engine calibration only and to understand if any changes in EGATS calibration were needed.

The engine efficiency was improved by phasing the combustion to a more favourable location (i.e. MFB50 closer to Top Dead Center (TDC) firing). This was done by advancing the injection timing. Advancing the injection timing provides more favourable combustion phasing and therefore improved fuel consumption, however it increases the engine out NOx. The base calibration was an EU 6 tailpipe compliant (with EU 5 engine out NOx emissions) consisting of two pilots and one main injection across the entire engine operating map. The advancing of injection timing was done for the entire injection train: the pilots and main injections were advanced by the same number of crank angle degrees, keeping the separation between them unchanged. The engine idle regime was not part of this optimization and had a different calibration. The Exhaust Gas Recirculation (EGR) of a hot exhaust gas was used to achieve EU5 engine out emissions levels and was kept unchanged.

4.1 Results and Discussion

Starting from the base calibration, three different calibration steps where made on the injection timing. The injection train was advanced (moving the injections closer to TDC firing) for 2 °CA, 4°CA and 5°CA. All other engine parameters were kept the same (Air demand, EGR rate, Rail pressure, etc.). Figure 6 shows the calibration results of combustion phasing optimisation. The fuel consumption was improved 3%, 4% and 7% respectively compared to the base calibration.

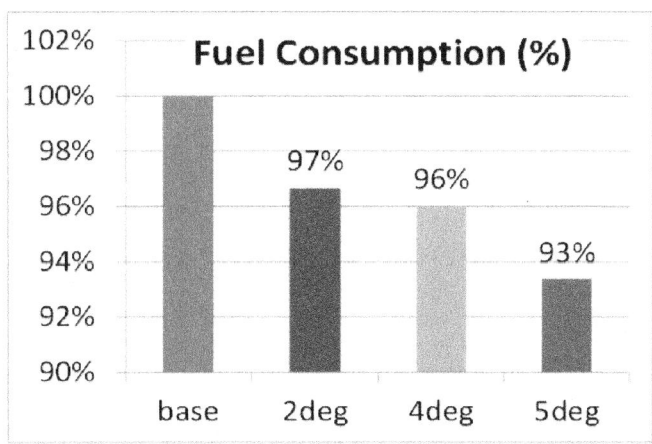

Figure 6. Fuel consumption benefit with different calibrations

Figure 7 shows the effect of advanced injection timing on the engine out NOx. As expected the more advanced injection timing produced more engine out NOx. The baseline calibration has an engine out NOx level of 180mg/km (1.98 g in total over the NEDC). To be EU5 NOx compliant the vehicle will only need a DOC and DPF EGATS. For the most advanced injection timing of 5°CA the engine out NOx was 450mg/km which was equivalent to the EU3 NOx tailpipe requirement. It is therefore necessary, for the EU6c limit, to use De-NOx systems to bring NOx emission below 80mg/km.

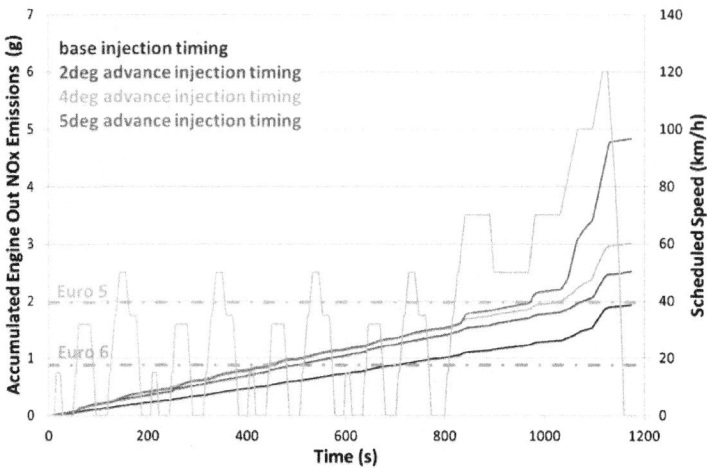

Figure 7. Engine out NOx for different calibrations over the NEDC

The SCRonDPF with the small uf-SCR described in Figure 1 was used to reduce the tailpipe NOx. The Adblue dosing strategy was a fixed ANR. The Adblue dosing was scheduled to start when the exhaust temperature reaches 180°C at the turbocharger exit (i.e. before the SCRonDPF). The Adblue doser used had a high injection pressure ~ 50 bar which allowed earlier dosing at lower exhaust temperatures compared to the usual 200 °C, without the risk of deposit formation. The Adblue droplets were smaller compared to the other competitor Adblue water and air cooled dosers (20um vs 100um) [4]. Figure 8 shows the tailpipe and engine out NOx values. Bby using the SCRonDPF + uf-SCR and early Adblue dosing strategy,for the case of 2°CA advanced injection timing, the EU6c requirement were met over the NEDC. For calibrations with 4°CA and 5°CA advanced timings, the selected EGATS and Adblue dosing strategy was not able to reduce the NOx emissions to meet EU6c levels.

The calibration of 4°CA advanced timing needs additional work to minimize the engine out NOx during the first acceleration in the extra-urban phase of NEDC (around 800 sec in Figure 9). These NOx peaks during the acceleration phases were caused by the reduction in cylinder filling due to the slow supply of air and EGR compared to the fuel supply. The delay in the air and EGR supply was caused by the turbocharger and EGR valve inertia to build up the required boost pressure and to supply required EGR amount. This led to the engine running with non-optimised cylinder filling (i.e.

air/EGR-to-fuel ratio) causing the increase in NOx emissions. Further calibration work is planned to improve air and EGR supply during accelerations.

The same applies for the case of the 5°CA advanced timing a more aggressive dosing strategy and/or a larger EGATS might also provide benefits and enable EC6c compliance.

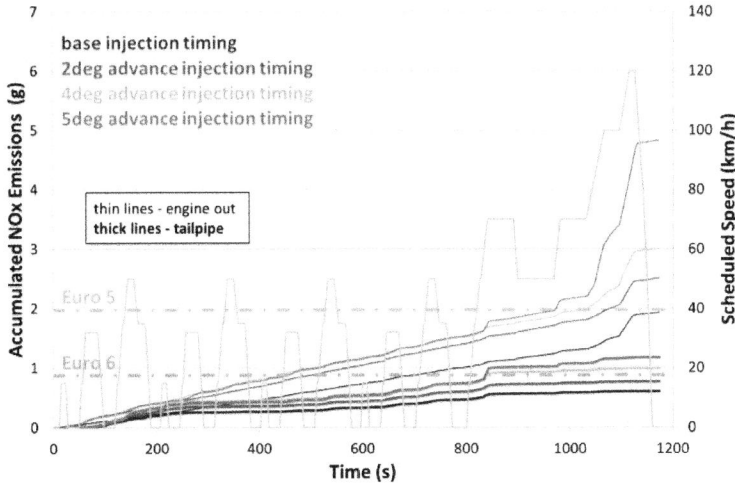

Figure 8. Tailpipe NOx for different calibrations over NEDC

A vehicle with the 2°CA advanced injection timing calibration, which showed the potential to provide 3% fuel consumption reduction and to meet the EU6c emissions, requires an increased Adblue consumption. A calculation was performed to understand the additional Adblue consumption, cost and how this compared to the fuel consumption saving and total fluid cost (Fuel+Adblue). The comparison was carried out for all four calibrations (Base, 2°CA, 4°CA and 5°CA advanced timing) and the following assumptions were made:

- EU6c emissions level were met either with the improved engine calibration or different EGATS
- NO2:NOx post DOC of 50:100
- ANR=1
- Baseline fuel consumption of 5L/100km.

The estimated Adblue, fuel and total fluid consumptions over 10,000 km of NEDC are shown in Figure 9.

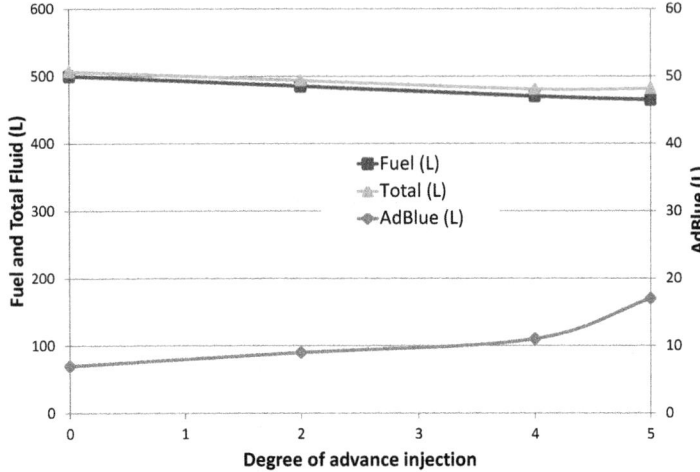

Figure 9. Fluids consumption (Fuel, Adblue, Fuel+Adblue)

In the UK currently, Adblue and fuel are closely priced (1L diesel=1.1GBP, 1L Adblue=1.1GBP) therefore the major cost is fuel used as the Adblue consumption is approx. 2% of the fuel consumption. In order to simplify comparison between the fuel and Adblue consumption for different calibrations, the total fluid consumption (Fuel+Adblue) was compared.

It can be seen (Figure 9) that calibrations with advanced timing reduces the total fluid consumption. It is also interesting to note that the current 4°CA and 5°CA calibrations have very similar total fluid requirements, so no additional gain in total fluid consumption can be achieved with more advanced timing. Moreover, using 5°CA calibration the Adblue fluid consumption is considerably higher (50%) as NOx engine out increases significantly to almost double the 4°CA calibration's value (Figure 8). The other disadvantage of using 5°CA calibration will be the interval of Adblue tank refill compared to the other calibrations.

5   The Passive NOx Adsorber PNA

The NOx Adsorber Catalyst (NAC) has the ability to store NOx during diesel engine "lean" operation, but due to its finite storage capacity it requires periodic regeneration.

The regeneration is achieved by running the engine in a "rich" mode for a short time, to increase the concentration of reductants in the exhaust gas (CO, HC and hydrogen). This creates a fuel consumption penalty and if this needs to be performed several times over a tested duty cycle, it can have a significant impact on the regulated CO_2 emissions.

One way to avoid the requirement for engine rich operation is to use a Passive NOx Adsorber (PNA). This type of catalyst stores NOx during engine "lean" operation at low exhaust gas temperature and releases it at a higher temperature (>190°C), when the SCR catalyst is more efficient. The PNA is capable of self-cleaning at higher temperatures without the need to run engine in a "rich" condition. The additional feature of the PNA is the ability to reduce HC and CO emissions, which provides opportunities to combine DOC and NAC functions. In this way a reduction in the thermal mass of the EGATS will enable faster warm up for downstream components, such as SCR and/or SCRonDPF.

The PNA prototype used for vehicle testing was optimised for NOx operations with sufficient HC and CO reduction. The PNA volume and PGM content was optimised for a D segment vehicle and NEDC duty cycle. The HC and CO cleaning performance is comparable to DOC, so a PNA catalyst is used alone.

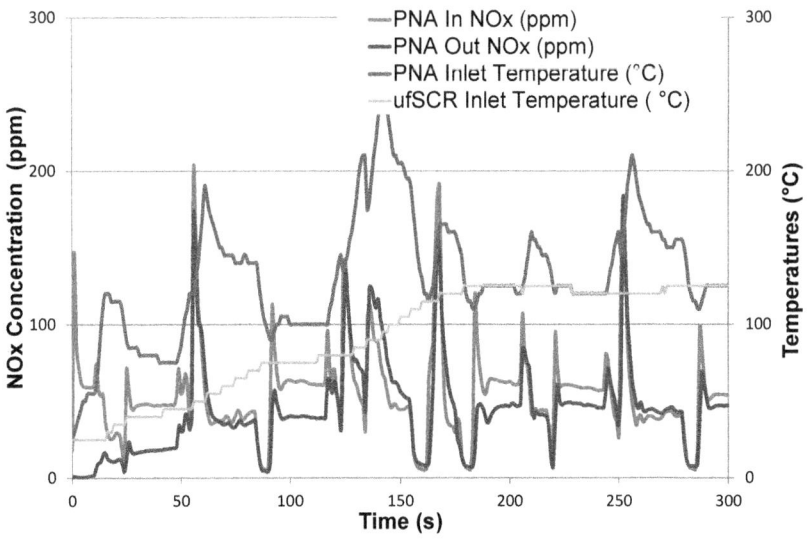

Figure 10. The PNA storage during low temperature operation at NEDC

Figure 10 shows PNA characteristics for the tested vehicle over the NEDC. It can be seen that the PNA stores NOx during low temperature operations where SCR catalyst is less effective. The most difficult NOx emissions to clean are those produced just after cold start (first 300 sec of the drive cycle) and the PNA has the ability to release them effectively in the extra-urban phase when SCR temperatures are above 200°C (Figure 11). The storage capacity of the PNA could be tuned to suit various vehicle segments and duty cycles as the amount of NOx produced during cold start is very different for the NEDC, WLTC, FTP 75 or RDE and also depending of the engine power and vehicle inertia weight.

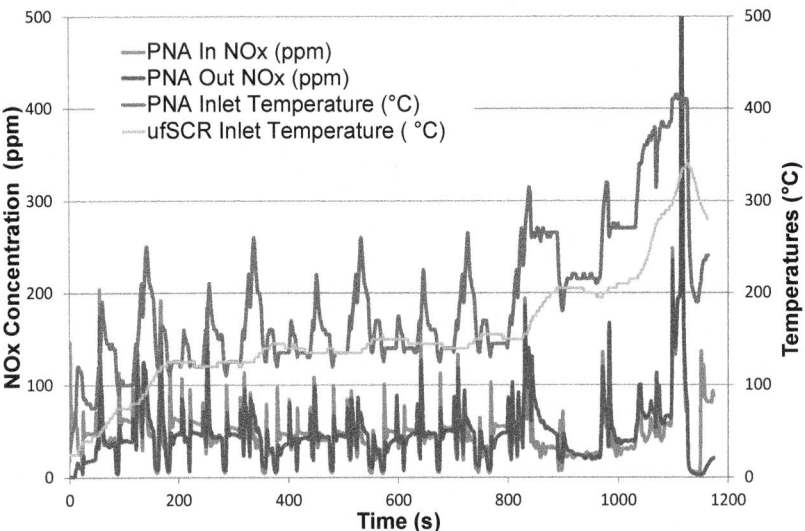

Figure 11. The PNA storage and release during NEDC extra urban phase

The NOx storage is available up until 200-220°C when the stored NOx starts to be released. The temperature when the stored NOx is released is a very important design feature of the PNA, as it needs to be higher than the SCR catalyst light-off temperature. In this case the ammonia in the SCR catalyst is able to react and reduce the released NO_X.

6 Summary/Conclusions

Diesel engine development focuses on reducing fuel consumption and pollutant emissions. Nevertheless, to meet future stringent EU6c NOx emissions legislation over more dynamic test cycles, such as WLTC and in-use compliance for RDE, diesel passenger cars have to be equipped with NOx aftertreatment technologies. Various SCR layouts were experimentally tested: under floor SCR -ufSCR, closed coupled SCRonDPF and the combination of SCRonDPF and small uf-SCR. The testing was carried out on a vehicle from segment D (>1600kg), over NEDC, WLTC, FTP-75 and HSC duty cycles.

The SCRonDPF+uf-SCR layout provided the highest NOx reduction efficiency on all cycles. This was achieved by the combed advantages of the SCRonDPF layout being closer to the turbocharger exit and having an additional small catalyst volume that enabled a high NOx efficiency at higher engine loads.

This layout was further used to carry out a fuel consumption study with the engine and Adblue dosing calibration changes. The engine efficiency and fuel consumption was improved by phasing the combustion to a more favourable location (i.e. closer to the TDC firing) by advancing the injection timing. The 2°CA, 4°CA and 5°CA advanced timing were evaluated.

The 2°CA calibration with the selected hardware and engine calibration change have the potential to reduce fuel consumption for 3% on the NEDC while still meeting EU 6c requirements. The total fluid consumption (Fuel+Adblue) was reduced by 2.5% compared to the baseline.

Utilizing the high NOx conversion potential of EGATS provides the possibility to calibrate the engine and EGATS individually for the optimum fuel consumption and the lowest emissions. The synergy between the engine and EGATS will be essential for future, more demanding emissions legislation, such as RDE and SULEV while meeting challenging CO_2 targets.

The Passive NOx Adsorber – PNA, with the capability of storing NOx at low engine temperatures and self-releasing them at higher temperatures, without a need to run in a "rich" mode, was tested with the same vehicle. The tested PNA had the additional ability to store and to oxidise HC and CO and therefore to combine two technologies: NAC+DOC.

The results demonstrate that PNA effectively stored NOx during cold start and in low load operation, releasing them at temperatures >200° C, when the SCR was more efficiently reducing NOx. The HC and CO reduction performance was comparable with the DOC performance. The PNA's storage capacity can also be tuned to suit various vehicle applications, to accommodate the varying amounts of NOx produced during cold start.

The combination of PNA with ufSCR or SCRonDPF has the potential to provide a very efficient NOx reduction system for more demanding emissions legislation, such as RDE and SULEV, across a wide range of passenger and light commercial vehicle (LCV) applications. To confirm this, further work is planned on a PNA system sized for WLTP, RDE and SULEV and for D class vehicles and LCV, together with the optimisation of CO_2 and NOx trade-off. The results will be published at a later date.

7 References

1. Johansen, K., Bentzer, H., Kustov, A., Larsen, K. et al., "Integration of Vanadium and Zeolite Type SCR Functionality into DPF in Exhaust Aftertreatment Systems - Advantages and Challenges", SAE Technical Paper 2014-01-1523, 2014, doi:10.4271/2014-01-1523.

2. Milovanovic, N., Hamalian, S., Tumelaire, C.F., Larsen, K. and Lewander, M., "The Novel SCR and PNA exhaust gas aftertreatmant system for future diesel passenger cars", IMeChe Internal Combustion Engines Conference, London, UK, December 2015.

3. Enderle, C., Vent, G. and Paule, M., "BLUETEC Diesel Technology – Clean, Efficient and Powerful", SAE Technical Paper 2008-01-1182, 2008.

4. Needham, D., Spadafora, P., Schiffgens, H.J., Kirwan., J.E., Cabush, D.D. and Kalina, A. "Delphi SCR Dosing System – An Alternative Approach for Close-Coupled SCR Catalyst Systems", 21st Aachen Colloquium Automobile and Engine Technology, 2012.

5. Gerhardt, J., Heiter, T., Kern, C., Maier, R., SAmuelsen, D., Strobel, M. and Welting, D. "Denoxtronic 5 and other Bosch System Solutions to meet "Post – EU6" Emissions Requirements", 34th International Wiener Motorensymposium, 2013. –

8 Definitions/Abbreviations

°CA	Degree crank angle
ANR	Ammonia to NOx ratio
CF	Compliance Factor
CO_2	Carbon Dioxide

DOC	Diesel Oxidation Catalyst
DPF	Diesel Particulate Filter
ECU	Engine Control Unit
EGATS	Exhaust Gas Aftertreatment System
EGR	Exhaust Gas Recirculation
EPA	Environmental Protection Agency
FTP-75	EPA Federal Test Procedure
GHG	Greenhouse Gases
HSC	High Speed Cycle
MFB50	50% of Mass Fraction Burned
MFE	MAHLE Flexible ECU
NAC	NOx Adsorber Catalyst
NEDC	New European Drive Cycle
NOx	Nitrogen Oxides
PNA	Passive NOx Adsorber
PEMS	Portable Emission Measurement Systems
RDE	Real Drive Emission
SCR	Selective Catalyst Reduction
SCRonDPF	SCR washcoat on a DPF
SULEV	Super Ultra-Low Emission Vehicle
TDC	Top Dead Center
uf-SCR	Underfloor SCR
WLTC	World Harmonized Light duty Testing Cycle

Experimental investigation of heat transfer characteristics of UWS spray impingement in diesel SCR

Yujun Liao, Panayotis Dimopoulos Eggenschwiler

EMPA, Swiss Federal Laboratories for Materials Science and Technology
Automotive Powertrain Technologies Laboratory
Überlandstrasse 129, CH-8600, Dübendorf, Switzerland

1 Abstract

To comply with the stringent regulations as stated in Euro6, the NOx emissions of heavy-duty vehicles have to be reduced by up to 80% compared to Euro5. Engine exhaust after-treatment SCR is a promising technique to reduce NOx emissions without sacrificing engine efficiency. The introduction of the reducing agent urea plays a significant role on the reduction reactions. The main challenges for the implementation of mobile urea-SCR systems include rapid decomposition and homogeneous distribution of urea and the mitigation of deposit formation. A key factor affecting these performances is the heat transfer characteristics of UWS spray under exhaust flow conditions. However, the heat transfer characteristics of UWS spray impingement have not been studied experimentally under exhaust flow conditions so far.

The present study is focused on the heat transfer characteristics of the impinging SCR spray in crossflow. The heat transfer characteristics of spray impingement are analyzed based on the temporal and spatial evolution of the wall temperature using infrared thermography. This work enhances the understanding of the wall impingement of UWS sprays.

Key words: Urea-SCR, heat transfer, UWS spray impingement, infrared thermography

2 Introduction

To comply with the stringent regulations as stated in Euro6, the NOx emissions of heavy-duty vehicles have to be reduced by up to 80% compared to Euro5. Exhaust SCR is a promising technique to reduce NOx emissions without sacrificing engine efficiency [1]. In most mobile cases, urea-water-solution (UWS) is used as a source of ammonia because of its non-toxicity, and convenience of storage. UWS is sprayed into the exhaust gas flow. The complete evaporation of water from spray droplets, thermal decomposition and hydrolysis follow. Water evaporation and thermal decomposition are endothermic processes, thus heat transfer issues are critical for the proper preparation of the reducing agent.

The main challenges for the implementation of mobile urea-SCR systems include rapid decomposition and homogeneous distribution of urea as well as the mitigation of deposit formation. Due to compact design requirement of the exhaust pipe and the relatively low efficiency of urea thermal decomposition, the spray impingement on the exhaust pipe or on the mixer is unavoidable. UWS spray impingement on the exhaust pipe wall or on a mixer on one side can assist liquid evaporation and urea thermal decomposition; on the other side can lead to deposit formation, since spray impingement results in local cooling. When the wall temperature drops below a certain threshold, liquid film starts forming. The spatial distribution of the reducing agent is strongly affected by the impingement characteristics which are determined by local temperature distribution. Evaporation from the wall film leads to further cooling and to increasing risk of deposit formation such as solid urea, biuret, cyanuric acid, ammelide, ammeline and melamine [2]. However, studies of UWS sprays under exhaust gas flow conditions are scarce. The lack of knowledge hinders the optimization of mitigation of deposit formation and homogenous distribution of urea.

Many studies have been done so far on urea-SCR. The fluid dynamic behavior of a 6-hole SCR injector spray was reported in [3]–[5]. UWS evaporation and spray/wall interaction were investigated in a hot air stream by Grout [6]. Musa [7] performed investigations on the evaporation characteristics of a single urea solution droplet. Dunand [8] specified the conditions of the impact regimes for water droplets onto a heated plate. Birkhold [9], [10] realized the importance of spray/wall interaction and included it into his systematic modelling of UWS injection. However, to the knowledge of the author, heat transfer characteristics of UWS spray impingement have not been studied experimentally under exhaust flow conditions so far. The complexity of the SCR system requires a close look at the transient heat transfer behavior of UWS spray impingement in hot crossflow.

In this work, infrared thermography was introduced to investigate the transient heat transfer behavior of the spray wall impingement.

3 Experimental setup

The measurements were conducted in the flow lab at Empa, Dübendorf. This lab was designed for the experimental investigation of UWS injection into the exhaust channel for SCR application. It allows the reproduction of engine exhaust like conditions. Fig. 1 shows an overview of the Empa flow lab. The compressed air is taken from Empa pressurized air network and fed into a large dampening volume to get rid of possible oscillations. The air flow is then led through three heating units to reach the target temperature. The prepared gas flow subsequently passes through the injection channel assembly which consists of two optically accessible measurement chambers. In this work, the small chamber with a square cross section of 80mm by 80mm and 200mm in length is used. The gas flow through the measurement chamber can be regulated up to 450kg/h and 500°C.

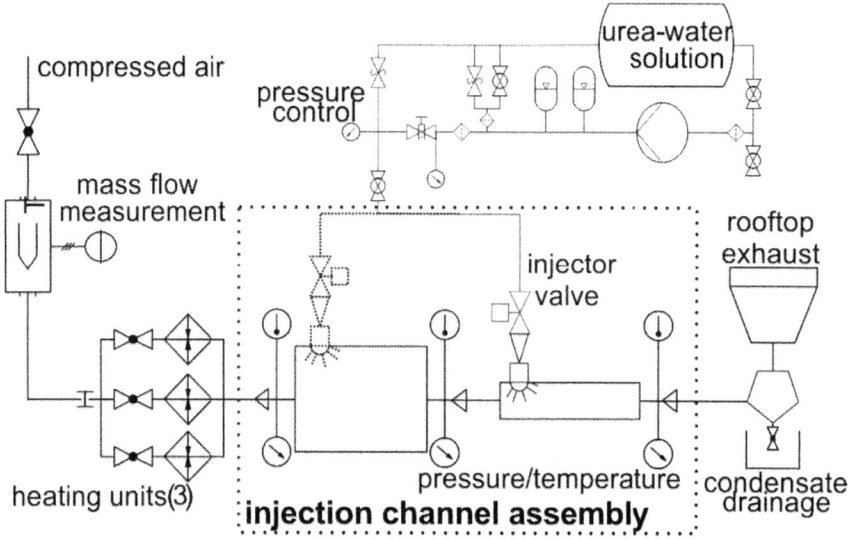

Figure 1: Schematic diagram of the Empa flow lab

The urea dosing unit consists of a high pressure system and a 3-hole commercial injector. The high pressure system can deliver fluid pressure of up to 15bar. The injector was mounted on the top of the measurement chamber, 50° inclined to the gas flow direction. The injector was operated with 9bar backpressure during experiments. The injection frequency is 1Hz with 60ms injection duration.

In this study, infrared thermography has been applied to investigate the heat transfer characteristics of the UWS spray/wall interaction. A stainless steel plate of 0.3mm was placed in the channel, 14mm above the bottom of the channel. The CaF_2 glass allows the optical access for the infrared camera, which has a transmissivity of 93% in the mid-infrared range of 3 to 5 um. The mirror which is coated by Gold allows very high portion of reflection in this range as well, about 90%. The IR camera used is a Cedip Jade III which can record up to 170 frames per second with a resolution of 320×240 pixels.

The camera calibration process was performed by exposing the camera detector to a black body of different known temperatures at both low and high camera housing temperatures. Later during the post-processing, the calibration points were linearly interpolated based on the housing temperature at which the measurements were made. The digital levels stored in the camera acquisition software are able to be converted to temperatures according to the calibration function, which is a polynomial fit of the calibration points. Since the IR camera transfers all the radiated energy it receives into digital levels. The relationship between the digital output of the camera and the object surface temperature should be determined under real experimental setup. Therefore, the CaF_2 glass transmissivity, mirror reflectivity and plate emissivity were determined using a black body in a hot environment. Based on these values, the measured temperatures were corrected towards the real surface temperatures.

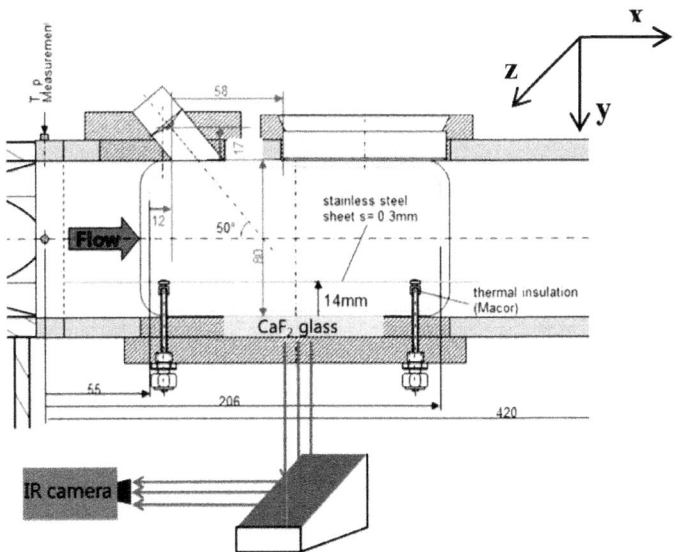

Figure 2: Schematic diagram of infrared thermography setup in the channel

4 Results and discussion

The measurements were done under various conditions typical for diesel exhaust. The results shown below are under gas flow conditions of 300°C and 200kg/h. During the experiments, the camera recorded 150 frames per second and the data were processed in Matlab. All the results shown in this section are the rear surface temperatures of the stainless steel sheet, which are averaged over five single injection events. Between two subsequent injections enough time elapsed so that the sheet reached every time thermal equilibrium with the surrounding gas flow.

Figure 3 shows the corrected rear surface temperature profiles at different time instants after the electronic start of injection (SOI). Rear surface temperatures on the plane which is y=83mm below the injector nozzle exit are shown. The x and z coordinates indicate the measurement locations with respect to the nozzle exit. Temperatures are coded according to the color scale beside the image. The spray from the 3-hole pressure injector has three distinct spray cones. The cooling regions are isolated from each other and concentrated in elliptical shapes. The footprints of the three spray cones are slightly asymmetric with respect to the channel centerline, which is attributed to slight misalignment of the injector mounting. As observed from Figure 3, the first impingement starts from the second frame which is 8ms after the electronic SOI. Before impingement the sheet temperature is uniformly distributed around 282°C. The cooling period (from SOI+8ms to SOI+75ms) lasts a bit longer than the valve actuation time of 60ms, which is similar to the injection open profile. The maximum temperature decline on the rear surface is about 141°C in the center of the impingement ellipsoids. As can be seen from Figure 4 the lowest temperature detected in the ellipsoid center is 141°C at 88ms after SOI.

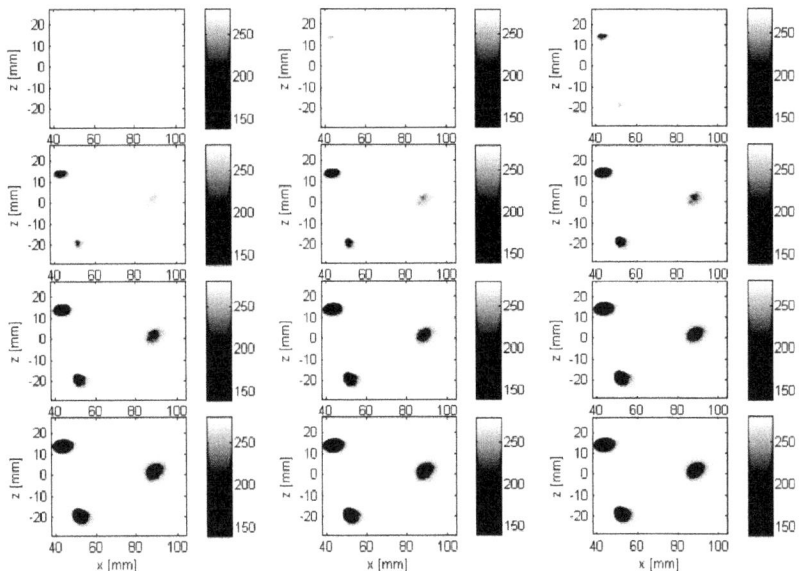

Figure 3: Spatial temperature distributions on the rear surface at different times after SOI

Figure 4 shows the temperature profile at the end of the injection which is 88ms after the electronic SOI. At this time instant, the minimum temperature on the rear surface is 141°C. The temperature footprints of the three spray cones behave somehow similar, showing the coldest temperature for a large part of the area and having a large gradient on the edge. This means that, the spray cools the area where there is a direct contact between the spray cone and the plate. Lateral heat conduction seems not contribute too much. The temperature profile is related to the local mass flux of the spray, which will be verified in future studies.

Figure 5 displays temperature profiles across the centerline of the front spray cone at different times after SOI.

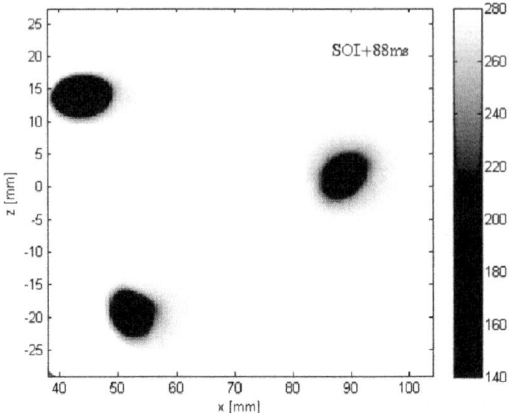

Figure 4: Detailed temperature distribution on the rear surface at the injection end

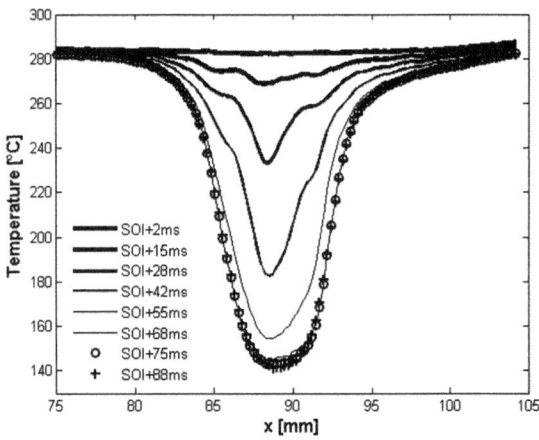

Figure 5: Temperature profiles across the centerline of the front spray cone at different times after SOI

15ms after the electronic SOI, the maximum temperature drop is about 13.5°C. The temperature gradient firstly increases and then decreases, reaching the maximum from SOI+28ms to SOI+42ms. After the electronic SOI, it takes some time for the spray to get fully developed and reach the maximum mass flow rate on the impinged plate. However, after certain time the temperature difference between the plate and spray gets less, thus the heat that can be transferred between the plate and the spray decreases. After SOI+68ms, the temperature continues to decrease but with a lower rate. This is either due to the decreasing liquid mass impinging on the plate because of the closing delay of the valve or the decreasing evaporation on the contact layer. Again it's evidenced, that the cooling effect remains relatively concentrated in the spray impinged area while the temperature of the non-impact area stays unaffected. The lateral heat conduction is not significant in this time scale.

5 Conclusion

The spray cooling effect on the channel wall is evidenced and quantified by infrared thermography. UWS spray impingement leads to substantial temperature drop on the exhaust pipe wall due to high heat transfer coefficients. In the case of this experiment, the spray cools the 0.3mm steel plate rear surface from 282°C to 141°C with injection duration of 60ms. The heat transferred from the plate to the spray can assist liquid evaporation and urea thermal decomposition. However the spray cooling effect on the wall leads to liquid film formation and thereafter deposit formation.

From the spatial temperature distribution, it's concluded that the spray from the 3-hole pressure injector has three distinct spray cones. The spray cooling footprints are isolated from each other and concentrated in elliptical shapes. The spray cools the area where there is a direct contact between the spray cone and the plate. The cooling effect on the exhaust pipe wall is local and the steel plate lateral heat conduction plays a minor role in this time scale.

The temporal temperature evolution on the plate depends on the injector opening profile, more in detail, the spray mass flux reaching the plate. However, the measured rear surface temperature evolution is somehow delayed in time compared to the front surface temperature due to the heat conduction across the thickness direction. The local heat flux firstly increases and then decreases, reaching the maximum during the injection.

References

[1]   M. Koebel, M. Elsener, and M. Kleemann, "Urea-SCR: a promising technique to reduce NOx emissions from automotive diesel engines," Catal. Today, vol. 59, no. 3, pp. 335–345, 2000.

[2]   A. M. Bernhard, D. Peitz, M. Elsener, A. Wokaun, and O. Kröcher, "Hydrolysis and thermolysis of urea and its decomposition byproducts biuret,

cyanuric acid and melamine over anatase TiO2," Appl. Catal. B Environ., vol. 115–116, pp. 129–137, 2012.

[3]   A. Spiteri and P. Dimopoulos Eggenschwiler, "Experimental fluid dynamic investigation of urea-water sprays for diesel selective catalytic reduction-denox applications," Ind. Eng. Chem. Res., vol. 53, no. 8, pp. 3047–3055, 2014.

[4]   A. Varna, A. C. Spiteri, Y. M. Wright, P. Dimopoulos Eggenschwiler, and K. Boulouchos, "Experimental and numerical assessment of impingement and mixing of urea–water sprays for nitric oxide reduction in Diesel exhaust," Appl. Energy, Apr. 2015.

[5]   A. Varna, K. Boulouchos, A. Spiteri, P. Dimopoulos Eggenschwiler, and Y. M. Wright, "Numerical Modelling and Experimental Characterization of a Pressure-Assisted Multi-Stream Injector for SCR Exhaust Gas After-Treatment," SAE Int. J. Engines, vol. 7, no. 4, pp. 2012–2021, 2014.

[6]   S. Grout, J. B. Blaisot, K. Pajot, and G. Osbat, "Experimental investigation on the injection of an urea-water solution in hot air stream for the SCR application: Evaporation and spray/wall interaction," Fuel, vol. 106, pp. 166–177, 2013.

[7]   S. Musa, M. Saito, T. Furuhata, and M. Arai, "Evaporation characteristics of a single aqueous urea solution droplet," ICLASS-2006, Kyoto, Pap. ID ICLASS06- 195, vol. 2, no. 1, 2006.

[8]   P. Dunand, G. Castanet, M. Gradeck, D. Maillet, and F. Lemoine, "Energy balance of droplets impinging onto a wall heated above the Leidenfrost temperature," Int. J. Heat Fluid Flow, vol. 44, no. August 2015, pp. 170–180, 2013.

[9]   F. Birkhold, U. Meingast, P. Wassermann, and O. Deutschmann, "Modeling and simulation of the injection of urea-water-solution for automotive SCR DeNOx-systems," Appl. Catal. B Environ., vol. 70, no. 1–4, pp. 119–127, 2007.

[10]   F. Birkhold, U. Meingast, and P. Wassermann, "Analysis of the Injection of Urea-Water-Solution for Automotive SCR DeNOx-Systems: Modeling of Two-Phase Flow and Spray / Wall-Interaction," SAE Int., vol. 2006–01–06, no. 724, 2006.

From laboratory to road – Real Driving Emissions

Helge Schmidt, Jens Badur

TÜV NORD Mobilität GmbH & Co. KG

1 Abstract

In European emission type approval during the past years the exhaust emission limits have been reduced significantly. Although air quality was improved from 1990 on, a high percentage of the European population is exposed to air pollutant concentrations above European limit values, mainly on particles, nitrogen dioxide (NO2) and ozone (O3).

Exhaust emissions of passenger cars and light duty trucks in Europe are measured by using the "New European Driving Cycle" (NEDC) under well defined ambient conditions in a laboratory. The NEDC represents only a small part of all driving conditions in real traffic. On 03.02.2016 the European Parliament decided that exhaust emissions in real traffic (Real Driving Emissions = RDE) shall be measured in Europe by using Portable Emission Measurement Systems (PEMS). Due to European air quality regulations NOx emissions are the main issue of RDE. European Commission is also interested in particle measurement especially on gasoline cars with direct injection.

When measuring emissions in real traffic numerous influencing factors have to be considered. Besides variable ambient conditions changing traffic situations affect the results of such measurements. This complex set of influencing factors has to be addressed by defining route requirements and boundary conditions. An elaborate data evaluation has been created. While the Moving averaging windows method (MAW; or EMROAD by JRC) is based on CO_2 emissions, for the Standardized wheel power frequency distribution method (SPF; or CLEAR by TU Graz) the wheel power is used for normalizing exhaust emissions.

Due to the fact that RDE are measured in real traffic, this new method will be challenging for all parties involved.

2 Introduction

The improvement of air quality is a priority for the European Commission. Road traffic is a major source of air pollution in Europe. In European emission type approval during the past years the exhaust emission limits have been reduced significantly. This decrease is yet not reflected by immission values in the same amount. Although major emission reductions were achieved from 1990 on, a high percentage of the Europeanpopulation is exposed to air pollutant concentrations above European limit values, mainly on PM, NO2 and O3.

Trend der Stickstoffdioxid-Jahresmittelwerte

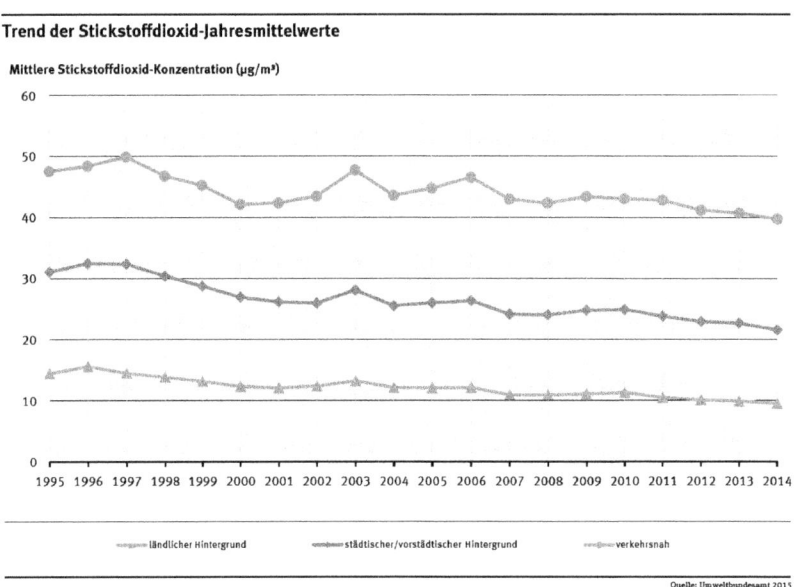

Figure 1: NO2 immissions, annual average values (source UBA)

According to the current European legislation for measuring exhaust emissions of passenger cars and light duty vehicles the "New European Driving Cycle" (NEDC) is applied. Exhaust emissions are measured under well defined ambient conditions in a laboratory.

The NEDC represents only a small part of all driving conditions in real traffic. To ensure that the emission type approval values represent real world emissions within Regulation (EC) No 715/2007 the Commission has been asked to „*keep under review the need to revise the New European Drive Cycle as the test procedure that provides the basis of EC type approval emissions regulations. Updating or replacement of the test cycles may be required to reflect changes in vehicle specification and driver behavior. Revisions may be necessary to ensure that real world emissions correspond to those measured at type approval. The use of portable emission measurement systems and the introduction of the 'not-to exceed' regulatory concept should also be considered*". /1/, /2/, /3/

3 New European Driving Cycle (NEDC)

The NEDC is defined by Commission Regulation (EC) No 692/2008 of 18 July 2008 implementing and amending Regulation (EC) No 715/2007 of the European Parliament and of the Council on type-approval of motor vehicles with respect to emissions from light passenger and commercial vehicles (Euro 5 and Euro 6) and on access to vehicle repair and maintenance information referring to ECE Regulation No. 83 and No. 101 of the United Nations Economic Commission for Europe (UNECE).

The "New European Driving Cycle" (NEDC) is a synthetic cycle which was created to determine exhaust emissions and fuel consumption of passenger cars and light duty vehicles in Europe. First of all the vehicle is soaked for at least six hours at temperatures of between 20 and 30 °C. The actual driving cycle begins with a cold start. The start is followed by the Urban Driving Cycle (UDC; duration: 780 seconds, distance: ca. 4 km) which was introduced first already in 1970. The UDC was designed to represent urban driving with a maximum speed of 50 km/h, with low engine loads and with a high share of idling.

The UDC is followed by the Extra-Urban Driving Cycle (EUDC; duration: 400 seconds, distance: ca. 7km) which was introduced in 1990. The EUDC was created to represent more aggressive driving with a maximum speed of 120 km/h. The emission values of both parts are combined to one final result. The total driving distance of the NEDC amounts to around 11 km, the average speed is 33.6 km/h and the maximum speed is 120 km/h. /3/, /4/, /5/, /6/

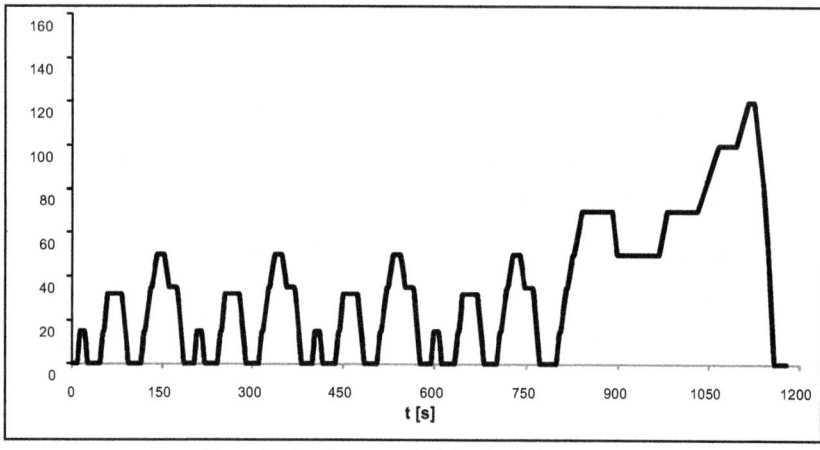

Figure 2: New European Driving Cycle (NEDC)

4 Real Driving Emissions (RDE)

According to the decisions of the European Parliament on the 03.02.2016 emissions will be measured in real traffic by using Portable Emission Measurement Systems (PEMS). PEMS are already used for measuring emissions of heavy duty vehicles according to ECE R49. The existing measurement systems had to be adapted to passenger cars and light duty vehicles. For this purpose the weight and the aerodynamic influence of the mobile analyzers were optimized.

In addition it was decided to omit HC measurement by FID due to safety aspects. Nitric oxides (NOx) and carbon monoxide (CO) will be measured on positive ignition and compression ignition vehicles. In addition carbon dioxide (CO_2) is needed as reference value for data evaluation. In a later stage Particle Number will be added. Figure 3 shows an example of a current PEMS for passenger car and light duty application. /7/, /8/

Figure 3: PEMS for passenger cars (source: Sensors)

In the current draft several parameters for the test trip are defined. The RDE trip has to include an urban part (34%) which is followed by a rural segment (33%) and by a motorway segment (33%) with a minimum distance of 16 km for each segment. The single segments are defined by the driven velocities. The trip duration shall be 90 minutes up to 120 minutes.

Comparing the different test methods it becomes obvious that RDE is going to include a much bigger variety of load points within the engine operation map than the NEDC and WLTP. Figure 4 shows a comparison of NEDC, WLTP and RDE load points.

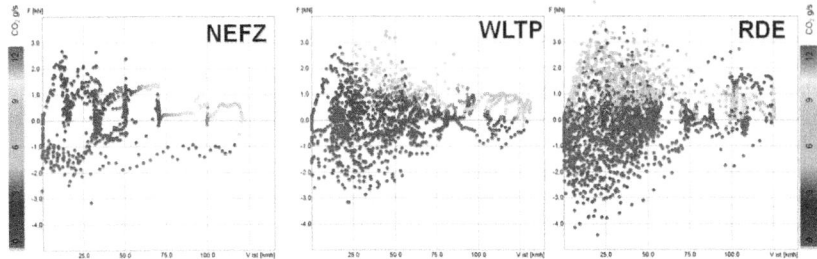

Figure 4: Comparison of NEDC, WLTP and RDE

Besides the engine load the results of RDE measurement are influenced by several parameters. Some of the influencing parameters are shown in Figure 5.

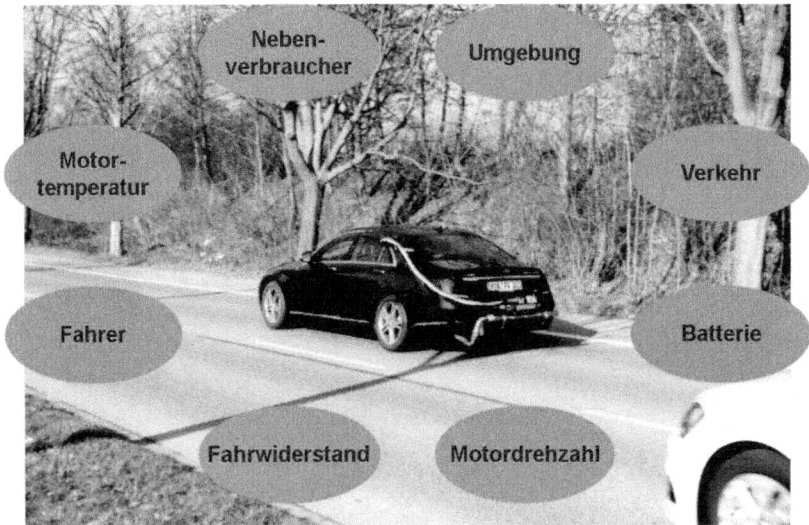

Figure 5: Parameters influencing RDE results

For RDE data evaluation at the moment two different tools are available. The Moving Average Windows method (MAW or EMROAD by Joint Research Center, JRC) is based on CO_2 values while the Standardized Power Frequency Distribution tool (SPF or CLEAR by TU Graz) is based on wheel power. Both tools shall ensure comparable emission results based on „normal driving". Figure 6 shows an example for a data evaluation by the Moving Average Window method (MAW).

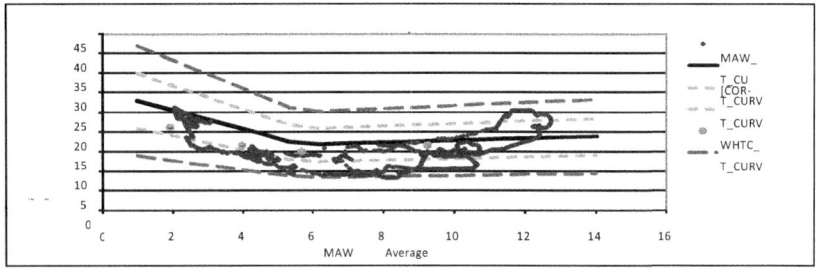

Figure 6: Data evaluation by Moving Average Window method (MAW)

The European Parliament decided to introduce a 'not-to exceed' regulatory concept. The „not-to-exceed" values (NTE) are generated by multiplying the Euro 6 limits by a conformity factor (CF).

NTE pollutant = CF pollutant x Limit Euro 6

NTE: Not-To-Exceed Limit for each pollutant

CF: Conformity Factor for each pollutant

It was suggested to specify a temporary CF for nitric oxides of 2.1. For 2020 the implementation of a final conformity factor of 1.0 plus a margin of 0.5 for considering measurement uncertainty of PEMS (CF = 1.5) was suggested. In a later stage a CF for particle number will be added. The conformity factors are shown in Table 1.

65

Table 1: NTE

Pollutant	NOx	PN	CO [1]	THC	THC+NOx
Temporary CF	2,1	to be determined	-	-	-
Final CF	1,0 + Margin 0,5 = 1,5	to be determined	-	-	-

[1]   CO emissions hall be measured and recorded

Due to the fact that RDE are measured in real traffic, this new method will be challenging for all parties involved.

5   Summary and prospect

Exhaust emissions of passenger cars and light duty trucks in Europe are currently measured using the „New European Driving Cycle" (NEDC) under well defined laboratory conditions. The NEDC represents only a small part of all driving conditions in real traffic. In Europe in the future exhaust emissions in real traffic (Real Driving Emissions = RDE) are going to be measured by using Portable Emission Measurement Systems (PEMS). Due to European air quality regulations NOx emissions are the main issue of RDE. European Commission is also interested in particle measurement especially on gasoline direct injection. RDE include a wide variety of boundary conditions. These influencing parameters have to be considered when evaluating RDE data and are addressed by route criteria, data evaluation tools and Compliance Factors (CF). The RDE Monitoring phase is going to start in 2016 for new vehicle types. It was suggested to specify a temporary CF for nitric oxides of 2.1. For 2020 the implementation of a final conformity factor of 1.0 plus a margin of 0.5 for considering measurement uncertainty of PEMS (CF = 1.5) was suggested. In a later stage a CF for particle number will be added. Due to the fact that RDE are measured in real traffic, this new method will be challenging for all parties involved.

Literature

1. Jan Cortvriend, DG Environment, European Commission. "Emission reductions resulting from the implementation of the Euro standards" 2nd International Conference, Real Driving Emissions; 17th September 2014

2. „RDE und wie es weitergehen könnte" Dipl.-Ing. Lars Mönch, Umweltbundesamt, Dessau; 27.01.2016

3. REGULATION (EC) No 715/2007 OF THE EUROPEAN PARLIAMENT AND OF THE COUNCIL of 20 June 2007 on type approval of motor vehicles with respect to emissions from light passenger and commercial vehicles (Euro 5 and Euro 6) and on access to vehicle repair and maintenance

4. COMMISSION REGULATION (EC) No 692/2008 of 18 July 2008 implementing and amending Regulation (EC) No 715/2007 of the European Parliament and of the Council on type-approval of motor vehicles with respect to emissions from light passenger and commercial vehicles (Euro 5 and Euro 6) and on access to vehicle repair and maintenance information

5. ECE Regulation No. 83 UNIFORM PROVISIONS CONCERNING THE APPROVAL OF VEHICLES WITH REGARD TO THE EMISSION OF POLLUTANTS ACCORDING TO ENGINE FUEL REQUIREMENTS

6. ECE Regulation No. 101 UNIFORM PROVISIONS CONCERNING THE APPROVAL OF PASSENGER CARS POWERED BY AN INTERNAL COMBUSTION ENGINE ONLY, OR POWERED BY A HYBRID ELECTRIC POWER TRAIN WITH REGARD TO THE MEASUREMENT OF THE EMISSION OF CARBON DIOXIDE AND FUEL CONSUMPTION AND/OR THE MEASUREMENT OF ELECTRIC ENERGY CONSUMPTION AND ELECTRIC RANGE, AND OF CATEGORIES M1 AND N1 VEHICLES POWERED BY AN ELECTRIC POWER TRAIN ONLY WITH REGARD TO THE MEASUREMENT OF ELECTRIC ENERGY CONSUMPTION AND ELECTRIC RANGE

7. European Parliament – Press release, Parliament decides not to veto car emissions test update, Brussels, 03.02.2016

8. COMMISSION REGULATION (EU) …/…of XXX amending Regulation (EC) No 692/2008 as regards emissions from light passenger and commercial vehicles (Euro 6), RDE draft

Li-ion batteries for automotive applications – Quo vadis?

Dr. Holger Fink

Robert Bosch Battery Systems GmbH and
Lithium Energy and Power GmbH

1 Abstract

Besides excellent user experience of electrical power trains, cost reduction of battery systems, legislative factors, and global urbanization will significantly drive electrification of automotive power trains within the next few years. The key success factor for a sustainable success of electrified power trains is the energy storage system. Using technical and commercial data currently available, a market prognosis for Robert Bosch for the years 2020ff is derived and is described in this paper. The most import key performance indicators (KPIs), such as safety, lifetime, and energy density, as well as specific price in EUR/kWh and the volumetric energy density in Wh/l for 2020 are presented. Besides Li-ion technology (LIT), current status and the roadmaps of the so called post Li-ion technologies (PLIT) are introduced. Post Li-ion battery systems are promising as increase in energy density and particularly lowering of cost can be foreseen. In this paper, Robert Bosch provides a possible scenario for the transition from LIT to PLIT.

2 Status of Li-ion Battery Technology

Currently Li-ion battery systems are used as high-voltage batteries in Hybrid-, Plug-In-Hybrid- and Electric Vehicles. These batteries attest that Li-ion batteries can fulfill the demanding requirements of automotive applications especially regarding safety, electrical power and lifetime.

Figure 1 shows the characteristics of Li-ion high voltage batteries currently produced by Robert Bosch.

	PHEV				EV
	Panamera	Spyder	Cayenne	Vehicle in China PHEV	F500
SOP	2013	2013	2014	2014	2013
Energy [kWh]	9,5	7,2	10,9	14,4	22,9
Power [kW]	84	240	90	134	125
Weight [kg]	134	139	136	217	272
Volume [l]	85	120	85	187	210

Figure 1: Examples of Bosch Li-ion high voltage batteries in the market.

The following KPIs are typically used to evaluate automotive batteries:

- Safety
- Lifetime
- Energy density
- Performance (electrical power related to the energy content)
- Cost
- Quality

The current status of Li-ion batteries with respect to the above mentioned KPIs is as follows:

Safety

Robert Bosch has developed a multi-stage safety concept for its automotive batteries. Dedicated selection of materials and integration of additional safety functions in the cell are combined with introduction of further safety measures in the battery system and in the vehicle. Using this multi-stage safety concept all safety tests and crash test criteria of the OEMs were fulfilled.

Lifetime

The battery systems shown in Figure 1 are designed such that a mileage of 150 thousand kilometers (based on reference driving profiles of the respective OEMs) and a lifetime in the range of 10 to 15 years are achieved. The lifetime of 10 to 15 years is characterized by a minimum of 80% of begin of life storage capacity and begin of life electrical power. After reaching this defined end of life criteria, the batteries could be continued to use.

Energy-density

The gravimetric energy density of automotive battery systems is currently in the range of 70 - 85Wh/kg, and the volumetric energy density is in the range of 110 – 140Wh/l. Technical improvement of battery cells will be able to double the energy density until 2020. The doubling of the energy density will be referred to in more detail in the later sections.

Performance

The characteristics regarding electrical power and energy of the Li-ion cells are indicated in Figure 2. The performance of the cells is described by relating the electrical peak power **P** to the nominal energy content **E**. The cells shown in Figure 2 cover the range from **P/E** = 6W/Wh (high energy cells for EV) up to **P/E** = 40W/Wh (high power cells for HEV). High electrical power is required for Plug-in-Hybrids. Today's battery systems can deliver electrical peak power of P = 90kW with a nominal energy content of E = 10kWh; the P/E ratio being P/E = 9kW/kWh. In Electric Vehicles the required performance is currently in the range of **P/E** = 5 - 6 and is expected to decrease in the future with an anticipated increase of the battery capacity to 50kWh and more.

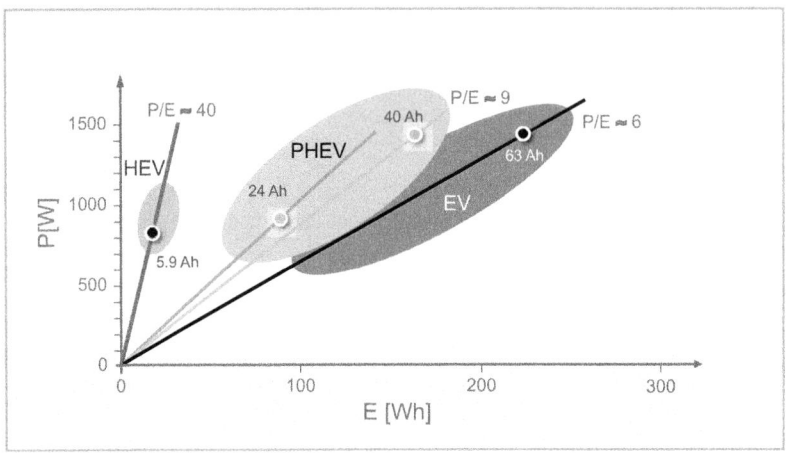

Figure 2: Electrical power and energy of Li-ion cells currently used in battery systems that are produced by Robert Bosch. The characteristic performance value P/E is the ratio of electrical peak power to the nominal energy content having a unit of W/Wh.

Cost
Price of Li-ion battery systems of the first generation are in the range of 400 – 600 EUR/kWh. A significant cost reduction of Li-ion battery cells will lead at least to half the price of Li-ion battery systems for automotive applications until 2020.

Quality
The first experience out of the field use shows that Li-ion batteries can fulfill the demanding quality requirements of automotive applications. The root cause for the field returns is mostly a faulty electromechanical component, such as a contactor. Optimization is underway in the next generation.

Summing up, it can be stated that Li-ion battery systems today already fulfills the automotive requirements regarding safety, lifetime, performance and quality. The energy density and the cost, however, needs to be improved significantly for mass adoption of Li-ion technology. The market requirements especially for these two factors (energy density and cost) will be explained in detail in the next section.

3 Market and market targets 2020

In addition to already established applications that are using high voltage batteries for HEV, PHEV and EV there will be, in the near future, new applications for Li-ion batteries in so called Boost-Recuperation-Systems (BRS) and Recuperation-Systems (RS) as described in Table 1. These new applications have battery voltages below 60V and are therefore referred to as low voltage systems.

Table 1: Field of automotive applications for Li-ion batteries

Automotive applications of Li-ion batteries					
Application	New applications		Already established applications		
	RS	BRS	HEV	PHEV	EV
Electrical driving range	-	Creeping / Parking	Up to 2 km	50km up to 100km	350km up to 500km
Energy content of the batteries	0,1kWh up to 0,3kWh	0,2kWh up to 1kWh	0,6kWh up to 1,5kWh	9kWh up to 20kWh	50kWh up to 100kWh
Power	4kW	12kW	25kW up to 50kW	70kW up to 140kW	100kW up to 350kW
Voltage	12V	48V	120V up to 300V	300V up to 450V	400V up to 750V
Specific success factors	Cost per reduced g CO_2	Cost per reduced g CO_2	Cost per reduced g CO_2 respective fuel consumption	Cost per electrical driving distance, performance	cost, energy density respective el. driving distance
Usual mounting locations	Motor compartement, cabin	Motor compartement, cabin, trunk	Motor compartement, trunk	Trunk	Under-floor
LV* / HV **	LV* battery systems		HV** battery systems		

*LV: Low voltage, that is a voltage level < 60V
**HV: High voltage, that is a voltage level > 60V

Table 1 shows that the requirements for the new applications differ significantly from the high voltage batteries. It is therefore apparent that different technical concepts are necessary. One the other hand, it is also obvious that in future the requirements in the already established applications will be different in comparison to today's technical requirements.

For the application in HEV, BRS and RS with high performance and high P/E ratio there are already very good technical concepts available for the Li-ion battery cell. Future improvement will be evolutionary via introduction of new product generations. However, there is extensive improvement required to reach the market targets for the PHEV and EV batteries that are optimized for high energy content. Figure 3 gives an overview of the cost targets as well as the targets for the volumetric and gravimetric energy densities. According to this prognosis the specific cost of batteries in EUR/kWh will decrease until 2020 by a factor of 2 to 3 and the energy densities will increase for PHEV applications a little less than a factor of 2, and for EV applications a little more than a factor of 2. To reach these demanding targets new technical concepts for battery cells and packs have to be developed. The next section is dedicated to future battery systems for PHEV and EV applications.

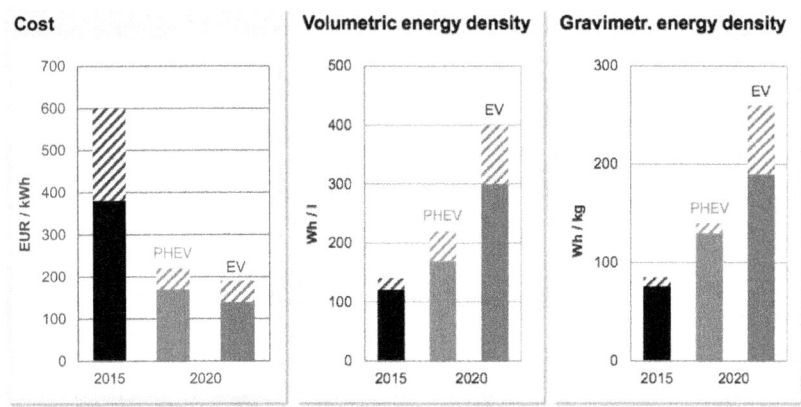

Figure 3: Current status and targets for 2020 for Li-ion batteries in PHEV and EV applications. The respective target region is the hatched area.

4 Technologies to reach the PHEV and EV targets 2020

The success in reaching the target shown in Figure 3 is very much related to the improvement of the battery cells. This is due to the fact that the cells are contributing to more than half of the cost, the volume and the weight of a battery system.

To reduce the specific cost (cost per kWh) of Li-ion battery cells both the material cost and the value add have to be considerably reduced. Starting from the current status of automotive Li-ion cells the biggest lever to reduce the cost is to increase the energy density of the cells. By doubling the energy density of the specific material and value add, cost can be reduced to half. By introducing improved manufacturing technologies the value add could be additionally reduced by 50% so that the cost until 2020 may reach a value close to the quarter of the current cost.

To increase the energy density by at least a factor of 2, improved electrode materials will be required. Figure 4 shows the capacity and the electrochemical potential of electrode materials. To reach high energy densities, the storage capacity and the cell voltage have to be increased.

In automotive cells, currently the cathodes are based on lithium-nickel-cobalt-manganese oxide or manganese spinel and in the anodes is a mixture of natural and artificial graphite is used (Gen 1 in Figure 4).

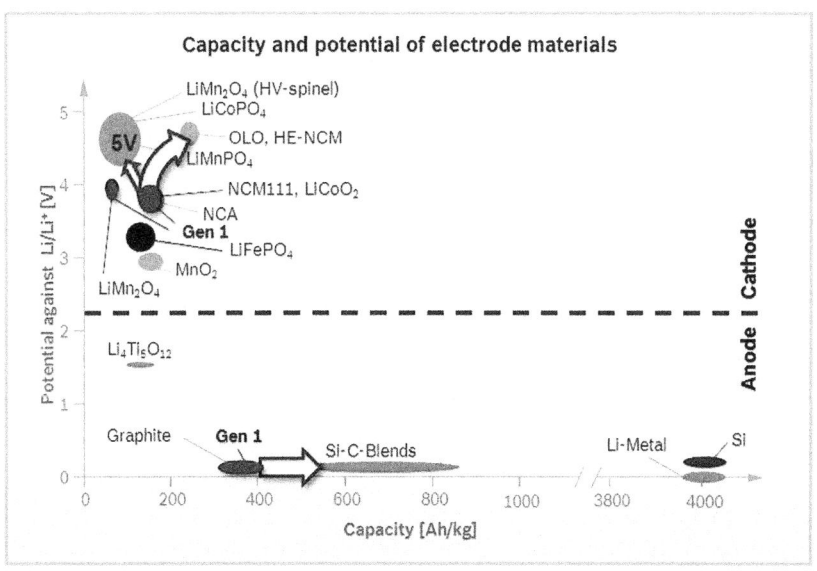

Figure 4: Capacity and electrochemical potential (against a lithium reference electrode) of cathode and anode materials for Li-ion cells (abbreviations: HV-Spinel (high voltage spinel), LiCoPO4 (lithium-cobalt-phosphate), OLO (overlithiated oxide), HE-NCM (high-energy NCM), LiMnPO4 (lithium-manganese-phosphate), LiMn2O4 (lithium-manganese-spinel), LiFePO4 (lithium-iron-phosphate), NCM111 (LiNi0,33Co0,33Mn0,33O2, lithium-nickel-cobalt-manganese-oxide), LiCoO2 (lithium-cobalt-oxide), NCA (LiNi0,8Co0,15Al0,05O2, lithium-nickel-cobalt-aluminum-oxide), MnO2 (manganese oxide), Li4Ti5O12 (lithium-titanate-oxide), Si (silicon), Li (lithium), C (carbon)).

Anode
The future development of anode materials for EV cells show a clear tendency towards silicon-graphite blends (Si-C). Silicon is used in the form of silicon oxide or silicon alloys. Figure 4 shows that the capacity of the anode materials can be more than doubled

without changing the electrochemical potential significantly. The challenges to introduce silicon based anodes are caused by the dramatic change in volume depending on the state of charge. Depending on the ratio of silicon to other components, the anode can change the volume in the range of 30% or even more with every charge and discharge cycle. This may, however, not reduce the lifetime of the cells to an unreasonable extent.

Cathode
The future development of cathode materials is currently undertaken in a manner that results in the increase of energy density and simultaneously reduces material cost. In today's Li-ion cells the most expensive component is typically the cathode.

In a first approach, the current used lithium-nickel-cobalt-manganese-oxides (NCM111, NCM523) are optimized with respect to reduction of nickel and cobalt content. This leads to so called "nickel rich" materials as e.g. NCM622 or NCM811. The series of digits represent the share of nickel, cobalt and manganese. Besides the cost reduction, the nickel-rich materials also have an increased capacity at almost the same electrochemical potential.

A second approach is trying to use materials that have a higher electrochemical potential of around 5V against lithium. Candidates of these so called "high voltage" materials are lithium-manganese-spinel, lithium-cobalt-phosphate and lithium-manganese-phosphate. The disadvantage of these materials is a reduced capacity compared to NCM111 which leads to a small increase of the energy density even in case of a significant higher electrochemical potential.

Therefore a third class of materials is currently in focus, which are called "lithium rich" and are based on lithium rich nickel-cobalt-manganese-oxides. High energy NCM is one example of these materials. The advantage of lithium rich materials is that they provide at the same time a higher capacity and a higher electrochemical voltage. Therefore the energy density can be increased considerably. Furthermore, the used raw materials are less expensive than the respective materials in today's cathodes.

The main challenges for introducing these new cathode materials are to reach the required lifetime and the required safety level, which is more difficult to provide with higher energy densities. Therefore the safety concepts have to be improved to use these new cathode materials without unreasonable safety risks.

Electrolyte
In Figure 4 it is indicated that there will be increase in the cell voltage from 4.15V up to around 4.6V. Although this seems to be only a minor change, this small voltage increase is one of the biggest challenges currently faced in the development of electro-

lytes. The availability of stable electrolytes for an increased cell voltage will most probably determine the point in time at which Li-ion cells with increased cell voltage can be introduced in automotive applications.

With the described improvements in the technology of the cell chemistry there will be automotive Li-ion cells feasible with up to 300Wh/kg gravimetric energy density and 700Wh/l volumetric energy density. The technical limit of Li-ion cells will be based on today's knowledge at around 370Wh/kg. Such high energy density values will not be reached only by improving chemistry. New mechanical concepts have to be developed as well. This will be explained in the next section.

Mechanics

In automotive applications prismatic hardcase cells and pouch cells are currently dominating the market. In consumer applications also cylindrical hardcase cells in 18650 format with capacities of up to 3.5Ah have a significant market share. Li-ion cells designed for automotive applications have current capacities up to 70Ah. The cell mechanics has to ensure that the requirements regarding safety, lifetime, performance and quality are reached in combination with the cell chemistry. The decisive criteria for different approaches are the cost and the energy densities that can be achieved. Table 2 shows an evaluation of the 3 mechanical concepts that are currently used.

Table 2: Evaluation of mechanical concepts for Li-ion battery cells.
Ratings: -- (very poor), - (poor), 0 (neutral), + (good), ++ (very good)

Criteria		Prismatic Hardcase	Pouch	Cylindrical Hardcase
Cost (per kWh)	Material cost mechanics	0	+	−
	Value add	0	−	++
Energy density (volumetric respectively gravimetric)	Utilized cell volume	−− ~65%	++ >90%	+ ~85%
	Weight	0	++	−
	Utilized module volume	+	−	−−

Cost: Because the manufacturing concept has been optimized over several product generations cylindrical hardcase cells show an advantage in the value add cost. This advantage is bought with a little higher specific material cost. Considered the overall specific cell cost per energy for the cell mechanics and assembly the cylindrical hardcase cell is cheaper than pouch and prismatic hardcase cells.

Energy density: Table 2 illustrates that prismatic hardcase cells have a very poor ratio of the volume utilized for the chemistry related to the whole cell volume. Only about 65% is utilized for the chemistry. Cylindrical hardcase cells can utilize about 85%, pouch cell can achieve a ratio of up to 90%. Looking on the efficiency of the cell mechanics regarding building battery modules consisting of several cells, prismatic hardcase cells have by far the highest efficiency. Overall the pouch cell technologies has the best efficiency of the currently used cell mechanics.

In essence, it can be stated that none of the current concept has significant advantages that it will dominate the future trend. It is assumed that the cell manufacturers will stick to their established concepts in order to amortize the high investment. To achieve the targets 2020ff it is most probably required to develop new mechanical concepts that are distinct from the current mechanics.

5 Post Li-ion Technologies

The separation between current Li-ion battery cell technologies (LIT) and so called post Li-ion technologies (PLIT) may lead to ambiguity and therefor a clarification of the terms seems to be appropriate.

In this article the term Li-ion technology is applied for Li-ion accumulators in which lithium is in both electrodes mainly intercalated, that is stored in the chemical compound of the electrodes without changing the structure of the electrodes during this process. Cathodes of such cells are based on lithium-metal-oxides. The anode materials are based on graphite, amorphous carbon or lithium-metal-oxides such as lithium-titanate-spinel. The ion transfer takes place in an electrolyte which allows free migration of lithium ions.

The term post Li-ion technology is used for Li-ion accumulators in which both electrodes are not performing intercalation of lithium. At least in one electrode a chemical reaction takes place which is correlated with a conversion of substances. So far the term post Li-ion technology was used synonymously with lithium-sulfur or lithium-air accumulators. The principle structure of such cells is described in Figure 5.

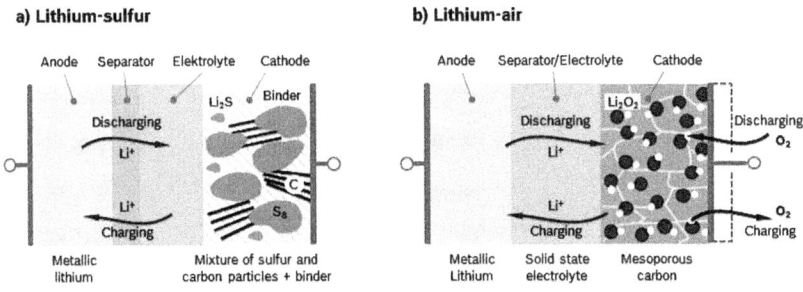

Figure 5: Typical structure of a) lithium-sulfur and b) lithium-air accumulators.

In lithium-sulfur accumulators the charge transfer in the electrolyte is performed via Li-ions. During discharging of the cell, sulfur is transferred in the cathode into lithium-sulfides. As intermediate products mixtures of Li_2S_8, Li_2S_6, Li_2S_5, Li_2S_4, Li_2S_3 and L_i2S_2 appear – so that while the sulfur content decreases, the lithium content keeps increasing. In a fully discharged cell the lithium-sulfide Li_2S is built, however, the opposite sequence takes place while charging the cell. Instead of using lithium, the anode can also be made of silicon or tin. Lithium-sulfur accumulators promise to achieve high gravimetric energy density but only moderate volumetric energy density. The cell is desirable as very low cost raw materials can be used compared to Li-ion technology.

In Li-air accumulators the charge transfer in the electrolyte is performed via Li-ions too. While discharging the cell oxygen is converted in the cathode to lithium-peroxide. This requires the presence of positive charged lithium ions. The conversion takes place in the opposite direction during charging the cell and oxygen is released. Usually the anode is based on metallic lithium, e.g. a lithium foil. Lithium-air accumulators can theoretically achieve a very high energy density. The realistic technical achievable values are depending very much on technical concepts for the exchange of oxygen in the cathode and on the availability of solutions that can avoid side reactions e.g. with H2O.

Both lithium-sulfur and Li-air accumulator technologies are currently in an early research state. Before starting the industrialization for automotive applications, many technical challenges have to be solved. The introduction into series production is not expected to happen before 2025. But there are other less prominent post Li-ion technologies available that might be ready for series production earlier. These technologies make use of the same cathode materials that are utilized in standard Li-ion technology and have anodes based on metallic lithium. Beside liquid electrolytes also solid state electrolytes are used in some of these technologies. The structure of such a battery cell with a polymer based electrolyte is shown in Figure 6.

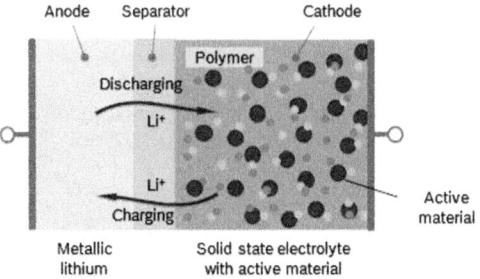

Figure 6: Typical structure of a PLIT battery cell with a solid state polymer electrolyte.

Some of the technical challenges with this technology are as follows:

- Lithium anode: During charging and discharging metallic lithium is removed and accordingly attached. Especially with high current densities there is a tendency to build lithium dendrites. It has to be ensured that these dendrites can't force through the separator in order to avoid internal shorts.

- Separator: The technical concept in Figure 6 needs a thin separator which is able to provide the above mentioned requirements at the layer interface to metallic lithium.

- Solid state electrolyte:

  - Performance: the required P/E ratio for automotive applications is in the range of 3 - 5kW/kWh. For this purpose, new materials with high conductivity and transfer rate are necessary. These materials must be chemically stable against lithium and against the active materials of the cathode.

  - Temperature: the ionic conductivity is usually strongly depending on the temperature. In the relevant temperature range of automotive application (-30°C up to 60°C), the required performance has to be achieved.

- Layer interface: The electrical resistance and the power performance is mainly determined by the ionic conductivity over the layer interfaces, which also requires to be optimized.

Lithium battery cells with solid state electrolyte may achieve in the described technology probably energy densities of up to 450Wh/kg (gravimetric) and 1000Wh/l (volumetric). High capacity as well as the electrochemical potential of metallic lithium result in large energy density of these cells. The forecasted values are above the technical limit of standard lithium ion technology which is expected to be at about 350Wh/kg and 900Wh/l.

Solid state Li-Ion cells are still in the phase of research. Prototype cells are, however, available which give a realistic forecast that the industrialization can start as early as 2017/2018. In case of success, this could lead to a first series introduction towards the end of 2020. From today's perspective the transition from standard Li-ion technology to a post Li-ion technology will not be disruptive but rather smooth using a step by step replacement process. Among other factors, this prognosis is based on the assumption that post Li-ion technologies will achieve in the first generation a P/E value of up to 3kW/kWh which are suitable for EV. The required performance for PHEV and HEV will not be achieved at the beginning. Therefore both technologies will be used at least in different applications in parallel for a length of time.

6 Challenges for Li-ion cell manufacturers

The article describes the technical measures that may provide until 2020 an improvement of the specific cost per kWh as well as the volumetric and gravimetric energy densities by at least a factor of 2. In addition to the technical challenges to achieve these targets, the battery cell manufacturers have to allocate extremely high capital expenditures to increase the manufacturing capacities. In parallel, large R&D costs are required to develop and industrialize post Li-ion technologies. In case the challenging cost targets are met, post Li-ion battery technologies will at the latest provide the tipping point for electrification.

How long will an electric car's propulsion battery last?

Detlef Hoffmann
Business Development Manager

SGS Germany GmbH, München, Transportation, Battery Test House
http://www.sgs-cqe.de/de/battery-testhouse.html
detlef.hoffmann@sgs.com

Introduction

Each rechargeable battery irretrievably loses storage capacity over time. A phenomenon that everyone knows from home electronics: For example, a tablet's usable capacity of 80 % after 500 cycles is considered normal, 80% after 1,000 full cycles is quite good. The lithium ion batteries in electric vehicles also lose capacity over time. The result is a permanent reduction of range and perceivable value loss for electric cars.

In contrast to vehicles with an internal combustion engine, electric cars do not need a gear box, clutch, converter, gas tank, dynamo, and many other components. Therefore, only little mechanical deterioration is possible. According to studies, repairing e-cars is on average up to 35% less expensive than the maintenance of cars with an internal combustion engine. Electric cars can have low maintenance cost and a very long life. However, the aging of the drive battery significantly influences the magnitude of this advantage.

Causes for aging of a battery

Over the course of time, the characteristics of a battery system change. Two effects can be observed during the aging process: on the one hand, the battery's capacity gradually drops whereby the range of the electric vehicle suffers. On the other hand, the internal resistance of the battery increases, what can lead to a loss in performance, for example during acceleration. The battery cell is made up of various materials that are in contact and can react with one another. Physical-chemical effects are responsible for the aging process of a battery cell – for instance, the loss of electrode surfaces and rechargeable electrode materials, the disconnection of electrical conduction paths or an increased charge transfer impedance. Many factors cause the aging of a battery: the loss of electrode surface area, the loss of rechargeable electrode material, the disconnection of electric conduction paths, and the increase of the charge transfer impedance.

Changes at the border surface between anode and electrolyte (the so-called Solid Electrolyte Interface, SEI) are to be emphasized in particular. Through chemical processes this layer grows more and more over the course of the lifespan. The capacity of the battery suffers as a result. The lithium ions that are converted into chemical compounds can no longer react electrochemically. In addition, the layer, through which the lithium ions in the electrolyte have to migrate, thickens. This can in turn increase the resistance within the battery. Beyond this, mechanical strains also lead to the aging of the battery for example, when lithium-ions are stored within the active materials. This can cause mechanical stress causing ultimately the formation of cracks within the particles, which then break apart. Additionally, a degeneration of the binder can cause the electrical disconnection between the individual particles of the active material.

Considering the aging of a battery module or pack, the effectiveness of the cooling of each single cell may be important. To have the maximum range for a battery electric vehicle, full charge-discharge cycles have to be used. To avoid overcharge and over-discharge of each single cell, there is a battery management system, which receives the cell voltages from the cell supervision circuit. Without active balancing, the charging process has to stop, when the first cell reaches its charge cutoff-voltage. Also the discharging must finish, when the first cell reaches its discharge-cutoff-voltage. This is why the weakest cell determines the overall capacity.

Impact on vehicle range, power, residual value

The battery is the decisive factor in terms of the car's range. The potential distance which can be travelled with one full charge is one of the most important arguments for or against the purchase of an e-car. If the already limited capacity of the traction battery decreases over the period of use, it has far-reaching consequences: Even with a fully charged battery the electric car is covering less distance and is not performing at maximum capacity, e.g. in the acceleration process. The result is a considerable loss in value. If the battery of a used e-car has to be replaced during resale or at the end of a leasing period, it may quickly become an expensive affair. For example, the replacement battery for 24kWh energy storage may cost more than 5.000 €.

What level of capacity loss is normal?

The usable capacity of rechargeable lithium ion batteries, the so-called State of Health (SOH), is subject to aging in calendar as well as cyclical terms. This applies for traction batteries of electric vehicles too. On the one hand, this means that the storage capacity of car batteries is lost all on its own over time, for example during storage or while the car is parked. On the other hand, with an increasing number of operational cycles consisting of charging process and driving, the batteries deliver less and less performance. The tolerable capacity loss in traction batteries of e-vehicles is a subject of controversial discussion among users. The US non-profit organization, "Plug in America" (PIA) examined the battery of the Tesla Roadster among others. The basis was a field test, in which anonymous reports and data from more than 100 Tesla owners were evaluated. The result: after 160.000 kilometers of driving distance, the Tesla battery, based on 18650 LCO cells, shows a capacity of 80 to 85%. Unlike results from PIA's LEAF-Battery Survey, no significant correlation was reported between climate and battery longevity. This may be a result of the active cooling system of the Tesla Roadster battery.

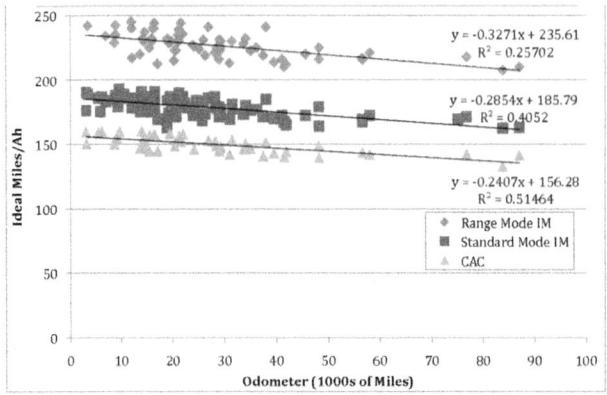

Figure 1: Capacity versus miles driven (Source: (1.) Plug in America)

The Tesla Model S battery is based on 18650 cells with 3.1 Ah capacity and a NCA cathode. Figure 2 shows field results of the car's range versus mileage as published by the Tesla Motor Club. There is a significant scattering of values and there is an average drop of approximately 5% range after 50.000km. However, there was only range 1% loss for the next 50.000km. This is quite different from the Tesla Roadster aging behavior. For a reliable prediction, it is important to gain field experience in order to develop a good understanding of the relevant aging processes during operation.

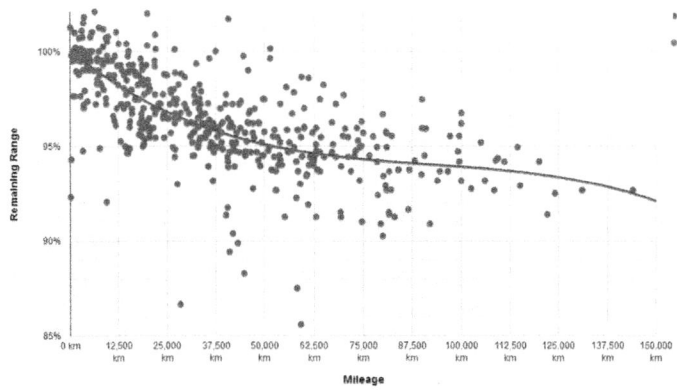

Figure 2: Tesla Model S Remaining Range as a function of mileage,
Source: (10.) "MaxRange Tesla Battery Survey" of the Tesla Motor Club

How to determine battery lifetime in the laboratory?

Measurement of calendar and cycle lifespan

For the vehicle as well as for the battery manufacturer it is important to be as familiar- as possible with the aging behavior of their the batteries – either, in order to prolong their lifespan through technical improvements or in order to be able to give reliable guarantee assurances. Practical experiences with the aging of batteries are still limited due to the relatively new application of lithium ion technology in vehicles. As an alternative, the industry is turning to accelerated aging processes in which the entire life cycle of a traction battery is simulated in the shortest time span possible but still within the operation limits. In test laboratories aging in terms of calendar and cycle time is separated, as they are based on different underlying physical laws. This way, a more accurate extrapolation of longer lifespans and driving performances is possible.

Calendar aging process is shown in figure 3. It may be accelerated by increasing the temperature. Parallel to this, a higher state of charge during storage can also be factored in as a further acceleration effect. Performance parameters including the actually useable capacity and resistance are measured periodically to monitor the aging. For many vehicle applications, end of life criterion is 20% capacity loss.

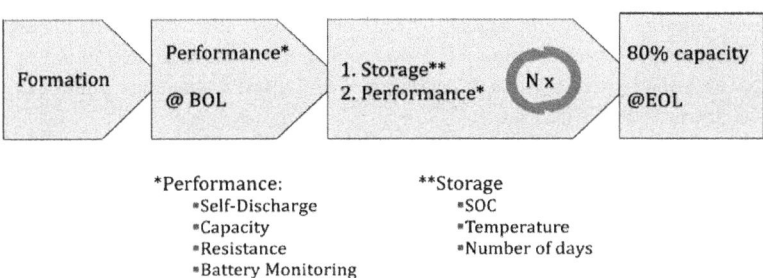

Figure 3: Determination of the calendar life time in the laboratory

Cyclic aging process is detailed in figure 4. In many cases, it can be accelerated by increased temperature. This may not work for every cells type, because some are designed for operation at higher temperature range (e. g. 40°C instead of 25°C). More charge throughput per period is another option for acceleration. The later can be achieved through a higher C-rate or an increase in the depth of discharge (DoD).

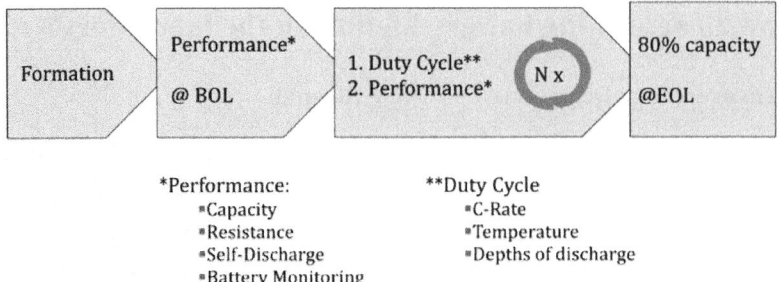

*Performance:
- Capacity
- Resistance
- Self-Discharge
- Battery Monitoring

**Duty Cycle
- C-Rate
- Temperature
- Depths of discharge

Figure 4: Determination of the cycle lifespan in the laboratory

In any cases, the aging procedure must keep carefully within the current, voltage and temperature limits defined by the cell manufacturer. This is also true for any individual cell in a pack.

Cycling a single cell

Figure 5 shows an example of a charge-discharge-cycle of a commercially available lithium-ion 18650 cell with NCA cathode and a rated capacity of 2.9Ah.

The cell surface temperatures were recorded. During the cycling, a change of cell surface temperature from Joule heating by the current flow and exo- and endothermic chemical reactions was observed. The cycle-process was programmed in a way to keep within the limits for voltages and currents at a given temperature as defined by the cell manufacturer's data sheet.

Voltage and current are shown in figure 5. First phase is charging to 100% SOC with constant current C/2, followed by a constant voltage of 4.2V, which is the charge cut-off. There is a charging-stop criterion of 59mA for this type of cell. After the zero current period, there is a phase of complete discharging which starts with a significant voltage drop and runs with constant current of 1C until the discharging cut-off of 2.5V is reached. There is a voltage increase step when the current is switched to zero. The charging is done with constant current of C/2 onto the charging voltage cut-off followed by a constant voltage phase with decreasing current.

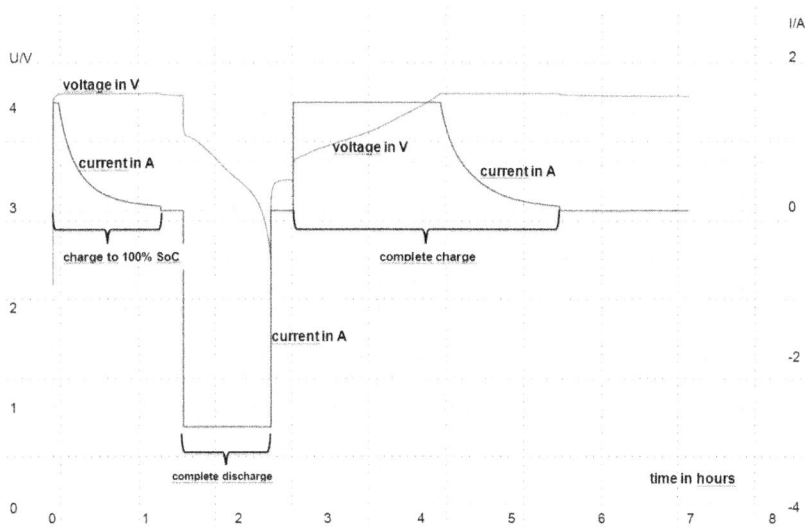

Figure 5: Charge to 100% SoC followed by a full cycle (charge, discharge) current and voltage as functions of time

The internal resistances were calculated from these voltage drops to be:

- 141mΩ for the 0,409V drop at start of discharge   (SOC=100%, I=2,90A)
- 295mΩ for the 0,856V step at end of discharge   (SOC=0% , I=2,90A)
- 167mΩ for the 0,242V step at the begin of charge   (SOC=0% , I=1,45A)

These resistances and the values determined by the electrochemical impedance spectrometry shown in figure 8 are comparable in magnitude.

The capacity was determined by measuring the current accurately and integrating it over the time. This procedure is known as Coulomb counting. Figure 6 shows the capacity decrease as given in the data sheet from the cell manufacturer. 80% SoH are reached after 400 cycles at 25°C.

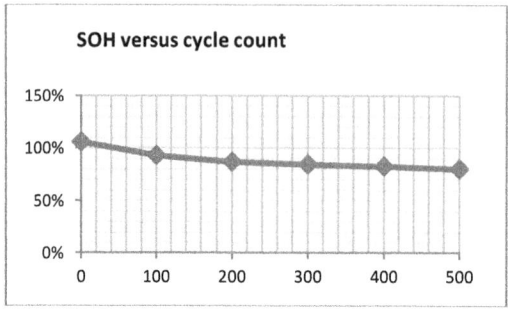

Figure 6: Remaining capacity in % as a function of cycle counts for the 18650 cell, temperature was 25°C, values are taken from the cell manufacturer's data sheet

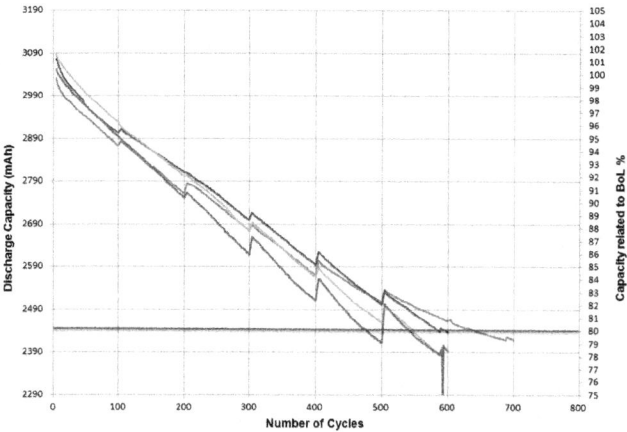

Figure 7: Remaining capacity as a function of cycle counts for 4 individual 18650 cells at 40°C

Figure 7 shows the capacity of 4 individual 18650 cells of the same type measured as a function of cycle number at 40°C with respect to their individually determined begin of life (BoL) capacity. The SoC was determined by Coulomb counting. 80% SoH is reached between 480 and 635 cycles. After each 100 cycles the cells were stored at 25°C and the impedance was measured. The usable capacity increased again during this storage (idle) time. The cyclic aging which was experienced by the cell before, leads to a faster decrease when further cycling is performed. This was observed also for several other cell types and can be explained as follows: During the first 100 cycles there is a loss of Lithium, which is available to take part in the cycling process.

During idle, some of the lost lithium becomes available again. However, some mechanism for lithium loss during cycling changes the cell permanently. This is why, in the subsequent cycles, the loss of lithium for cycling progresses faster than before.

Impedance increases with aging

The aging of the cells depends, among others, on the choice of the active material, the electrolyte, and the binder. With respect to materials and cell design, the aging mechanisms were examined directly on the cells. For this, the electrochemical impedance spectroscopy (EIS) was used. A four- point measurement setup was found to be essential to avoid the impact of voltage drops from current flow via cell external contacts. The cell impedance depends on temperature and came out to be significant lower for low SoC – both for new and old cells. Therefore, to monitor the aging, the impedance was be measured always at a well defined SoC. Figure 8 shows results for 50% SoC and 25°C.

The so called high frequency intercept with the real axis as shown in figure 8 is representative for resistive losses and a moderate increase of this resistance was observed, mainly for the first 300 cycles. Reasons for the overall resistance growth may be an increase in electrolyte resistance due to lithium loss or an increase in the resistance of electrode particles or of the binder. The intercept frequency resistance depends on a combination of several cell parameters. It came out to be of limited value for age indication for the considered cell type. The semi-circle at medium frequencies shows the behavior of a capacity in parallel with a resistor. This comes from the electrode-electrolyte interface, e. g. from a double-layer capacitance of the SEI-layer in parallel with its resistance. With the increase in the number of cycles, there is a significant continuous, approximately linear, growth of the diameter of the semi-circle. This makes it a good parameter to monitor cyclic aging for this cell type.

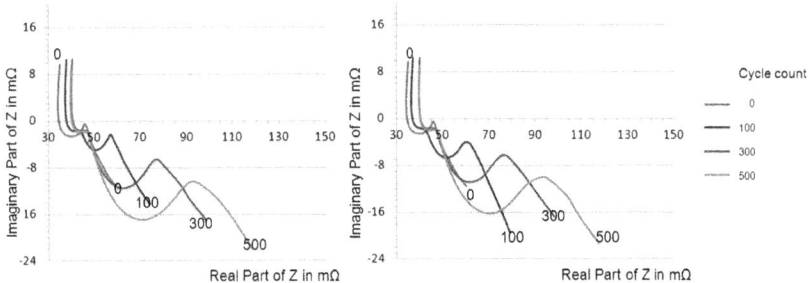

Figure 8: EIS-Measurement results shown as Nyquist plots, the number of cycles is given as a parameter, measurements were performed at 50% SoC and 25°C

Figure 8 shows the individual Nyquist plots from two cells from the same sample. The two additionally measured cells showed very similar results. From this, we conclude, that impedance increase can be used as an indicator for aging, at least for certain cell types.

Battery pack state of health

For the entire battery system, a two-step approach is generally used for determination of lifetime: first, the aging of the cells is determined. In a second step, a lifespan examination at the battery system level is essential. It is the only way to determine the effect of the pack-specific arrangement of the battery cells and the influence of the battery management system, which performs cell balancing and thermal control.

Critical during the testing procedure is the consideration of all factors causing battery aging as well as the provision of realistic battery loads – in other words that the strains of a battery are simulated as in actual driving operation. Examples of such driving cycles are available, e. g. EUCAR-HEV Specification 2005 and FreedomCar DOE/ID-11069. Measured, vehicle-specific driving profiles can also be tested. Dynamic thermoregulation for the liquid temperature control has to be implemented in the test setups to simulate the thermal management in the vehicle. Figures 14 and and 15 show measurement results for a module consisting of high capacity automotive pouch cells.

How to determine battery state of health in short time?

For a used electric car, the batteries state of health is an important factor of the value of the complete car. As battery state of health depends on the way the car is handled and driven, it is also an important factor for leasing inspections. With knowledge from the performance of the battery at begin of life, it is possible to determine the state of health without Coulomb counting during a complete discharge – charge- cycle. As demonstrated in (4.), the information on the SoC (state of charge) is also included in the open circuit voltage (OCV) of the cell.

Cell level SoH determination

For the cell considered here, figure 9 shows the details of the voltage drop after switch-off the discharge current. The open circuit voltage approaches a value, which represents the SoC of the cell. The time needed to reach a stable OCV-value depends on the SoC itself. The comparison of figure 8 and figure 9 shows significant differences between the new and aged cell. When the discharge current is switched to zero, the aged cell shows a higher voltage drop than the new one. This corresponds to the increase of the internal resistance which was also measured with the impedance spectrometry shown in figure 8. The comparison of figure 9 and figure 10 indicates that the open circuit voltage of the old cell increases more slowly to a stable value than the new one. The short time open circuit voltage (OCV) drop seen directly after current switch-off is significantly higher for the old cell than for the new one.

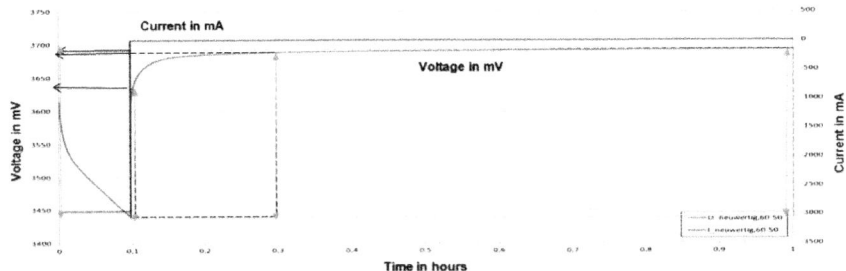

Figure 9: Cell at BoL, Voltage response to a current pulse applied to discharge he cell from 60% to 50% SoC

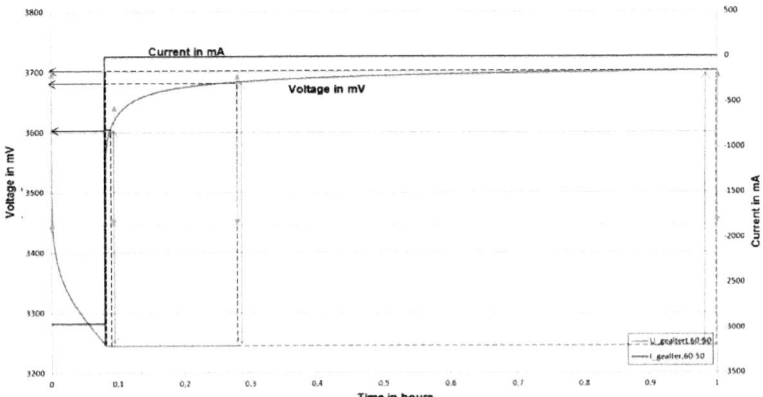

Figure 10: Cell at EoL(79.9% SoH): Voltage response to the same current impulse as in figure 9

The OCV of the considered cell in the final state (e. g. 3 hours after the switch) shows a monotonous increase with the SoC. If the capacity is measured in Ah by current counting, the OCV for a cell with the same amount of charge differs for a new and the old cell as shown in figure 11.

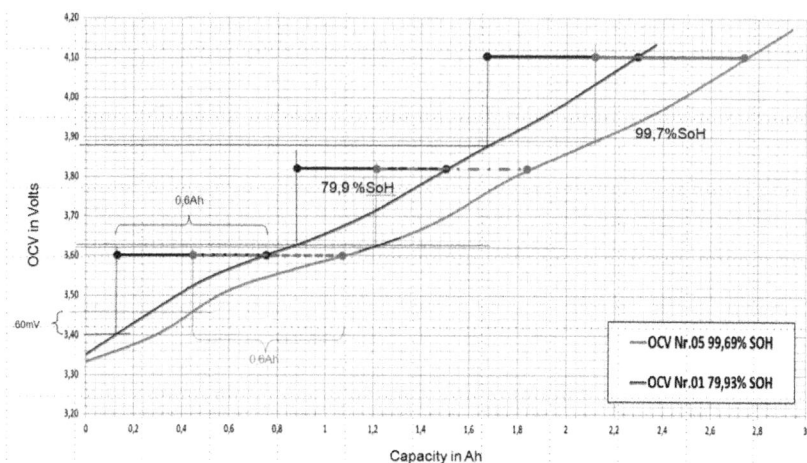

Figure 11: OCV as a function of cell capacity for a BoL Cell (lower curve) and an EoL cell (upper curve)

As seen in figure 11, in the lower SoC range (3,6V), the same amount of discharge leads to a significant OCV difference between the old cell and new one. This procedure is applicable for a quick measurement only, if the cell is already in the region of 3,6 V. In this case, a comparison can be done after approx. 12minutes discharge leading to a difference of 60 mV during the full life time. Discharge the same amount from the high or medium OCV range gives a very small OCV difference between old and new cell, for the type considered here.

If the SoC is scaled to 100% for the fully charged cell at EoL, both OCV curves fit together quit well, as shown in figure 12. There are SoC deviations of maximum 4% SoC in the low and high SoC regime. Best coincidence of the OCV-curves are between 60 and 76% SoC. For this example, a voltage drop of 8.6mV per 1% SoC is determined at room temperature. Within the reported accuracy, the SoC can be determined from OCV independent of the batteries age. Similar results were also found by the author for other lithium-ion cell types and are also reported in (4.), together with a correction of the moderate temperature dependence of the OCV.

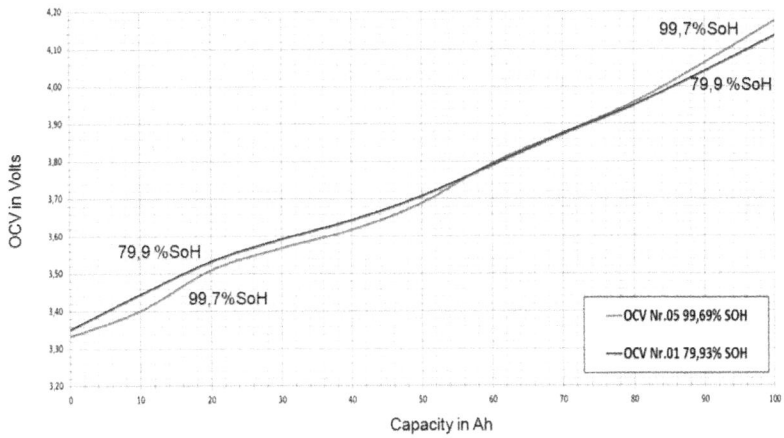

Figure 12: OCV as a function of the SoC for a new and an aged 18650 cell, EoL cell scaled to 100% SoC when fully charged

In order to allow for an age measurement from any state of charge and from any state of health on cell level, the following procedure was used:

1   Charge the cell until the charge cut-off current is reached, this determines 100% SoC

2   Discharge the cell for a certain Δ SoC, e. g. -10%, which can be determined by current counting.

3   The OCV must be recorded until there is no further increase. Otherwise the remaining increase must be extrapolated.

4   Then repeat the discharge (No 2) of -10% SoC until discharge cut-off is reached.

To find out any hysteresis, the analog procedure was also performed in the charge mode. In this case, for the high OCV-levels near the charge cut-off voltage, the current must be reduced in order to charge with constant maximum voltage. The C/100 constant discharging and charging OCV was also measured for comparison. Figure 13 gives an overview on the voltages measured during charging and discharging the cell at BoL. The nominal current for charging is 1,45A, for discharging it is 2,9A. This current relation shows itself in the different voltage drops of the charging and discharging curves. There is a significant increase in the voltage drops for all SoC levels with aging, as shown in figure 14 (aged cell) in comparison with figure 13 (new cell). This is also due to the increase of the internal resistances with aging.

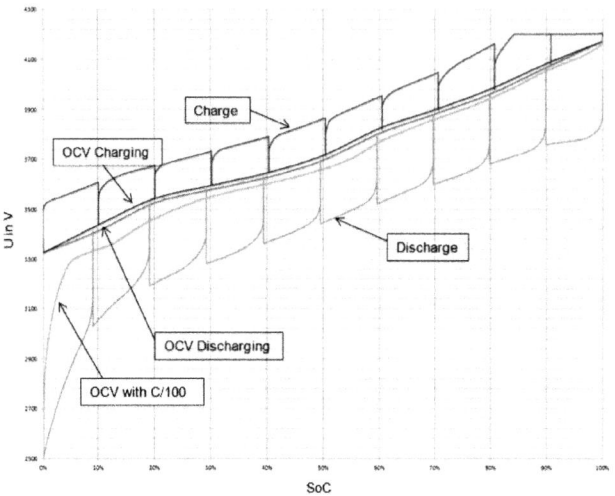

Figure 13: Cell at BoL: Voltage drop as a function of the SoC

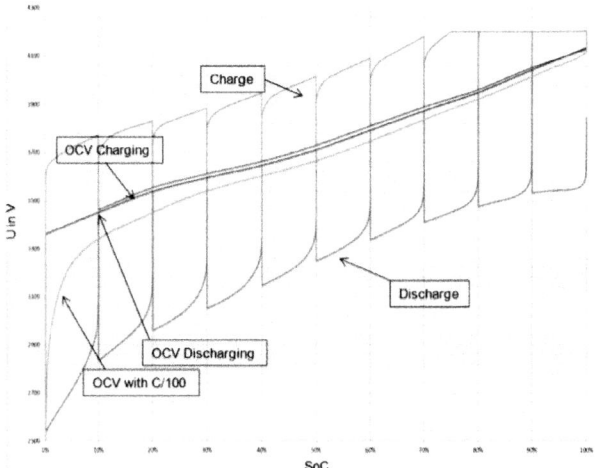

Figure 14: Cell at EoL: Voltage drop as a function of the SoC

The curves given on top of the figures have been recorded during charging. After three hours of idle without external current flow, the OCV values were measured and are represented by the OCV-Charging curve. The lowest curve was measured during discharge. The enclosing curve above "OCV discharging" shows the OCV values after three idle hours. These are close to the OCV steady state values for charging. The result of the measurement with constant C/100 is also given for comparison. The diagrams in figure 13 and 14 cover the complete voltage range between 2,5 and 4,2 V. In the range below 30% SoC, the voltage drops are bigger. This is due to the impedance of the cell, which increases with decreasing SoC.

The complete 10% SoC step by step discharge measurement took 31 hours. If this results are available for a new and an accelerated aged 80% SoH cell, the unknown age of any cell of this type can be determined in short time, by measuring the OCV after an appropriate idle time at the reference temperature. After this, a 1C discharge impulse is applied for 5 minutes on the cell under tests and the voltage is recorded as a function of time for 10 minutes. The voltage drops and steps are determined at current switch-off and after 20s and 5 minutes. As shown above, the magnitude of the voltage drops are the measure for the aging. If needed, the SoH-interpolation between 80% and 100% can be improved by the known dependence of the impedance on the number of cycles at low frequencies or by calibration with the help of intermediate SoH voltage drops for the initial data set.

Module level SoH determination

The measurement principle explained above was applied to a module consisting of 40Ah low resistance cells, the aging characteristic of those had been measured before. A cyclic aging was performed also on the module and the voltage drops were measured with 3C discharge current pulses. Results are presented in figure 15 and show an almost monotonous increase up to at least 2000 cycles for voltage drop measurements after 60s and more.

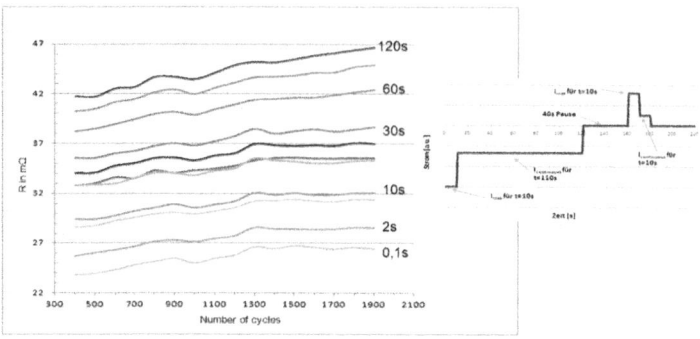

Figure 15: Change of different dynamic resistances for a 3C discharge of a module with the number of Cycles

Also the OCV was measured on this module. For this, the module was fully discharged and after this it was charged with 20Ah. The OCV was measured after this. Figure 16 shows the dependence of the OCV-level from the number of cycles. It is clearly demonstrated, that aging has an impact on the OCV. 20Ah for a new module means 50% SoC. For the aged one, the OCV starts with a higher value due to the capacity loss. The module results can be traced back to the single cell results, which are not published in this article.

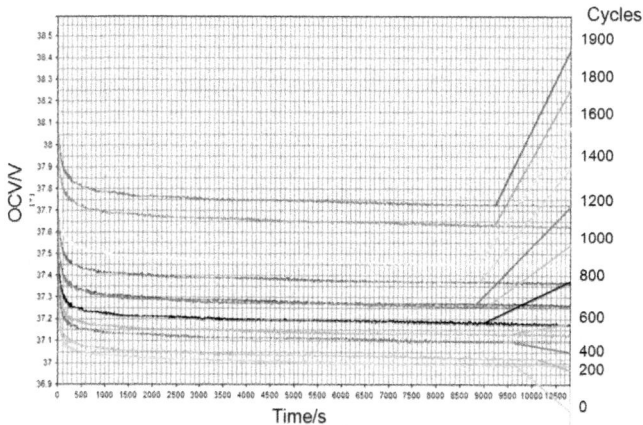

Figure 16: OCV of a module, measured for different states of cyclic aging, 1000 cycles represent 93,7% SoH (determined by Coulomb counting)

Conclusion and Outlook

Battery aging behavior in the field has significant impact on customer satisfaction, no matter whether mobile, automotive or stationary applications. It is not easy to predict, which makes it an issue for further evaluation and research. If the aging behavior of the cell and module type has been determined in the laboratory in advance, there is the possibility for an independent short time determination of SoH on cell and module level. This will help to understand individual scattering of aging and range reduction in the field and allows for a profound and fast evaluation of the SoH and residual value of drive batteries in used cars.

The aging of a battery cannot be prevented. However, there are promising approaches with which the industry wants to achieve a significantly longer lifespan for lithium ion batteries. The question of battery durability is not only of major significance for the application in e-mobility. Lithium-ion batteries are also used as buffer for wind and solar energy. In that area, the expected lifespan may be as high as 7.000 cycles and is an important calculation factor as investment decisions have to be made over long periods of time.

Modern batteries are complex systems. All progress that can be made in terms of lifespan, for example through improved battery chemistry, possibly has other disadvantages. The use of new materials, for instance, could increase cost or engender additional safety risks. For this reason special, comprehensive test series are of major importance during the market launch of any new energy storers.

Bibliography

1. Tom Saxton, Plug In America's, Tesla Roadster Battery Study, July 13, 2013

2. P. H. L. Notten, D. L. Danilov, „Battery Modeling: A versatile Tool to design advanced battery management systems" Advances in Chemical Engineering and Science, 2014, 4, 62-72

3. Y. Zhang, C.-Y. Wang, X. Tang, „Cycling degardation of an automotive LiFePO$_4$" Journal of Power Sources 196 (2011) 1513-1520

4. S. Takenaka, Y. Kanai, N. Yamashita, „Real-time and highly accurate SOC estimation method for management system of lithium-ion batteries", Posterpresentation AABC 2014, Mainz

5. T. Mezger, P. Nobis, Analysezentrum für Elektromobilität (AZE), „Batteriemessungen im Rahmen des Projekts AZE", BMWi Förderkennzeichen: 03KP202

6. N. D. Williard, "Degradation Analysis and Health Monitoring of Lithium ion batteries" University of Maryland, Master Thesis Fall 2011

7. J. Groot, "State of Health Estimation of Li-Ion Batteries; Cycle Life Test Methods" Division of Electric Power Engineering, Department of Energy and Environment, Chalmers University of Technology, Goeteborg, Sweden 2012, Thesis for the degree of licentiate of engineering

8. R. P. Ramasamy, R. E. White, B., N. Popov "Calendar life performance of pouch lithium-ion cells", Journal of power Sources 141 (2005) 298-306

9. Uwe Tröltzsch, "Modellbasierte Zustandsdiagnose von Gerätebatterien", Doktorarbeit an der Universität der Bundeswehr München, Neubiberg, 7.12.2005

10. Max Range Tesla Battery Survey of the Tesla Motor Club, access date 20.01.2016

Increased efficiency in the calibration process of automotive Li-ion battery systems

Dipl.-Ing. Lukas Behr, Dr. Ulrich Zimmermann, Dipl.-Ing. Stefan Trinkert
Robert Bosch Battery Systems GmbH

Dr.-Ing. Thomas Kruse
ETAS GmbH

Dr.-Ing. Stephan Rees, Dipl.-Ing. (Fh) Friedhelm Bröckel
Robert Bosch GmbH

Prof. Dr.-Ing. Jian Xie
Institute of Energy Conversion and Storage, Ulm University

Abstract

This paper discusses the use of model-based calibration for Li-ion batteries. The scope is set on the modeling algorithm, in this case the Gaussian Process Regression (GPR) in combination with an external dynamic structure (NARX), and its capability to describe dynamic battery behavior. A data-driven model is generated based on measurement of a Stuttgart Cycle. Model quality is evaluated on an Artemis Cycle. The achieved results show that the GPR can be used in a future model-based battery calibration process for dynamic applications in order to increase calibration efficiency.

1 Introduction

More stringent CO_2 and exhaust emission standards that will come into force in the future are driving a rising trend towards hybrid and electric vehicles. This demands the development of advanced lithium-ion battery systems. Challenges are posed by ever-shorter development cycles and the need to further optimize in terms of range, performance, and cost [1]. Mastering these challenges is crucial for the success of these vehicles in the market.

1.1 Calibration of automotive Li-ion battery systems

The Battery Control Unit (BCU) for lithium-ion battery systems controls the system's states and guarantees a safe and optimal operation. The calibration of the BCU software is a tedious and time-consuming process demanding for more efficient solutions. Similar to the calibration of an ECU for internal combustion engines, multiple parameters for complex models, functions, and maps for adaption to each project specific battery have to be determined. Due to the electrochemical nonlinear behavior, the challenging automotive requirements, and increasing complex software structure the calibration process today requires very extensive, cost- and time-consuming testing procedures.

1.2 Model-based calibration process

In order to increase the efficiency of the battery calibration process, a model-based calibration methodology is proposed. Figure 1 shows a model-based calibration workflow.

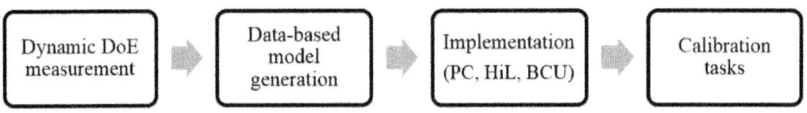

Figure 1: Model-based calibration workflow for battery systems

For internal combustion engines, model-based calibration is already state-of-the-art [2]. Within this methodology, the Gaussian Process Regression (GPR) has been successfully established as a highly performing regression algorithm. In use cases without time-dependent effects, very promising results have been achieved, see [3-5]. A first investigation of model-based battery calibration for static parameter maps resulted in a reduction of 75 % of the required measurement points [6]. Taking into account researches on dynamic ECU calibration [7] and identification of dynamic engine behavior with GPR [8] the next step is to consider modeling the dynamic battery behavior with regard to battery specific characteristics and dynamics based on real life data.

As shown in Figure 1 the workflow starts with a dynamic identification measurement based on an optimal Design-of-Experiment (DoE) providing a maximum system information with a minimum amount of measurements. The main step is to build an accurate data-based model from this data. Afterwards this model can be used for calibration tasks and function development as offline PC-model, in Hardware-in-the-Loop systems (HiL) or even as reliable models on the BCU, as proposed for ECU purpose in [9], and thus increase overall efficiency in terms of time, cost and quality.

1.3 Scope and structure

The scope of this paper is the data-based model generation part. In a first step, existing measurements are used instead of an optimal DoE (will be considered in future work). The goal is to show that the GPR as an alternative to classical battery modeling is capable of describing dynamic battery behavior up to a required accuracy.

Starting with section 2 battery fundamentals and important characteristics are introduced and a real life data set is discussed in respect to the dynamics. Section 3 will give a more detailed perspective on the GPR algorithm used for identification. For modeling dynamic behavior the time horizon and features have to be respected, thus the GPR has to be adapted by an external nonlinear autoregressive structure (NARX), described in section 4. Consequently the data-based model will be trained, section 5, and the results discussed, see section 6.

2 Battery fundamentals

The battery system can be considered on different scales due to its design. The pack level is the highest level of system integration and is equipped with the BCU. It consists of several modules, while each module contains a number of single cells. In general, the behavior observed on a single cell can be scaled up and vice versa.

2.1 Battery modeling

This section will give a brief overview of modeling techniques usually applied for characterizing battery behavior in comparison to the approach presented in this paper. Modeling techniques can be classified in white, grey and black box modeling [10] referring to their physical interpretability [11]. White box models are complex and detailed physical or electrochemical descriptions of internal battery characteristics as shown in [12], [13]. Based on differential equations they require a high computational effort [14] but also result in high accuracy. The grey box approach is used for more practical applications. Popular are descriptions based on electrical Equivalent Circuit Models (ECM) [15] and are already in series in the automotive field [16]. These models consist of a voltage source, serial resistors, different number of parallel circuits with resistors and capacitors and other circuit elements depending on the modification, calculation time and accuracy required [17]. Data-based models consider the observed system, battery or cell, as a black box. They describe the input-output relationship with a mathematical algorithm [17] such as the GPR employed in this presentation. Thus the information about the system is fully delivered by the measured data. Advantages of data-based models are high accuracy with small complexity [18].

2.2 Battery dynamics

Figure 2 shows the dynamic behavior of a battery cell on the basis of a current pulses with the resulting voltage responses. The voltage responses are *not constant but change during the course of the discharge pulses* [19]. The instantaneous drop is characterized by the inner ohmic resistance followed by charge transfer and diffusion processes [19].

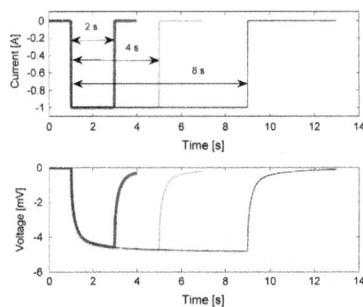

Figure 2: Current pulses and voltage responses [20]

These electrochemical and also thermal processes have complex and nonlinear effects that are mostly interacting with each other and showing different reaction times within some milliseconds (current), from minutes to hours (relaxation of voltage after the pulse or temperature adaptations), up to several hundred hours (aging processes) [21]. Influencing parameters are current, State-of-Charge (SoC), temperature, and State-of-Health.

2.3 System definition

Due to these complex dependencies, modeling the dynamic behavior accurately marks a challenge. For the sake of simplicity of the first approach, the investigated system is reduced to two inputs and one output, see Figure 3. As inputs the current and SoC and as output the resulting voltage response is regarded. The voltage is considered on cell level.

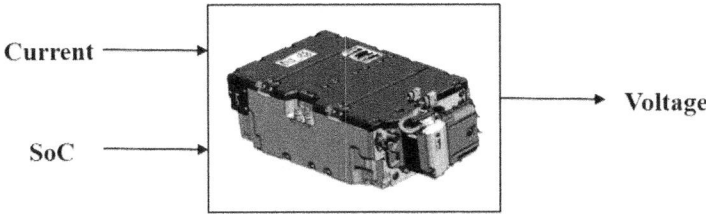

Figure 3: Definition of systems in- and output for the battery

2.4 Measurement for training and validation

Data is available for two driving cycles, the so-called "Stuttgart" and the Artemis Cycle, and is used for the subsequent analysis, see Figure 4. The measurements were generated with a battery pack containing 104 Li-ion cells in series. The single cell capacity is 28 Ah. Both driving cycles have been measured in the laboratory at a battery pack test bench at the temperature T = 25 °C. The chosen sampling rate of 1Hz is sufficient to identify battery behavior in this first approach. To identify specific local effects a higher sampling rate should be considered for future investigations. For an optimal data-based model it is important that the training data contains a maximum of information within the operating boundaries. Comparing both measurements, the Stuttgart Cycle covers a larger current range, while the Artemis Cycle is operating in far smaller boundaries. The SoC boundaries are comparable for both cycles, but the Stuttgart Cycle shows a higher SoC gradient in the beginning due to higher discharge current. The first part of this cycle until approx. to the time t = 1500 s a full electric operation is simulated by the current profile, while for the second part t > 1500 s hybrid operation is simulated. The Artemis Cycle shows a full electric operating current profile.

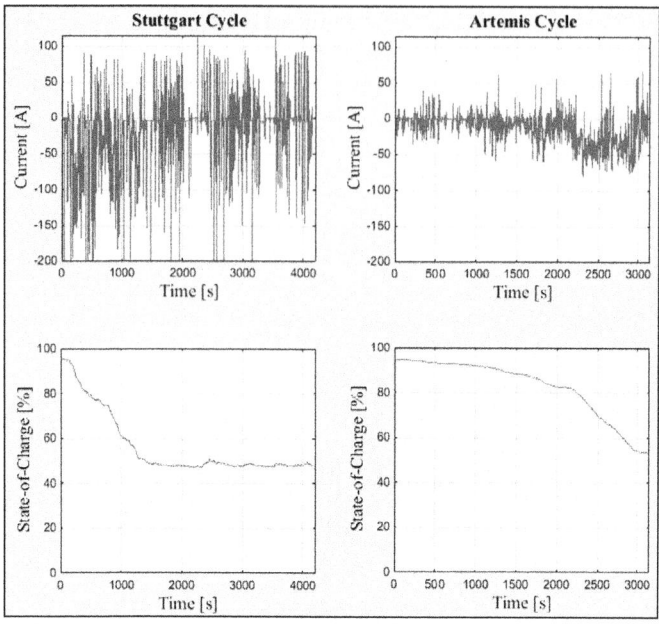

Figure 4: Comparison of the Stuttgart and the Artemis Cycle in respect to current (top) and State-of-Charge (bottom)

According to this comparison, Figure 4, it is obvious that the Stuttgart Cycle is more suitable for training the data-based model while the Artemis Cycle can be used for model validation. For future work the use of synthetic excitation signals should be investigated for optimal dynamic DoE with a maximum of information within the boundaries and dynamic gradients.

3 Gaussian Process Regression

Contrary to standard parametric regression models as polynomials or neural-nets, Gaussian Processes are able to model even strongly nonlinear systems with high accuracy. In addition, the modeling is easy to apply since the model parameters are determined automatically by probabilistic principles. A Gaussian Process Regression (GPR) is a Bayesian and nonparametric modeling approach that replaces the dependent variable y by a stochastic process, as opposed to the concept of parametric models, where prior knowledge about the functional relationship is included into the model description [22]. For an input vector x_* the GPR gives a probability distribution p for the output

y_*. This probability distribution follows a normal distribution $\mathcal{N}$ with a mean value m_* and a variance σ_*.

$$p\big(y_*(x_*)\big) \sim \mathcal{N}(m_*, \sigma_*) \tag{1}$$

$$m_* = c_*^T (C + \sigma_n^2 I)^{-1} y \tag{2}$$

$$\sigma_* = C(x_*, x_*) - c_*^T (C + \sigma_n^2 I)^{-1} c_* \tag{3}$$

$$C = C(X, X), c_* = C(X, x_*). \tag{4}$$

C is the covariance matrix that defines the stochastic process. Typically, squared exponentials are used as positive definite covariance function. X is the matrix with N multidimensional training samples, σ_n^2 the noise variance and I the identity matrix. The model output includes besides a mean value m_* (the model prediction for y_*) also the variance value σ_*, which can be used as validity measure of the prediction. In its standard implementation, the GPR framework does not require any parameterization by the user and is, therefore, appealing in the practice for nonlinear identification tasks.

A major drawback of the GPR described above is its computational complexity of cubic order $O(N^3)$ due to the inversion of the covariance matrix C of size $N \times N$, see Equation (2) and (3), where N indicates the number of training samples. Thus today's PCs can handle only data sample sizes up to a couple of thousands [23]. Over the last decade several fast and sparse approximation schemes emerged to approximate the standard GPR based on a reduced subset of M samples ($M \ll N$), which usually scales the complexity down to $O(M^2 N)$, see e.g. [24]. This gives the advantage to cope easily with large data sets, even with $N > 10^5$. Practical applications have shown that those approximations are still able to compete with other nonlinear regression algorithms, while maintaining the benefits of the standard GPR. The only free parameter to be adjusted is the subset size M. The Model-based calibration tool ETAS ASCMO, applied in this paper, is using empirically optimized values for M

4 Dynamic description with NARX

To consider dynamic effects with a data-based modeling algorithm a superordinate model structure is used: The system input space is expanded with the feedback of past input and output values up to a certain time horizon (Figure 5). This is known in the literature under the term nonlinear autoregression with exogenous inputs (NARX), see, for instance, [25] or [26]. In the following, the feedback values are referred to as features. The NARX approach transforms the dynamic identification problem into a quasi-stationary relationship, with the new input vector $\tilde{x}(k)$, where k indicates a discrete time-step, see Equation (5):

$$y(k) = f_{NL}(\tilde{x}(k)) = f_{NL}(x_1(k), x_1(k-1), \ldots, x_2(k), x_2(k \qquad (5)$$
$$-1), \ldots, y(k-1), \ldots),$$

Based on an available data set of measured input-output values, any data-based regression can be used for modeling the nonlinear relationship $f_{NL}(\tilde{x}(k))$. In ETAS ASCMO a sparse GPR regression as described in section 3 is used in the NARX structure, which enables to handle even big training data sets with ten thousands data points and a high number of inputs.

After model training, two scenarios are possible for the model's application: one-step ahead and multi-step ahead prediction. In case of a one step-ahead prediction, the past real system outputs are known and given during the run time by actual measurements. The model has to predict just the upcoming time step. This could potentially be used for an online controller.

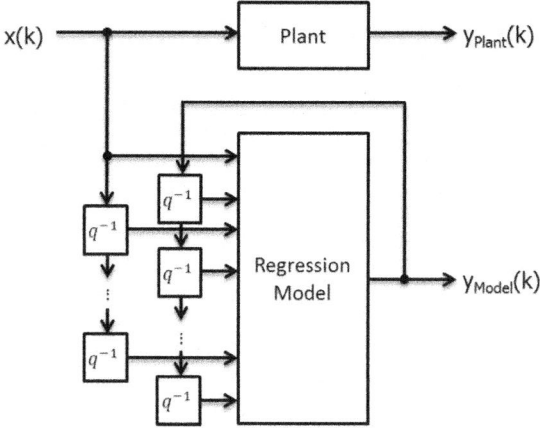

Figure 5: Nonlinear autoregression with exogenous inputs (NARX) and its application for multi-step ahead prediction (simulation)

At the multi-step ahead prediction, the past system outputs are replaced entirely by the model's predictions, see Figure 5. This equals to an offline simulation and is seen as the standard use case for ECU calibration, where the user wants the model's response to a given data stream of input variations. Multi-step ahead prediction is also used for this research task.

Feature selection

Often, the time dependencies of the system to be identified are not known in advance and a main question remains: which features need to be fed back for the NARX structure to support a reliable model quality? Two common implementations are forward selection (FS) and backward elimination (BE) to automatically determine the appropriate feedback structure, as described in the following:

Forward selection

1. Start with an empty feature set.

2. Train for each feature, which is not yet included in the feature set, an individual model and validate it.

3. Select the feature that caused the largest increase in model quality and include it into the feature set. As long as the increase in model quality is significant, go on with step 2.

Backward elimination

1. Model training and validation with all features.

2. Remove one feature at time from the feature set, conduct the model training and validate its importance in terms of model quality.

3. Select the feature whose removal caused either the maximum increase or the least decrease in the model quality and finally discard it from the feature set. As long as the decrease in the model quality is less than a predefined threshold, go on with step 2.

The forward selection usually converges much faster to a solution, whereas the backward elimination allows for considering inter-dependencies among the features. For practical applications, the forward selection is normally sufficient to get good modelling results.

5 Data-based GPR model with NARX

For determining an optimal GPR model with NARX, the maximal time lag for the feedback values has to be chosen and features have to be selected.

The appropriate time lag can be identified with the help of the Inverse Autocorrelation Function (IACF), Figure 6. It indicates which time lags are significant. As proposed in [27], the use of the 95% confidence interval, marked by the dotted lines in Figure 6, as a target value shows which time lags to choose.

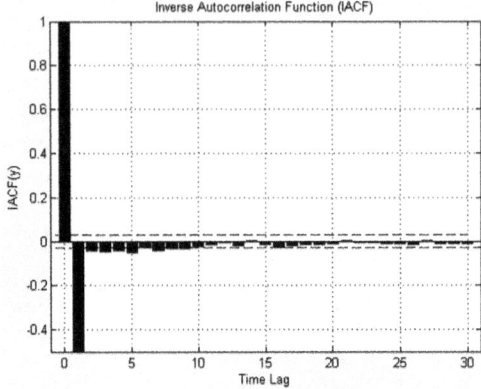

Figure 6: Inverse Autocorrelation Function (IACF) of an observed time lag of 30 for determination of time horizon

In this case maximal time lags to be considered are determined for two scenarios:

- Time lag (1): A compromise of small computational effort and selection of most significant time lags.

- Time lag (10): After the 10th time lag the influence of the other time lags are no longer significant while within the required 95% lines, thus a maximum of model quality can be achieved.

The different feature selection strategies were already introduced in section 4. For the optimal model determination the following approaches are performed:

- AF(1) / (10): The simplest approach is the selection of All Features (AF) for time lag (1) and (10) without any reduction of NARX input space.

- FS(1) / (10): Forward Selection (FS) for time lag (1) and (10), as introduced.

- BE(1) / (10): Backward Elimination (BE) for time lag (1) and (10), as introduced.

For model quality evaluation the Root Mean Square Error (RMSE), the Normalized RMSE (NRMSE) and the R^2, all based on independent validation data, were chosen. The required quality that is sufficient for calibration tasks is considered to be at a NRMSE < 5 % and a $R^2 > 0.9$. In Figure 7 the Forward Selection process for time lag (10) is visualized, showing the best combination of features in respect to RMSE.

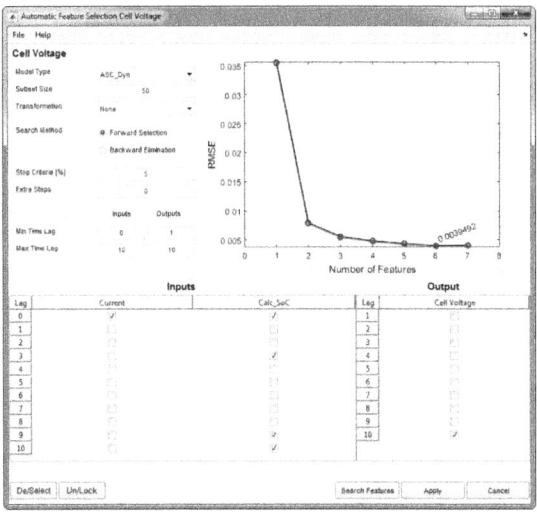

Figure 7: RMSE over number of used features (top). Selection of features (bottom)

6 Results

The results of the applied strategies for determining the optimal set of features are listed in Table 1. The best result in respect to the validation/test data achieves the Forward Selection with 10 observed time lags FS(10). From the 32 maximal available features 6 are selected.

Table 1: GPR model quality for the training and test data

	Max. no. feat.	No. feat. used	Training (Stuttgart)			Test (Artemis)		
			RMSE $*10^{-3}$ [V]	NRMSE [%]	R^2	RMSE $*10^{-3}$ [V]	NRMSE [%]	R^2
AF(1)	5	5	30.78	4.99	0.898	130.63	28.16	0.004
FS(1)		4	6.90	1.12	0.995	19.94	4.30	0.977
BE(1)		4	2.95	0.48	0.999	307.95	66.39	-4.54
AF(10)	32	32	15.39	2.50	0.975	19.68	4.23	0.977
FS(10)		6	3.95	0.64	0.998	8.60	1.85	0.996
BE(10)		31	10.83	1.76	0.987	13.39	2.89	0.990

The combination of the chosen features is shown in Figure 7 and yields in Equation (6) whereas f_{GPR} is the sparse Gaussian Process Regression as described in chapter 3:

$$y_{Voltage}(k) = f_{GPR}\left(\begin{array}{c} x_{Current}(k) \\ x_{SoC}(k), \ x_{SoC}(k-3), \ x_{SoC}(k-9), \ x_{SoC}(k-10) \\ y_{Voltage}(k-10) \end{array} \right) \quad (6)$$

The FS(10) result fulfills with a NRMSE = 1.85% the quality requirement of NRMSE < 5 % respectable. The R^2 value of 0.996 shows that this model is well usable for modeling battery dynamics.

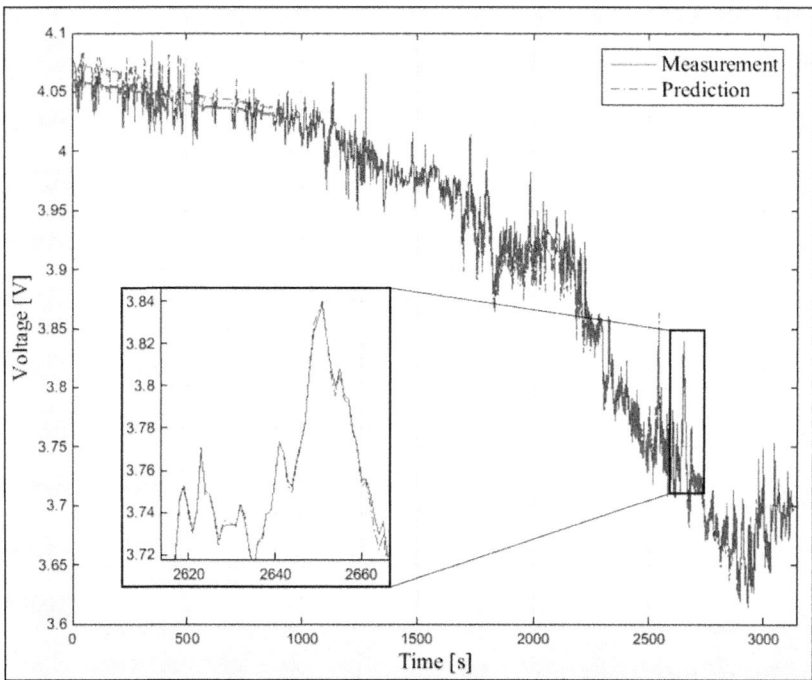

Figure 8: Model validation on the Artemis test data

In Figure 8 the validation on the Artemis Cycle is shown. The model prediction follows the global trend given by the measurement very well and indicates the strong global relationship between the input SoC and the output voltage. Observing local peaks the scope within Figure 8 demonstrates that also local dynamics are well represented by the model. Between the beginning of the graph and time t = 1000 s some offset between prediction and measurement can be recognized. An explanation for this could be that the Stuttgart Cycle as training data does not cover this voltage area well enough: as seen in the comparison in Figure 4, the SoC curve of Stuttgart Cycle decreases much faster compared to the Artemis Cycle due to higher current rates. This fact should be checked on again in upcoming investigations. From a calibration engineer's perspective the results of this first approach are very promising, because they demonstrate that already on basis of one driving cycle it is possible to establish with GPR in combination with NARX a data-based battery model that can predict another driving cycle with respectable quality.

7 Summary and Outlook

This paper discusses the use of model-based calibration for Li-ion batteries, whereas the scope was set on the appropriate modeling algorithm.

An optimal data-driven model to describe the dynamic battery behavior could be determined with a Gaussian Process Regression combined with an external dynamic structure (NARX), as implemented in the tool ETAS ASCMO. As training data, measurements from a Stuttgart Cycle were used while the model quality was evaluated on the Artemis Cycle. The model shows good results considering the NRMSE of 1.85 % and R^2 value of 0.996. The model prediction followed the global trend given by the measurement well and modeled local behavior in acceptable quality. The achieved results showed that the Gaussian Process Regression can be used in a future model-based battery calibration process for dynamic applications in order to increase calibration efficiency.

The next steps to be considered in future investigations will be:

- Determination of an appropriate sampling rate for considering detailed local effects
- Determination of an optimal DoE for generating training data with maximum amount of information and select a general set of features
- Comparison of the data-based model to benchmark battery models
- Expansion of input space

References

1. Fink, H.; Rees, S.; Fetzer, J.: Generation 2 Li-ion battery systems – Technology trends and KPIs. In: 10. MTZ-Fachtagung, Wolfsburg, 2015

2. Roepke, K.; Guehmann, C. et al.: Design of Experiments (DoE) in Powertrain Development. Expert Verlag, Renningen, 2015

3. Klar, H.; Klages, B.; Gundel, D.; Kruse, T.; Huber, T.; Ulmer, H.: New processes for efficient, model-based engine calibration. In: 5th International Symposium on Development Methodology, Wiesbaden, 2013

4. Cornetti, G.; Huber, T.; Kruse, T.: Simulation of diesel engine emissions by coupling 1-D with data-based models. In: 14th Stuttgart International Symposium, Stuttgart, 2014

5. Hoffmann, S.; Schrott, M.; Huber, T.; Kruse, T.: Model-based Methods for Calibration of Modern Internal Combustion Engines. In: MTZ (2015), No. 4, p. 46 - 51

6. Behr, L.; Kruse, T.; Rees, S.; Xie, J.: Increase Efficiency in the Calibration of Lithium-ion Battery Systems. In: ATZelektronik 06/2015, 2015

7. Tietze, N.: Model-based Calibration of Engine Control Units Using Gaussian Process Regression. PhD thesis, TU Darmstadt, 2015

8. Gutjahr, T.; Ulmer, H.; Kruse, T.; Markert, H.; Ament, C.: Advanced Approaches for the Identification of Dynamic Engine Behavior with Probalistic Modeling. In: Roepke, K. (ed.): Design of Experiments (DoE) in Engine Development, Expert Verlag, Renningen, 2011

9. Lang, T.; Diener, R.; Hanselmann, M.; Markert, H.; Ulmer, H.: Data-based Models on the ECU. In: 6th International Symposium on Development Methodology, Wiesbaden, 2015

10. Schmidt, J.P.: Verfahren zur Charakterisierung und Modellierung von Lithium-Ionen Zellen, PhD thesis, KIT University of Karlsruhe, 2013

11. Ljung, L.: System identification: theory for the user. Volume 2. Upper Saddle River, NJ, Prentice Hall, 2009

12. Tippmann, S.; Walper, D.; Balboa, L.; Spier, B.; Bessler, W.G.: Low-temperature charging of lithium-ion cells part I: Electrochemicalmodeling and experimental investigation of degradation behavior. In: Journal of Power Sources 252 (2014) p. 305-316, 2014

13. Schmidt, A.P.; Bitzer, M.; Imre, A.W.; Guzzela, L.: Experiment-driven electrochemical modeling and systematic parameterization for a lithium-ion battery cell. In: Journal of Power Sources, 195 (2010) p. 5071 – 5080, 2010

14. Rausch, M.; Klein, R.; Streif, S.; Pankiewitz, C.; Findeisen, R.: Model-based state estimation for lithium-ion batteries, In: at-Automatisierungstechnik 62(4), 296-311, 2014

15. Andrea, D.: Battery management systems for large lithium-ion battery packs. Artech House, Boston, 2010

16. Cois, O.; Fink, H.: Battery Management System – Concepts to reduce Battery specific costs. In: Battery Conference 2015, Aachen, 2015

17. Keil, A.; Jossen, A.: Aufbau und Parametrierung von Batteriemodellen. Download: https://mediatum.ub.tum.de/doc/1162416/1162416.pdf, time stamp: 2016-01-18, 09:48, University of Munich

18. Buchholz, M.; Remmlinger, J.; Dietmayer, K.: Genau und einfach – Datenbasierte Modelle für das Batteriemanagement. In: Technische Akademie Esslingen, Ostfildern, 2014

19. Waag, W.: Adaptive algorithms for monitoring of lithium-ion batteries in electric vehicles. Phd thesis, RWTH Aachen, Aachen, 2014

20. Waag, W.; Käbitz, S.; Sauer, D.U.: Application-specific parameterization of reduced order equivalent circuit battery models for improved accuracy at dynamic load. In: Measurement 46 (2013) 4085 – 4093, 2013

21. Quasthoff, M.; Unger, J.; Jakubek, S.: Entwicklungsmethodik eines generischen Batterie-Simulationsmodells und dessen Einsatzmöglichkeiten. In: 5. Fachtagung Baumaschinentechnik, Dresden, 2012

22. Rasmussen, C. E.; Williams, C. K. I.: Gaussian Processes for Machine Learning. MIT Press, Boston, 2006

23. Berger, B.; Rauscher, F.; Lohmann, B.: Analysing Gaussian Processes for Stationary Black-Box Combustion Engine Modelling. In World Congress, volume 18. IFAC, Milano, 2011

24. Quiñnonero-Candela, J. ; Rasmussen, C. E.; Williams, C. K. I.: Approximation Methods for Gaussian Process Regression. Technical report, Microsoft Research, May 2007

25. Sjöberg, J.; Zhang, Q.; Ljung, L.; Benveniste, A.; Delyon, B.; Glorennec, P.-Y.; Hjalmarsson, H.; Juditsky, A.: Nonlinear Black-Box Modeling in System Identification: a Unified Overview. In Automatica, volume 31, Elsevier, 1691–1724, 1995

26. Nelles, O.: Nonlinear System Identification: From Classical Approaches to Neural Networks and Fuzzy Models, Springer, 2001

27. Gutjahr, T.: Dynamic System Identification with Gaussian Processes in Model-Based Engine Development. PhD thesis, University of Ilmenau, 2012

HEV concept with lean operated SI-engine optimized for fuel consumption and emissions

Jing Cheng, Prof. Dr.-Ing. Michael Bargende
Universität Stuttgart, Institut für Verbrennungsmotoren und Kraftfahrwesen
(IVK)

Dr. Frank Altenschmidt, Christoph Ley
Daimler AG

Abstract

The integration of a lean operated Spark-Ignition (SI) engine into a Hybrid Electric Vehicle (HEV) is rarely considered as a suitable combination to improve the fuel efficiency. The main argument is that both technologies are aiming at the improvement of fuel economy in the partial load area by means of the lean operation and electric drive, which are mutually exclusive. With the help of overall-system simulation, this paper investigates the fuel saving coming from the combination with an optimized operating strategy. This research shows that the electric motor assists the engine in lean operation to reach the optimal operating points rather than meeting the propulsion requirement with electric drive. Driving cycles with different dynamic characteristics are considered to study the influence of operating strategy on fuel consumption under real world driving conditions. In addition, the impact of the engine displacement is also analyzed.

1 Introduction

The tightening legislations for CO_2 and exhaust gas emissions have major influence on the technical development in the automobile industry. Nowadays the main purpose for the development and production of hybrid vehicles is to introduce a means of cleaner transport. Through hybridization the vehicles retain the performance and driving range of conventional cars, while improving the fuel economy due to the assistance of an electric motor.

On the other hand, there have been a series of technological improvements of conventional combustion engines, the stratified-charge combustion of gasoline engines being one of them. The fuel saving potential comes primarily from the reduced pumping losses in the low load area, by extension of the unthrottled operating range. Restrained by the engine load and emission limits, the engine must be operated in different operating modes according to the operating areas. For the spray guided combustion process used in this paper, the three main operating modes can be seen in figure 1: the stratified mode (STRAT), the homogenous-stratified mode (HOS) and the homogenous mode (HOM).

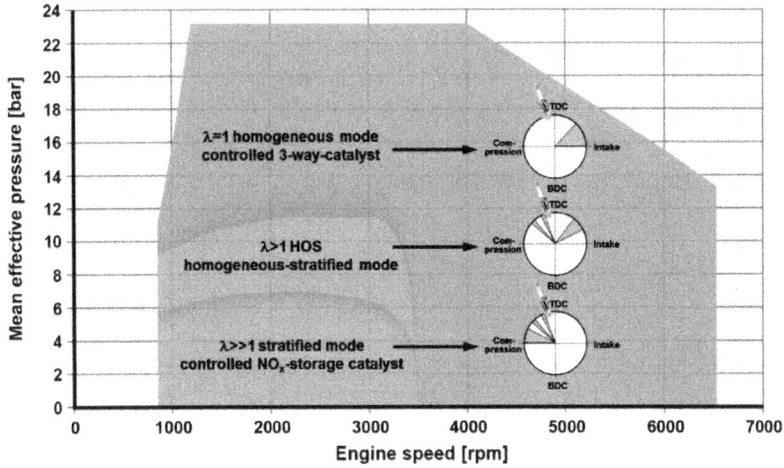

Figure 1: Operating modes of a SI engine with lean operation [1]

In a previous research the synergy has been identified between the gasoline engine with direct injection and the electric motor of a hybrid powertrain [2]. However, the utilization of a rigid rule based energy management strategy (seen section 4) and the simulation performed only in the New European driving cycle (NEDC) may undermine the fuel saving potential of the combination, as the NEDC may not be most representative of real life driving scenarios. Hence, in this paper an optimization algorithm is applied for the optimal power allocation between the combustion engine and electric motor, and driving cycles with different characteristics are adopted for the simulation.

2 Simulation System Configurations

The target powertrain is a P2 plug-in hybrid middle class sedan with an automatic gearbox. The maximum power output of the engine and the electric motor is 155kW and 60kW, respectively. In order to obtain accurate predictions of fuel consumption and emissions of the vehicle in near real time conditions, an overall system model is built with focus on engine modeling.

2.1 Data Collection

For both modeling and validation of the simulation, measurements have been collected on a stationary test bench from an SI engine with a exhaust aftertreatment system (three-way catalytic converter and NOx trap). The test carrier is a four cylinder gasoline engine with direct injection. Its peak power and torque are 155 kW and 350 Nm, respectively.

The measurements are collected in the following stages:

Stage 1: Complete stationary emission maps for all operation modes (HOM, HOS and STRAT) are measured. In every operating point the emission data, the fuel consumption, the set values from the Engine Control Unit (ECU), the sensor values in air path as well as the signals from the exhaust gas aftertreatment are collected.

Stage 2: A reduced collection of stationary operating points are measured with varied ECU set values like ignition timing, injection offset, EGR rate, coolant temperature. With such variations the raw emissions model is supposed to make better emission prognosis in case of a transient engine operating state, by taking the fluctuation of the other operating parameters into consideration.

Stage 3: For the aim of the validation for the simulation models, various driving cycles with varied starting temperature are measured.

2.2 Simulation Model Structure

The basis of the whole system model is a zero-dimensional feedforward simulation model of the vehicle in Matlab/Simulink®, in which detailed plant models with the combustion engine's features are implemented. The whole system takes the vehicle speed profile as an input; a PID controller serves as a driver and the control variable is the accelerator and brake pedal position. The vehicle body and accessories are simplified as basic physical models, as well as some components of powertrain like the transmission and the clutch. The electricity storage system including the battery cells and the Battery Management System (BMS) are delivered in form of s-function from the manufacturer. The electric motor is described by characteristic maps.

The ICE plant models in the overall system are built in a highly elaborate way. The structure can be seen in figure 2 (a detailed demonstration in [2]). It comprises the following components:

- Engine Control Unit (ECU) model: Emission-relevant ECU functions are imported as s-functions directly into the ECU model in Simulink. The functionalities cover operation modes control, air control, fuel control, boost control, ignition control, injection control and camshaft control. The set points and the operation data generated by the ECU model are to be used in the MVM and REM.

- Mean Value engine Model (MVM): This model adopts the MVM approach based on the engine library from Swiss Federal Institute of Technology (ETH) Zurich [3]. After modification for gasoline engines with lean operation, the MVM is calibrated with the measurements from the engine test bench. The outputs of the MVM serve as the sensor signals to the whole engine model, such as information like mass flow, temperature and pressure of the air path and combustion chamber.

- Raw Emissions Model (REM): The raw emission model is built in a quasi-static approach. Based on the transient boundary conditions, the utilization of an Artificial Neural Network (ANN) enables the REM to capture the non-linear behavior of the emission formation. The test plan in section 2.1 is specifically designed for training and validation of the ANN. The raw emissions outputs are fed to the exhaust gas aftertreatment system.

- Exhaust Aftertreatment Model (EAM): The main purpose of the EAM is to model the NOx trap which the engine uses for lean operation. The catalyst efficiency is related to the raw NOx mass flow, the exhaust gas mass flow and the gas temperature before the catalyst. The trapped NOx is to be accumulated over time in lean operation until it reaches a threshold and triggers regeneration in engine control unit.

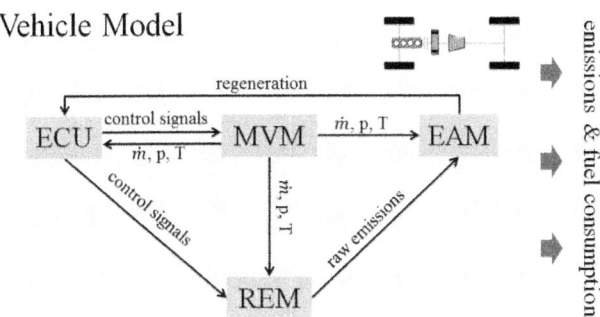

Figure 2: Overall system model

3 Optimization Strategy

3.1 State of the Art

With two different propulsion sources, the hybrid electric vehicle faces a challenge with respect to power distribution and energy management. The aim is to provide comfort and agility for the customer, while keeping the fuel consumptions and emissions as low as possible. Meanwhile the vehicle is supposed to stick to a set of boundary conditions like the state of charge of the battery. The task of the supervisory energy control strategy can be generalized in two kinds of decision: when to turn the engine on and off, and how to allocate the torque between the electric motor and the engine.

For the mass-production HEVs nowadays, the supervisory energy control mostly follows a heuristic or rule based strategy. This approach provides a quick and intuitive solution but it is not an optimal strategy to reduce the fuel consumption. However, there are varieties of different optimization approaches studied and applied in the development of HEVs in car manufactures and universities. They can be divided into causal (heuristic, online Equivalent Consumption Minimization Strategy (ECMS)) and non-causal (dynamic programming, ECMS/ Pontryagin's Minimum Principle (PMP)); or optimization theory based (dynamic programming, ECMS) and not optimization theory based (heuristic, fuzzy logic) [4].

The ECMS approach is adopted in this paper, considering its real time calculation, feasibility and the compatibility to the existing system.

3.2 ECMS

The basic problem of HEV fuel consumption and emission optimization of HEVs can be summarized with following cost function [5]:

$$J = \int_{t_0}^{t_{end}} [\dot{m}_f + a_{NO_x}\dot{m}_{NO_x} + a_{PM}\dot{m}_{PM} + a_{HC}\dot{m}_{HC}]\, dt \tag{1}$$

J: cost of the fuel consumption and pollutant emissions

t_0, t_{end}: begin and end of the time horizon

$\dot{m}_f$: fuel mass flow rate

$\dot{m}_{NO_x}, \dot{m}_{PM}, \dot{m}_{HC}$: mass flow rate of nitrogen oxide, particulate matter and hydrocarbons

a_{NO_x}, a_{PM}, a_{HC}: weighting coefficients of nitrogen oxide, particulate matter and hydrocarbons

The aim is to minimize the integral value of the weighted sum of fuel consumption and pollutants over the driving time. ECMS can be used for offline optimization (also known as PMP) as well as online optimization. Since the main purpose of this paper is to evaluate the fuel saving potential of a lean burn engine in a hybrid powertrain, it is reasonable to assess the variations offline. ECMS uses the equivalency factor s_0 to minimize the virtual sum of the real fuel consumption of the engine and the equivalent fuel consumption of the electric motor at each time step. When only the fuel consumption is considered, the equation for optimization can be simplified in form of power as shown in (2) and s_0 is assumed to be constant during the drive cycle [6].

$$P_H = P_{fuel} + s_0 P_{ele} \tag{2}$$

P_H: virtual cost of total power;

P_{fuel}: power combustion engine;

P_{ele}: power electric motor;

During the repartition of the power demand, the system is subject to local constraints such as the limits of the mechanical power output of the engine and the electric motor, and the electric power and state of energy of the battery [5].

In this paper charge only sustaining operation is considered, which means that the State Of Charge (SOC) of the battery at the beginning equals the SOC at the end. This imposes a global constraint for the offline optimization. This also facilitates comparison of fuel consumption from different simulations.

4 Simulation

In May 2015, the European Union has decided the NEDC should be complemented by Real Driving Emission (RDE) test procedures, which better reflect the emissions of vehicles in normal driving conditions. The laboratory test must be confirmed by a real road driving car equipped using a Portable Emission Measuring systems (PEMS). However, the RDE test procedures are in still in discussion. For the simulation in RDE context under different driving conditions, two variations of the RTS95 cycles (random test sequence 95 %) with high and low dynamics (velocity profile in figure 3) are considered in this work.

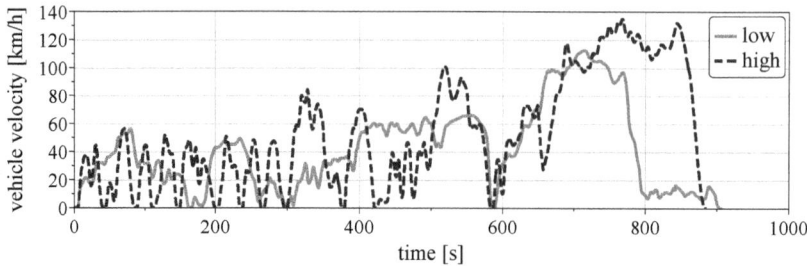

Figure 3: Low and high dynamic real driving cycle RTS95

Two sets of simulations are carried out: one focuses on the fuel saving potential of the lean operating mode; the other investigates the influence of engine displacement on the total fuel consumption of the engine with and without lean operating mode. For all simulations the equivalency factor s_0 is determined iteratively, such that the initial SOC of the battery equals the final SOC.

5   Results and Analysis

Firstly, it is important to consider the impact of the dynamic characteristics of the driving cycles on the control strategy (charge sustaining operation). Figure 4 (left) illustrates the distribution of the operating points of the vehicle in different driving cycles. It can be seen clearly that the NEDC operation points cover only a narrow area of the whole engine operating map. The difference between the two driving cycles used in this work is like the name implies: the RTS95 high dynamic cycle has much more operating points on the high positive or negative load areas compared to the mild dynamic cycle. This difference also results in the recuperable energy in the driving cycle. As an indication of recuperable energy the brake energy is shown in figure 4 (right). No matter if brake energy in total or divided by distance is considered, the high dynamic cycle possesses about three times of the brake energy of the low dynamic cycle. This means that if a SOC balance has to be achieved, the electric energy term in the cost function in the high dynamic cycle is "cheaper". Mathematically, this is directly reflected on the equivalency factor s_0. When other boundary conditions stay the same, the iteratively determined s_0 in the high dynamic cycle tends to be moderately higher than the one in low dynamic cycle. That is to say, the high dynamic cycle possesses more electric energy for electric drive or for operating points shifting. If the assumption that the use of the electric motor undermines the fuel saving potential of lean operation is valid, then in the high dynamic cycle the fuel saving from the variation from a homogenous engine to an engine with lean operation may be less compared to the saving from the same variation in the low dynamic cycle.

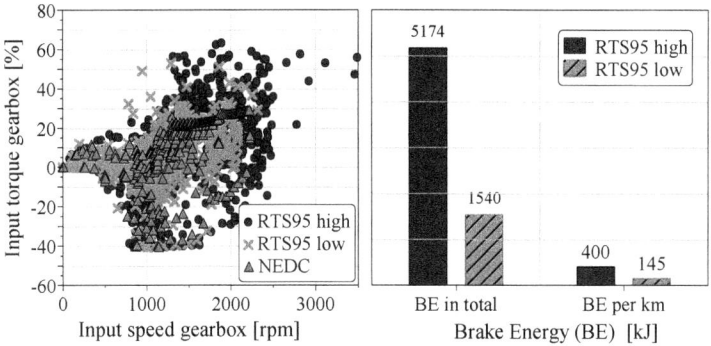

Figure 4: left: Operating points distribution; right: Brake energy

The following figure 5 illustrates the operating point shift of a hybrid powertrain in different driving cycles. Firstly, if the simulation is carried out in the high dynamic cycle, the engine with lean operation tends to have more downshifting of the operating points, which means the available electric energy may be used rather for the load point shifting than for electric drive. The statistics of electric motor operation in figure 6 illustrate this point. The simulation for the homogenous engine tend to have more electric drive, while the simulation with the engine in lean operation uses the electric motor more often to reduce the engine load. The same pattern is also observed in the low dynamic cycle, although in this cycle the required torque is mostly within the lean operation load limit. Secondly, the resulting operating points of the engine with lean operation concentrate more in the same load at each engine speed. The observation of the specific fuel consumption in figure 7 reveals that this arises from the consumption curve gradient. First of all, it is obvious that the introduction of SCH and HOS operation reduces the specific fuel consumption at low load. If the HOS operation is available, there is a significant optimum load point at the red arrow. In the ECMS optimization the electric motor tries to move the load point of engine toward this point, which causes the uniform engine load. The operating points gather also at the load limit of lean operation at higher engine speed, because the ECMS algorithm uses characteristic maps for fuel consumption. The differences between the HOS and HOM fuel consumption maps at the transition boundary lead to a shift into the HOS operating area.

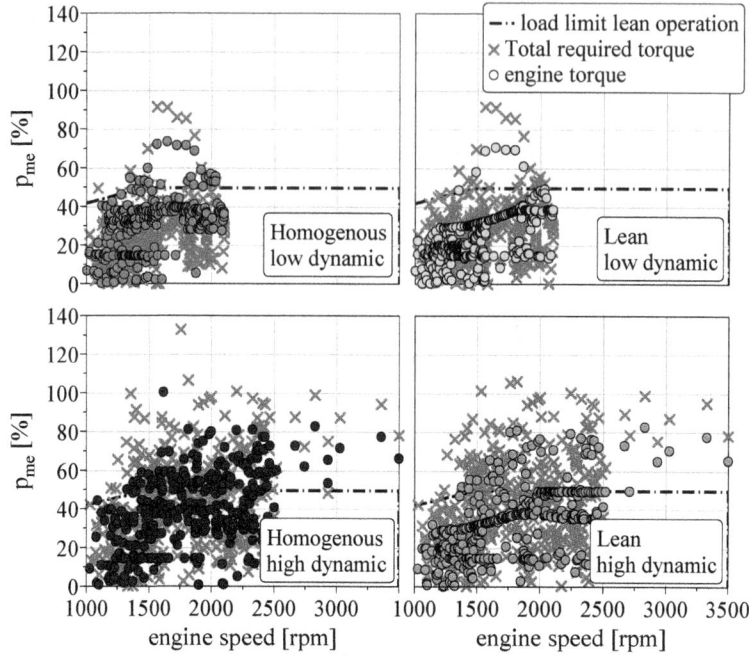

Figure 5: Distribution operating points

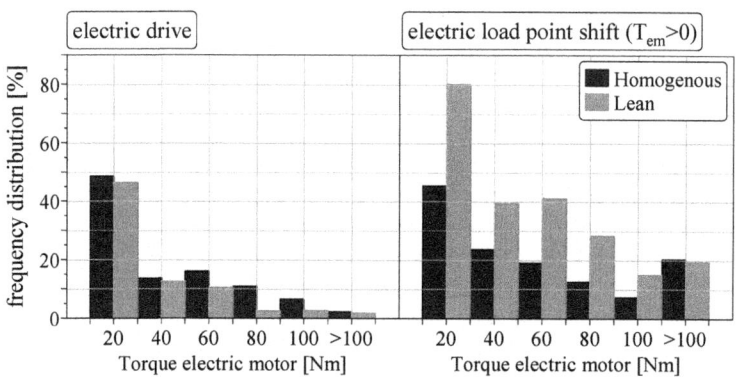

Figure 6: Electric motor operation with different engine features in high dynamic cycle

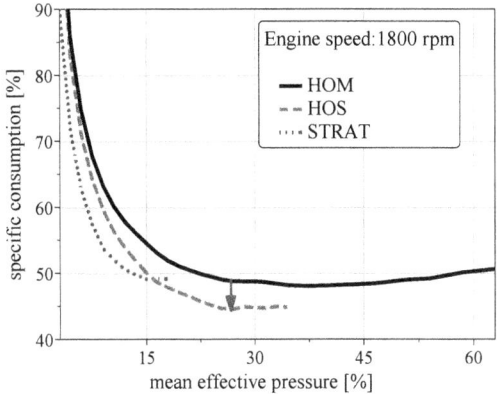

Figure 7: Specific fuel consumption in three operation modes

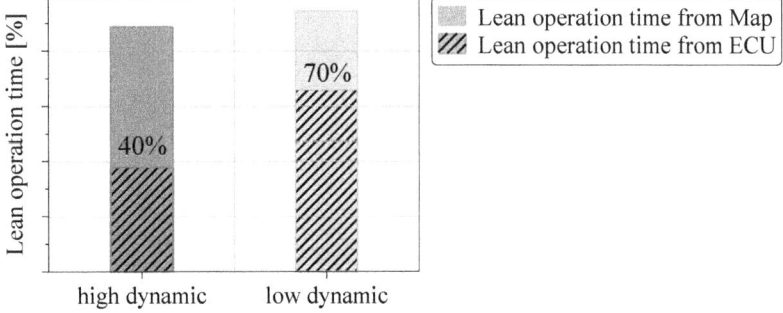

Figure 8: Comparison of lean operation duration

In addition it must be mentioned that not all operating points within the lean operation load limit are actually in lean operation. The decision of the ECU also depends on various transient boundary conditions like engine air charge, temperature of engine coolant or temperature of the exhaust manifold. Multiple hystereses are also implemented in the ECU to prevent frequent shifting between lean and homogenous operation. These conditions are included in the ECU simulation model which governs the operation modes. However, in this research these conditions are not fed into the ECMS optimization for the determination of an efficient operating point. This procedure leads to the fact that the ECU forbids lean operation for some operating points shifted into the lean operation area by the ECMS. Figure 8 shows that in the low dynamic cycle the ECU only allows for 70% of the lean operation according to the ECMS outcome is actually carried out by ECU, and naturally for the high dynamic cycle this quote is much less (40%) because of the drastically changing requirement for the powertrain and the corresponding dynamic operating state of the combustion engine.

The influence on the total fuel consumption for the two driving cycles can be seen in figure 9. First of all, there is no doubt about the advantage of usage of the lean operation. It saves 3% and 2% of the total fuel consumption compared to the powertrain with the homogenous engine in high and low dynamic cycles, respectively. This fuel saving potential is higher than the one obtained in the last research, where only rule based control strategy was used and the results were computed for the NEDC [2]. It also shows that the adoption of lean operation on a hybrid powertrain not only improves the total fuel efficiency, this improvement even grows when the control strategy is optimized. Furthermore, despite the fact that the character of the high dynamic cycle reduces the instances of lean operation of the engine, lean operation still leads to a higher fuel saving figure than in the low dynamic cycle.

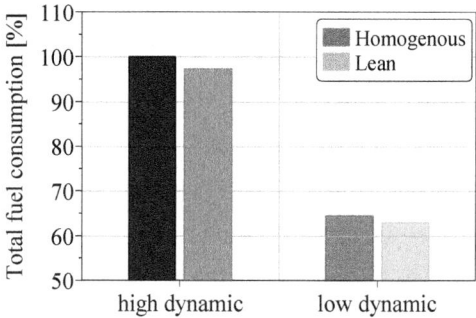

Figure 9: Fuel consumption under different conditions

In the following the influence of engine displacement on fuel consumption is considered.

The engine torque is proportional to the product of mean effective pressure (p_{me}) and engine displacement. Hence, the bigger the engine displacement, the smaller the resulting mean effective pressure (p_{me}) of the engine operating point [7]:

$$p_{me} = \frac{2\pi n_R T}{V_d}, \tag{3}$$

n_R being the number of crank revolutions for each power stroke (two for four-stroke cycles; one for two-stroke cycles), V_d the engine displacement, and T the effective engine torque.

For the SI engine with lean operation this means that the maximum torque for lean operation can be expanded by raising the engine displacement as the load limit for lean operation is defined by p_{me}. It also means that for the engine with larger displacement the fuel saving potential of lean operation is increased with respect to homogeneous operation. Simulations conducted in the two driving cycles prove this assumption (figure 10). The tendency is more distinctive in the high dynamic cycle because there are more operating points above the load limit of lean operation. Another interesting result is that for the engine with lean operation a reduction of engine displacement from 2 liter to 1.8 liter does not lead to a decline of total fuel consumption. This is due to many engine operating points being raised above the load limit of lean operation during the downsizing, and throttling losses exceed the fuel benefit from downsizing.

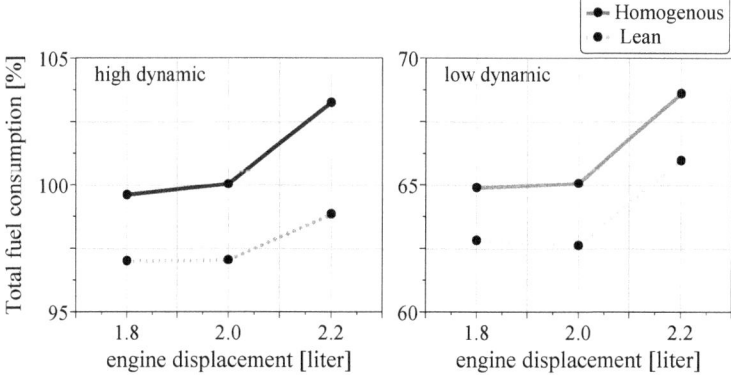

Figure 10: Influence of engine displacement on fuel consumption

6   Summary and Outlook

This paper presents an overall system model with focus on an SI engine with lean operation to study its behavior in a hybrid powertrain. As a solution for the energy management optimization problem an ECMS algorithm is implemented to coordinate the operation of the combustion engine and electric motor. The first results confirm the fuel saving potential of the SI engine with lean operation in a hybrid powertrain. The utilization of the ECMS further improves the fuel efficiency. The work also reveals that the amount of time in electric drive of the hybrid powertrain with the lean operation engine is much smaller when compared to the homogenous engine as the electric energy is used rather for the load point shifting. Under the same driving conditions the benefit of the adaption of lean operation increases for engines with bigger displacements. Hence, this study provides an approach to reduce the fuel consumption of the engine without downsizing and the corresponding loss of maximum power. However, the pollutant emissions of SI engine with direct injection are not negligible; lean operation makes the emission of nitrogen oxide even more critical. It is necessary to take pollutants, especially nitrogen oxide, into consideration when conducting the optimization in the further research.

References

[1]   A. Waltner, F. Altenschmidt, and U. Schaupp, "Magerbrennverfahren – Die Zukunft für Ottomotoren," in *Internationaler Motorenkongress 2014*, pp. 433–445.

[2]   J. Cheng, F. Altenschmidt, C. Ley, and M. Bargende, Eds, *The Hybrid Powertrain: A Challenge for the Simulation*, 2015.

[3]   M. Benz, T. Nüesch, M. Hehn, and S. Zentner, "Engine Library Reference Book," Institute for Dynamic Systems and Control, Swiss Federal Institute of Technology (ETH) Zurich, Swiss, 2010.

[4]   L. Guzzella and A. Sciarretta, *Vehicle propulsion systems: Introduction to modeling and optimization,* 3rd ed. Heidelberg, New York: Springer-Verlag, 2013.

[5]   L. Serrao, S. Onori, and G. Rizzoni, "ECMS as a realization of Pontryagin's minimum principle for HEV control," in *2009 American Control Conference*, pp. 3964–3969.

[6]   A. Sciarretta and L. Guzzella, "Control of hybrid electric vehicles," *IEEE Control Syst. Mag,* vol. 27, no. 2, pp. 60–70, 2007.

[7]   J. B. Heywood, *Internal combustion engine fundamentals*. New York: McGraw-Hill, 1988.

Software tool to create a hybrid operation strategy for simulations

Dipl.-Ing. Dan Keilhoff, Prof. Dr.-Ing. Hans-Christian Reuss

Institute of Internal Combustion Engines and Automotive Engineering, University of Stuttgart

Abstract

In this paper a software tool to create a hybrid operation strategy is introduced. It is supposed to be used in simulations for the generation of load spectra. The tool was designed with Matlab/Simulink ® and bases on the tool "Common Powertrain Control" (CPC module) published in [1].

Compared to the functionality in [1] the tool was extended massively. It now supports seven different hybrid topologies, i.e. four parallel and three axle-split configurations. The high number of parameters makes it possible to adapt the virtual powertrain to several boundary conditions. Furthermore the modular character of the model allows modifications if necessary.

It must be noticed that the CPC module is not a complete simulation tool for powertrains. It only offers the possibility to create a hybrid operation strategy. For this reason, a demanded torque or a demanded traction force must be provided to the tool. It then calculates the torque distribution considering limitations like max possible torques or max possible currents. To match the requirements of the designed powertrain, up to 20 operation modes are available. Based on different signals like battery state of charge (SOC) and acceleration pedal the user defines which mode is to be used.

1 Introduction

Among others simulations are used to create load spectra to design powertrain components. In order to be able to match the expected loads, especially for hybrid vehicles an operation strategy is essential. Since the mentioned simulations are executed at a very early stage of the development process many conditions are not fixed yet. The data of the internal combustion engine (ICE), the electric machine (EM) and gearbox will change as well as the strategy itself. Assuming that the changes are not too big, the simulations are supposed to be still valid.

With the CPC module the authors introduced an approach to reach that goal in an earlier paper [1]. The suitability of this abstract hybrid strategy was proven. However, it worked only for a so called P2-hybrid, where the EM is located at the gearbox entry.

Below the most important features as well as the mentioned extensions of the tool are explained.

2 The "CPC module"

2.1 Defining the virtual powertrain

The CPC module supports seven hybrid topologies. An example of each topology is shown in figure 1. They differ from each other in terms of number and location of the EM as well as the existence of a clutch between the ICE and the EM.

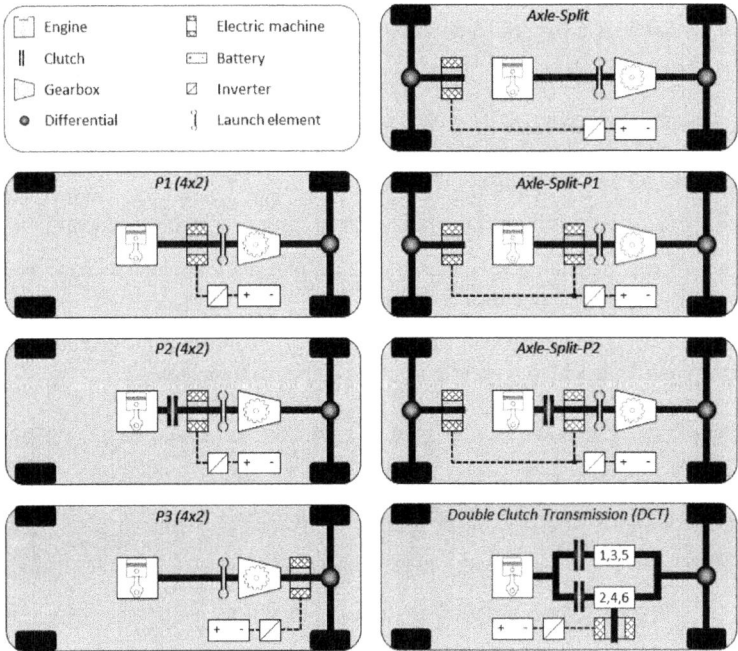

Fig. 1: Supported hybrid topologies

Available are four parallel structures:

- P1: EM is located at the gearbox entry w/o clutch between ICE and EM
- P2: EM is located at the gearbox entry w/ clutch between ICE and EM
- P3: EM is located at the gearbox exit
- DCT: EM is located at a countershaft of a double clutch transmission

The common property of these four topologies is that the torque addition of ICE and EM takes place before the first torque split occurs. It is not relevant, whether this torque split is the front/rear distribution for a 4WD or the right/left distribution in a differential gearbox.

In the group of the axle-split configurations three different versions can be calculated:

- AS: axle 1 is driven by the ICE; axle 2 is driven by an EM
- AS-P1: axle 1 is driven by a P1-hybrid; axle 2 is driven by an EM
- AS-P2: axle 1 is driven by a P2-hybrid; axle 2 is driven by an EM

Once the topology is selected the details of the powertrain must be defined. First of all it must be clarified, how the EM is connected to the powertrain. The CPC module supports two options, see figure 2:

- Coaxial: EM is located on the shaft; no additional gearbox needed
- Axially parallel: EM is located on a parallel shaft; additional connection element is needed; clutch is optional

Furthermore an additional clutch can be defined. This allows it to separate the EM from the powertrain and to prevent it from too high revolution speeds.

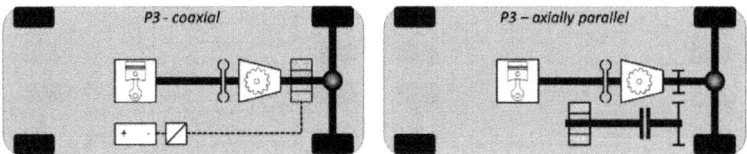

Fig. 2: Supported options to connect the EM to the powertrain

To describe the components of the powertrain the CPC-own parameters and calibratable curves can be used, such as current limits or maximum torques. As an alternative some information can be calculated in the simulation environment and then be provided to the CPC module. If, for example, the information "maximum torque of the ICE at the actual speed" is calculated somewhere outside of the CPC module, the CPC module can be ordered to use the external data source. By doing so it is ensured that consistent data are used.

2.2 Types and selection of operation modes

To match the requirements of the user the CPC module offers a number of operation modes, see table 1.

Tab 1: Available operation modes   modified after [2]

Operation mode	Machine states		
	ENG	EM1	EM2
Standstill			
Idle (IDLE)	pos	0	0
Charge during standstill (CDS) [1]	pos	neg	0
Engine Stop-Start (ESS)	0	0	0
Driving			
Pure engine driving (ENG)	pos	0	0
Pure electric driving (EM) [2]	0	pos / 0	pos / 0
Hybrid driving (HYB) [2]	pos	(pos)	(pos)
Load point increase (LPI) [1,2]	pos	- / 0	0
Launch procedures (LNCH)	pos / 0	pos / 0	pos / 0
Overrun / Rolling			
Engine overrun (OVR_ENG)	neg	0	0
Electric overrun (OVR_EM)	0	neg	0
Torque-free rolling (OVR_ZERO)	0	0	0
Breaking			
Friction braking (FRIC)	neg / 0	0	0
Recuperation (RECUP) [2]	neg / 0	neg / 0	neg / 0

[1] Not available for all topologies	pos	positive torque
[2] Sub-states available for selected topologies	neg	negative torque
	0	disengaged / off
	(...)	if necessary

Not all of them are available for all hybrid topologies. It is, for example, not possible to run the EM during vehicle standstill with a coaxial P3 topology. In theory a lot more states are possible, especially for the axle-split topologies. But all modes, where one axle generates a positive force and the other one a negative force, were ruled out since these modes are expected to show a very low overall conversion ratio as well as high tire wear.

Based on the velocity and the actuation of the acceleration and the braking pedal four major states are distinguished: "standstill", "driving", "rolling/overrun" and "braking". Additionally the values of the SOC, the acceleration pedal and the velocity are separated into different levels. The combination of the SOC and the acceleration pedal levels leads to a matrix, see figure 3. For the major state "driving" two matrices are used: one for the low and one for the high velocity level. Obviously the usage of the acceleration pedal does not make any sense during "standstill" and "braking". For every field of the matrices the user can decide which operation mode to use. Figure 4 shows a possible calibration.

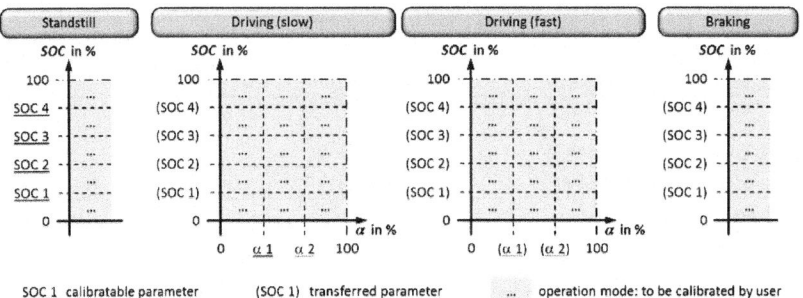

Fig. 3: Selection of operation modes

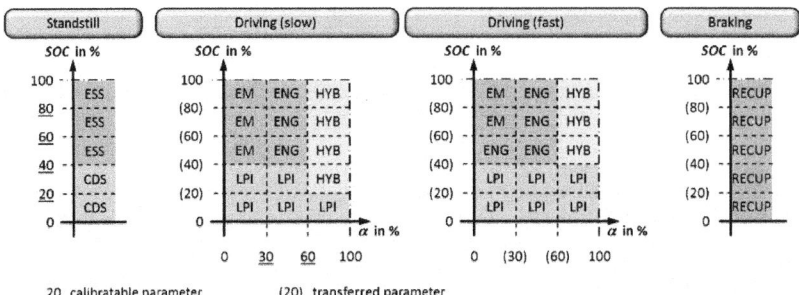

Fig. 4: Example of a calibration

3 Calculation of a parallel topology

Once the state has been selected, the torque distribution is calculated. For the parallel topologies an abstract powertrain is used. Using the P2-hybrid topology as an example figure 5 shows both the schematic and the abstract version of a powertrain.

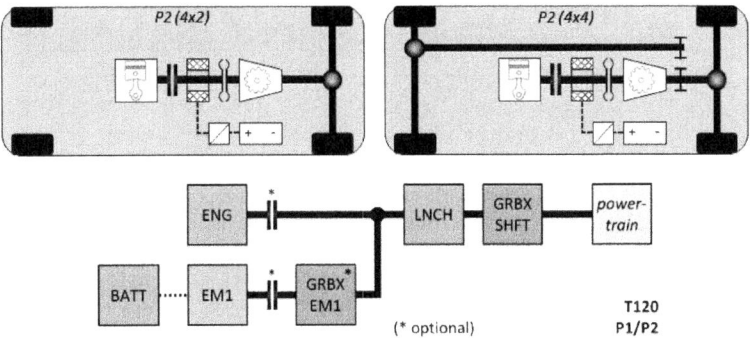

Fig. 5: Schematic and abstract version of a P2-hybrid

Starting point is a torque demand at the gearbox output side. This demand is required as an input signal, so it must be calculated outside of the CPC module. Based on this demand the powertrain is calculated backwards until the ICE and/or the EM are reached. During the calculations not only transmission or conversion ratios but also torque limits are considered. By doing this it is ensured that for example a maximum allowed torque at the gearbox input is not exceeded under any circumstances.

To evaluate this approach a large number of measurements were compared with their corresponding simulations. Three routes were driven, each with a mid-size and a luxury passenger car. To cover as many driving situations as possible a city cycle, an interurban cycle and a highway cycle were selected. Every measurement was done both in Sport and Eco mode, which leads to six datasets per vehicle. The results for the luxury car show a very good conformity between the measured and the simulated data [1]. But also the simulation of the mid-size car with less weight, less ICE and less EM power provides meaningful data. The figures 6, 7 and 8 show this exemplarily for the interurban cycle in Eco mode.

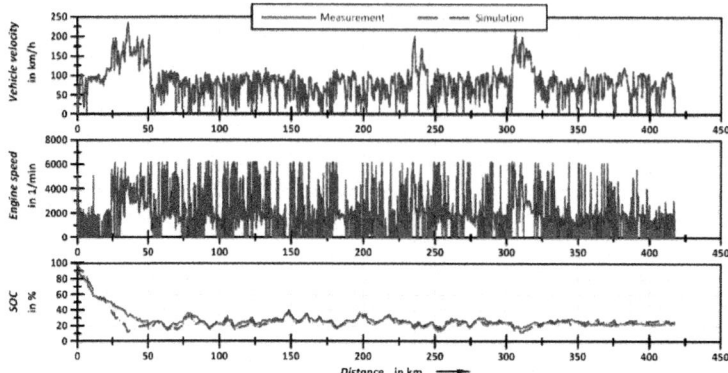

Fig. 6: Comparison between measured and simulated data (1)

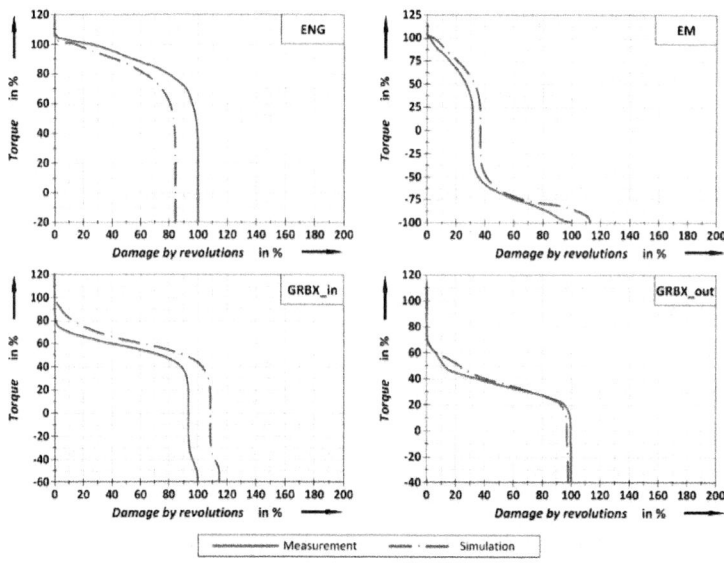

Fig. 7: Comparison between measured and simulated data (2)

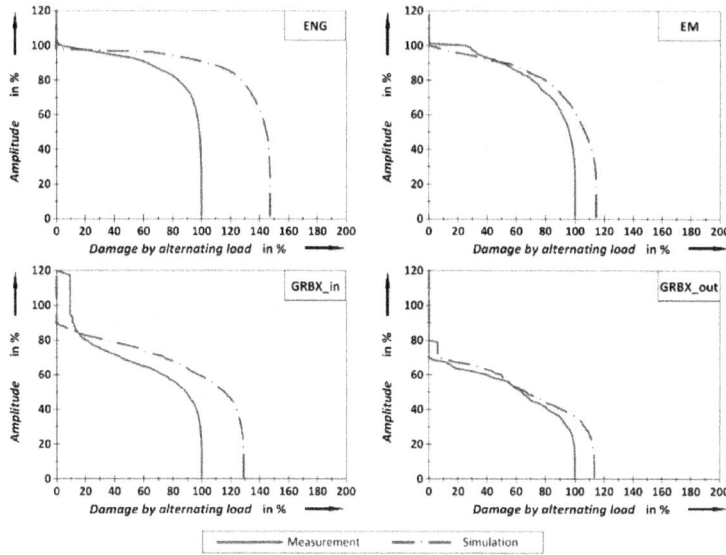

Fig. 8: Comparison between measured and simulated data (3)

Since the calibration sets were optimized for the simulated vehicle, the next step was to compare the data used for the mid-size and the luxury car. The idea was to run the mid-size simulation with the luxury data set as well as the other way round. The results were supposed to give some information about the deviation between measurements and simulations if a "wrong" calibration set is used. However, it appears that almost all parameters were identical. As a consequence the mentioned simulations were skipped. The conclusion is that the strategy – together with the calibration – is very robust as long as the "hybrid philosophy" stays the same. On the other hand the calibration sets are expected to be different, if a "power hybrid" is compared with a "consumption hybrid".

4 Calculation of an axle-split topology

Again an abstract version of the powertrain is used, see figure 9. Due to the higher number of electric machines and the different locations, it is a lot more complex than the parallel version.

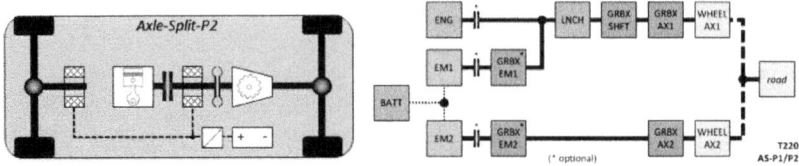

Fig. 9: Schematic and abstract version of a AS-P2-hybrid

The two powertrains do not have a mechanical link indeed. But since there is just one battery an electrical link exists. And this has a very important consequence: as soon as the battery limit is reached any raise of the current for one EM reduces the current for the other one and thus the max possible force. In other words: the max possible forces of the two powertrain are affected by each other.

In contrast to the parallel structures the starting point for the calculations is a total force demand. The CPC user defines a ratio how the force is split between the two axles. The CPC module tries to fulfill the force demand and to maintain the requested split at the same time. If this is not possible the user can prioritize the two options "split" and "force". If the option "split" is chosen, the calculated torques stick to the split ratio even if the force demand is not reached. The reverse behavior is caused by choosing the option "force". Now the calculations leads to torque values that provide the demanded force or at least as much as possible. In this case it is accepted that the force split differs from the requested value.

The figures 10 and 11 show exemplarily the results for the operation mode "hybrid". In this mode all machines are enabled. The requested force split between powertrain 1 (P2-hybrid: ENG + EM1) and powertrain 2 (EM2 only) is 60 % to 40 %.

To demonstrate the principle, the following situation is assumed. All machines are running at a constant speed. At this speed they have a certain torque limit, the EMs additionally a current limit. All values are fictitious and do not represent an existing powertrain. The total force demand is raised from 0 up to 10000 N in steps of 100 N to evaluate the behavior at different driving situations.

In figure 10 the option "split" has been prioritized. Up to a force demand of approximately 5300 N both the demand and the requested force split ratio can be fulfilled. The engine can provide the percentage of powertrain 1 on its own so the EM1 is not needed so far.

If the demand exceeds the value of 5300 N, the EM2 reaches its max allowed current. From this moment the total force stays constant since the powertrain 2 cannot provide more torque and the split ratio must be met.

To choose the option "force" leads to a different behavior, figure 11. Again the EM2 hits its current limit at a force demand of approximately 5300 N. But now the force demand for powertrain 1 is raised to provide the total force. As a consequence the split ratio is moved to powertrain 1.

At a force demand of approximately 6800 N, the engine reaches its limit and the EM1 is needed. Therefore the battery current is getting higher until the max allowed battery current is reached. That happens at a force demand of approximately 8300 N.

From this moment every raise of the current of one EM leads to a reduction of the current of the other EM. In other words: the calculation of the torque distribution becomes a calculation of a current distribution. To decide which EM is prioritized a factor is calculated for both powertrains. This factor contains the information how much force is gained out of a current of 1 A. The powertrain with the higher factor gets more current. In the shown example powertrain 1 produces more force with the same current. As a consequence the current – and therefore the torque – of EM2 is reduced while the current of EM1 is raised.

At a force demand of approximately 8700 N, the EM1 reaches its max allowed current. Now all machines are running at a limit and the total force stays constant.

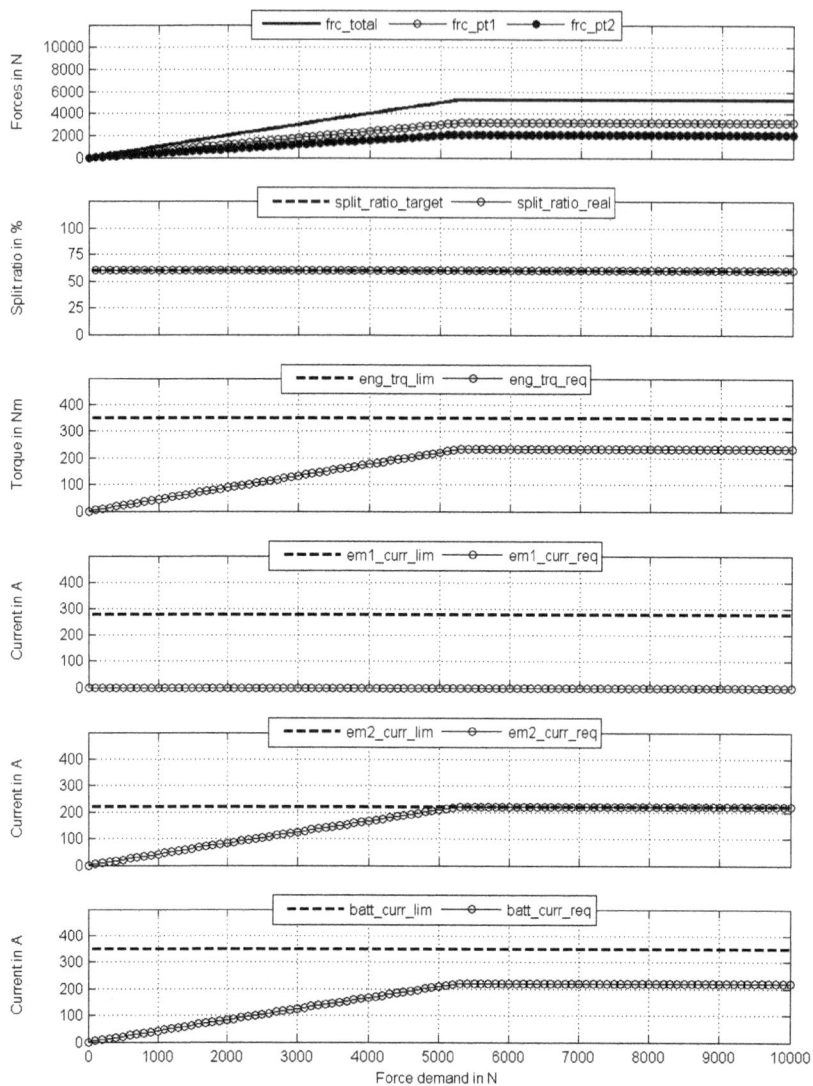

Fig. 10: Results for operation mode "hybrid" and the option "split"

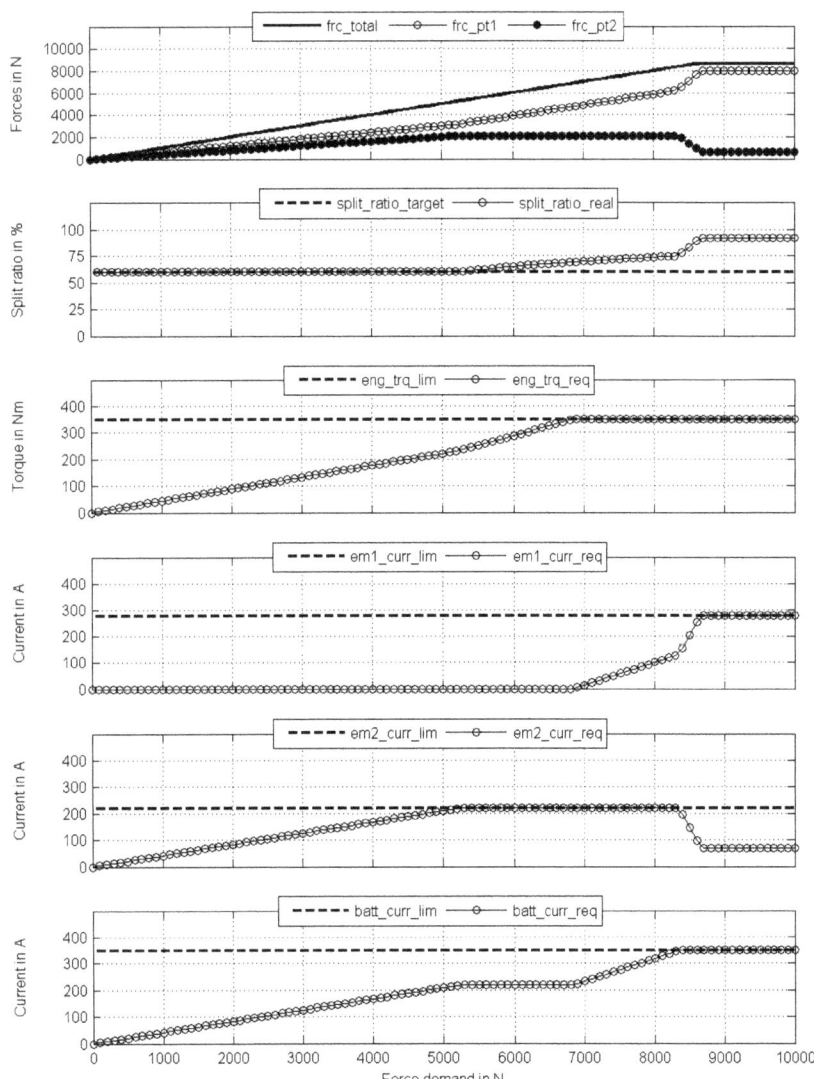

Fig. 11: Results for operation mode "hybrid" and the option "force"

5 Outlook

The CPC-module is available as a Simulink model and as a C-Code file. Since it was designed to be included in existing simulation environments, the interface contains a lot of parameters and calculations. By offering a lot of options a high number of topologies are supported. All these features compromise the calculation time. If a certain environment is chosen and/or some powertrain topologies are ruled out, the model could be cleaned. This is expected to improve the performance of the model.

To support future users a separate documentation of the CPC module will be written.

Furthermore vehicle measurements with an axle-split hybrid would allow an additional validation of the calculation approach.

6 Bibliography

1. Keilhoff; D.; Prof. Reuss, H.-C.; Modulare Hybrid-Betriebsstrategie für die Simulative Ermittlung von Lastkollektiven; Tag d. kooperativen Promotionskollegs Hybrid; 11.11.2014, Stuttgart

2. Kozuharov, A.; Transfer of a modular hybrid operation strategy to an axle-split-powertrain for the simulative generation of the load spectra; Master thesis; October 2015, Stuttgart University

Iterative refinement of the discretization of the Dynamic Programming State Grid

Andreas Haag, Promotionskolleg Hybrid

Peter Antony, Daimler AG

Prof. Dr. Ferdinand Panik, HS Esslingen

Prof. Dr. Michael Bargende, Universität Stuttgart

1 Abstract

The dynamic programming method is well known and widely used for the calculation of the optimal operating strategy of hybrid electric vehicles. In this paper the challenge of large grid sizes due to growing battery capacities and complex powertrain concepts and the need for fine discretization is met by an iterative approach which has the chance to reduce memory usage and computing time by reducing the SOC range with each iteration.

2 Introduction

Dynamic Programming (DP) is a well-known mathematical optimization method developed by Bellman [1]. It is also very common in calculating the operating strategy of hybrid electric vehicles (HEV) and plug-in-HEV (PHEV) [2,3] as it is able to come up with a global optimal solution while its disadvantages are the offline-usage, the computing time and the high memory usage, also known as the "curse of dimensionality" [4]. According to its global optimal solution the DP method is widely used for the comparison of different powertrain configurations or architectures [5] to avoid the influence of different operating strategies on the result or tuning the strategy to each powertrain concept. The DP is also used as a benchmarking tool to compare different operating strategies that might be used online [6], e.g. rule based strategies.

The inability of the DP to be used online as mentioned above has its reason in the fact that the DP needs to know about the whole driving cycle as it is calculated backwards in time to calculate the optimal cost-to-go matrix and then forward in time to find the optimal solution. Its second disadvantage is the computation time, as each control and state vector, whether continuous or not, has to be discretized, and the combined grid of all vectors has to be computed at each time step. Along with this comes the fact that the whole state grid has to be saved as a cost matrix together with a matrix of the optimal control for each time step which causes severe memory usage. For the sake of optimality the discretization of the battery's state of charge (SOC) has to be chosen as a very fine grid as it influences the DP's result [7]. If the grid of the cost matrix is coarse, the cost interpolation to the grid can lead to non-optimal behavior of the control strategy.

Discretization and interpolation of the cost matrix lead to another issue with numerical DP. If there are undesired states e.g. an SOC below a predefined minimum at the end of the driving cycle, the cost matrix will show large numbers at these points of the grid to avoid the simulation from accessing this area. This is typical for charge-sustaining operation strategies of HEVs. Interpolation between feasible and infeasible states leads to incorrect cost calculation. Some methods provide approaches to overcome this issue

[8]. The so-called Boundary Line Method for state grids with a single dimension calculates the extremal points of the feasible region and adds them to the grid while the Level Set Method works with multi-dimensional state grids and uses a final cost term instead of the higher value for the infeasible state region at the end of the driving cycle. Both methods show good results with final SOC state values very close the lower constraint. This usually means a reduction of the fuel consumption as the energy in the battery can be used for electrical driving. For the use of a final cost term the relationship between additional cost and the distance between final SOC and constraint should be known. Due to interpolation in the forward part of the DP simulation a coarse cost matrix may lead to suboptimal solutions which may not be met by the methods mentioned above.

When comparing different operating strategies the DP can be used as a benchmarking tool which makes it necessary to have an optimal solution. By comparing complex powertrains with multiple gear sets and more than one electric machine, which result in large multi-dimensional control grids, the DP can give an optimal solution for any given concept as it is independent and does not need special tuning. Finally the development of PHEVs with large batteries and multiple driving cycles in charge-depleting mode to compute makes it necessary to have a DP method capable of large grid sizes within a small computing time.

Luus [9] proposed an iterative DP method called "region contraction" already in 1990. In this method the control space being discretized is being narrowed with the help of previous iteration's result leading to a significant improvement of the simulation's result with constant grid size. As a model he used a system of 8 ordinary differential equations with 4 control variables and 8 state variables.

Other solutions to increase the DP's speed stay with suboptimal results, like Approximate Dynamic Programming (ADP) [10]. The combination of Dynamic Programming with suboptimal methods to approximate a cost function are summed up in [11].

In [4] an iterative approach is used for optimizing the operating strategy of an HEV with the vehicle speed as part of the state grid. The authors propose a fixed reduction factor of about 0.6 applied to the control and state grid to avoid the solution from being locked up in suboptimal regions due to smaller reduction factors.

3 Simulation Setup

3.1 Vehicle Model Setup

In the following simulations we use a vehicle model with an internal combustion engine (ICE) of 140 kW, an electric machine of 80 kW and a battery with a capacity of 9 kWh. The vehicle mass is 1750 kg, the WLTP is used as driving cycle.

The road load is calculated with the road load equations as shown in [12].

All machines use static loss or consumption models while the battery model is of 0th order. For the purpose of creating a benchmark we do not use any restrictions of a powertrain. Instead the two-dimensional static models of the machines are being reduced to 1-D models in which the machine's mechanical power corresponds to its lowest possible loss or consumption [13]. Note that the use of these simplified models reduces the computational effort, as only one degree of freedom has to be controlled by the DP.

3.2 DP Model Setup

For the following simulations we use the DP MatLab function provided in [14] as it is free and easy to use.

The DP is set up with the vehicle model described above. As the static models of the machines are reduced to 1-D for benchmark purposes only one control vector is needed to control the electric machine's power.

In the simulation model a restriction to the total number of grid points in state and control space is being used. This ensures control of memory usage. The total grid size is the product of the size of all control vectors and all state vectors. In this case a state vector for the battery SOC and for the engine state is used. So the number of elements g of the SOC state is given.

As the control vector's limits are constant, its range is defined at each time step as following:

$$c_{max} = \min(P_{Batt,max}, P_{EM,max}, -P_{Road})$$
$$c_{min} = \max(P_{Batt,min}, P_{EM,min})$$

$$(1)$$

Thus the road load can be met without interpolating the control vector. The ICE power results from the road load P_{Road} and the electric machine's power P_{EM}.

3.3 Iterative DP Setup

The iterative DP is set up according to the following steps:

An initial DP is being calculated with the chosen grid sizes and the bounds of the battery as limits of the SOC state grid with the bandwidth $\Delta SOC_1 = SOC_{max} - SOC_{min}$. The battery limits can be chosen according to the driving cycle or to physical limitations. A starting $SOC_{1,0}$ and a range for the final $SOC_{end,min}$ and $SOC_{end,max}$ are defined. This results in the discretization:

$$f_1 = \frac{\Delta SOC_1}{g - 1} \tag{2}$$

with the number of SOC grid elements g.

The resulting SOC of the initial DP is used as the center of the grid created for the following iteration:

$$X_2 = SOC_1 \tag{3}$$

A contraction factor c is chosen. It defines the size SOC range which is discretized for the next iteration:

$$\Delta SOC_2 = \Delta SOC_1 \times c \tag{4}$$

As the number of elements in the grid is constant, the discretization is refined:

$$f_2 = \frac{f_1}{c} \tag{5}$$

As time is discretized in the DP as well, every time step is calculated individually. With the SOC changing at every time step for each step an initial and a final value $X_{2,in}$ and $X_{2,out}$ is needed with $X_{2,t+1,in} = X_{2,t,out}$. The initial point $X_{2,0,in} = SOC_{1,0}$ equals the predefined value. With the value X as its central point and the ΔSOC_2 as bandwidth the upper and lower limits X_{min}, X_{max} of the discretization range are yielded (see Fig. 1).

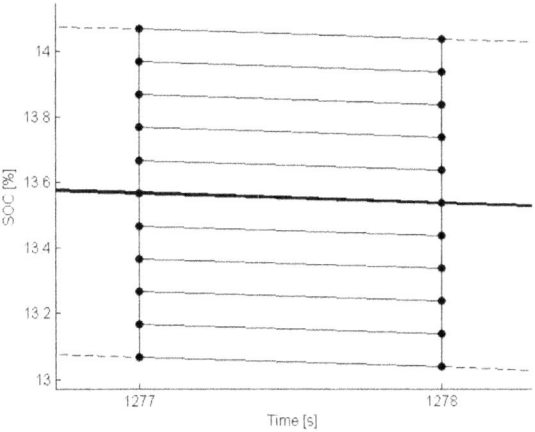

Fig. 1: Example of a contracted grid X (markers) based on the SOC (thick line) and the limits X_{min}/X_{max} (dotted lines) of a preceding iteration

If the discretization range exceeds the limits of the SOC range given in the initial DP, the central points are being shifted in a way that $X_2 \pm \frac{\Delta SOC_2}{2}$ is within the limits. The benefit is that the whole discretization range can be used by the optimization instead of being marked as infeasible. At the end of the driving cycle a specified SOC range between $SOC_{end,min}$ and $SOC_{end,max}$ is permitted in the initial DP. As the contraction of the region may not include the whole range, the lower SOC bound is considered as relevant for minimizing fuel consumption. So the contraction factor c must be increased to allow $X_{2,end,out} - \frac{\Delta SOC_2}{2} \le SOC_{end,min}$. To compare different simulation setups a common final SOC discretization is needed, and for the final DP iteration the contraction factor is adapted to adjust the discretization to its final value. Fig. 2 shows the contracted SOC bandwidth for SOC near the lower boundary (left) and with the final minimum SOC (right).

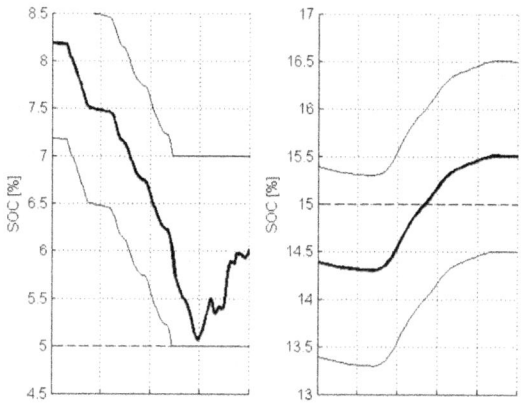

Fig. 2: Example of shifting the X grid near the lower limit (left) and keeping the desirable final SOC within compression range (right)

The creation of the SOC grid as an equally spaced linear grid with the number of elements g is conducted as following:

$$[X_{2,lin,min}, X_{2,lin,max}] := \left\{ -\frac{g-1}{2} : 1 : \frac{g-1}{2} \right\} \tag{6}$$

Note that by making sure g is an odd number the values of X_{lin} are integers. The relation to the SOC is made by projecting the linearized grid $X_{2,lin}$ onto the SOC grid of each time step:

$$SOC_{2,t,in} = X_{2,t,in} + X_{2,lin,in} * f_2 \tag{7}$$

Within the time step the calculations are being made with the SOC values of the battery. To return the time step's result to the DP the projection is being reversed with the new grid center $X_{2,t,out}$ at the end of the time step:

$$X_{2,lin,out} = \frac{SOC_{2,t,out} - X_{2,t,out}}{f_2} \tag{8}$$

This makes the DP work with an equally spaced linear grid while the vehicle model can work with the SOC range provided by the projection.

The iterative DP continues with the result of the preceding iteration until the final discretization is reached.

To avoid locking up the simulation in suboptimal regions of the SOC the result of each iteration is tested upon the distance of the SOC towards the bandwidth bounds. In this case the iteration can be repeated with less contraction.

As the use of parallel computing is not influenced by the method provided in this paper, its influence on computing time is not considered.

4   Results

Fig. 3 shows the results in fuel consumption in relation to the lowest value that occurred in all simulations for total computing grid sizes between 10^3 and 10^5 and contraction factors between 0.1 and 0.9.

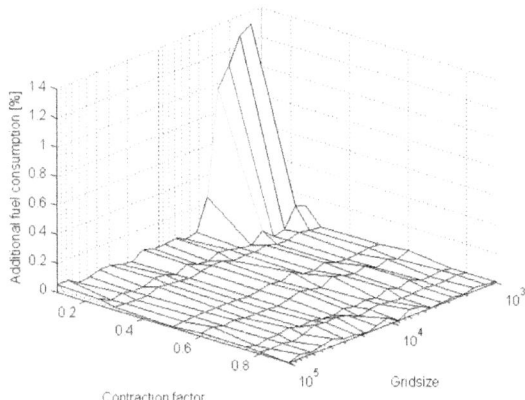

Fig. 3: Relative fuel consumption of different contraction factors and grid sizes

It can be seen that for contraction factors above ~0.25 and total grid sizes above ~10^4 all simulations show quite similar results which makes the approach seemingly robust with 155 out of 160 total simulations within a range of 0.2% of additional fuel.

For comparison a single DP with the same discretization was conducted with a result of 0.05% and a total grid size of 7.5×10^5. The computing time of 149 simulations was below the computing time of the single DP.

Fig. 4 shows the results of each iteration. It can be seen that a finer discretization leads to an improvement of the result. As with coarse start grids the results vary, the few results above an additional fuel consumption of 0.5% show that the contraction factor is crucial for the success of an iterative DP.

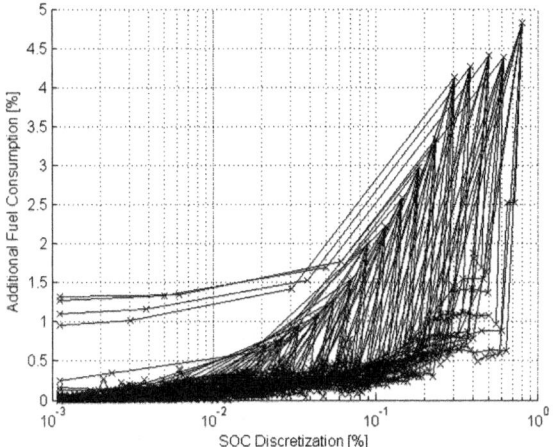

Fig. 4: Relative fuel consumption vs. discretization of SOC

For the comparison of computing times a single DP was conducted with a total grid size of 7.3×10^6, as the computing time increases with its grid size. For the iterative DP a contraction factor of 0.25 and a total grid size of 1.2×10^4 were used as these showed the best results in the computations shown above. The iterative DP took 6 iterations with a total computing time 144 s while the single DP took 155 min, so computing time was improved by a factor of 64. The results vary by 0.04% with a negligible advantage for the iterative DP. While saving a veritable amount of computation time the total grid size could be reduced by a factor of 600.

5 Conclusion

The approach of an iterative DP with a contraction factor regarding the desired final SOC state, the final SOC and the absolute bounds of the state system improves the capability of the DP to cope with the need for benchmarking to compare different strategies for long driving cycles or large batteries as well as the need to compute optimal operating strategies for complex powertrain concepts with control vectors.

6 References

1. Bellman, R. E.; Dreyfus, S.: "Applied Dynamic Programming", Princeton University Press, 1962

2. Cipek, M.; Skugor, B.; Kasak, J.; Deur, J.: "Dynamic Programming-based Optimization of Control Variables of an Extended Range Electric Vehicle", SAE Technical Paper 2013-01-1481, 2013

3. Kum, D.; Peng, H.; Bucknor, N.: "Optimal Energy and Catalyst Temperature of Plug-In Hybrid Electric Vehicles for Minimum Fuel Consumption and Tail-Pipe Emissions", IEEE Transactions on Control Systems Technology, Vol. 21, 14-26, 2013

4. Wahl, H.; Gauterin, F.: "An Iterative Dynamic Programming Approach for the Global Optimal Control of Hybrid Electric Vehicles under Real-time Constraints", IEEE Intelligent Vehicles Symposium, 2013

5. Serrao, L.; Onori, S.; Rizzoni, G.: "A Comparative Analysis of Energy Management Strategies for Hybrid Electric Vehicles", Journal of Dynamic Systems, Measurement and Control, Vol. 133, 2011

6. Ravey, A.; Blunier, B.; Miraoui, A.: "Control strategies for fuel-cell-based hybrid electric vehicles from offline to online and experimental results", IEEE Transactions on Vehicular Technology, Vol. 61, 2452-2457, 2012

7. Onori, S.; Serrao, L.; Rizzoni, G.: "Dynamic Programming" in "Hybrid Electric Vehicles", Springer London, 41-49, 2016

8. Elbert, P.; Ebbesen, S.; Guzzella, L.: "Implementation of Dynamic Programming for –dimensional Optimal Control Problems with Final State Constraints", IEEE Transactions on Control Systems Technology, Vol. 21, 924-931, 2013

9. Luus, R.: "Application of dynamic programming to high-dimensional non-linear optimal control problems", International Journal of Control, 52:1, 239-250

10. Larsson, V.; Johannesson, L.; Egardt, B.: "Cubic spline approximations of the Dynamic Programming cost-to-go in HEV energy management problems", IEEE European Control Conference, 1699-1704, 2014

11. Bertsekas, D.; Tsitsiklis, J.: "Neuro-Dynamic Programming: An Overview", Proceedings of the 34th Conference on Decision & Control, 560-564, 1995

12. Guzzella, L., & Sciarretta, A. (2007). "Vehicle propulsion systems", Vol. 1, Springer-Verlag Berlin Heidelberg

13. Haag, A.; Antony, P.; Panik, F.; Bargende, M.: „A Comparison of Varieties of Powertrains for Hybrid Electric Vehicles", 12th Symposium Hybrid- und Elektrofahrzeuge, Braunschweig, 2015

14. Sundstrom, O.; Guzzella, L., "A generic dynamic programming Matlab function," Control Applications, (CCA) & Intelligent Control, (ISIC), 2009 IEEE , vol., no., pp.1625,1630, 8-10 July 2009

Fast predictive burn rate model for gasoline-HCCI

Mahir Tim Keskin, Michael Grill, Michael Bargende

IVK (Universität Stuttgart), FKFS

1 Abstract

Operating gasoline engines at part load in a so-called Gasoline-HCCI (gHCCI) combustion mode has shown promising results in terms of improved efficiency and reduced emissions. So far, research has primarily been focused on experimental investigations on the test bench, whereas fast, predictive burn rate models for use in process calculation have not been available. Such a phenomenological model is henceforth presented. It describes the current burn rate as the sum of a sequential self-ignition process that is modeled using a temperature distribution and an Arrhenius equation on the one hand and a laminar-turbulent flame propagation described by a modified entrainment model on the other hand. The newly developed model correctly predicts burn rates for a wide range of variations of control parameters, including PTDC combustion and operating mode switches, using a single set of tuning parameters, while requiring very low computational times.

2 Introduction

The ongoing trend of enforcing lower fleet consumption and tightening emission standards, combined with ever-increasing customer demands and an increased competitive pressure on a globalized market, is a driving force for further improvement of the internal combustion engine regarding efficiency and emissions. In this regard, gHCCI seems to be a promising approach for gasoline engines. Used for part-load operation – pressure gradients become to steep at higher loads – it enables considerable improvements in both fuel economy (up to 15 % [1] over the NEDC) and engine-out NO_x emissions (up to 99 % [2]). This in turn creates the opportunity to increase AFR and to benefit from dethrottling while still using a cost-efficient three-way-catalyst.

This enormous potential triggered extensive research work on the combustion process, starting as early as the 1970s [3]. However, up to today, no commercial application has been introduced within the automotive industry. One of the main difficulties is the complexity of combustion control [4-6] due to the fact that – unlike for conventional gasoline or Diesel engines – control parameters which have a direct impact (i.e. with a very short dead time) on combustion phasing are lacking. Although experimental researchers have repeatedly demonstrated that these drawbacks can be overcome, controlling combustion in gHCCI mode remains a demanding task that is associated with high complexity and corresponding costs, e.g. for sensors. Whereas the automotive industry has hence currently focused on alternative means to reduce fuel consumption, gHCCI remains an interesting option for the near future:

- Costs for sensor technology required for combustion control can be expected to decrease due to the general progress in electronics.

- Tightening fleet consumption targets will intensify the need to achieve further improvement and will probably require using new technologies.

- Likewise, tightening emission legislation such as "Real Driving Emissions" [7] will make scavenging and mixture enrichment hardly affordable in future type approval procedures, reducing attractiveness of concepts such as downsizing.

Thus an increasing interest in gHCCI can be expected. As soon as commercial applications are contemplated, tasks such as developing operating and control strategies or assessing potential and hardware requirements will become important. A fast predictive burn rate model, such as the one presented here, would then be a useful tool to reduce the need of cost-intensive test bench investigations.

3 Measurement Data

Measurement data from the research project [4] has been used as the basis of model development. Therein, a one-cylinder unit with direct injection and fully variable electro-hydraulic valve train served as test engine, which was derived from a V6 gasoline engine from Daimler (internally labeled M 272 DE), see [4]. On the test engine a wide range of parameter variations was conducted, see [4] or [8]:

- Residual gas strategy: In order to reach a temperature level that is sufficiently high for self-ignition, a certain amount of hot residuals has to be stored in the cylinder. This can be either achieved by means of a negative valve overlap, partly combined with PTDC combustion ("Combustion Chamber Retention"), or by performing an exhaust valve double lift ("Exhaust Port Recirculation").

- Injection strategy: Either a single injection (well before FTDC) or a pre-injection shortly before PTDC and a main injection well before FTDC was performed.

- Valve timings/throttle position: Variations of these parameters were used to create different residual gas contents.

- Engine speed: 2000 rpm or 3000 rpm, load (IMEP) 2 bar or 3 bar.

- Ignition angle: Some operating points were conducted with spark assist, with the ignition angle varying between 40°CA and 0°CA before FTDC.

A thoroughly conducted measurement data analysis revealed several hints at the occurrence of a flame propagation mechanism. For instance, an earlier ignition angle lead to an earlier start of combustion, but only if the boundary conditions (AFR, residual gas content) allowed for flame propagation: otherwise no change could be observed. Likewise, a variation of the residual gas content at nearly stoichiometric operating points showed a behavior in which the early phase of combustion was accelerated by low residual gas content, whereas a higher residual gas content (and thus a higher temperature) was beneficial for the combustion later on. Last but not least it could be observed during an operating mode switch from conventional spark ignition to gHCCI that there are cycles with a "hybrid" characteristic, starting like conventional spark ignited combustions and showing a steep increase in the burn rate afterwards.

4 Model Description

4.1 General approach

As shown in the measurement data analysis and in optical investigations [9], combustion progress during gHCCI –under certain boundary conditions – strongly resembles the one observed in conventional spark ignition engines, which is characterized by the propagation of a flame front starting at the spark plug and can be described using an entrainment model in working process calculation. It thus seems obvious to include the entrainment model in the overall model, albeit adjustments accounting for the different boundary conditions have to be considered. Besides, a second mechanism has to represent the thermo-kinetically dominated part of combustion, corresponding to the idea of a sequential self-ignition process. In order to describe this volume reaction, the whole timeline since the SOI as well as possible inhomogeneities in temperature or mixture composition have to be taken into account. This is managed by means of a distributed ignition integral, in which ignition delay is monitored separately for groups of differing temperature in the combustion chamber. The overall model comprises thus two different parts *flame propagation* and *volume reaction*, see Figure 1, with the total burn rate being calculated as the sum of the two partial burn rates.

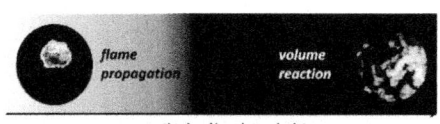

flame propagation   volume reaction

pre-reaction level in unburned mixture

Figure 1. Comparison of the two basic mechanisms of combustion (visualizations from [5])

4.2 Volume reaction calculation

The self-ignition process itself depends on reaction kinetics, which can be modeled by means of an ignition integral as the most basic approach. The sequential progress can be secured by imposing inhomogeneities by means of a normal distribution. Finally, the influence of mixture formation has to be taken into account for late injection timings close to TDC.

4.2.1 Mixture formation calculation

Measurement data analysis revealed that a significant influence of mixture formation on combustion can only be observed for PTDC combustion (where fuel was injected relatively close to TDC), whereas nothing of this kind could be seen in the main combustion around FTDC (where fuel was injected at BDC at the latest). Mixture formation was thus only modeled for injection events occurring less than 70°CA before TDC. For doing so, several approaches are known from phenomenological Diesel combustion models. As relatively simple approaches based on turbulence showed satisfying results, a formulation similar to [10] has been chosen, see [8].

4.2.2 Description of inhomogeneities

The modeling of the volume reaction as a sequential self-ignition process is based on the idea of a distributed ignition integral. According to this, a temperature distribution is assumed for ignition delay calculation in such a way that there a groups of higher and lower temperature to which a certain mass fraction is assigned. When the ignition integral in a certain group reaches a predefined limit value, immediate combustion of the assigned mass fraction is assumed. It is notable that this partition in several groups is only relevant for ignition delay calculation, whereas the thermodynamic calculation itself unvariedly follows the two zone approach apparent in the entrainment model. The mathematical description of temperature inhomogeneities is based on a normal distribution function, which is widely used in general as it represents the one distribution to which every distribution converges if it is subject to a large number of independent influencing parameters (central limit theorem). However, using a "simple" normal distribution does not allow for asymmetrical distributions which have to be expected due to wall heat losses. Therefore the use of a "contaminated" normal distribution, a linear combination of any desired number of normal distributions, is useful:

$$f_c(x) = \sum_{i=1}^{n} \varepsilon_i \frac{1}{\sigma_i \cdot \sqrt{2\pi}} \cdot e^{-\frac{1}{2}\left(\frac{x-\mu_i}{\sigma_i}\right)^2} \tag{1}$$

$f_c(x)$ probability density function of contaminated normal distribution [1/unit of random variable]

ε share of a single normal distribution [-]

i index of summation [-]

n	number of normal distributions [-]
σ	standard deviation [unit of random variable]
μ	mean [unit of random variable]
x	random variable [any desired unit]

The single normal distributions within the "contaminated" one can be interpreted as distinctive areas in the combustion chamber that are characterized by a different temperature distribution. Considering the aforementioned wall-heat influence, it seems natural to distinguish two such areas: a "normal" area and a "wall-influence" area. The latter allows accounting for the facts that the average temperature close to the cylinder wall is lower and that the temperature difference between the two areas increases during compression.

4.2.3 Ignition delay calculation

A simple approach widely used (e.g. in Diesel combustion modeling or knock modeling) is based on the idea to sum up the whole reaction kinetics in a single "substitute" reaction and to describe its reaction rate by means of the Arrhenius equation. In several phenomenological models, the original form of the latter has been enhanced by several chemically plausible factors. During model development, it become evident that using a formulation such as

$$r = A \cdot c_{Rad} \cdot c_{O2} \cdot p \cdot e^{-\frac{E_A}{R \cdot T}} \tag{2}$$

r	reaction rate [1/s]
c_{O2}	oxygen concentration [mol/m³]
c_{Rad}	radical concentration [mol/m³]

yields very good results. The only amendments to the original formulation therein are the terms for pressure, oxygen and radical concentration. For calculating the latter, proportionality to the stoichiometric residual gas fraction is assumed, with an additional tuning parameter allowing to distinguish between residuals retained in the combustion chamber and residuals recirculated via the exhaust port. This parameter also accounts indirectly for stratification, as a lesser degree of stratification can be associated with better mixing and thus reduced radical activity.

4.3 Modifications to the entrainment model

In order to account for the different boundary conditions, two adjustments have been made to the entrainment model: Firstly, flame surface calculation has to be modified due to the fact that more than just one ignition point is possible, and secondly flame speed calculation has to be reconsidered because the flame front will reach mixture regions in which pre-reactions have taken place to a non-negligible extent. Both modifications will only have a considerable impact at high pre-reaction levels, allowing for a consistent transition to the original entrainment model.

4.3.1 Flame surface calculation

As shown in [11] for conventional spark ignition, the current flame surface can be calculated at any crank angle for any burned volume share based on spherical shells if combustion chamber geometry and spark plug position are known. In order to ensure a smooth transition to conventional spark ignition, it is convenient to account for the increase in flame surface occurring with multiple ignition points by means of a multiplier, assuming thus implicitly a similar influence of cylinder walls on flame surface. The calculation of this multiplier follows a step-by-step approach. Firstly, the number and the volume share of the respective ignition centers are derived from the distributed ignition integral. Then the increase in surface for a certain burned volume fraction can be calculated geometrically unambiguously. Subsequently, an overlap probability can be determined using an analogy to an urn problem and the resulting reduction in surface is deducted, so that the sought surface multiplier can be finally computed, see [8].

4.3.2 Laminar flame speed calculation

It can be expected that an increase in the pre-reaction level of the unburned mixture will lead to an increase in flame speed, as less heat has to be supplied to the pre-heating zone to initiate an inflammation, with heat transfer being the dominating parameter for flame propagation. However, wide-spread flame speed correlations do not account for the influence of pre-reaction levels, as for practical reasons, measurement of flame speed is typically conducted at near-ambient temperatures where pre-reactions are negligible. In order to estimate the qualitative impact of pre-reaction levels on laminar flame speed, the following thought experiment can be considered:

A perfectly homogeneous mixture is conditioned for a certain time span at a certain temperature T_{ub}. If T_{ub} is sufficiently high, self-ignition will occur after a certain period of time. The temperature which is leading to self-ignition after a short, defined period of time t^* is to be called activation temperature T_{act}. Henceforth, for various starting temperatures $T_{ub} < T_{act}$ a spark ignition can be conducted after the previously defined period of time t^* has lapsed, leading to a flame propagation. Describing pre-reaction levels based on an Arrhenius integral and assuming

- that the amount of energy that has to be applied to the pre-heating zone is proportional to the difference between the ignition integral limit value and the actual value,

- that the period of time necessary to supply this energy is proportional to its amount

- and that flame speed is inversely proportional to the necessary period of time,

a modified laminar flame speed correlation can be formulated as follows [8]:

$$s_{L,mod} = s_{L,0} \cdot \frac{R(T_{act}, t*) - R(T_0, t*)}{R(T_{act}, t*) - R(T_{ub}, t*)} \tag{3}$$

$s_{L,0}$ laminar flame speed at reference conditions [m/s]

T_0 reference temperature [K]

The fraction in equation (3) can be interpreted as a pre-reaction factor. It can be simplified and supplemented by a damping term that reduces the impact of the simplified assumptions used for derivation, especially at high pre-reaction levels, see [8].

4.4 Interaction between the two combustion mechanisms

As already discussed, the general model approach allows for both flame propagation as well as volume reaction to occur simultaneously, thus leading to interaction. From an entrainment model point of view, three changes due to the volume reaction have to be considered:

- Flame propagation can start from multiple locations.

- The flame front may reach regions where the mixture is already burned.

- Unburned mass in the flame zone can still burn directly via the volume reaction mechanism.

The impact of flame propagation on the volume reaction mechanism is twofold:

- Ignition delay is shortened due to the heat release from flame propagation.

- Mass that is burned via flame propagation cannot be burned a second time.

All of the mentioned aspects are accounted for in the model implementation [8]. Regarding the last aspect, it is assumed that mass burned via the entrainment model is taken evenly (i.e. according to their mass proportion) from all of the temperature groups that have not reached their self-ignition point yet, see Figure 2.

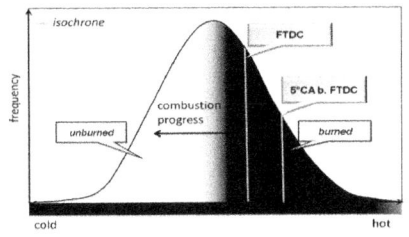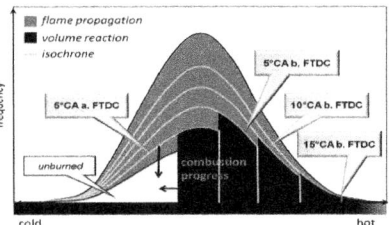

Figure 2. Combustion progress without (left) and with (right) flame propagation

5 Simulation Results and Discussion

For practical reasons, the examples discussed here are limited to the CCR strategy. However, it is noteworthy that results for EPR are similarly promising, see [8].

5.1 Variation of residual gas content

A variation of residual gas content, which is basically achieved by means of an exhaust valve timing variation (EVC), is analyzed in Figure 3. The comparison between simulation and experiment shows very good agreement: Combustion phasing as well as maximal burn rate are correctly reproduced for all operating point, the influence of residual gas content is displayed properly. Apart from minor errors, the model overall predicts burn rates correctly from both a quantitative and a qualitative point of view.

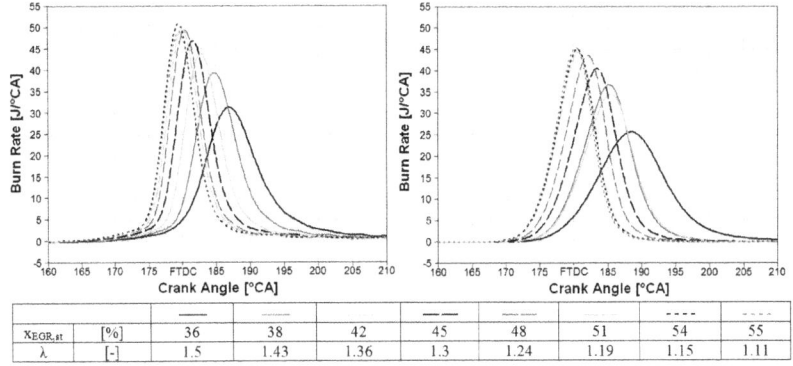

$x_{EGR,st}$	[%]	36	38	42	45	48	51	54	55
λ	[-]	1.5	1.43	1.36	1.3	1.24	1.19	1.15	1.11

Figure 3. Burn rates from experiment (PTA, left) and simulation (right) for the residual gas content variation (eng. speed 2000 rpm, IMEP 2 bar, single injection 350°CA b. FTDC, no spark)

5.2 Variation of main injection

In this variation the SOI is shifted from 335°CA b. FTDC to 270°CA b. FTDC, with the EGR rate and AFR held nearly constant. When comparing simulation and experiment in Figure 4, good agreement can be stated again, with the only notable difference being a slightly faster combustion in simulation. It is remarkable, however, that the counter-intuitive effect of a later injection leading to an earlier combustion (see the two operating points with the latest SOI) can be reproduced in simulation. This can be attributed to two reasons: firstly, the ignition integral hardly progresses in the time span between the two injection timings due to the comparatively low temperatures, secondly, the latest injection timings changes AFR and residual gas content slightly, leading to boundary conditions beneficial for auto-ignition.

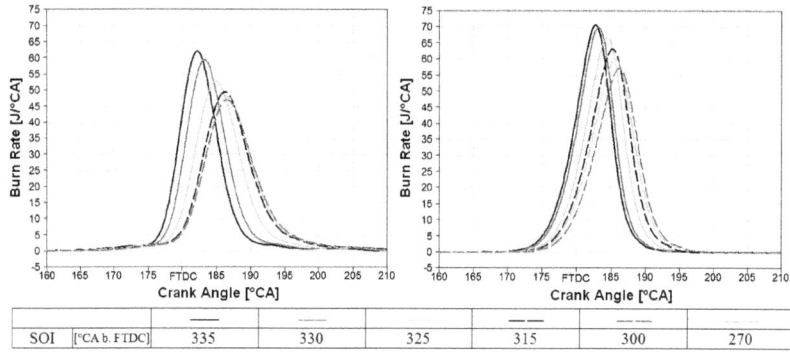

SOI	[°CA b. FTDC]	335	330	325	315	300	270

Figure 4. Burn rates from experiment (PTA, left) and simulation (right) for the main injection variation (eng. speed 3000 rpm, IMEP 3 bar, λ 1.45 - 1.50, st. EGR rate 31 % - 33 %, no spark)

5.3 Variation of residual gas content and main injection

When a combined variation of the two parameters already discussed is performed, maintaining combustion phasing approximately constant requires earlier injection timings when the residual gas content decreases. If additionally, the AFR is kept nearly stoichiometric by a simultaneous throttle variation, an interesting behavior can be observed: Burn rate traces show a "crossing characteristic" where low residual gas content leads to a faster flame propagation in the early combustion phase, but at the same time also to a delay in auto-ignition, resulting in a lower maximum burn rate and a later end of combustion. The model predicts this effect correctly, see Figure 5.

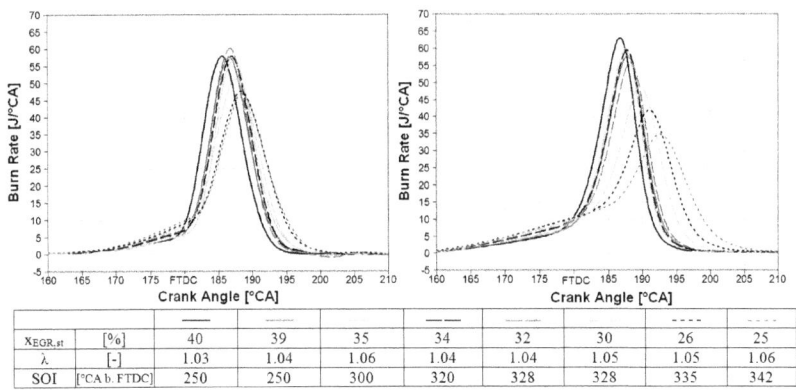

$X_{EGR,st}$	[%]	40	39	35	34	32	30	26	25
λ	[-]	1.03	1.04	1.06	1.04	1.04	1.05	1.05	1.06
SOI	[°CA b. FTDC]	250	250	300	320	328	328	335	342

Figure 5. Burn rates from experiment (PTA, left) and simulation (right) for the variation of EGR rate, SOI and throttle position (eng. speed 3000 rpm, IMEP 3 bar, ign. angle 30°CA b. FTDC)

5.4 Variation of ignition angle

As already mentioned, an earlier ignition angle leads to an earlier combustion phasing without significant changes in maximum burn rate, provided that residual gas content and equivalence ratio are not excessively high. Simulation results show the same behavior, see Figure 6. The heat release from flame propagation obviously triggers earlier auto-ignition in this case.

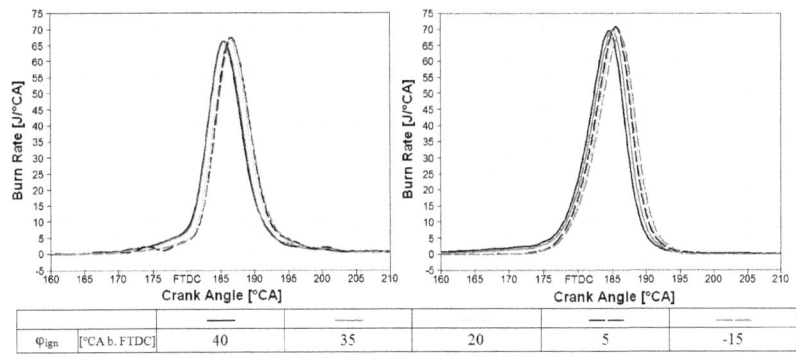

φ_{ign}	[°CA b. FTDC]	40	35	20	5	-15

Figure 6. Burn rates from experiment (PTA, left) and simulation (right) for the ign. angle variation (eng. speed 2000 rpm, IMEP 3 bar, λ 1.21, st. EGR rate 40 % - 41 %)

5.5 Variation of pre-injection

A variation of pre-injection timings may serve as an example to investigate PTDC combustion during the negative valve overlap. Figure 7 shows again a comparison between experiment and simulation. Whereas combustion phasing is retarded in simulation for earlier injection timings, the general shape of the burn rate traces as well as the grading of the single operating points is displayed fairly well. In particular, the steep increase at the beginning, reminiscent of Diesel pre-mixed combustion can be reproduced. When assessing model quality, one has to consider that especially for PTDC combustion numerous uncertainties arising from measurement and PTA will affect the boundary conditions of the simulation; for instance, the cylinder mass during the negative valve overlap eventually results from a gas exchange analysis. Sensitivity calculations showed that a mere increase of 6 mg in trapped mass is sufficient to fully compensate the aforementioned retarded combustion phasing [8].

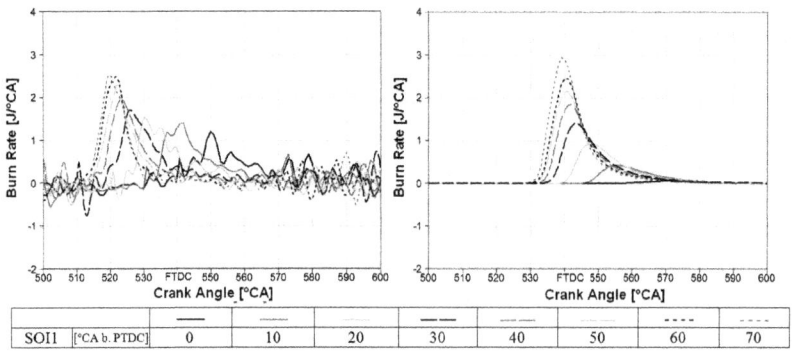

Figure 7. Burn rates from experiment (PTA, left) and simulation (right) for the pre-injection variation (eng. speed 2000 rpm, IMEP 3 bar, SOI 180°CA b. FTDC, λ 1.57 - 1.60, st. EGR 37 %)

5.6 Operating mode switch

Last but not least an operating mode switch from conventional spark ignition to Gasoline-HCCI is investigated. The changes in operating parameters occurring during this switch are broadly discussed in [4]. Figure 8 shows a compact overview of the change in burn rate from cycle to cycle. It is evident that the transition from conventional flame propagation to auto-ignition is predicted correctly in simulation. This applies in particular to the "hybrid" combustion occurring in working cycle 7.

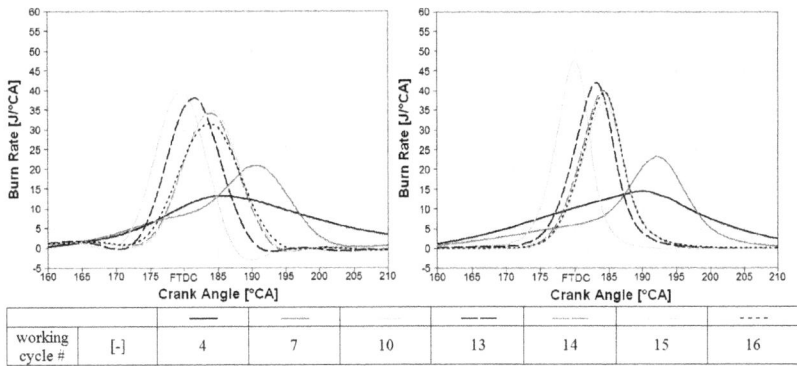

working cycle #	[-]	4	7	10	13	14	15	16

Figure 8. Burn rates from experiment (PTA, left) and simulation (right) for an operating mode switch

6 Conclusion

A fast, predictive burn rate model for gHCCI has been developed. It is based on the idea to see gHCCI combustion as a combination of laminar-turbulent flame propagation on the one hand and a sequential self-ignition process on the other hand. In order to model the first mechanism, an entrainment model, which has convincingly proved itself in practice for conventional spark ignition engines, is included. Several amendments and modifications have been made to account for the different boundary conditions in gHCCI mode, such as a term describing the increase of laminar burning speed at high pre reaction levels in the unburned charge. The second mechanism is essentially represented by ignition delay calculation, in which the reaction rate is computed separately for some hundred groups of different temperatures based on the Arrhenius equation. Thermal inhomogeneity is described by a contaminated normal distribution which accounts for the influence of wall temperature on mixture close to the cylinder wall. This allows restricting the number of thermodynamic zones to two only, a burned and an unburned zone. Both mechanisms for the combustion progression run simultaneously and can influence each other, leading to a dynamic partitioning between the two modes depending of the boundary conditions. All in all, a relatively low number of tuning parameters is sufficient for model calibration, with all of them being intuitively comprehensible. Once calibrated, the model is able to predict burn rates for a wide range of variations of control parameters, including PTDC combustion, different operating strategies and an operating mode switch, using a single set of tuning parameters. Comparisons for more than 150 operating points show good agreement between simulation and experimental data.

The model's characteristics make it thus highly suitable for daily use in engine development processes, potentially providing stimulus for the further development and application of gHCCI. Besides, the basic conception of the model to cover flame propagation and self-ignition at the same time may unlock the potential for the development of a comprehensive model of combustion in gasoline engines, which could possibly depict all of the phenomena occurring in gasoline engines including spark ignition, gHCCI, knock and pre-ignitions.

References

1. Lang, O., Salber, W., Hahn, J., Pischinger, S. et al., "Thermodynamical and Mechanical Approach Towards a Variable Valve Train for the Controlled Auto Ignition Combustion Process", SAE Paper 2005-01-0762, 2005.

2. Zhao, H., Ma, T., Jiang, X., Cao, L. et al., "Combined Experimental and Modelling Studies on CAI Combustion Engines" presented at Haus der Technik Kongress Controlled Auto Ignition, 2005, Germany, October 20-21, 2005.

3. Onishi, S., Jo, S., Shoda, K., Jo, P. et al., "Active Thermo-Atmosphere Combustion (ATAC) - A New Combustion Process for Internal Combustion Engines", SAE Paper 790501, 1979.

4. Babic, G., "Betriebsstrategien Benzinselbstzündung," Final Report on FVV project no. 883, 2010.

5. Herrmann, H., Herweg, R., Karl, G., Pfau, M. et al., "Regelungskonzepte in Ottomotoren mit homogen-kompressionsgezündeter Verbrennung" presented at Haus der Technik Kongress Controlled Auto Ignition, 2005, Germany, October 20-21, 2005.

6. Kulzer, A., Hathout, J., Sauer, C., Karrelmeyer, R. et al., "Multi-Mode Combustion Strategies with CAI for a GDI Engine", SAE Paper 2007-01-0214, 2007.

7. Kühlwein, J., Rexeis, M., Luz, R., Hausberger, S., "Update of Emission Factors for EURO 5 and EURO 6 Passenger Cars for the HBEFA Version 3.2" http://ermes-group.eu/web/system/files/filedepot/10/HBEFA3-2_PC_LCV_final_report_aktuell.pdf.

8. Keskin, M., "Brennverlaufsmodell Benzinselbstzündung". Final Report on FVV project no. 1109, 2015.

9. Sauter, W., Hensel, S., Schubert, A., "Benzinselbstzündung". Final Report on FVV project no. 831, 2008.

10. Rether, D., "Modell zur Vorhersage der Brennrate bei homogener und teilhomogener Dieselverbrennung," Ph.D. thesis, University of Stuttgart, 2012.

11. Grill, M., "Objektorientierte Prozessrechnung von Verbrennungsmotoren," Ph.D. thesis, University of Stuttgart, 2006.

Acknowledgments

The model presented in this report is the result of a research task defined by the Forschungsvereinigung Verbrennungskraftmaschinen e. V. (FVV) and conducted at the Institut für Verbrennungsmotoren und Kraftfahrwesen (IVK) of the University of Stuttgart. It was financed within the framework of a program to promote cooperative industrial research (Industrielle Gemeinschaftsforschung und -entwicklung, IGF) by the Bundesministerium für Wirtschaft und Technologie (BMWi) of the Federal Republic of Germany by means of the Arbeitsgemeinschaft industrieller Forschungsvereinigungen e. V. (AiF), IGF-Nr. 16835 N, based on a decision of the German Federal Parliament. A working group under the direction of Dr. André Kulzer, Dr. Ing. h.c. F. Porsche AG, accompanied the research work. The authors would like to thank this working group and the companies involved for their support and the BMWi and AiF for providing financing.

Definitions, Acronyms, Abbreviations

AFR	Air-Fuel-Ratio	**ign.**	ignition
BDC	Bottom Dead Center	**IMEP**	Indicated Mean Effective Pressure
CA	Crank Angle	λ	Equivalence Ratio
CCR	Combustion Chamber Retention	**NEDC**	New European Driving Cycle
eng.	engine	**PTA**	Pressure Trace Analysis
EGR	Exhaust Gas Recirculation	**PTDC**	Pumping Top Dead Center
EPR	Exhaust Port Recirculation	**SOI**	Start of (main) Injection
EVC	Exhaust Valve Closing	**SOI1**	Start of Pre-Injection
FTDC	Fired Top Dead Center	**st.**	stoichiometric
gHCCI	Gasoline Homogeneous Charge Compression Ignition	**TDC**	Top Dead Center
HCCI	Homogeneous Charge Compression Ignition		

Transient simulation of nitrogen oxide emissions on diesel engines

Benjamin Kaal, Michael Grill, Michael Bargende

FKFS

1 Introduction

Using the data from multiple temperature and pressure measurement points, high and low pressure indication and high-speed exhaust measurement devices on a turbocharged automobile diesel engine as a basis, a new transient capable 0D/1D model for the prediction of nitric oxide formation in diesel engines is developed.

2 On the Importance of Transient Emissions

Upcoming legislation, including more stringent emissions limits, more dynamic driving cycles and Real Driving Emissions, set a focus on transient emissions. Meanwhile, the ever increasing development efforts necessitate the massive use of simulations to keep costs low and development time reasonable short. However, the increasing importance of transient emissions has so far been neglected, both in the study and the simulation of emissions. To address these points, transient emissions on a passenger car diesel engine were measured using fast emissions measurement instrumentation to develop a new emission model for the prediction of transient nitric oxide emissions.

3 Transient 0D/1D-Simulation: From Air Path over Injection and Combustion to Emissions

The successful simulation of transient emissions depends on more than just a transient capable emissions model. Since the simulation of nitric oxide emissions is one of the last steps in the simulation chain when simulating a complete combustion engine and is therefore dependent on numerous input values from other simulation models it is very important that those other models are transient capable as well.

This transient capable simulation chain normally starts with the air path, which is fed with more or less constant boundary conditions, for example the ambient conditions of a test bench. The 1D flow model then predicts a pressure curve which is produced by the cyclic operation of the engine. For static simulations the flow model usually is given some cycles until it reaches a steady state and meaningful results can be retrieved. For transient simulations, after a short startup phase, the flow model has to exhibit the same transient behavior as the actual system since every cycle represents a discrete time step of the transient measurement. Therefore, the requirements on a transient flow model are much higher than those on a static one. One part where this is especially true is the turbocharger. In static simulations the turbocharger only needs to deliver the correct boost pressure for a given set of boundary conditions. In transient simulations it is also important how it reaches that boost pressure. As an example, this means, that the inertia of turbocharger becomes very important, a value which is not crucial for static simulations. This makes tuning the turbocharger more difficult as well, since often, for static tuning, the efficiency or mass flow will be adjusted to reach measured boost pressures for given conditions: An approach which is acceptable for static simulations but can change the transient behavior of the turbocharger in an unwanted manner.

The transient capable flow model then delivers the boundary conditions necessary for the burn rate model to simulate the in-cylinder combustion. Since this in turn supplies the emissions model with the boundary conditions to calculate the nitric oxide emissions, the burn rate model needs to be transient capable as well. Only a transient capable burn rate model can simulate the continually changing burn rate and thus every required intermediate step of the in-cylinder parameters in order to represent the real transient behavior of an engine. In this paper the quasi-dimensional CI-combustion model described in [1] was used, which simulates the diesel combustion based on the injection rate.

This again necessitates a transient capable injection model which is able to calculate the injection rates for arbitrary operating conditions of the engine. For this purpose, the injection model uses a number of fix points to describe the injection as a polygonal line in the style of [2]. After calibrating the injection model to a measured reference point, this allows for a sufficiently accurate prediction of transient injection rates for a modern common rail injection system with pilot and post injections.

The burn rate model then uses the information from the injection rates to create slices of fuel in constant time steps. The slices are then subjected to an admixture of combustion chamber gas which introduces air into the slices and thus starts to form a combustible air-fuel mixture. Ignition occurs after the calculated integral reaction rate, as the inverse of the ignition delay, exceeds a specific limit. The ignition delay is calculated for every time step based on a combined Arrhenius and Magnussen approach. Prior to ignition the already injected fuel, which is subjected to the admixture of gas, is divided into two pools. The premixed pool is assumed to be sufficiently mixed with air to allow for a rapid combustion immediately after the ignition delay. The combustion of the diffusion pool however is controlled by the ongoing admixture of air and therefore occurs much slower. After the start of the combustion fuel can only be assigned to the diffusion pool. With this structure the model can simulate arbitrary burn rates of diesel combustion with pilot and post injections while capturing the typical diesel burn rate shape with a fast premixed combustion immediately after ignition delay and a slow diffusion controlled burnout. Figure 3.1 shows simulated injection rates while Figure 3.2 shows the corresponding simulated burn rates for a transient load increase calculated with this quasi-dimensional model.

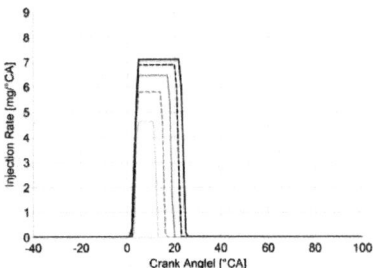

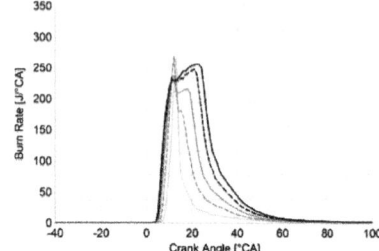

Figure 3.1: Simulated injection rates for a transient load increase

Figure 3.2: Simulated burn rates for a transient load increase

Another important factor for transient simulations, which hasn't been discussed so far, is cylinder wall temperatures. While for steady-state simulations, wall temperatures can easily be imposed as constant boundary conditions, transient simulations necessitate a different approach: wall temperatures have to change in accordance with operating conditions. To this end, cylinder wall temperatures have to be simulated using a simplistic modelling of the cylinder structure and cooling via a FEM-model. For practical applications a simple tuning of these FEM-models using few measured wall temperatures or empirical values usually yields satisfying results.

4 The Transient NO-Model

The transient emissions model used in this paper is based on the phenomenological model proposed by Kožuch in [2] which calculates nitric oxide emissions by means of an admixture from the unburned zone into the burned zone. The model according to Kožuch relies on a 2-zone process calculation. The combustion occurs stoichiometrically and the additional admixture from the unburned zone occurs past the flame directly into the burned zone, whereby the temperature and composition can be adjusted in the burned zone.

The admixture is phenomenologically modeled according to Eq. (1). It is calculated using a turbulent velocity $u_{turb,g}$ on the basis of a k-ε model, the density of the unburned zone ρ_{ub}, the volume of the burned zone V_b and the number of nozzle holes N_N. The dimensionless constant c_g is used for tuning.

$$g = \frac{dm_{ub,ad}}{dt} = c_g \cdot \rho_{ub} \cdot u_{turb,g} \cdot V_b^{\frac{2}{3}} \cdot N_N \tag{1}$$

Eq. (1) thus supplies a phenomenologically determined admixture mass flow which introduces unburned mixture into the burned zone. This mass flow lowers the temperature in the burned zone and thereby the admixture provides for a reduction of simulated NO. Figure 4.1 shows a schematic view of the complete emissions model according to Kožuch, with nitric oxide formation in the burned zone.

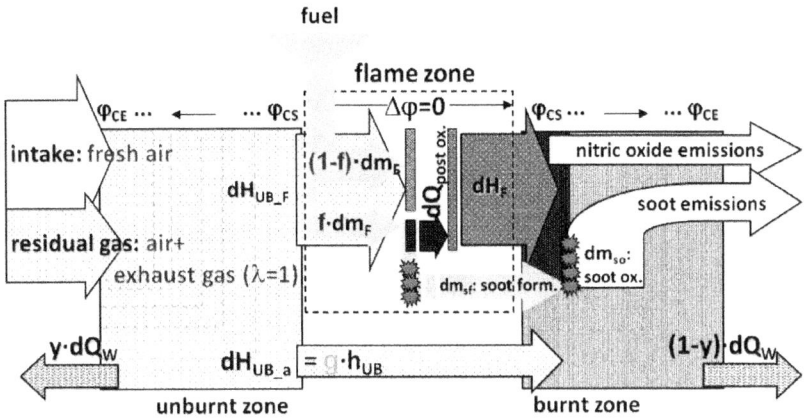

Figure 4.1. Diagram of emissions model according to Kožuch [3]

The specific additions to the model for fully transient capable simulations are related to two effects. First, measurements have shown, that there is a direct influence of the combustion chamber wall temperature upon the nitric oxide emissions of a diesel engine, which are not covered by the transient responses of the other models involved. This is an important factor, since during transient operations the wall temperature can change significantly. As the original emissions model by Kožuch has no direct influence of the combustion chamber wall temperature integrated, the model was expanded to allow for it.

For this purpose the 2-zone approach chosen by Kožuch was upgraded to a 3-zone approach. The former burned zone in which the nitric oxide formation occurs was divided into two separate pseudo-zones: the cool peripheral zone and the hot core zone. Following the basic theory behind wall heat losses with a flow and temperature boundary layer, the peripheral zone is postulated to have a temperature too low for nitric oxides to form. Since the peripheral zone is modelled to be directly adjacent to the combustion chamber wall it encloses the hot core zone which retains the temperature of the original burned zone und thus still shows nitric oxide formation.

The size of the peripheral zone is calculated depending on the current combustion chamber wall temperature. As shown in Figure 4.2, the three boundaries of the combustion chamber, created by the demarcation of cylinder head, piston and liner each span the peripheral zone perpendicularly in the direction of the combustion chamber. The size of the peripheral zone (and consequently the core zone) is thus completely described by the (momentary) burned zone size and one thickness for each of the three boundaries. These thicknesses are calculated depending on the corresponding wall temperatures using an exponential correlation as shown in Eq. 2.

Figure 4.2: Model representation of combustion chamber wall temperature influence

$$b = F \cdot \left(\frac{T_{wall}}{400}\right)^E + C \qquad (2)$$

The division of the burned zone into a core and peripheral pseudo-zone in the described way enables the model to predict the direct combustion chamber wall temperature influence which was identified in measurements.

The second effect which needs to be incorporated into the model for full transient capability is the effect of locally rich conditions. Since the simulation of nitric oxide formation in the model according to Kožuch is completely dependent upon the admixture of unburned mass into the burned zone to reduce the temperature, the model only works as long as there is enough unburned mass to continue the admixture throughout the relevant NO formation period. If global conditions approach stoichiometric gas composition an ever increasing amount of the total available air mass is needed for the combustion of the fuel and less air mass is available for admixture. This can lead to a sudden breakup of the admixture accompanied by a massive increase in NO formation due to abruptly increasing temperatures in the burned zone.

Even for static conditions this represents a problem, since modern diesel engines are operated close to stoichiometric conditions at full-load and even at partial-load if high

176

EGR rates are applied. During transient operations this becomes even more relevant since the slow reaction time of the air path compared to the fuel injection system leads to conditions even closer to stoichiometric air fuel ratios. Combined with planned new and more dynamic test cycles, this will also become relevant for certification.

To address this problem another division into two further pseudo-zones was integrated into the model, see Figure 4.3. The core zone with the nitric oxide formation from the wall temperature influence was divided into two additional pseudo-zones: the rich and the "stoichiometric" pseudo-zone. The "stoichiometric" pseudo-zone keeps the temperature and composition of the core pseudo-zone and thereby the original burned zone. The rich pseudo-zone in contrast is modelled with a rich composition with the corresponding lower oxygen concentration and temperature which leads to greatly reduced nitric oxide formation.

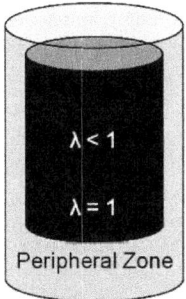

Figure 4.3: Model representation, air deficiency influence

The division of the core into the "stoichiometric" and rich pseudo-zone is modelled with a trigonometric approximation as described in Eq. 3.

$$a = cos(A \cdot \lambda \cdot \pi + P \cdot \pi) + O \qquad (3)$$

For a detailed description of the used emissions model please see [4].

5 Calibration and Application of the Transient NO-Model

For tuning of the new model the base model according to Kožuch must first be tuned as usual [3], [4]. This means that the admixture has to be tuned to the measurements by means of the tuning constant c_g. All other parameters of the emissions model usually can remain at their default values. The potential tuning of the wall temperature as well as the air deficiency influence are independent of this and of each other. In comparison to the emissions model according to Kožuch, however, a significant improvement to predictability results via the new model even without explicit tuning, because the wall temperature influence was not considered and the predicting of emissions at operating points with low global λ values was not possible at all with the base model. The new model can thus be used advantageously even without additional tuning effort or in the case of missing corresponding measurement data. For the best possible forecasting quality, an explicit tuning of the wall temperature and air deficiency influence is nevertheless required. Both a simplified as well as a detailed tuning is possible for both sub-models. In the following, however, only the simplified tuning is presented. For the detailed tuning, see [4].

A simplified tuning of the wall temperature influence can be carried out with a single tuning parameter at only one measurement point. Parameter F can be tuned to an operating point with very low coolant temperatures once the admixture parameter cg has been tuned on the hot engine. With this, the progression of the approximation over the temperature can be scaled without too severely changing the basic form. The parameter C should not be adjusted here. It describes the minimum wall thickness at high wall temperatures and would always have to be balanced by an adjustment of the parameter cg to the admixture in the base model according to Kožuch. Parameter E should only be carefully changed or not changed at all in the case of only little applicable measurement data, since it can heavily influence the form of the approximation over temperature.

A simplified tuning of the air deficiency influence can also be carried out with only one tuning parameter. In this case the tuning occurs via the parameter P, which adjusts, in a certain range, the lambda value from which the model has effect. In this case, a single operating point with a sufficiently low air ratio, so that the model already takes effect, suffices for tuning of the air deficiency influence. The parameter O describes the stoichiometric remainder at maximum effectiveness of the model and should not be adjusted, because there is normally no data for such low air ratios without special measurements. Parameter A causes a deformation of the tuning curve and should not be adjusted, because this deformation must be balanced via the other parameters, which requires more measurement data.

6 Simulations

Since calibration of models for transient simulations still occurs at steady-state operating points it is prudent to first analyze the behavior of the models at those conditions. After a good agreement between simulation and measurement is reached for steady-state operating points via tuning of the involved models the switch to transient simulations should only necessitate a few changes in tuning usually for facets of the models irrelevant for steady-state simulations, for example, the inertia of the turbocharger, the wall temperature model and so on.

6.1 Steady-State

In this chapter, a comparison is made between the emissions model according to Kožuch, the new model and measurement data. Figure 6.1 shows the simulation results of the two models together with the measurement data.

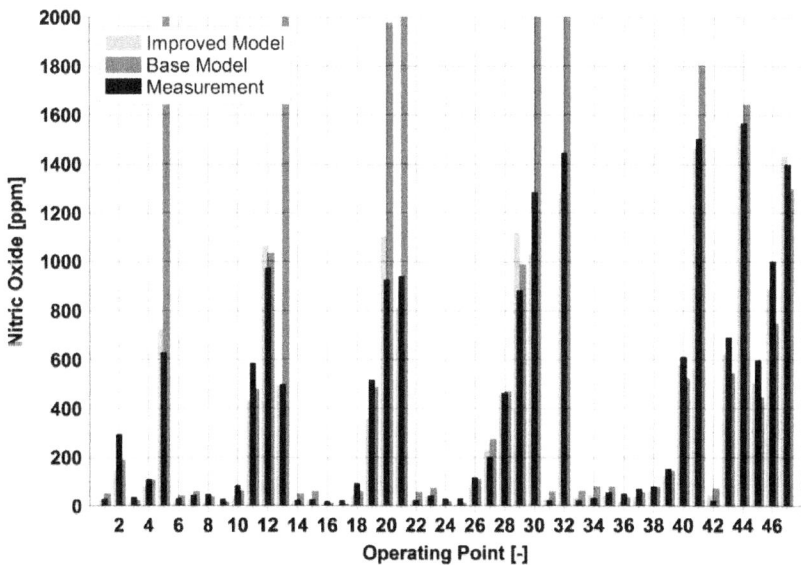

Figure 6.1: Model validation on the performance map

For the model according to Kožuch, at first, operating points 5, 13, 20, 21, 30 and 32 stand out, since, due to their low air ratios, they show a breaking admixture with the effects described in Chapter 4. The simulation results of the new model, however, show that it is capable to predict emissions even for the operating points with the lowest global λ values. Agreement with the measured values has also mostly improved for the remaining operating points. This can be best determined based on the results from operating point 33 and higher. The highest speeds of 2000, 2400 and 3000 min[-1] are found here, at which the temperature influence is strongest due to the increase in load between 1 and > 20 bar IMEP and the λ values are still sufficiently large so that the combustion chamber wall temperature influence is not overlapped by the air deficiency influence. At these operating points, the new model shows a significantly better correspondence with the measurement values than the model according to Kožuch, thanks to the consideration of the combustion chamber wall temperatures.

In summary, it can be said that the new model exhibits overall better forecasting quality. This particularly applies when the operating points with low global air ratios are included in the consideration, since a prediction of the nitric oxide concentrations for such operating points only becomes possible with the new model in the first place.

6.2 Transient

6.2.1 Temperature Load Steps

The new emissions model is first validated based on two identical load steps at different engine temperatures. The load steps each occur from approximately 1 bar to 6 bar IMEP at 850 min⁻¹. One load step occurs with the engine cold, at a coolant and oil temperature of 35 °C each; the second with the engine hot at 85 °C each. The two load steps thus have differing combustion chamber wall temperatures at the beginning and likewise exhibit differing steps for those temperatures. The models were each tuned to the best possible correspondence with the measured values after both load steps. The results of the simulation, together with the measured nitric oxide concentrations, are shown in Figure 6.2.

Particularly in the case of the "cold" load steps, saw-tooth-shaped spikes are recognizable in the simulated NO curves. These originate from the combustion model, whose ignition delay reacts too sensitively to the modified temperature level for these load steps. The corresponding jumps in the simulated nitric oxide emissions in Figure 6.2 are thus not due to the emissions models, but rather the combustion model: a good example for the need of a fully transient capable simulation chain.

In comparison with the measured data, the emissions model according to Kožuch shows a spread of results for the hot and the cold load step which is far too small. For the concrete load steps, even a false tendency is returned: the model according to Kožuch simulates higher nitric oxide concentrations for the cold load step than for the hot, in the case of a common tuning for both load steps to the measured nitric oxide concentrations after the load steps. The differences in the ancillary conditions of the two load steps caused by the different temperature levels thus result in a trend which is contrary to the measured nitric oxide concentrations in the case of the model according to Kožuch.

In the case of the new model, the trend corresponds to the measured behavior: the hot load step returns higher simulated nitric oxide concentrations than the cold. While the emissions after the cold load step are well on target, they are just slightly under the measured values for the hot load step. Likewise, the strongly pronounced emissions peaks directly after the load steps simulated by the model according to Kožuch are extensively eliminated with the new model.

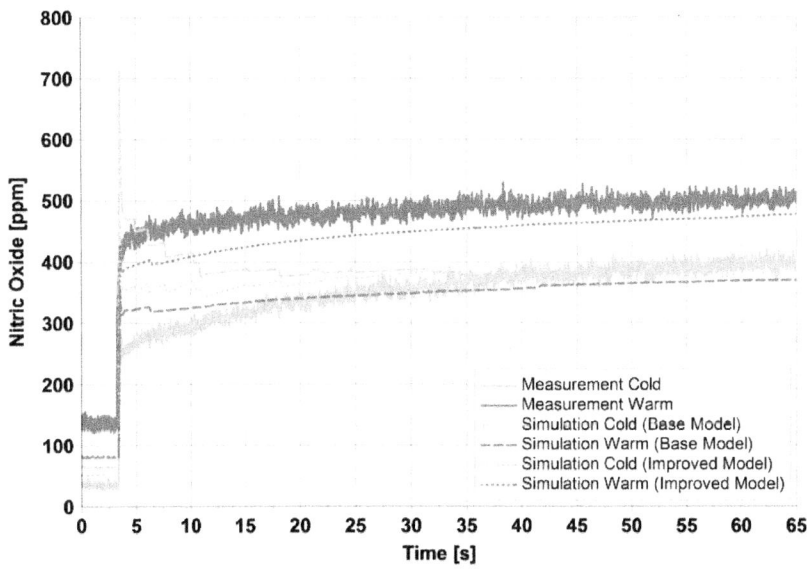

Figure 6.2: Model validation based on temperature load steps – cold and hot

6.2.2 Lambda Load Step

For validation of the representation of the air deficiency influence while transient, a load step from zero load to full load is considered (1...12 bar IMEP, 850 min[-1]) with a correspondingly large spread of the combustion air ratio: ideal prerequisites for validation of the new sub-model.

Figure 6.3 shows the nitric oxide concentrations over time simulated with the two models compared with those measured. Whereas the new model can follow the measured values during the load step very well, except for a minor temporal delay, and also reproduces the progression and the absolute values after the load step well, the model according to Kožuch reacts too slowly to the load step and especially does not exhibit the drop in nitric oxide concentrations after reaching a maximum value, which is visible in the measurement.

An alternative depiction with the simulated and measured nitric oxide concentrations over the air ratio is shown in Figure 6.4. In this depiction, the wrong prediction of the nitric oxide emissions for low air ratios by the model according to Kožuch becomes particularly clear. For falling λ values, the emissions first rise for both models, just as for the measurement. At λ = 1.58 for the measurement or λ = 1.46 for the new model, the nitric oxide emissions decrease again. The model according to Kožuch does not exhibit this behavior, but rather continually and even increasingly steeply rising nitric oxide concentrations.

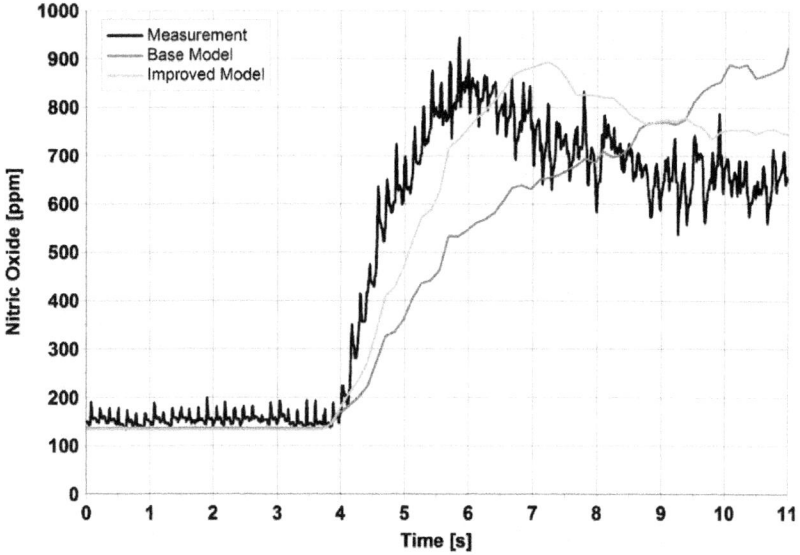

Figure 6.3: Lambda load step – NO over time

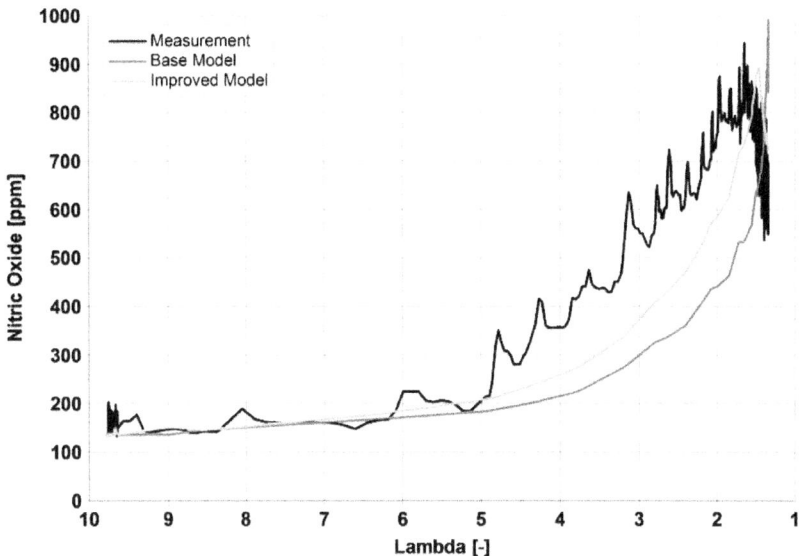

Figure 6.4: Lambda load step – NO over combustion air ratio

7 Summary and Outlook

For transient simulations of nitric oxide emissions it is essential to only use models which are transient capable. A transient capable emissions model alone is not enough, since it always has to depend on input values from other models. For a completely transient simulation this normally means that at least a transient capable flow model, burn rate model (for diesel engines this necessitates an injection model), wall temperature model and emissions model is needed.

With the described burn rate model and an appropriate flow as well as wall temperature model transient boundary conditions can be created to be used by an emissions model. However, additional research has shown that there are some direct influences on the nitric oxide formation which are especially important for transient simulations and need to be directly integrated into an emissions model.

Therefore, an existing phenomenological model with good prediction accuracy proposed by Kožuch was expanded to allow for those direct influences. The new model showed better predictions for steady-state and especially transient simulations. With the new model and a fully transient capable simulation chain good agreement with measured transient emission was achieved.

Acknowledgments

The model presented in this report is the result of a research task defined by the Forschungsvereinigung Verbrennungskraftmaschinen e. V. (FVV, Frankfurt) and conducted at the Institut für Verbrennungsmotoren und Kraftfahrwesen (IVK) of the University of Stuttgart. It was financed within the framework of a program to promote cooperative industrial research (Industrielle Gemeinschaftsforschung und -entwicklung, IGF) by the Bundesministerium für Wirtschaft und Technologie (BMWi) of the Federal Republic of Germany by means of the Arbeitsgemeinschaft industrieller Forschungsvereinigungen e. V. (AiF), IGF-Nr. 17372 N, based on a decision of the German Federal Parliament.

A working group under the direction of Dipl.-Ing. Jürgen Münzenmaier, Daimler AG, accompanied the research work. The authors would like to thank this working group and the companies involved for their support and the BMWi and AiF for providing financing. The authors are grateful for the possibility to perform this project.

References

1. D. Rether, M. Grill, A. Schmid, M. Bargende, "Quasi-Dimensional Modeling of CI-Combustion with Multiple Pilot- and Post Injections", SAE Technical Paper 2010-01-0150, no. 10.4271, 2010

2. C. Barba, „Erarbeitung von Verbrennungskennwerten aus Indizierdaten zur verbesserten Prognose und rechnerischen Simulation des Verbrennungsablaufes bei Pkw-DE-Dieselmotoren mit Common-Rail-Einspritzung", Dissertation: ETH Zürich, 2001

3. P. Kožuch, „Ein phänomenologisches Modell zur kombinierten Stickoxid- und Rußberechnung bei direkteinspritzenden Dieselmotoren", Dissertation: Universität Stuttgart, 2004

4. B. Kaal, M. Sosio, „Instationäre Emissionsmodellierung am Dieselmotor", Vols. 1062 - 2015, Frankfurt am Main: Forschungsvereinigung Verbrennungskraftmaschinen e.V., 2015

Renewables in transport 2050 – Empowering a sustainable mobility future with zero emission fuels

Patrick R. Schmidt
LBST – Ludwig-Bölkow-Systemtechnik GmbH
Munich/Ottobrunn, Germany
E: patrick.schmidt@lbst.de

Dietmar Goericke
FVV – Forschungsvereinigung Verbrennungskraftmaschinen e.V.
Frankfurt a.M., Germany

Dr. Werner Zittel, Werner Weindorf, Tetyana Raksha
LBST

Contents

Abstract

Presented are results of the second FVV Fuel Study done by expert consultants Ludwig-Bölkow-Systemtechnik GmbH (LBST) on behalf and in cooperation with the Forschungsvereinigung Verbrennungskraftmaschinen e.V.

Keywords: Renewable Electricity, Power-to-Gas, Power-to-Liquid, Transportation, Europe

1 Introduction

Post 2020 climate targets and the energy transition in the transport sector are presently drivers for discussions for more sustainable fuels and drive systems. With energy transition in the electricity sector in Germany, and increasingly in the EU, electricity-derived liquid fuels (power-to-liquids, PtL) enter strategy discussions. Gasoline, diesel, and kerosene from renewable electricity are promising insofar as they can be seamlessly integrated into already established infrastructures and drive systems. Furthermore, PtL expands the basket of options for electricity-derived fuels – like power-to-hydrogen (PtH$_2$) for fuel cell vehicles and power-to-methane (PtCH$_4$) for combustion engines – into the domains of high-performance transportation applications. Characteristic for this kind of applications are long distances, high energy demands paired with high requirements concerning the energy density of the fuel/drive system. Typical high-performance applications are to be found in long-distance truck transportation, international aviation, and not least with non-road mobile machinery.

Moreover, compared to biomass-based fuels, the production of electricity-based fuels is area-efficient, also resulting in a higher production potential while maintaining the highest sustainability requirements. The need for additional renewable electricity systems essentially depends on the future transportation demand as well as the fuel/drive systems of the different transportation modes.

The focus of this paper is laid on the resulting electricity demands, fuel costs, and cumulated investments for the EU-28. In a boundary approach, a 100% energy transition was modelled for all motorised transport modes till 2050 on the basis of exclusively renewable electricity.

2 Two transportation demand scenarios

To explore the range of presently discussed and possible future transportation demands, two distinctly different transportation demand scenarios («HIGH», «LOW») were selected from a roaster of studies of the last years. Changes in transportation demand are summarised in the following table, comparing transportation demand as used in the FVV Study for Germany.

Transportation demand development (including international transport) 2010 to 2050 in the high (HIGH) and low (LOW) scenario for Europe and Germany, respectively:

Transportation demand scenario	Sector	Change from 2010 to 2050	
		DE	EU
HIGH	Passenger	+30%	+50%
	Freight	+60%	+80%
LOW	Passenger	-25%	+10%
	Freight	+20%	+50%

For the EU-28, transportation demands were derived from scenarios compiled by AEA[1] on behalf of the European Commission and published in 2012 as 'EU Transport GHG-Routes to 2050 II'. For the EU HIGH transportation demand scenario the AEA scenario 'BAU-a' (business-as-usual) was chosen, and for the LOW transportation demand scenario AEA scenario 'C5-b'. The inner-EU passenger and freight traffic

---

[1] Today part of the Ricardo Group

correspond with the EU reference scenario which the European Commission had published in 2013. In addition, AEA scenarios also include international transportation demands[2] between the EU and non-EU countries. International transportation has been taken into consideration in the FVV Study in order to fully satisfy all energy demands in transportation.

3 Three fuel/drive scenarios

The development of the specific fuel emission factor follows a target scenario assuming that the primary energy basis of each fuel will be successively converted to renewable electricity by 2050. The following figure shows this energy transition with the example of gasoline/kerosene/diesel[3] from crude oil to synthetic fuel from renewable electricity.

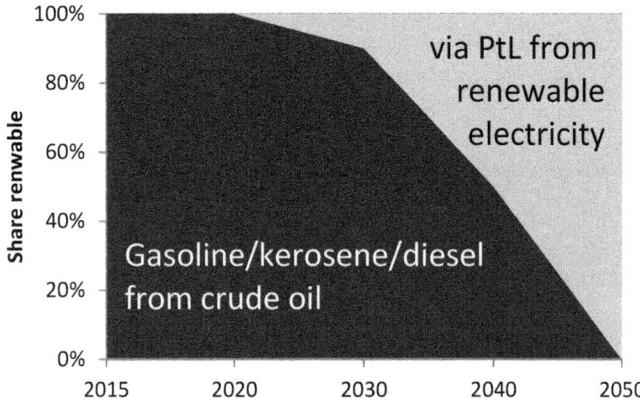

Target scenario for energy transition in the primary energy basis,
here for the example in the case of gasoline/kerosene/diesel

---

[2] That is especially international air transport for passengers and international cargo shipping, based on the territorial principle, i.e. all international outbound traffics up to the first destination are taken into account.

[3] In the FVV Study two routes for the production of synthetic liquid fuels were examined, namely via the Fischer-Tropsch route and via the methanol route. Synthesis processes always produce a mixture of fuel products comparable to refineries today.

Analogous to gasoline/kerosene/diesel, the same approach was taken for the other assumed fuels of the FVV Study, which are methanol, hydrogen and charging/overhead line electricity.

To supply the different transportation demands, three distinctive scenarios were defined to assign drive systems and fuels in accordance with the FVV Working Group 'Future Fuels':

The «**PTL**» scenario corresponds to a 'business-as-usual' development and as synthetic boundary scenario it is deliberately conservatively designed. Already prevailing drive systems in transport modes are maintained here. Combustion engines are further optimised with regards to fuel consumption, however, there are not fundamental changes in the technology bases (evolutionary development). All means of transport that are operated by combustion engines use power-to-liquids from renewable electricity instead of fossil fuels. For passenger cars the ICE hybrid is the predominant propulsion technology operated by PtL gasoline/diesel till 2050.

The mixed scenarios «**FVV**» display a portfolio of currently discussed and to some extent already existing fuel/drive systems, although one still adheres to the basic technology of the combustion engine. For a maximum reduction of fuel consumption, combustion engines are embedded into advanced electric propulsion systems. For the passenger car category this e.g. means that in 2050 range-extender hybrid electric vehicles (REEV) will dominate the new registrations.

The «**eMob**» scenario focuses on the 'Regional' scenario of the 'eMobil 2050' study of the Öko-Institut published in 2014. Predominant are electrical drives on the basis of batteries and fuel cells, for the passenger car in particular battery-electric vehicles (BEV).

The roll-over of the vehicle stock takes place in form of percentage shares of new vehicle registrations in one year till 2050 and in the lifetime defined in the model. As an example for the assumed fuel/drive scenarios, the shares of new registrations of passenger cars for the three fuel/drive scenarios are depicted in the following tables. Grey coloured cells highlight the assumed shift of emphasis for alternative fuels and drives.

Share of new passenger car registrations in the three fuel/drive scenarios:

% new registrations	ICE Gasoline/Diesel	ICE Methane	Hybrid Gasoline/Diesel	Hybrid Methane	REEV Gasoline/Diesel	BEV	FCEV
«PTL»							
2010	100%	0	0	0	0	0	0
2020	80%	0	20%	0	0	0	0
2030	40%	0	60%	0	0	0	0
2040	10%	0	90%	0	0	0	0
2050	0	0	100%	0	0	0	0
«FVV»							
2010	100%	0	0	0	0	0	0
2020	36%	5%	45%	2%	6%	4%	1%
2030	0	0	55%	5%	25%	10%	5%
2040	0	0	37%	2%	45%	16%	9%
2050	0	0	0	0	70%	20%	10%
«eMob»							
2010	100%	0	0	0	0	0	0
2020	86%	5%	3%	0	3%	3%	0
2030	68%	5%	6%	0	9%	12%	0
2040	0	0	10%	0	17%	72%	0
2050	0	0	5%	0	12%	82%	0

The three fuel/drive scenarios are convoluted[4] with transportation demand scenarios HIGH and LOW in a fleet stock roll-over model developed by LBST. The resulting fuel, emission, and costs quantity structures are then determined. The results are described in the following chapters.

---

[4] Whereby «eMob» for reasons of coherence between the scenarios only with «LOW»

4 Greenhouse gas emissions

The resulting greenhouse gas emissions are depicted hereinafter taking the «FVV» scenarios for high (left) and low (right) transportation demand as an example.

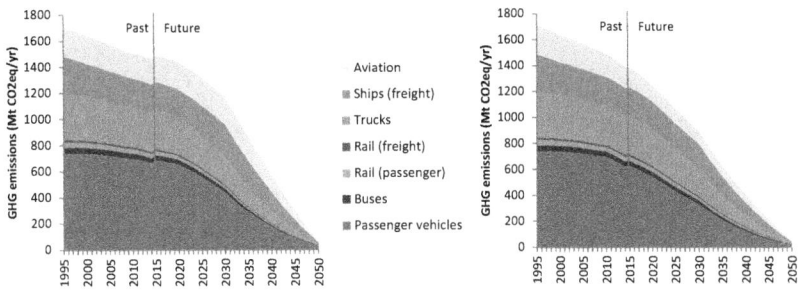

| Greenhouse gas emissions «FVV» at high transportation demand | Greenhous gas emissions «FVV» at low transportation demand |

On the basis of the target scenario of 100% renewable electricity and power-based fuels the greenhouse gas emissions will decrease to zero for all transport modes till 2050 as expected. Climate effects by emissions from aviation in high altitudes are not taken into account here.

5 Resulting energy demand

5.1 Fuel demand (final energy)

Resulting energy demands are depicted hereinafter in TWh/a final energy using the «FVV» scenario for high (HIGH, figure left) and low (LOW, figure right) transportation demand as an example.

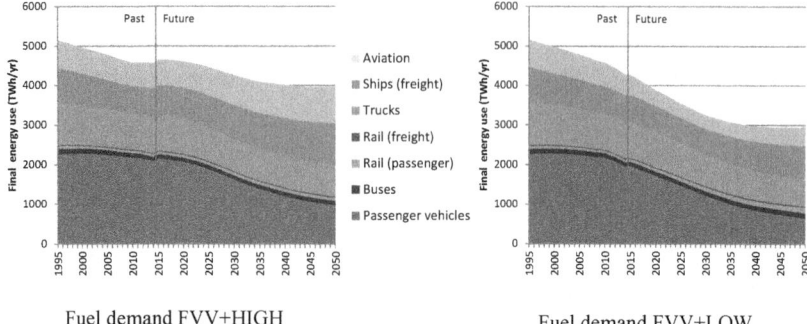

Fuel demand FVV+HIGH   Fuel demand FVV+LOW

From the development of the fuel demand both at HIGH and LOW transportation demand it becomes apparent how fuel consumption in the passenger car segment decreases in real amounts on the basis of assumed specific passenger car consumption
development (in the FVV scenario 70% represent REEV, 20% BEV, and 10% FCEV
of new passenger car registration in 2050). The relative importance of the other
transport modes like trucks, ships and aircrafts increases. This development can be
observed in all scenarios, least distinct in the PTL scenario at HIGH transportation
demand.

5.2 Renewable electricity demand

The subsequent figure shows the development of the renewable electricity demand
from transportation assuming energy transition in all modes of transport by 2050.

The base electricity demand of some 2800 TWh$_e$ represents the electricity demand
without transport of EU-28 in 2013. This electricity demand could increase in the future, on the one hand by new electricity consumption for heat supply (power-to-heat,
e.g. via direct heating or heat pumps) or for the production of basic chemicals (power-
to-chemicals, e.g. for methanol, ethylene, hexane). On the other hand power consumption could also decrease in the future through efficiency measures in established applications – targets for this purpose exist. A final discussion of sectorial development
paths outside the transport sector had not been subject of the FVV Study.

For comparison, in the following figure the range of technical production potentials
for renewable electricity in EU-28 is depicted (upper bar, 9000-12000 TWh$_e$/a). This
classification is based on the evaluation of potential assessment studies, complemented by own calculations.

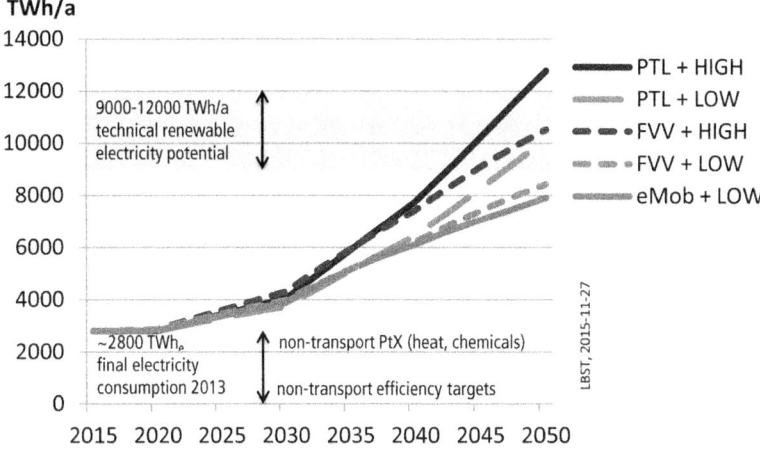

Development of renewable electricity demand by transportation scenario (lines)
compared to the technical renewable electricity potentials in EU-28 (upper bar)

The comparison of the renewable electricity demand with the technical renewable electricity production potentials in the EU-28 shows, that all energy transition scenarios in transport – with the exception of the PTL scenario at HIGH transportation demand – could be supplied by domestic renewable electricity.

To give an impression of the dimension of possible electricity demands: to supply these scenarios in 2050, the present electricity sector would have to increase by a factor 3 to 4.5.

For the provision of the necessary amounts of renewable electricity for the EU-28, renewable power plants of some 364 GW installed capacities[5] in 2014 would have to be increased to 2200 to 4300 GW. Depending on the scenario, this corresponds to deployment rates of 60-220 GW per year in the PTL+HIGH scenario or 50-120 GW per year in the eMob+LOW scenario.

---

[5] Thereof 150 GW hydropower, 121 GW onshore wind, 8 GW offshore wind, and 85 GW PV

6 Fuel costs

Largest single item for the determination of full cost of electricity-based fuels are the electricity costs, particularly in the case of the synthesis fuel pathways (PtG methane, PtL gasoline/kerosene/diesel). The development of the renewable electricity mix as assumed in the FVV Study was extrapolated by means of an S-shaped curve of the renewable power deployment rates of the last years till the technical potentials in 2050. This output of renewable power plants and PtX facilities results in a cost reduction for the production of renewable electricity and the production of electricity-based fuels.

A significant influence on the full costs of synthetic fuels is the energy efficiency of the fuel production process. For the synthesis pathways, this can be significantly improved by suitable heat integration of high-temperature electrolysis (SOEC) and CO_2 extraction. For a conservative estimation, exclusively CO_2 extraction from air was assumed in the FVV Study and the use of high-temperature electrolysis from year 2040 onwards.

The subsequent figures illustrate the total cost of renewable electricity and electricity-based fuels in 2015 and 2050. The comparison shows that particularly synthetic fuels have a significant potential for cost reduction, however in the long term may remain above the 100 US\$/barrel crude oil assumed by the IEA for 2030.

The costs for import PtL from areas with preferential solar and wind conditions, such as in North Africa, could be 20% more favourable than PtL production in the EU.

If PtL gasoline/diesel even without taxes is to compete with a gasoline/diesel price of 1.10 € per litre (including energy and value added tax) the renewable electricity price would have to be around 5 €ct/kWh$_e$ and 4000 equivalent annual full load hours using high-temperature electrolysis and CO_2 from a concentrated source.

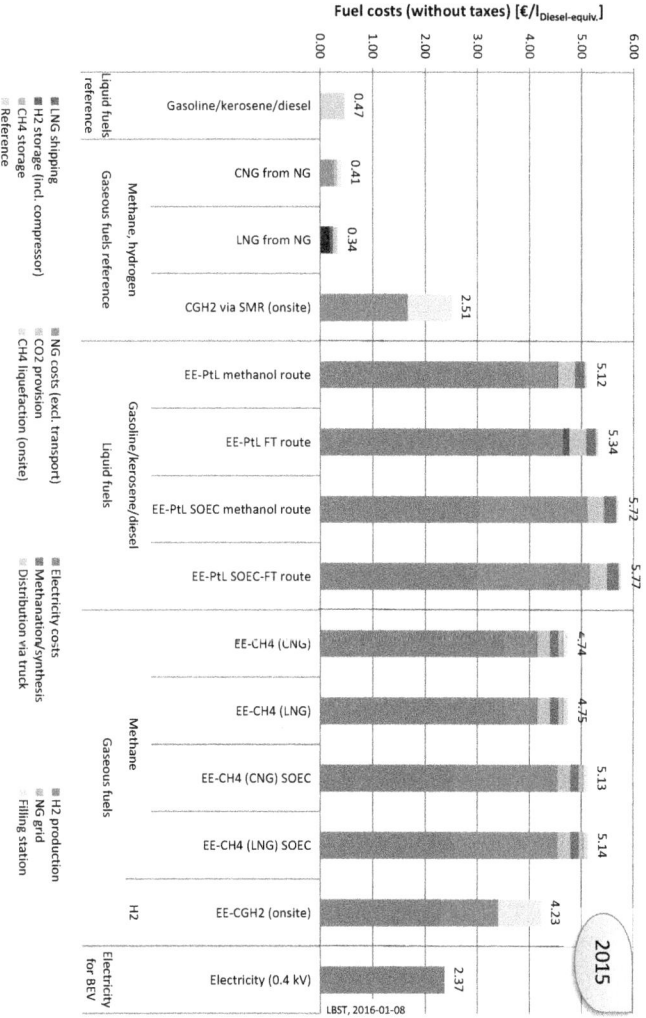

Fuel costs for fossil reference fuels, PtL, PtCH₄, PtH₂ and BEV charging
ex gasoline pump / charging socket for EU-28 in 2015

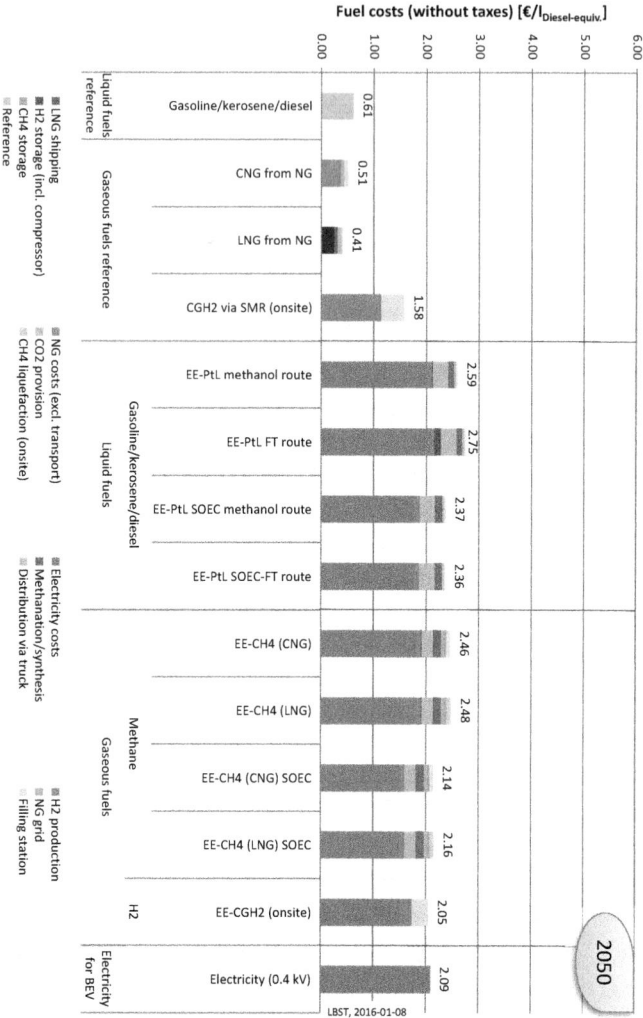

Fuel costs for fossil reference fuels, PtL, PtCH₄, PtH₂ and BEV charging
ex gasoline pump / charging socket for EU-28 in 2050

7 Cumulated investments for energy transition

Cumulated investments resulting from the scenarios include all investments in renewable power plants, PtX plants as well as distribution infrastructures for hydrogen and methane. In the context of the FVV Study home-charging was assumed for battery-electric vehicles. For the provision of electricity for overhead lines a storage share of 50% in conjunction with an electricity storage efficiency of 75% was assumed for matching the (fluctuating) renewable electricity supply with the electricity demand from electric locomotives. The vehicles were not subject of the FVV Study. In the calculation of the cumulated investments the initially higher costs and reductions in specific investments due to capacity and learning curve effects have been fully taken into account per integral.

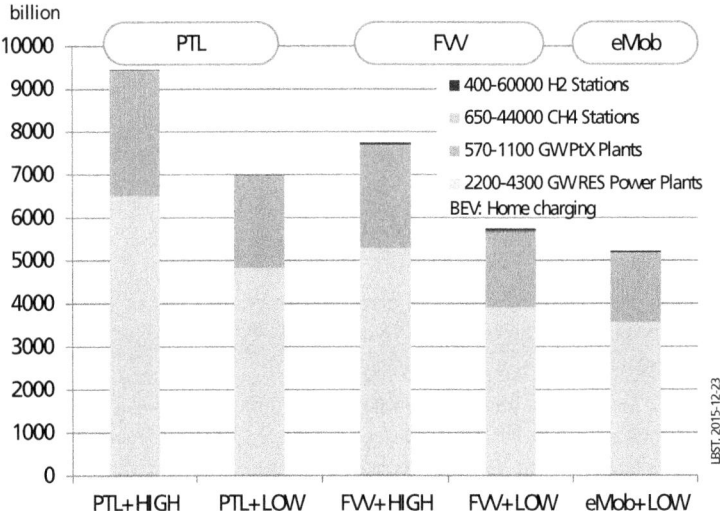

Cumulated investments for an energy transition in transport (EU-28) by 2050 for renewable power plants, PtX plants, and new filling stations (without vehicles)

Depending on the scenario, cumulated investments till 2050 amount to between 5200 and 9500 billion €. Investments into renewable power plants prevail, followed by investments into PtX plants. From a macro-economic point of view investments for methane and hydrogen filling stations are nearly negligible. From a business point of view, the deployment of alternative fuel distribution structures are a significant hurdle, especially in the initial phase with often low utilisation.

In the year 2014 the gross domestic product of the EU was 13,920 billion €. The linear mean value of cumulated investments over a period of 35 years results in average investment needs of 150 to 270 billion € per year. This corresponds to about 1 to 1.9% of the gross domestic product in Europe. By comparison, the annual oil expenses of the EU amounted to about 290 billion € in 2014. Annual investment needs for a complete energy transition in transportation roughly corresponds to or is even significantly lower than current annual expenditures for fossil fuels. The potential for regional added value and energy security in Europe is thus significant.

8   Conclusions

The following main conclusions can be drawn from the results of the FVV study and the scenarios analysed therein:

- The future transportation demand (in terms of passenger-km and tons-km, respectively) is the most sensitive parameter for the development of the fuel or electricity demand in the transportation sector.

- All scenarios except PTL at HIGH transportation demand could technically be satisfied with domestic renewable electricity in the EU.

- Depending on the fuel/drive scenario the electricity market would increase to the 3 to 4.5-fold by 2050 (electricity demand at present plus new electricity demand from transportation).

- PtX costs are mainly driven by the electricity production and supply costs, which again depend on the choice of the PtX fuel (PtH$_2$, PtCH$_4$, PtL) and respective system efficiencies.

- The costs for PtX fuel production in the EU could halve between 2015 and 2050. PtL imports are some 20% lower in cost compared to PtL production in Europe. Further opportunities for PtX cost reduction are project-specific (business-case).

- The investments in the fuel distribution infrastructure are negligible compared to the investment needs in renewable power capacities and PtX plants – this result applies to all analysed scenarios.

- From a macro-economic point of view, the cumulated investments for an energy transition in transportation seem manageable for all scenarios analysed.

- Transportation has to become more electric, both with regard to the fuel production and the ICE embedding in propulsion systems. This way the significant renewable electricity potentials can be exploited; lower electricity demands reduce the amount of renewable power plant capacities.

- For cost reasons PtX fuels are no fast-selling items in the foreseeable future. Existing regulatory frameworks should allow for the accountability of PtG and PtL towards environmental targets for transport as well as for the deployment of renewable capacities in order to match renewable electricity demands from transportation. In addition, incentives and regulatory support is needed to facilitate the process.

- Roadmap development and a discussion of suitable instruments to support the introduction of PtX fuels are fields for further research – the FVV Study provides a starting base for this.

Acknowledgement

The background study to this paper was financed by the FVV and supported by members of the FVV Working Group «Future Fuels».

Bibliography

Schmidt, P., Zittel, W., Weindorf, W., Raksha, T. (LBST – Ludwig-Bölkow-Systemtechnik GmbH): RENEWABLES IN TRANSPORT 2050 – Empowering a sustainable mobility future with zero emission fuels from renewable electricity – Europe and Germany; FVV – Forschungsvereinigung Verbrennungskraftmaschinen e.V. (Hrsg.), Munich/Frankfurt a.M., 2016

Acronyms and abbreviations

BEV   battery-electric vehicle

eMob   electric mobility (here: scenario name)

FCEV   fuel-cell electric vehicle

FVV   Forschungsvereinigung Verbrennungskraftmaschinen e.V. /
Research Association for Combustion Engines e.V.
(institution and scenario name)

ICE   internal combustion engine vehicle

PTL   power-to-liquid

REEV   range-extender battery-electric vehicle

Quality management of CFRP-components in the automotive production

Dipl.-Ing. Manuel Schuster, M.Sc. Markus Soutschek, M.Sc. Anna Hansmersmann

Fraunhofer IPA – Department „Lightweight construction technologies"

1 Introduction

Due to the outstanding mechanical properties and the possibility of near-net-shape fabrication the amount of components which are made from carbon fibre reinforced plastic (CFRP) is highly increasing. Based on rising requirements concerning emission levels of vehicles, CFRP is also becoming the focus of attention in the automotive industry. Higher material costs and the need of additional processing technologies are raising the costs of the whole CFRP process chain. Using this kind of material in batch production requires a significant reduction of costs along the production of CFRP components themselves and the machining of CFRP which represents an important step of the mentioned process chain. It is necessary to adapt conventional tools and process parameters due to the different material behaviour of CFRP and metal (anisotropic and isotropic) during the machining. Also the damage patterns differ between CFRP and metals, as a result of the typical characteristics of fibre materials. Because of this, new strategies of quality management are necessary in the production and processing of CFRP.

Until now, no standards in many sectors of quality management could be established. Hence the processing of these already expensive materials become frequently uneconomic.

Figure 1: Drilling of a CFRP component (Source: Fraunhofer IPA)

2 Damage mechanism in CFRP-processing

2.1 Influence of cutting tools

Besides the adjustment of process parameters, the tool wear is enormously influencing the expected quality when machining CFRP components.

The structure of CFRP-suited tools will always have to find a compromise between the processing of the soft matrix material on the one hand and the abrasive carbon fibre on the other hand. Soft matrix material requires a sharp cutting edge to enable a low heat input. The sharp edge whereas will be strongly stressed by the very abrasive working carbon fibre (see figure 2). This leads to an abrasion of the blade which influences the processing of the matrix in return. The lifetime of CFRP-suited tools is therefore not reaching the lifetime of metal-suited tools. The higher process forces due to abrasion are also increasing the possibility of occurring defects.

Figure 2: Abrasion of a HSS-drill after 15 drilled holes (left: new, right: worn) – (Source: Fraunhofer IPA)

2.2 Damage Mechanism

Incorrect machining leads to damaged components. These damages are explained by the following damage mechanisms.

2.2.1 Delamination

It is possible for the layers of CFRP-components to detach from each other while beeing machined. Usually delamination occurs in the layers at the entering zone (Peek-up) and the exit zone (Push-out) of the tool (see figure 3).

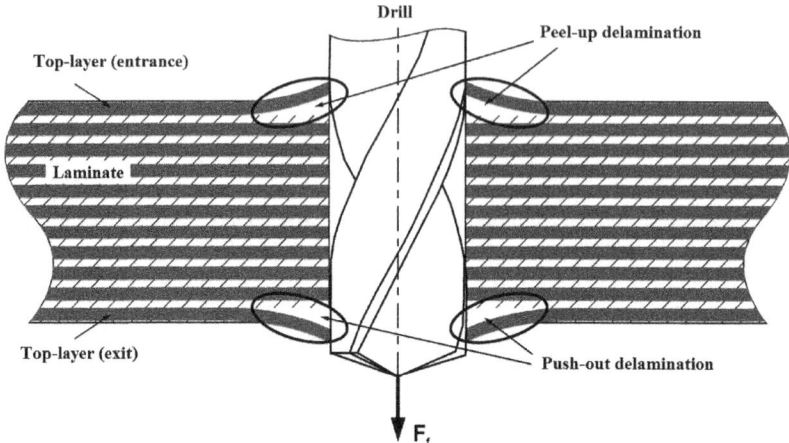

Figure 3: Delamination of fiber composite components – (Source: Fraunhofer IPA, T. Pfeifroth)

2.2.2 Fraying

Fraying is a damage which occurs if not all fibres in the processing area are properly separated and therefore protruding into the processing area (see figure 4). These protrusions mean problems for following processes, because they are an interfering contour which affects e.g. inserting a dowel pin.

However, frayings do not influence the mechanical properties of a component negatively.

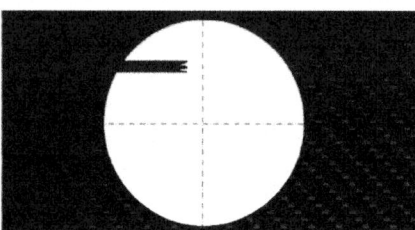

Figure 4: Fraying at a drill hole

2.2.3 Feathering

Feathering is caused by deposits of molten matrix material in the outer edge of the processing area (see figure 5 and figure 6). The deposits result from a too high heat input and from worn tools. This damage mechanism often goes along with delamination. The uprising gap from delamination is filled with molten matrix material.

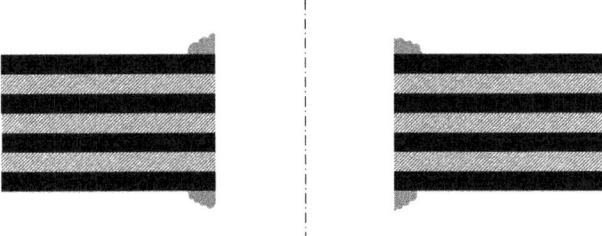

Figure 5: Feathering at a cross section of a drill hole

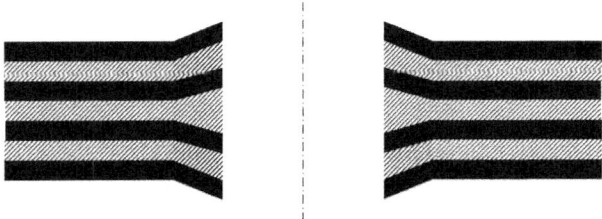

Figure 6: Feathering overlain by delamination at a cross section of a drill hole

2.2.4 Fibre breakage and fibre rupture

Fibre breakage is caused by the breakdown of single fibres outside of the processing area. If these broken fibres are pulled out of the laminate by the processing forces, it is called fibre rupture (see figure 7). Fibre rupture lowers the interlaminar properties and therefore also the mechanical properties of the component.

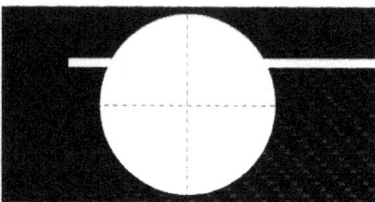

Figure 7: Ripped out fibre

2.2.5 Geometrical defects

Geometrical defects are defects which occur while processing and lead to a difference to the intended geometry (see figure 8).

Figure 8: Geometrical change

3 Quality control

The special structure and damage mechanisms of fibre composite components lead to new challenges in quality management. Especially finding a suitable test procedure in quality management for safety critical components is an important task. Comparing metal-components to CFRP-components, the reinforced plastic is vulnerable to damages in the machining process thus strict test procedures are needed. Because of high test-rates destructive test methods are in many cases uneconomical. Therefore different non-destructive testing methods are available for CFRP-components. These include for example thermography, ultrasound or testing with computer tomography (CT). Due to little investment costs of measuring equipment, visual inspection is still one of the most popular methods for quality control in industrial processing. This chapter gives an overview about different possibilities of non-destructive test methods of CFRP-components and their advantages and disadvantages.

3.1 Visual inspection

The human eye is used as a tool when inspecting components by vision (see figure 9). Visual perception of the human being is affected by many different subjective factors. That is why just a limited reproducibility and reliability of measuring data is able to be guaranteed. Catalogues of limiting samples are used for evaluating processing errors, especially in the automotive industry. The inspector has to compare the quality of the component with the mentioned catalogue to ensure its quality. If the component does not match the requested quality, post processing or waste disposal is necessary.

Figure 9: BMW employee inspecting the CFRP roof frame by vision – (Source BWM Group)

Defects have to be at a minimum size and located on the surface to be clearly identified by eye. Helping devices with zooming function may help to detect smaller errors. Based on the given massive disadvantages of visual inspection, more objective but also more expensive test methods are getting in focus due to the reproducibility of the results.

3.2 Ultrasound

Ultrasound testing captures the expansion of sound waves in components (see figure 10). Defects in components result in a change of expansion which can be measured and help to detect the defect. A couplant is necessary to transmit the sound pulses. Therefore the tested components need to be wetted or dipped into water. Ultrasound technologies are successfully used for indicating defects especially inside monolithic and primarily plane-parallel components. In principle scanning curved surfaces is possible, but the evaluation of the test results is complicated due to diffuse spreading of pulses. In comparison to conventional materials, inspection of CFRP-components with ultrasound technology is more challenging to the testing devices. Damping effects of the ultrasound signals by the laminar structure as well as diffusion phenomenons caused by inhomogeneity have to be considered when using this technology. Especially if applied to sandwich structures, the classical impulse-echo technology does not meet the requirements.

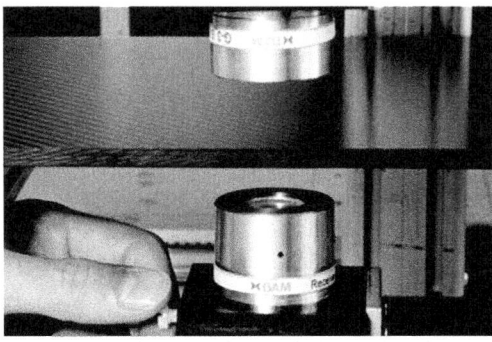

Figure 10: Inspection of a CFRP-panel with an ultrasonic-converter (Source: Bundesanstalt für Materialprüfung)

3.3 Thermography

It is possible to measure heat flow with thermography. Each object is emitting a characteristic spectrum of electromagnetic waves which is traditionally used to locate people or detect loss of heat from building sand pipes. Thermography is also applicable to control temperatures in industrial plants.

This technology is not common yet for testing semi-finished products and components. In this process, the surface of a CFRP-component is warmed up through a short impulse caused by an infrared radiator, UV light or ultrasonic (active thermography). From the surface the heat is transferred to deeper layers of the component by thermal conduction. Defects like delamination, pores or cracks disturb the heat transfer and thus keep the layers above on a higher temperature level.

The development of precise, rapid, robust and low-cost sensors makes thermography more and more attractive. However, the individual calibration to each test object still is a challenge. But once the calibration is successful, defects on the inside are reliably, rapidly and automatically detectable. Inline-control while producing and processing is possible due to good possibility to automation.

Besides defects, several other structure characteristics are indicated which complicates the evaluation. Corresponding to that, the requirements to the qualification of the inspector and the used software are high. This is why thermography is still uneconomic for small and medium enterprises (SME).

The Institute for Plastics Technology (IKT University of Stuttgart) shows this CFRP-roof as an example for thermography measurement (see Figure 11): On the left hand side, no visible errors are detectable. A thermography measurement shows damages caused by hail (a, b, c) on the right hand side. The visible square (d) is a sticker set as reference.

 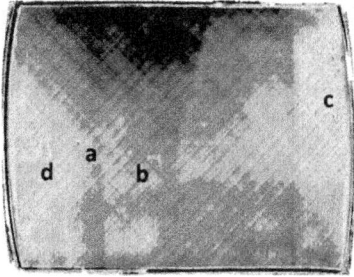

Figure 11: Inspecting a CFRP-roof for damage by hail – (Source: IKT University of Stuttgart)

3.4 Computer tomography (CT)

X-radiation enhances an absorption profile of the inspected component. Thereby the tube and the detector or the component itself is rotating so that the data can be set to a 3D picture of the component (see Figure 12). In addition to the fibre direction, the porosity of the component is an important part of the analysis.

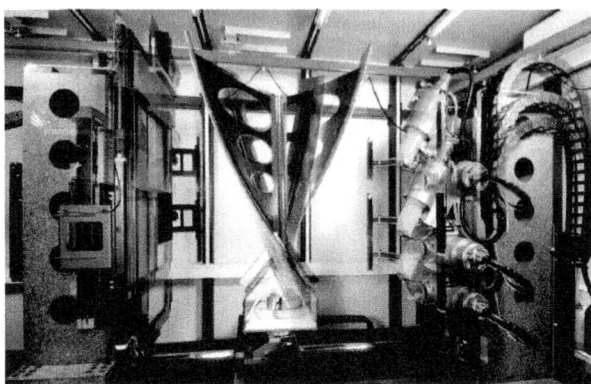

Figure 12: CT construction with CFRP structure – (Source: DLR, T. Ullmann)

Pores are located between fibre layers and immaterial transitions. They occur during the production of CFRP-components. Pores strongly weaken the whole structure and lower the requested strength parameters. The analysis of porosity of complex CFRP-structures layer by layer is often not sufficient. Therefor computer tomography enables the inspection of the whole volume. It is possible to locate pores independent of their position in the 3D body (see figure 13). The visible size of these pores is depending to the CT-resolution.

Figure 13: CT of a tube: 3D view on the left, view of all pores on the right. Ratio of pores: 1,71% (Source: QualiFibre Project Fraunhofer IPA)

3.5 Partly automated optical test method "AICC" by Fraunhofer IPA

All in chapter 3.1 to 3.3 presented measuring methods are primarily used for detecting defects inside of the structure. The limited use of these methods in line production is based on the high investment costs and long duration of each test run. Additionally due to their lack of mobile use, these methods are not flexible usable. The Fraunhofer IPA is therefore dealing with an optical solution for automated quality management of drill holes and edges. As a result, the Fraunhofer IPA designed a manual device which is able to inspect and approve the machining quality fast and simple (see figure 14). No special qualified worker is needed to judge the quality. Furthermore the mobile device guarantees maximum flexibility in quality control. The „Automatic Identification of Cut Carbon" (AICC) makes often used catalogues of limiting samples unnecessary and hence improves the quality of testing significantly. The AICC is also applicable as a stationary test bench for measurement laboratories. AICC enables measurements of free-form shapes of nearly any orientation. The evaluation of the component happens after machining or even at the same time as machine-integrated solution. The evaluation also works with quantitative figures as well as categorizing the components in usable and waste parts. The results are filed in a database which are accessible for complete documentation and process monitoring. These features make AICC practicable for manual control of incoming goods, in-house documentation at the supplier as well as automated process monitoring and optimization. A high-resolution camera, which is controlled and electrically supplied by an ethernet hub, is used for testing the components. The camera has a high-resolution CCD-sensor and a short focal length objective. The surface of the workpiece is illuminated by a ring of dark field lights.

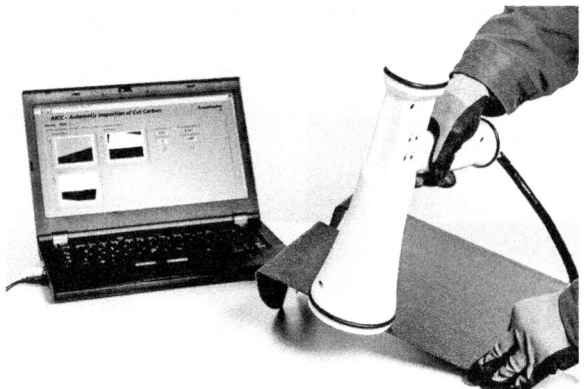

Figure 14: Evaluating with AICC – (Source: Fraunhofer IPA)

This consistent and homogenous illumination shows irregularities on the surface which can be detected by the caused differences in brightness. Reflections on glossy components like carbon- or fibreglass-reinforced plastics have to be avoided by selecting a suitable light intensity. The housing of AICC was developed by arranging the single components in multiple iterations so that ergonomic working is enabled. When testing the quality, the manual device is deployed over the test point followed by starting the measurement with a manual switch. Afterwards the photo material is evaluated by an algorithm developed at the Fraunhofer IPA. The footage is tested regarding delamination, fraying and feathering. Furthermore, differences in the shape of drill holes can be detected. The end user is able to determine individual threshold values of each quality attribute which can be adapted to different requirements at will. Several status displays (see Figure 16) simplify the evaluation. The numeric test results are clearly presented to the user. Afterwards the date can be archived. According to the predefined tolerance value, the quality of the tested spot is illustrated by different colour codes. Good parts are marked green, bad parts red.

However, AICC only detects defects between layers partly. Therefore previous processes need to be reliable to avoid these defects. Nevertheless AICC replaces current catalogues of limiting samples used in the industry followed by striking improvement of testing quality.

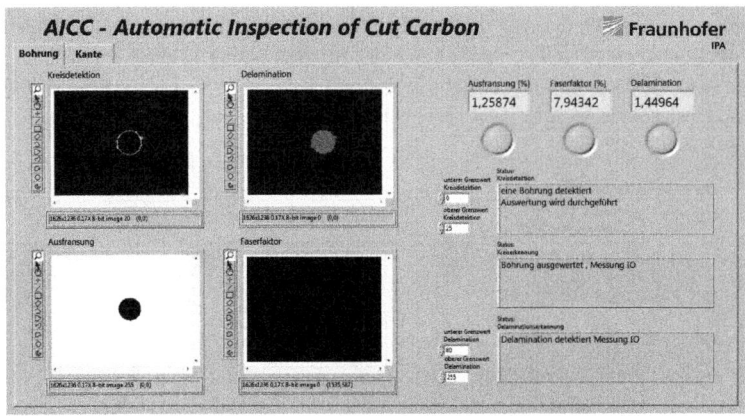

Figure 15: Self developed AICC software – (Source: Fraunhofer IPA)

High volume production of lightweight automotive structures

Prof. Dr.-Ing. Christian Brecher, Clemens Buschhoff, Dr.-Ing. Michael Emonts

Fraunhofer Institut für Produktionstechnologie IPT

1 Motivation

In times of globalization and simultaneous worldwide resource depletion energy-efficient mobility is increasingly becoming the focus of social and economic interest. The transformation of conventional drive concepts with gasoline and diesel engines via plug-in hybrid solutions is going to fully electrically driven Urban Vehicles, as the BMWi series, Tesla Motors and many more demonstrate.

In particular the relatively young species of electric vehicles is dependent on obtaining a market acceptance by realizing a certain minimum distance. Hence, a reduction in weight of the vehicle in general and the structural components in particular is mandatory in the sense of increased energy efficiency. But even for vehicles with conventional internal combustion engines the lightweight has a more important role: With the EU Directives becoming effective in 2020, European car manufacturers need to reduce the average CO_2 fleet emissions below 95 g/km [FRIE13].

One promising solution to accomplish these ecological and political requirements represents the structural and particularly the material lightweight. Hence the fiber-reinforced plastics (FRP) could lately be established in the market with its excellent mechanical properties combined with low density. By the use of modern high-performance materials such as unidirectional carbon fiber reinforced plastics (UD-CFRP) the same functionality of the components, i.e. same stiffness and strength, is realized with weight savings compared to steel up to 75 % and to aluminum by 60% [FRIE13]. Such potential can however only be exploited if it is possible to reduce manufacturing costs of CFRP components dramatically. For example, the mass-related costs for CFRP elements in automotive applications is up to 70 €/kg, whereas they amount to about 3 or 6 €/kg for steel and HS steel and about 7 €/kg for aluminum [JAHN12].

2 Fiber reinforced plastics in the automotive industry

Fiber reinforced plastics (FRP), a subgroup of composite materials, are generally characterized by a combination of at least two significantly different, not in each soluble phases: the fibers and the matrix surrounding them. By combining the materials the resulting properties of the composite material can be far superior. [EHRE06] The mechanical-properties such as strength and stiffness of the composite material are largely determined by the high-strength fibers. The matrix meanwhile fixes the position of the fibers in the intended position. Furthermore, it supports the fibers, induces loads into the fibers, and carries loads transversely to their direction. [WITT10] Figure 1 shows the relative importance of fiber and matrix for various properties of the composite material.

	Fiber	Matrix
Mechanical properties		
⌐ Stiffness	◕	◔
⌐ Strength	◕	◔
⌐ Fatigue	◕	◔
⌐ Damage tolerance	◔	◕
⌐ Impact behavior	◕	◔
⌐ Thermomechanical properties	◕	◔
⌐ Fiber-matrice bonding	◑	◑
Physical properties		
⌐ Corrosion behavior	◔	◕
⌐ Temperature stability	○	●
⌐ Chemical stability	○	●
⌐ Electrical stability	◑	◑
Processing properties	○	●

○ Low importance   ● High importance

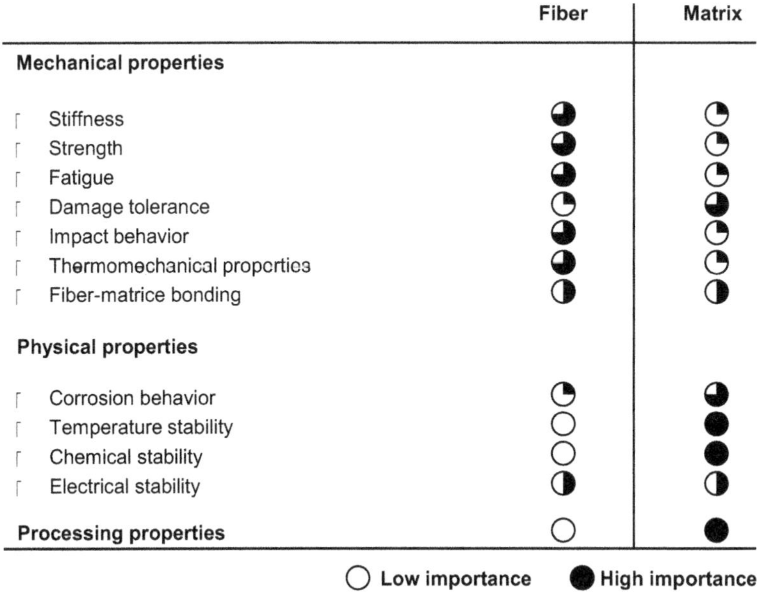

Figure 1: Influence of fiber and matrix material on selected properties

The majority of FRP produced today is based on thermosetting matrix materials, such as epoxy resins. The manufacturing processes vary from hand lay-up for large-scale wind turbine blade elements to resin transfer molding (RTM) of automotive component like boot lid reinforcements. The greatest disadvantage of thermoset matrices comparing to thermoplastics are the relatively long chemical cross-linking from several minutes to hours, and complex automation. The integration of such technologies in existing process chains is therefore challenging.

In FRP with thermoplastic matrix, the fiber impregnation and the molding process can be separated in two steps. Similar to the steel sheet processing thermoplastics can be formed by melting them up and reshaping a blank into complex geometries in cycle times of less than one minute [NEIT04]. Such combination of material and manufacturing process has therefore great potential in automotive applications with its short cycle times and efficient value chains [EICK13]. Also for further sectors dependent on lightweight, such as aviation and space-travel industry is the development of an economic and high volume production chain of great interest.

The fibers are in industrial applications mainly described by its material (glass, carbon or aramide fibers) and by its length (short, long and continuous fibers). Short and long fibers can be processed for instance via injection molding or extrusion. They are aligned randomly, or easily aligned in thin-walled components by the melt flow and provide isotropic properties. Correspondingly low is the attainable reinforcing effect of the fibers, which is why these components are mainly used in low-load applications, such as cladding [SCHÜ07].

In Figure 2 stiffness, strength and impact resistance are shown of an exemplary FRP with epoxy matrix along fiber direction. The achievable values are normalized to properties of fibers of 100 mm length. Although the shown curves are only qualitatively they demonstrate very well that the stiffness has a relatively low dependency on the fiber length and thus appealing rigidity values can be achieved with short fibers. Strength and particularly impact behavior however correlate clearly with fiber length.

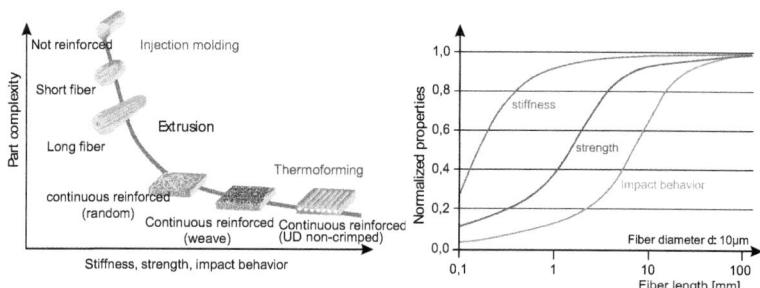

Figure 2: Influence of fiber length on part properties; left: feasible part complexity and mechanical properties; right: normalized mechanical properties in correlation to fiber length

Currently in automotive industry mainly thermosetting matrix systems are used for crash-relevant body parts made of continuous fiber reinforced plastics. These require long cycle times and are at present only suitable for small lots in the high price segment. Thermoplastic systems, however, solidify in comparatively short periods of time and are therefore used in the large-scale production, such as injection molding or extrusion.

3 Production technologies for fiber reinforced thermoplastic

In serial manufacturing of thermoplastic FRP glass mat-reinforced thermoplastics (GMT) and long-fiber reinforced thermoplastics (LFT) are established. With the LFT injection molding, the semi-finished material is melted in special extruders and injected under high pressure in the mold. Semi-finished materials are usually pultruded pellets. For the GMT process tangled fiber mats are inserted in the mold before injection molding. The fibers used for this purpose usually have a length of 20-30 mm, in products obtained by LFT extrusion process 12-25 mm and about 2-6 mm for the LFT injection method. [WITT10]

Since the production of Class A surfaces is only partially possible, the pressed components are usually used as non-visible, flat reinforcement structures. They are widely used for example in the automotive industry, coupled with the possible design options with good weight-specific stiffness, good crash properties and a high potential for function integration [EHRE06]. Examples of applications are a tailgate of GMT and a dashboard support from LFT. Further possibilities are underbody components, spare wheel wells and front-end carriers.

Currently there are various efforts to enhance the mechanical properties with GMT and LFT components, such as the load-optimized application of UD tapes [GRAU12].

The industrial processing of continuous fiber reinforced thermoplastics is usually realized by forming organo blanks. These usually woven or non-crimped fabric reinforced semi-finished products are already fully impregnated , non-porous and consolidated, by which for the further shaping processing only a melting and solidification is required.

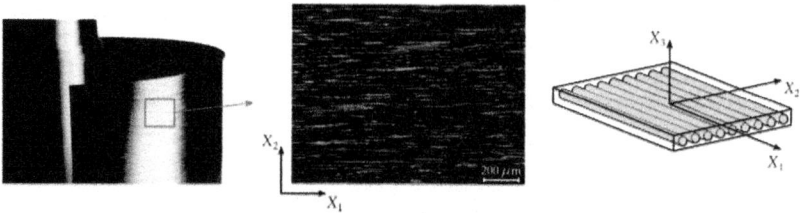

Figure 3: Setup of UD tapes: macroscopic (left), microscopic (center); schematic (right)
[HAAN13]

Unidirectional fiber reinforced tapes are also semi-finished products similar to organo blanks and are processed by tape laying or winding. These have very good mechanical properties in the fiber direction, but perpendicular to fibers the matrix dominates the properties. However the fibers are aligned only in one direction and the tape tends to be thinner (0.15-0.3 mm) as seen in figure 3. Thereby high fiber volume content is re-alized (up to 60 % or even higher). By applying the tapes in the area and along the di-rection of loading the tapes enable a resource-conserving and waste-optimized pro-cess. This ensures manufacturing of thinner and lighter semi-finished products and thus components. Furthermore, it is possible to use in less stressed areas significantly more favorable glass fiber-reinforced tapes rather than carbon fiber-reinforced tapes, in order to reduce component costs. In figure 4 a general overview of production technologies for fiber reinforced thermoplastics is given.

Material type	Process technology	Process costs per kg reinforcement	Isotropic properties	Key facts
UD tape	Tape winding / placement	high	low	• Outstanding mechanical properties • Low throughput / costly
Organosheet	Thermoforming (overmolding)			• Increased mechanical properties • Easy integration in existing process chains
Long fibers	Injection molding (GMT/LFT)			• Relatively low priced • Well established
Short fibers	Injection molding	low	high	• Low priced • Well established

GMT: glass mat reinforced thermoplastic
LFT: long fiber reinforced thermoplastic

Figure 4: Overview of technologies to process fiber reinforced thermoplastics

The end user is mostly interested for its application in generally good mechanical properties combined with low cost. By selecting only one processing technology described above the requirements probably cannot be realized. Either the mechanical properties are too low (i.e. part weight in combination with stiffness, strength manufactured by injection molding) or the costs are too high or with the tape placement. However, this can be realized through a combination of several materials and processes.

4 Process chains for high volume production of fiber reinforced thermoplastics

Within the Fraunhofer project FSEM 2 (Fraunhofer System Research for Electromobility 2) technologies and process chains for automated mass production of thermoplastic fiber composites are developed at the Fraunhofer Institute for Production Technology IPT. Exemplary of an automotive application a floor panel for an electric vehicle is realized. Therefore the battery packs is integrated within the floor panel in accordance with the vehicle concepts of the Tesla Model S or BMWi3 (see figure 5).

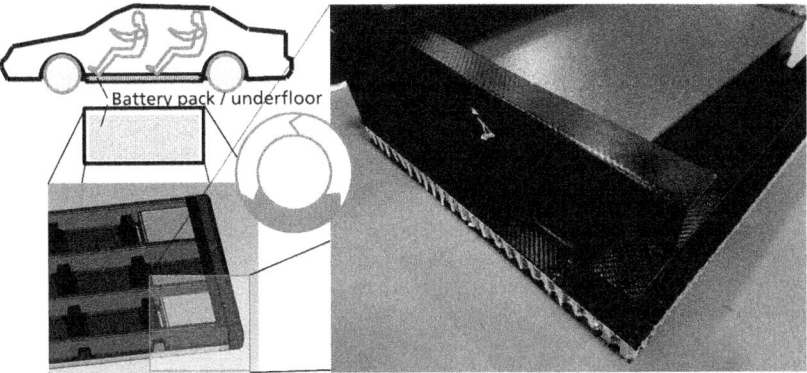

Figure 5: Demonstrator for fiber reinforced thermoplastics in automotive applications

The demonstrator is manufactured by following technologies and covers the entire process chain of processing thermoplastic composites and shows the potential of material as well as the technologies used.

1. Thermoforming of profiles made by organo sheets

2. Tape placement and winding of side skirt

3. Local reinforcement of honeycomb sandwich structure with thin organo sheets as top layers, as well as the formed profiles and organo sheets

4. Machining of fiber reinforced thermoplastic

5. Joining technologies (thermos-drilling of holes / ultrasonic welding)

Within this paper the thermoforming as well as the joining technologies are presented in more detail.

4.1 Materials used

Two different Organosheets were used within this study. A continuous glass fiber reinforced polyamide 6 (Tepex dynalite 102-FG290(8)/45%-2 mm; black) GF/PA6 and a carbon fiber reinforced polyamide 6 (Tepex dynalite 202-C200(9)/50%-2 mm) CF/PA6 from Bond Laminates (Lanxess).

The honeycomb panel is a ThermHex 25 mm PA honeycomb with TEPEX dynalite vm skins provided be econcore.

The tapes are Celstran CFT PA6-CF60 endless fiber reinforced UD tapes by Celanese with a width of 12 mm, a thickness of 0.15 mm and a fiber volume content of 49 %.

4.2 Thermoforming

The thermoforming technology has the potential to process fiber reinforced blanks with reproducibility and short cycle times beneath one minute. Similar to metal forming the material is heated and subsequently formed to final shape as shown in figure 6.

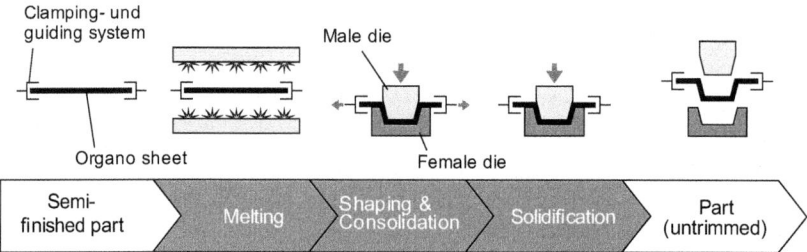

Figure 6: Process steps for thermoforming

In a first step, the organo sheet is clamped between two guiding rails and heated from both sides in an infrared oven. Once the optimal forming temperature is reached, the sheet is transferred in the press and formed to its final shape. To conserve the shape the blank needs to be solidified in the mold. Figure 7 demonstrates the process steps as well as material temperature and process times for the individual steps.

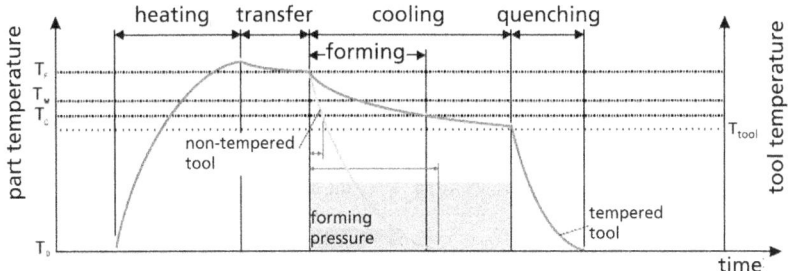

Figure 7: Part temperature of organo sheet along thermoforming process

Regarding process times the heating as well as the cooling are the steps which take the longest. However the clamping and unclamping are not considered. The actual forming process is realized within seconds.

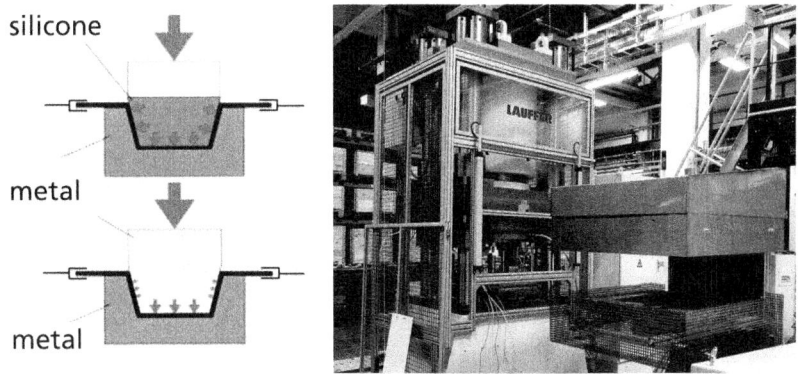

Figure 8: Schematic of silicone and metal tool concepts (left); thermoforming press and infrared oven (right)

Within the project aluminum tools are used for the production of S- and hat-shaped profiles and silicone tools for U-shapes to apply homogenous pressure during forming. As it can be seen in figure 8 a pressure cannot be applied or only in fraction on areas perpendicular to direction of motion with rigid forms. Silicone forms though enable homogeneous pressure over the whole stamp. In addition the applied pressure can be reduced as seen in table 1. On contrary heat conduction of silicone is much lower compared to aluminum and therefore cooling time of material in the press is much longer (see table 1). The values are based on prior internal processing studies with similar tool geometries. In general the parameters need to be simulated for the specific set of part geometry, and materials used.

Table 1 Process parameters for thermoforming

Tool	Material	Heating Time [s]	Pressing Time [s]	Applied Pressure
Hat-; s-shape	aluminum	60	120	500 bar ≙ 6000 kN
U shape	silicone	60	300	200 bar ≙ 1500 kN

The heating is mainly dependent by the material used and within this project kept constant at 60 seconds to a temperature of 260°C.

Figure 9 shows three already machined profiles. The parts are straight over the length. However the U-profile has on the bottom side a slight deflection due to pressure variation over the profile of the tool. Furthermore the hat-profile has a minor spring back. In the case of inaccuracies of the tools or also thickness variation of the organo sheets also pressure variations occur and result in warpage.

Figure 9: Machined U-, hat- and S-shaped profiles

4.3 Joining technologies

To exploit the maximum lightweight potential of multi-material solutions joining technologies need to be developed further and optimized. For FRP several joining techniques are used, such as adhesive bonding, welding or bolts and rivets. The welding enables the joining of (fiber reinforced) thermoplastics basically without additional weight, continuous as well as locally and with the ability of automation.

4.3.1 Ultrasonic welding

The ultrasonic welding process joins two parts using heat due to high frequency friction. A generator sends electric high frequency signals to multiple piezo elements which convert them into mechanical vibrations. The amplitude is subsequently intensified when passing the booster and sonotrode section. The sonotrode incorporates the tool in contact with the parts. Heat is created at the interface where the vibration antinode is located and melts the material of both parts. To ensure a stable and reproducible process, energy directors have to be placed at the interface between both parts. These energy directors concentrate the ultrasonic energy due to their tapered design and enable a controlled melting of the joining area. However, currently the application of energy directors is time consuming and not automated. By using narrow unidirectional fiber reinforced tapes as energy directors on contrary already existing material, machinery and processes can be used. Therefore the tape is in-situ welded on thermoplastic composite sheet with the laser assisted tape placement process and subsequently ultrasonically welded to a second thermoplastic composite sheet. The results show that the tapes as energy directors enhance strength of the joint significantly.

On figure 10 a fiber reinforced thermoplastic tape has been applied on the part surface by laser-assisted tape laying to operate as energy director.

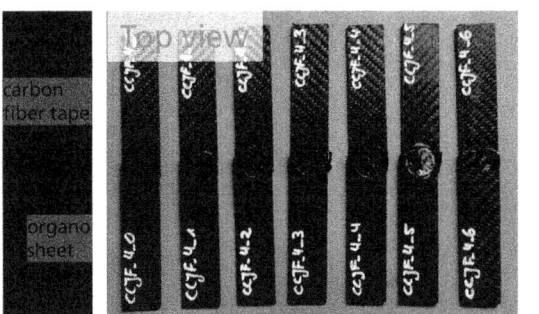 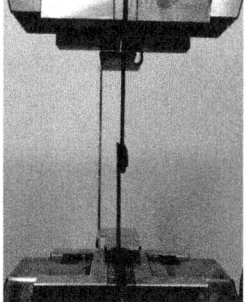

Figure 10: Organo sheet sample with carbon fiber tape as energy director (left); ultrasonic welded organo sheets with energy director (center); sample clamped in lap shear test

Lap shear welded samples were produced to investigate the influence of fiber direction in the tape as well as tape width on the mechanical strength of the bond. The results show that the tapes as energy directors enhance strength of the joint significantly. The welded samples were mechanically tested following the ASTM D 3163-01 standard (five samples per set of welding conditions) with a cross-head speed of

0.13 mm/min. The apparent single-lap shear strength (LSS) of the welds was calculated as the maximum load divided by the total overlap area.

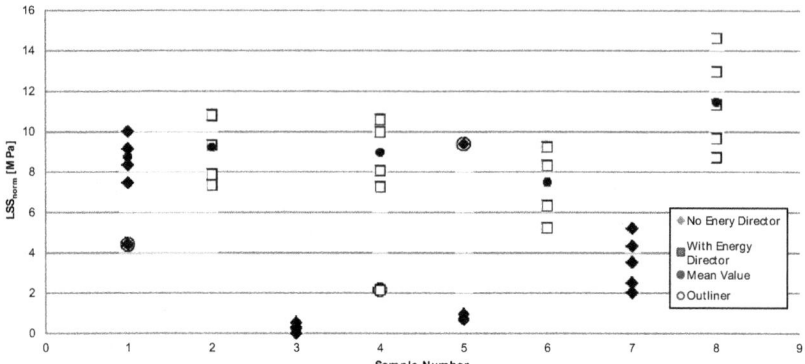

Figure 11: Results of lap shear test and improvements due to energy directors

The bonding strength could be increased significantly by using tapes as energy directors (figure 11). Therefore, the complex application of energy directors to the bonding area could be simplified by the laser-assisted process.

4.3.2 Mechanical bonding – Inserts

Bolts enable demountable joints and the joining of different materials. However, drilling holes into FRP destructs the fibers resulting in an interrupted flux and thus lower stiffness and strength. By forming the hole and implementing the bolt in the part within a thermos-drilling process a fiber fair and bionic structure is realized.

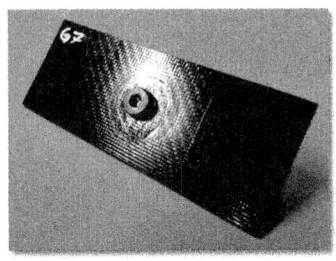
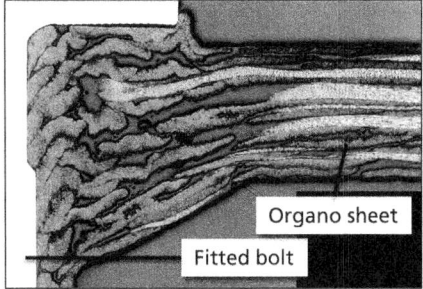

Organo sheet

Fitted bolt

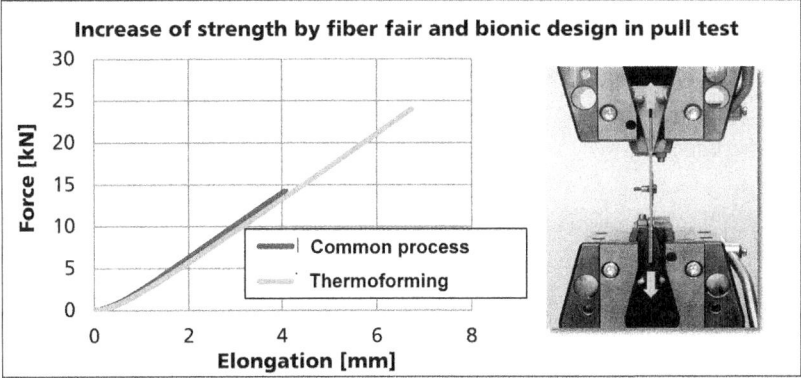

Increase of strength by fiber fair and bionic design in pull test

Common process

Thermoforming

Figure 12: Organo sheet with insert (top left); microscopic cross-section of insert (top right); comparison of thermo-drilled hole with insert and conventional process in pull test (bottom)

Figure 12 demonstrates very well the principle of the thermoforming. The material is not removed, but rather moved around the fitted bolt. Therefore the fibers are not destroyed and in close contact to the bolt. Undercuts can thereby be realized (see figure 12 top right). Furthermore the strength within a pull tests increases by 62.5 % compared to conventional drilling.

5 Conclusion

Within this paper process technologies for thermoplastic composites are presented. The economic processing of organo sheets has been shown exemplary along the process chain by the thermoforming and the joining. All technologies have the general capability of short cycle times and the integration in existing production environments.

Bibliography

1. [FRIE13] Friedrich, H.E.: Leichtbau in der Fahrzeugtechnik. 1. Auflage, Springer, 2013, ISBN 978-3-8348-1467-8

2. [JAHN12] Jahn, B. Karl, D.; AVK – Industrievereinigung Verstärkte Kunststoffe e.V.; Carbon Composites e.V. (Hrsg.): Composites-Marktbericht 2012: Marktentwicklungen, Trends, Ausblicke und Herausforderungen. 2012

3. [EHRE06] Ehrenstein, W.W.: Faserverbund-Kunststoffe: Werkstoffe, Verarbeitung, Eigenschaften. 2. Auflage, Hanser, 2006, ISBN 978-3-446-22716-3

4. [WITT10] Witten, E.; AVK – Industrievereinigung Verstärkte Kunststoffe e.V. (Hrsg.): Handbuch Faserverbundkunststoffe: Grundlagen, Verarbeitung, Anwendungen. 3. Auflage, Vieweg+Teubner, 2010, ISBN 978-3-8348-0881-3

5. [NEIT04] Neitzel, M.; Mitschang, P.: Handbuch Verbundwerkstoffe: Werkstoffe, Verarbeitung, Anwendung. 1. Auflage, Hanser, 2004, ISBN 978-3-446-22041-0

6. [EICK13] Eickenbusch, H.; Krauss, O.: Kohlenstofffaserverstärkte Kunststoffe im Fahrzeugbau – Ressourceneffizienz und Technologien. VDI Zentrum Ressourceneffizienz GmbH, 2013

7. [SCHÜ07] Schürmann, H.: Konstruieren mit Faser-Kunststoff-Verbunden. 2. Auflage, Springer, 2007, ISBN 978-3-540-72189-5

8. [GRAU12] Grauer, D.; Hangs, B.; Reif, M. Martsmann, A.; Jespersen, S.T.: Improving Mechanical Performance of Automotive Underbody Shield with Unidirectional Tapes in Compression-Molded Direct-Long Fiber Thermoplastics (D-LFT). In: SAMPE Journal, Volume 48, No.3, Mai/Juni 2012, S. 7-15

9. [HAAN13] Haanappel, S.: Forming of UD fibre reinforced thermoplastics: A critical evaluation of intra-ply shear. Dissertation, Universität Twente, 2013

Thermoplastic composites for high volume production

Dipl.-Ing. Stefan Epple, Prof. Dr.-Ing. Christian Bonten

Institut für Kunststofftechnik

Abstract

Fiber reinforced plastics are often used as lightweight materials where conventional materials, especially metals find their limits. Up to now, a thermosetting resin is often used as matrix even though using thermoplastics brings some advantages like a higher impact resistance to comparable thermoset composites, weldability and recyclability. In addition, it is possible to reshape or to weld the thermoplastic composites. To show some of the possibilities, some recent developments are shown. Especially highly filled thermoplastic composites produced by In-Situ-Pultrusion are demonstrated, because they are supposed to be used as local reinforcement of injection molded plastic parts. For this purpose, the pultruded parts are overmolded after being inserted into the injection mold.

Introduction

In high volume production, especially of structural parts, there are often two main requirements: High mechanical properties at low cost. In the second half of the past century, often a third requirement has been added: high design freedom.

In the last decade, a fourth requirement became more and more important: lightweight.

To reach all those requirements, there is a large toolbox of industrial processes available. Components that have e. g. to bear only minor mechanical strains, e. g. in cars, are usually made of short fiber reinforced plastic. For parts requiring better mechanical properties, longer or preferably continuous fibers are needed.

Following, some possibilities to produce parts that have to bear higher mechanical strains are discussed.

State of the Art

To reach the requirements high mechanical properties at low cost and high design freedom, the Ford Focus front end was built as a metal-thermoplastic hybrid part (Figure 1). The good mechanical properties at low cost come from the sheet metal, while the injection molded plastic brings the design freedom.

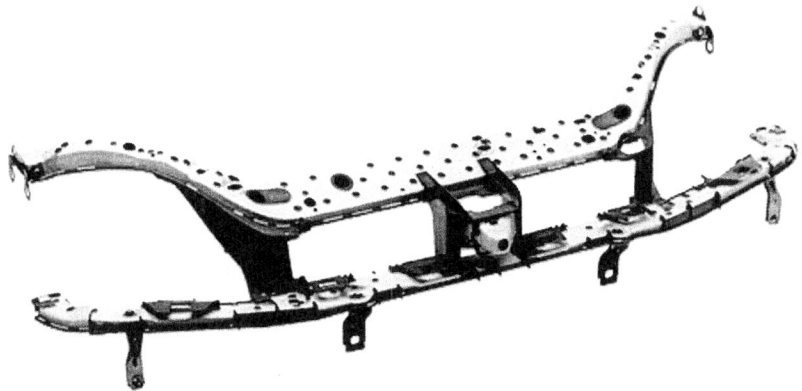

Figure 1: Hybrid front end for Ford Focus [Source: Dynamit Nobel Kunststoff GmbH]

The challenges in this design have been, that adhesion agent is required between metal and plastic, metal has a relatively high density and thermomechanical failure can occur, especially after temperature changes

Facing the challenges, another current development in Audi A8 front end is catching the requirement of lightweight (Figure 2). Instead of sheet metals, continuous fiber reinforced plastic sheets are used. This brings high specific strength with low density and hardly any thermomechanical failure caused by temperature changes while maintaining the high design freedom from injection molding.

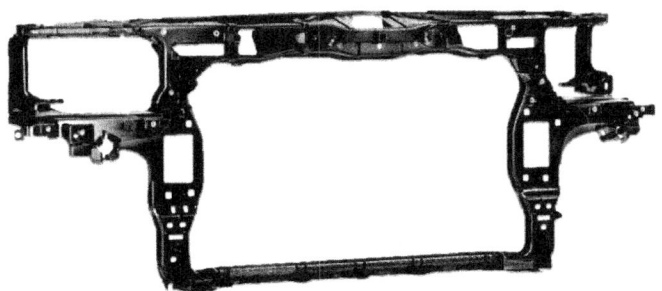

Figure 2: Hybrid front end for Audi A8 [Source: Lanxess AG]

The challenges in this design are the risk of fiber damage while reshaping the composite sheet, the mostly unfavorable fiber orientation and the limited composite sheet thickness.

One way to handle those challenges is not to use the consolidated composite sheets but the very thin composite tapes, made of unidirectional fibers and thermoplastics (Figure 3).

Figure 3: Tape demonstrator [Source: BASF SE]

Another way to produce lightweight parts with high mechanical properties at low cost and with high design freedom was developed at Institut für Kunststofftechnik (IKT).

To bring the advantages of continuous fiber reinforced plastics also into parts with complex geometries, it was put into effect a process based on the pultrusion process combined with reaction injection molding (RIM). In reaction injection molding, two components are mixed together. The mixture is then injected into the mold, where it reacts to the plastic. The RIM process offers the advantage of a very low viscosity of the mixture before the reaction starts. In this state, the two component mixture can impregnate the fibers very well.

It was the objective of the development to produce thermoplastic parts with a high amount of continuous glass fibers in various shapes and thicknesses. Adhesion between fibers and matrix and fiber coating were optimized [3].

In traditional fiber reinforced plastics, a thermosetting resin is used as matrix. These fiber reinforced thermosets have disadvantages, especially in batch production because e. g. the cycle times are too long, they are not weldable and poorly recyclable.

In batch productions it is common to use fiber reinforced thermoplastics. Often, short fibers are used as reinforcement. By using glass fibers, mechanical properties increase, however, highly-stressed components needs continuous fiber reinforced plastics.

The two major advantages of thermoplastic composites are that they have a higher impact resistance to comparable thermoset composites and can be welded and recycled.

One objective of the current research is to produce a pultruded thermoplastic composite that could be heated and remolded to have complex shapes. This is not possible with thermosetting resins. It will also be possible to recycle the thermoplastic composites in contrast to thermoset composites.

Thermosetting resins are of low viscosity and thereby can easily impregnate the reinforcing fibers before they react. In contrast, thermoplastics are usually of high viscosity, even in the molten state. It is not easy to impregnate reinforcing fibers.

To impregnate reinforcing fibers with a thermoplastic matrix, plastic films and woven fabrics must be laminated. High pressure and high temperature are required. These fiber-reinforced thermoplastic laminates are called organo-sheets. Fiber mat or long glass fiber reinforced thermoplastics are also used to produce rugged parts. The problem of these parts is their limited shape variety.

In-situ-pultrusion is one way to produce thermoplastic FRP. Like in conventional pultrusion with a thermoset matrix, continuous fibers are impregnated with the monomeric precursor of the matrix material, which then reacts and forms the polymer. When using ε-caprolactam as a monomer, cast PA6 will be formed and act as the matrix [1].

Different to conventional pultrusion, the fibers cannot get impregnated by pulling them through a resin bath, as the surrounding humidity would disturb the chemical reaction. Thus, the reaction has to take place in an inert gas atmosphere inside the pultrusion die.

The products from in-situ-pultrusion can e. g. be used as reinforcements for injection molded parts. One example is the addition of rip structures to bending beams composed of two FRP profiles via back injection molding. For this purpose, the profiles are inserted into an injection mold and overmolded with a thermoplastic. Thus, cycle times like in conventional injection molding processes are possible, combined with excellent mechanical properties and high component complexity. [2]

Two principles are coupled in the In-Sizu-pultrusion process. First, the two reactive components are mixed together. Then the mixture is injected into a mold. This process is often used to produce parts made of polyurethane. In pultrusion processes, fibers are pulled through a bath of thermosetting resin. Afterwards, the impregnated fibers are pulled through a heated die, where the resin reacts to the formed composite.

A puller behind the die is pulling the material through the process at a consistent rate. Behind the puller, the parts are sawn off to the desired length.

In current researches, the matrix of the composite was polyamide 6 (PA 6). That is why ε-caprolactam was used in the activated anionic polymerization process from monomer to polymer. The resulting PA 6, also known as cast PA 6, has good properties like dimensional accuracy, low moisture adsorption and a good creep resistance.

The RIM pultrusion process used for the production of continuous glass fiber reinforced thermoplastics is schematically shown in. Figure 4.

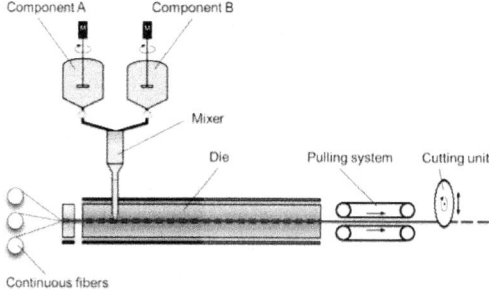

Figure 4: Scheme of the new in situ pultrusion process

The glass fibers are pulled through a pre-warming and sorting tool before they are entering the die. In the die, the fibers are impregnated with the monomer. In the following zones of the die, the monomer ε-caprolactam reacts to PA 6. The first part of the die is only heated while in the second part of the die, cooled or heated, depending on the process behaviour. This is important because in the process heat can be generated by the reaction which has to be removed from the die.

The puller and the saw are also shown in the scheme. Providing the mixtures of ε-caprolactam and catalyst as well as ε-caprolactam and activator in the correct way is very important for the process. Protective gas has to be used to reduce the influence of moisture on the reaction.

The activated anionic polymerization of PA 6 can occur within minutes. In this polymerization, the monomer respectively the growing macromolecule is an anion. An activator to accelerate the catalytic reaction is also used in this system.

Especially the low temperatures below the melting temperature of PA 6 which are needed for this reaction (130 to 170 °C) allow its use in the present process.

Usually glass fibers are delivered in wound bundles called rovings. These rovings usually have tex numbers of 300, 600, 1200, 2400, 4800 or 9600. A small beam diameter is characterized by a low tex number.

While it is easier to handle glass fibers with higher tex numbers, the possibility to impregnate the fibers decreases. A roving with the tex number of 2400 was selected for the tests. In Figure 5, the rovings are shown. Fiber diameters can be adjusted to the production process. Typical fiber diameters are 16 respective 20 μm. The mechanical properties of the fiber reinforced products depend largely on the ratio of fiber diameter to fiber length. If profiles are produced continuously, the fiber length in the product is virtually endless. Therefore, the fiber diameter plays a subordinate role in the mechanical properties of the final product. The finer fiber diameter of 16 μm was chosen.

Figure 5: Glass fibers in the sorting tool

To ensure the fiber-matrix adhesion, as well as the handling of the fibers, a sizing is applied to the glass fibers during the process. A number of chemical compounds are known for preventing or at least affecting the anionic polymerization. For the first experiments an already available sized fiber was selected.

At IKT, an In-Situ-pultrusion facility with two component units, puller, saw, die and mixer was installed.

To design the In-Situ-pultrusion die (Figure 6), preliminary examinations were done and the reaction behavior of ε-caprolactam was described.

It is important for the process that the material does not polymerize in the nozzle. The nozzle opening was therefore separated from the rest of the die and can be cooled. The die length was chosen in a way that in the times known from the preliminary studies, polymerization can take place in the die. It should be noted that the die should not be unnecessarily long. Otherwise the puller reaches its performance limits. On the other hand the die must not be too short because then the material would have too few time to polymerize.

Figure 6: In situ pultrusion facility

To mix the two components (ε-caprolactam with activator and ε-caprolactam with catalyst), a mixing head was developed, which has been especially adapted to the requirements of the process. Below about 80 °C, the mixture is in a solid state. Above about 120 °C it begins to polymerize.

This means that the part of the mixer outside the die, in which a static mixer is located, has to be heated, whereas the nozzle through which the monomer is passed into the die must be cooled so that the material does not polymerize prematurely [2].

Experimental

For the in situ pultrusion experiments activator concentration of 2.5 % and catalyst concentration of 3.75 % were chosen. The temperature in the tanks and the mixer was set to 100 °C. The temperature in the first part of the die was adjusted to 160 °C. In the second half of the die, the temperature was regulated to 140 °C by a temperature control unit. To achieve a high amount of glass fibers, 114 glass fiber rovings were used. The glass fibers were sorted in a box before entering the die. In this box, the fibers also get dried by dry and hot air, to prevent the influences of moisture to the activated anionic polymerization.

It was possible to produce a continuous glass-fiber-reinforced PA 6 part with a sufficiently good surface in case of specimens (Figure 7).

Figure 7: Pultruded parts

To evaluate the quality of the specimens, microsections of the samples were made. Using these images, on macroscopic scale, first statements about the distribution of fibers in the component can be made.

Figure 8 shows the straight areas of the specimen. The fibers which can be identified as black spots are well distributed and surrounded by the PA 6 matrix.

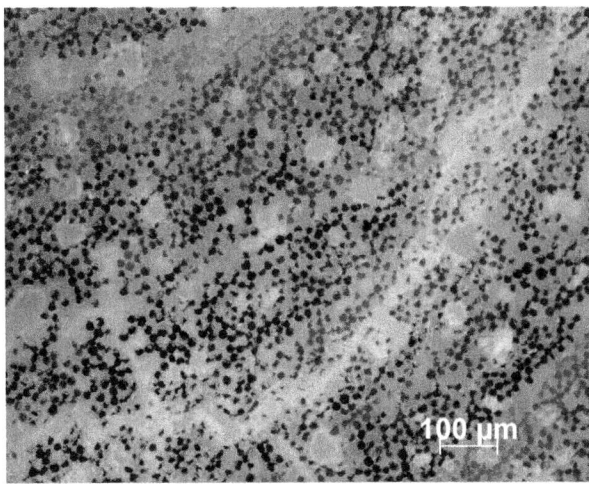

Figure 8: Straight area of the specimen

To study the adhesion between FRP profiles with a cast PA6 matrix and regular PA6, a specimen shape was used, which allowed to vary the effective adhesion surface without applying a bending moment (as it would occur in a conventional three-point flexural test). Therefore a PA6 overlap on two sides of the profile was realized (Figure 9), which allowed for pull-out tests on a tensile-testing machine.

Figure 9: Side view of the test specimen

The overlap length can be varied by choosing different lengths of the pultruded parts.

The test specimen were produced on an injection molding machine of the type Arburg Allrounder 520S 1600-400. The PA6 used in the tests was a Lanxess Durethan B 30 S.

Before injecting the PA6 into the mold, the pultruded profile was heated under defined conditions. For this purpose, the ceramic heater was placed 17 mm in front of the profile. Then the profile surface temperature was measured with the pyrometer to ensure that the profile temperature had reached 230 °C. The heating and measuring equipment was then pulled out of the mold and the injection molding cycle was started. The injection pressure was 900 bar, the dwell pressure was 525 bar and the cooling time was 30 seconds at a mold temperature of 80 °C. The test specimen (Figure 10) were sealed within airproof bags to later be tested as-molded [3].

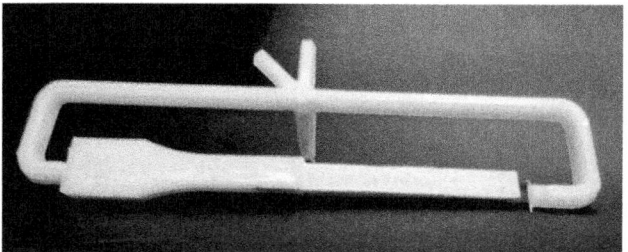

Figure 10: Test specimen as-molded

After the results of these tests have shown, that the bonding strength between the pultruded profile and the PA6 was sufficient, a bending girder has been designed (Figure 11). This bending girder has the pultruded parts in the highest stressed surface areas and is overmolded with short glass fiber reinforced PA6 (PA6GF30).

Figure 11: bending girder

Through three-point-bending-testing, there was no delamination between the PA6GF30 and the PA6GF30 with pultruded parts. In-Situ-pultruded parts therefore can be used as local reinforcement for injection molded complex parts.

Summary

By using overmolded in situ pultruded profiles, it is possible to produce parts with special requirements in mechanical properties, low cost, lightweight and design freedom. A high amount of glass fibers can be realized with a very good fiber matrix adhesion and distribution.

Acknowledgements

The authors thank the companies Lanxess Germany, Brüggemann Chemical and Johns Manville for the provided material.

References

1. NING, X.; ISHIDA, H.: RIM-Pultrusion of Nylon-6 and Rubber-Toughened Nylon-6 Composites. In: Polymer Engineering and Science, Vol. 31, No. 9 (1991)

2. EPPLE, S.; BONTEN, C.: Production of Continuous Fiber Thermoplastic Composites by In-Situ Pultrusion. In AIP Conference Proceedings (2014)

3. EPPLE, S.; BONTEN, C.: In-Situ-Pultrusion – Bonding of FRP-Parts to PA6 (PPS Europe/Africa Regional Conference 2015, Graz, Austria, 21. – 25. September 2015). Graz, 2015

Mercedes-Benz diesel technology OM654 near-engine-mounted SCR system for WLTP and RDE

Tillmann Braun, Peter Lückert, Dr. Frank Duvinage, Dr. Alexander Mackensen

Daimler AG

1 Introduction

With the launch of the new 4-cylinder engine OM654 Mercedes-Benz continues to build on the advantages of the diesel combustion process and is opening up new potential in the area of emissions reduction and variant reduction.

The diesel engine has firmly established itself in all segments of the Mercedes-Benz product portfolio and thus the breadth of vehicle requirements, and application costs have increased substantially. To deliver innovations as quickly as possible in a host of variants, a standardized strategy is required which decouples the engine from the vehicle variance as effectively as possible.

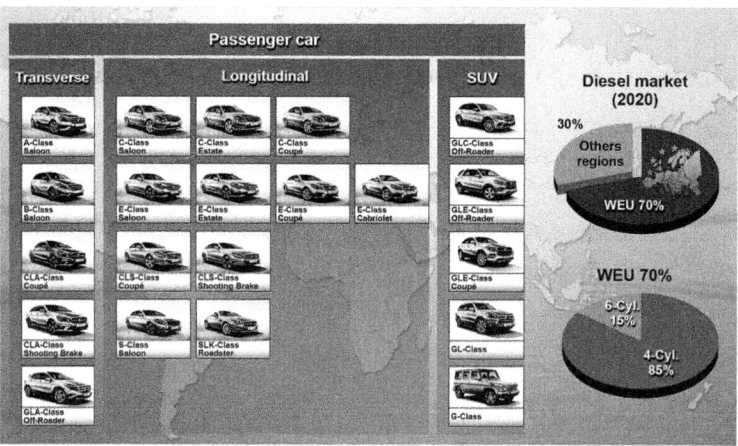

Fig. 1: Mercedes-Benz diesel vehicle portfolio

The OM654 ushered in a brand-new engine platform with a characteristic feature: all the key operating components are arranged directly on the engine:

- Engine with multiway EGR

- Charge air and EGR cooling

- Exhaust system (hot end) with all the components required for emissions reduction

As the first member of the "Mercedes-Benz Powertrain Architecture" the OM654 is thus compatible with all vehicle installation spaces in the Mercedes-Benz longitudinal architecture and can be integrated into all vehicle variants with minimal interface modifications.

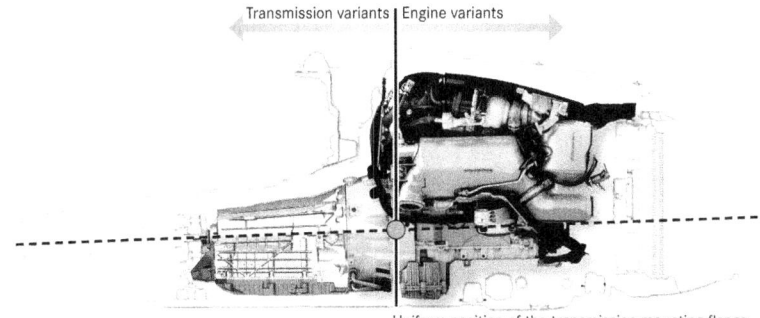

Fig. 2: Compact standardized engine OM654 with uniform vehicle interfaces

2 Key Development Areas: Functional Characteristics

In addition to a tangible improvement in agility and noise comfort, the reduction in fuel consumption and emissions constituted a particular development focus in a bid to make the diesel engine more compelling while also reinforcing its growing importance as a sustainable drive type.

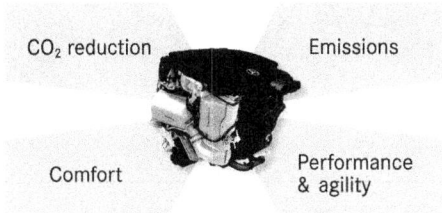

Fig. 3: Key development areas from a functional perspective

The OM654 in the presented variant features single-stage VNT turbo-charging and a displacement of 1950 cc. In conjunction with its new combustion process (stepped recess and 8-hole piezo-servo injector with up to 2050 bar injection pressure), an engine output of 143 kW and maximum torque of 400 Nm were achieved. For the E220 (213 series) with OM654 this means an extra 18 kW of power output compared with the OM651 (E220-212 series) and a power output per liter of 73 kW/l.

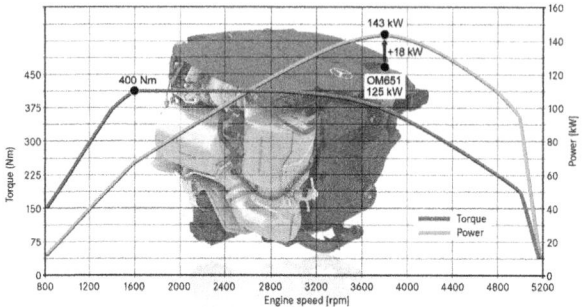

Fig. 4: Engine data

Despite the tangible increase in agility, fuel consumption was reduced significantly compared with the very good figures for the OM651 (E220-212 series). In-house measurements for the new E-Class with OM654 on the "AMS test circuit" reflected a reduction in test fuel consumption in excess of 10% compared with the outgoing model with OM651. This difference is comparable with the fuel consumption difference measured in the NEDC.

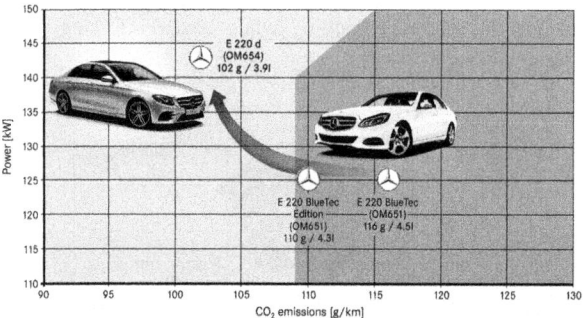

Fig. 5: CO_2 optimization

The presentation focuses on the emissions reduction of the new engine family, taking the single-stage launch variant of the OM654 by way of example. Engine and exhaust gas aftertreatment were designed as a single unit right from the requirement specifications stage of the new engine generation. The aim from the outset was to provide a future-proof, integrated system which delivers minimal emissions in virtually all operating spectrums.

Key components of this all-new emissions concept include:

- Low engine raw emissions through cooled high-pressure and low-pressure EGR
- Rapid light-off by integrating the particulate filter and all catalytic converters in the engine compartment
- Rapid AdBlue metering readiness after cold start through injection after DOC
- Dual swirl mixer with far better AdBlue evaporation and uniform distribution
- Rapid SCR implementation through SCR coating on DPF (sDPF)
- Sufficiently large DPF volume for high DPF regeneration intervals and, in turn, improved fuel consumption and catalytic converter ageing
- Additional SCR volume for high NO_X conversion at high engine load
- Complex model-based SW functions for maximum NO_X conversion and reliable prevention of NH_3 slip when operating at the performance limit of the catalytic converters
- Sensor-based NO_X conversion measurement with adaption of the metering quantity to offset tolerances
- Integrated control and monitoring functions for much better reliability and differentiated fault analysis in the event of error detection (OBD)

3 Emissions Reduction: Integration of Engine and Exhaust Gas Aftertreatment

3.1 Multiway EGR for Low NO_X Raw Emissions

The OM654 was systematically optimized for low NO_X and PM raw emissions. The engine features a multiway EGR, i.e. a cooled high-pressure EGR with switchable bypass and a cooled low-pressure EGR. The exhaust is removed downstream of the SCR catalytic converter, and, as such, the compressor is constantly fed almost entirely cleaned exhaust.

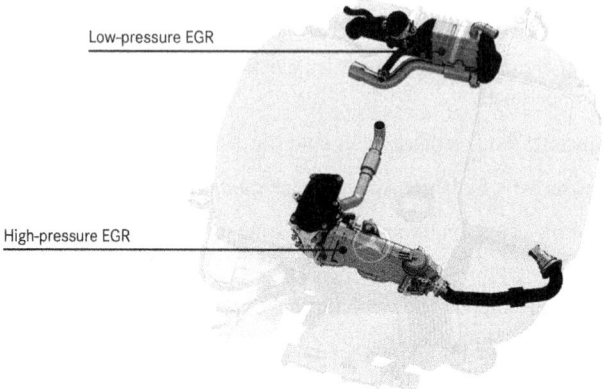

Fig. 6: High-pressure and low-pressure EGR system

Mounting the exhaust gas aftertreatment directly on the engine produced very short pipe routing in the low-pressure EGR path, see Fig. 6. In addition to low pressure losses, the EGR control quality and speed benefit in particular from the compact arrangement. The EGR rate is controlled on the high-pressure and low-pressure sides by means of electrically operated valves and can also be increased in the low-load range by an exhaust flap installed in the cold end.

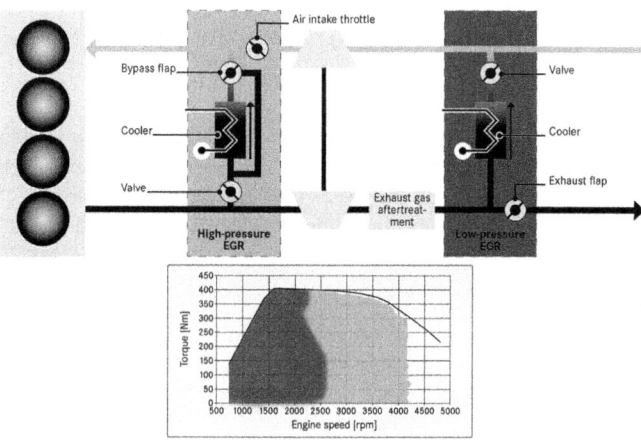

Fig. 7: Operating ranges of the high-pressure and low-pressure EGR

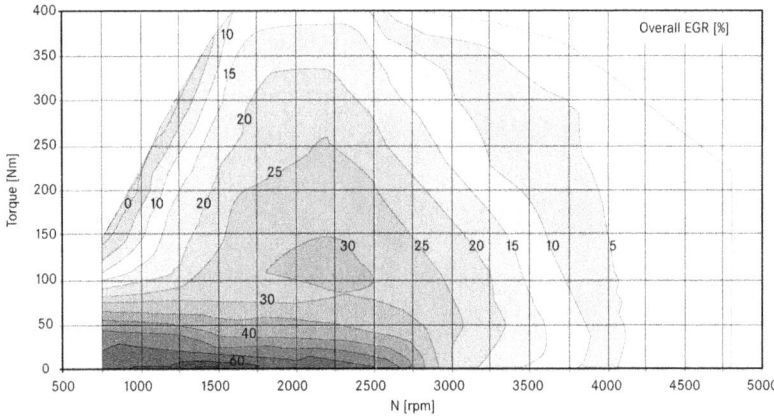

Fig. 8: EGR rate (combined value from high-pressure and low-pressure EGR) in the performance map

The combination of both paths allows the EGR region to be extended to nearly the entire performance map, see Fig. 7 and 8, as long as function and durability over lifetime are not affected (component protection). In this way the NO_X raw emissions through to the region close to full load (FL) are reduced substantially and can be adjusted very precisely even in dynamic operation thanks to the short paths and a model-based control strategy.

The low NO_X raw emissions of the OM654 thus make a substantial contribution to achieving low NO_X tailpipe emissions across a broad operating range.

3.2 Engine-Mounted Exhaust Gas Aftertreatment for Effective Emissions Reduction

On the OM654 all components relevant to emissions reduction are located directly on the engine. This minimizes heat loss and optimally facilitates rapid response of the catalytic converters. In order to accommodate the requisite substrate volumes in the available installation length, a very compact arrangement with fairly large substrate cross-sections was used and, in turn, the pressure loss reduced by over 40% compared with the previous engine OM651.

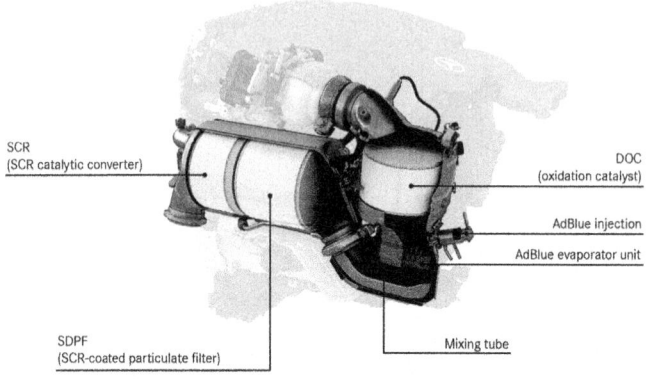

Fig. 9: Compact, engine-mounted exhaust gas aftertreatment

A SCR-coated particulate filter is used for the first time on the OM654. This arrangement facilitates AdBlue metering straight after DOC and requires very compact AdBlue treatment in the available installation space.

To ensure deposit-free AdBlue conversion and even distribution of the formed NH_3 across the fairly large substrate cross-sections, an all-new treatment concept was developed and successfully tested. Key components include a plate evaporator, which is arranged parallel to the flow due to pressure loss constraints, along with a downstream overlaying dual swirl, which provides thorough mixing and acts evenly on the sDPF cross-section. As part of extensive optimization steps, the NH_3 uniform distribution was increased to over 98% across a wide operating range.

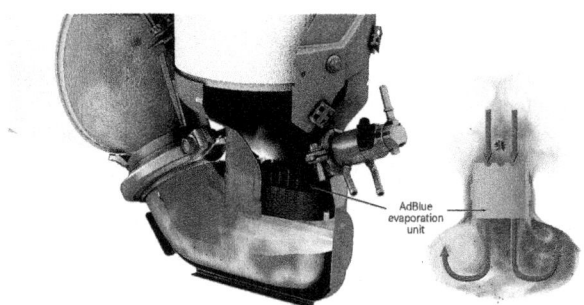

Fig. 10: AdBlue treatment

The homogeneous provisioning of high quantities of reduction agent in combination with the SCR catalytic converter arranged downstream of the sDPF facilitates high NO_X conversion rates across a wide operating range and, in turn, provides a standardized system for all model series in the Mercedes-Benz longitudinal architecture, i.e. from the lightweight vehicles (C-Class) to the large SUVs (e.g. GLS-Class).

4 Exhaust Gas Aftertreatment: System Design

4.1 Diesel Oxidation Catalyst (DOC)

The DOC in the OM654 constitutes the only oxidizing component in the exhaust system. Since the particulate filter of the OM654 is also used for the SCR function and, as such, has no oxidizing coating, the DOC must ensure the virtually complete HC and CO conversion as well as a sufficient quantity of NO_2 for an efficient SCR reaction at low exhaust temperatures.

The aim was to prevent as far as possible excess NO_2 (e.g. at low volume flows when new), and insufficient NO_2 (drawback for SCR conversion). To this end, a volume design with just under 1.95 l was chosen so that the volume flow influences are minimized and the full conversion capability is available across a wide massflow range. Through extensive screenings of different volumes and cross-sections as well as coatings and precious metal content, a design with high conversion performance (HC and CO) and stable NO_2 formation over the lifecycle was implemented, see Figure 11.

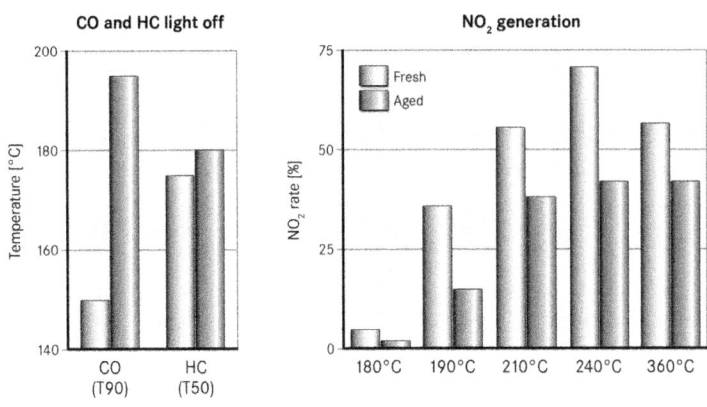

Fig. 11: DOC conversion and NO_2 formation

In addition to the aforementioned chemical conversions, the DOC coating also provides NO_X storage properties. In this respect, unlike with the classic NO_X adsorber converter, the NO_X is adsorbed in the catalytic converter structure and released at higher exhaust temperatures (from approx. 250°C). The stored NO_X is therefore not converted via rich combustion, but via an SCR reaction in the sDPF or SCR catalytic converter after desorption. In this way, the cold-start emissions are already effectively reduced before the start of the AdBlue metering. In the NEDC the total contribution of this effect is approximately 5%, in relation to the total raw emissions in the cycle.

4.2 SCR-Coated Particulate Filter (sDPF)

The relatively low ageing factor illustrated in Figure 11 with the example of the DOC is a basic requirement for good emissions across the vehicle's entire mileage. In addition to the catalytic converter design, ageing is determined substantially by the temperature and frequency of the DPF regeneration. With in-house measurements on the "AMS test circuit" a DPF regeneration interval of over 1000 km was established with the new E-Class and OM654; this means an increase of over 30% compared with other measured vehicles.

This example illustrates the consistent implementation of an integrated development of engine and exhaust gas aftertreatment in order to create an integrated system. Important decisions were made right from the concept phase regarding how to achieve high DPF regeneration intervals. In addition to increasing the particulate filter volume to 3.7 l, the new stepped recess combustion system significantly reduced particulate raw emissions.

With a focus on further improving urban emissions, Mercedes-Benz has been working for several years with substrate and coating manufacturers on optimizing and qualifying SCR-coated particulate filters. The following aspects needed particular attention:

- Particulate reduction at least on a par with previous cDPFs

  (challenge: in addition to the substrate, the deposition is influenced in particular by the coating)

- Maximum soot loading similar to existing cDPFs

  (challenge: the greater porosity of the substrate gives rise to a change in heat transfer and thermal mass)

- High SCR effectiveness

  (challenge: trade-off between washcoat mass and pressure loss)

Figure 12 shows the benefits of the sDPF concept under cold-start conditions and low engine load, taking the NEDC cycle by way of example. Through the AdBlue injection downstream of the DOC (on conventional systems it takes places downstream of the DPF) the required exhaust temperature of 150°C (sDPF average temperature) already exists after the first 2 to 3 km.

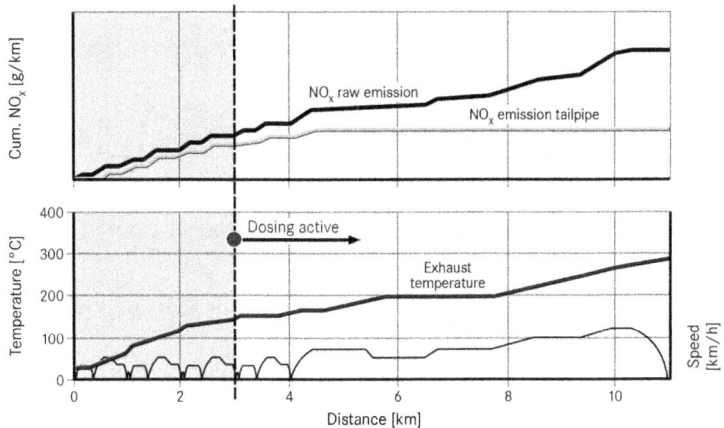

Fig. 12: NO$_X$ conversion of sDPF and SCR in the NEDC

The NO$_X$ reduction takes place straight after the start of the drive cycle and reaches a combined figure of 60%. In the ECE component of the NEDC, the NO$_X$ is only reduced at the sDPF. The particulate filter can be operated with fairly high NH$_3$ levels and, in turn, good NO$_X$ conversion, since desorbing NH$_3$ from the downstream SCR catalytic converter can be captured and converted fully with a sudden increase in temperature and throughput (e.g. transition to EUDC). This option allows for fairly high NO$_X$ conversion rates right from the sDPF and prevents NH$_3$ slip.

4.3 SCR Catalytic Converter (SCR)

The importance of the downstream SCR catalytic converter becomes even clearer taking the WLTC as an example. The sDPF accounts for the bulk of the NO$_X$ reduction in the low-load phases; the SCR conversion increasingly switches to the downstream SCR catalytic converter during acceleration and with high exhaust mass flows. With a volume of 2,8 l the SCR catalytic converter is sufficiently large to reduce significantly the NO$_X$ emissions across a broad operating spectrum. In the WLTC the OM654 achieves a NO$_X$ conversion rate of 86%.

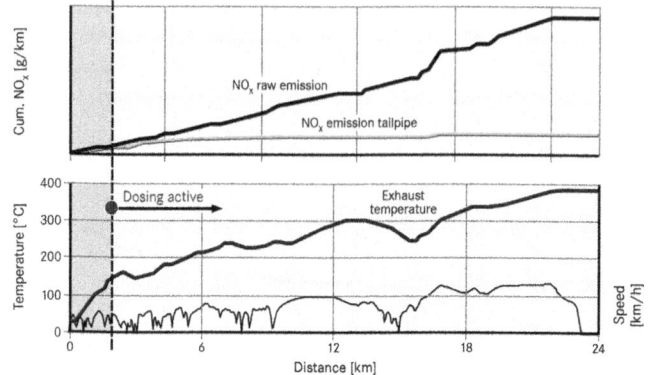

Fig. 13: NO$_X$ conversion of sDPF and SCR in the WLTC

5 Exhaust Gas Aftertreatment: Sensor System and Functions

5.1 Tolerance Analysis and Sensor Concept

In addition to the system design, the control of the SCR system and the function algorithms and models in particular are decisive for achieving high NO$_X$ reduction rates. In this respect, tolerances and ageing influences must also be taken into account as well as the large spectrum of possible usage scenarios and constraints. Achieving NO$_X$ conversion rates of well above 80% means that the SCR system must be operated permanently and in all operating conditions at the performance limit and false diagnoses must also be avoided on the OBD side, as well as AdBlue deposits or NH$_3$ slip.

At the start of system development, the key component and model tolerances were analyzed. Initially, the individual influences of the key parameters on emissions behavior and system stability were examined. The following parameters were identified as key influencing variables:

- Input value NO$_X$ raw emissions (model value vs. NO$_X$ sensor)

- Accuracy of exhaust mass flow

- Accuracy of AdBlue metering

- AdBlue treatment and uniform distribution

- Accuracy of temperature values at various points of the catalytic converters

In the second step, the tolerance bands and statistical frequencies of the key individual influences on measurements were determined. Finally the various influences were combined using the "Monte Carlo method" in accordance with their statistical frequency and assessment of the overall influence on system behavior.

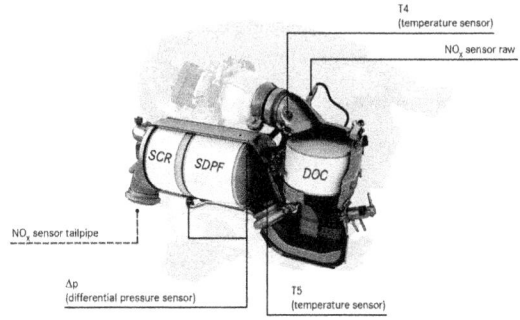

Fig. 14: Sensor concept of the exhaust gas aftertreatment unit

The following decisions in the concept phase resulted from these tests:

- Precise measurement of the NO_X raw emissions using NO_X sensor upstream of DOC
- Implementation of lower tolerance limits for AdBlue metering system
- Model-based conversion control with cyclically triggered model comparison
- Monitoring of the NO_X conversion by two NO_X sensors
- Adaption of the AdBlue metering quantity on the basis of the aforementioned NO_X conversion measurement
- Temperature sensor upstream of DOC and upstream of sDPF
- Modeling the surface temperatures inside the sDPF and SCR

5.2 Model-Based Metering Quantity Control and OBD

To achieve high NO_X conversion rates, the adsorption of ammonia in the catalytic converter is crucially important. Since the adsorption capacity of the SCR coating is a function of exhaust temperature and volume flow, the stored quantity ("NH_3 level") must be controlled as precisely as possible and the AdBlue metering varied accordingly. During acceleration where the adsorption capacity falls due to the increase in temperature and throughput, NH_3 emissions (NH_3 slip) must be avoided without reducing the NO_X conversion. These driving maneuvers are highly relevant to emissions behavior since very high NO_X raw emissions can be produced in these phases.

A model-based NH_3 level control system was developed for the OM654 which calculates the NO_X conversion and the adsorbed NH_3 quantity on the basis of physical, chemical models. The calculations are performed locally resolved over the length of the catalytic converter (1D model). In order to calculate the complicated SCR reaction kinetics in the control unit, the calculation algorithms were modified and simplified accordingly.

The key input variables for the model calculation are:

- NO emissions
- NO_2 emissions
- NH_3 concentration
- O_2 concentration
- Exhaust gas mass flow rate
- Exhaust gas temperature

In this way, both the NH_3 level and its local distribution are calculated. On the basis of this data, the AdBlue metering quantity and timing for NH_3 level build-up and reduction are determined exactly and hence "running empty" or "overflowing" of the catalytic converter prevented.

Intelligently linking sensor and model values allows the system to be operated stably at high conversion rates, while also monitoring its function. The same functions that control the system also form the basis for the OBD concept. Thus the OM654 already fulfills the stringent EU6-2 OBD limits at the start of series production.

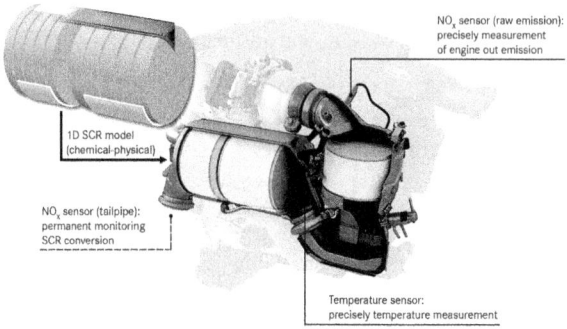

Fig. 15: Model-based NH_3 level control with overlaid adaption and OBD

6 Emissions Results

Through the multiway EGR in conjunction with the engine-mounted exhaust gas aftertreatment and the systematic further development of AdBlue treatment and control algorithms, NO_X reduction efficiency was increased substantially compared with the first- and second-generation BlueTEC systems.

The initial prototype vehicles were already subjected to extensive PEMS measurements and each development step systematically assessed on a test circuit with a hybrid usage spectrum. Known as the "Stuttgart PEMS circuit" the circuit used includes a 48% city, 21% intercity and 21% highway component. The vehicles are started cold in Untertürkheim and initially drive through Stuttgart city center in low-load operation and stop-and-go traffic for approximately 20 to 25 minutes. Starting with a low exhaust temperature level, a climb follows within the urban area ("Weinsteige") at fairly high load and followed straightaway by renewed cooling of the exhaust system by a downhill section back into the city. On the rural section, the vehicle is driven on a hilly section at an average speed of approximately 77 km/h and then accelerated very quickly on the highway up to the maximum speed of approximately 140 km/h. Before the transition back into urban mode, there is a lengthy downhill section where the exhaust system cools down once again. The route ends after driving once more through Stuttgart city center back to Untertürkheim.

Overall the route covers 84 km. Depending on the traffic density and driving style, the average speed is approximately 45 to 50 km/h. The combined uphill climb is approximately 1050 m.

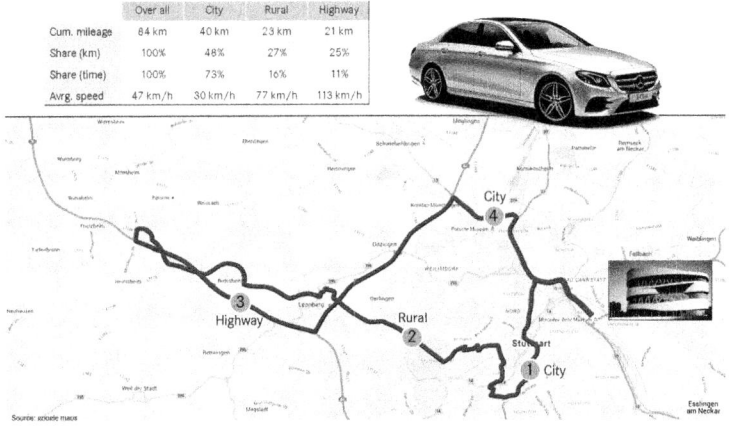

Fig. 16: "Stuttgart PEMS circuit" for assessing off-cycle emissions

Figure 17 illustrates the speed profile and the exhaust temperature curve during measurement on a "Stuttgart PEMS circuit" with a moderate driving style, along with NO_X raw emissions and NO_X tailpipe emissions. An overall good result of 89% NO_X conversion was achieved. If you only examine the city component, 84% NO_X conversion was achieved for this section.

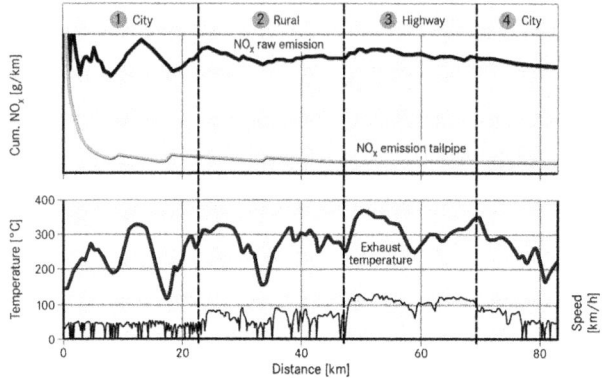

Fig. 17: Emissions on "Stuttgart PEMS circuit", E220 with OM654

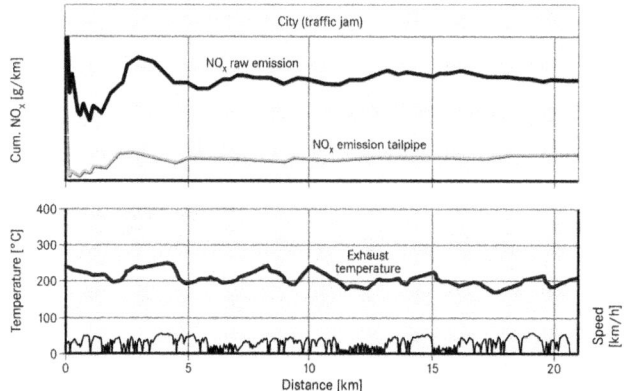

Fig. 18: Emissions for urban traffic, flat topography, stop-and-go, E220 with OM654

In addition to the "Stuttgart PEMS circuit", various route scenarios were completed with different driving styles and traffic density. Figure 18 illustrates by way of example an urban journey with almost flat topography in stop-and-go traffic. The average speed was less than 8 km/h. Despite the low engine load and numerous stationary phases, the AdBlue metering was achieved throughout the entire route and the NO_X total conversion rate was approximately 74%.

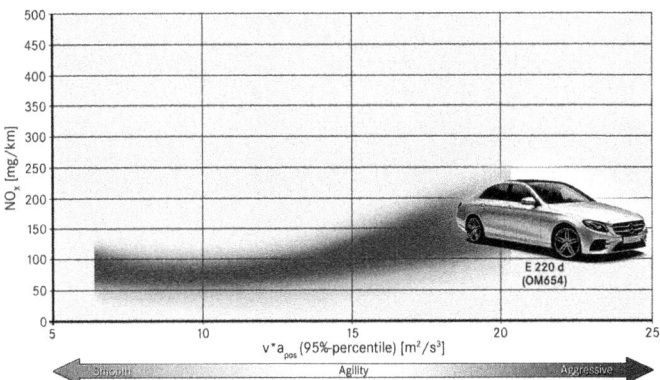

Fig. 19: Emissions from various scenarios with different driving styles, E220 with OM654

Figure 19 illustrates the emissions behavior of the OM654 in various route scenarios and with different driving styles along with the two measurements explained in Figure 18 and 19 by way of example. This produces an emissions range since the emissions are determined by the criterion "v x a_{pos}" as well as the topography, the vehicle weight and the environmental conditions (outside temperature, air humidity…) in particular.

7 Summary and Outlook

A host of measurements under different test conditions, scenarios and driving styles confirm the low emissions of the OM654 and the advantages of the all-new overall concept. In addition to the hardware configuration and the integration of engine and exhaust gas aftertreatment, the model-based instrumentation and control algorithms with overlaid metering quantity adaption as well as the integrated OBD functions form the basis for robust emissions reduction in a host of usage profiles.

The first member of the new engine generation at Mercedes-Benz will be launched with the OM654 in the new E-Class. The standardized concept lays the foundations for the rapid rollout of the new engines in a host of vehicle variants.

The gasoline engine and RDE challenges and prospects

G. Fraidl, P. Kapus, K. Vidmar

AVL LIST GMBH
Hans-List-Platz 1
A-8020 Graz

Abstract

Whereas in the past decade, the CO_2 reduction was the major driver for technology development with Gasoline en.gines, especially after the "Dieselgate" an "Extended Emission Compliance" including Real Drive Emissions is taking significant impact on future technology routes. Whereas the RDE legislation initially was targeted primarily towards the Diesel-NOx and the Gasoline-PN emissions, now also the NOx emissions of some Gasoline engine concepts have to be reduced.

In a very simplified way, the most significant RDE risk areas of Gasoline engines can be seen in following areas: scavenging at low engine speeds, enrichment and /or exceeding favorable catalyst space velocities at high engine speeds / loads, high load dynamics, in-field stability of PN emissions, insufficient catalyst temperatures and start / restart strategies (esp. Hybrid).

As most measures improving the RDE emission show a trade off with CO_2 and / or cost, the future RDE legislation is both a technical but also a commercial challenge. On the other hand, RDE also opens opportunities for the Gasoline engine to increase market shares as with the Diesel engines in some vehicle categories the effort for RDE compliance will be even higher than with Gasoline engines.

It is expected that the RDE requirements will enhance the trend from "Extreme Downsizing" towards "Rightsizing", which was already initiated by modern fuel economy technologies like Miller or Atkinson Cycle. In a next step, also the combination of 48Volt systems with the next generation of Gasoline engines will become a highly competitive technology combination.

From a cost / CO_2 reduction perspective, however, both CNG and LPG are most attractive technology approaches to reduce CO_2 and pollutants simultaneously. LPG might be considered as the more simple short-term approach, CNG as the more sustainable long-term solution. With both approaches, the key challenges are not at the pure technical side, but primarily with marketing to gain sufficient customer acceptance.

To balance RDE, CO_2 requirements and cost will become one of the most challenging engineering tasks and will require new development and validation approaches. Thus, RDE will not only have a significant impact on powertrain technology, but might also initiate a paradigm shift in powertrain development.

1 The starting point

Comparing the development trends for passenger car powertrains e.g. in Europe and in US in the last decades, quite different priorities are obvious, Fig.1.

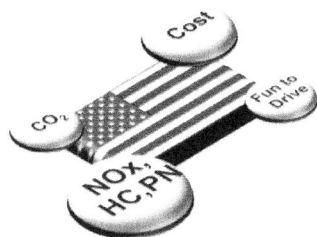 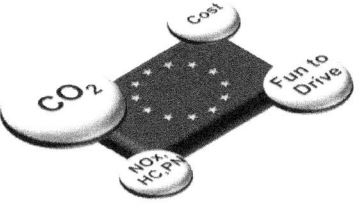

PAST
- Focus on POLLUTANTS
- Compliance = lowest emission under all conditions
- Sufficient real world emission

FUTURE
- "RDE" just to identify defeat devices"

PAST
- Focus on CO_2
- Compliance = fulfillment of test cycle
- Increasing discrepancy between Certification and real world emission

FUTURE
- Complex RDE legislation to improve the air quality issue

Fig. 1: Main Technology Drivers in US and Europe

For a long period, in the US the powertrain development was clearly focused on the pollutant emission to provide lowest emission under a broad range of operation conditions even compromising CO_2. This resulted in improved air quality, however also in a poor CO_2 scenario with a fleet average significantly higher than in Europe.

In the same time frame Europe showed a nearly diametrically opposed orientation: clear priority on aggressive CO_2 reduction, however, the improvement of pollutant emission more concentrated on a specific test cycle. Recognizing an offset between aggravated certification limits and actual improvements in air quality, the EU started to formulize quite comprehensive RDE requirements that, at least in view of development efforts will even exceed the US legislation. Recent irritations originating from the "Diesel-gate", however, pushed the focus of the discussions from a realistic technical basis more towards "political requirements" potentially imposing unreasonable risks on the automotive industry.

2 RDE Challenges

With Heavy Duty vehicles, RDE is already a broad standard since several years. However, the boundaries are completely different from passenger cars. Whereas with Heavy Duty vehicles, the mean power with RDE driving is quite equivalent to the test cycle and in the order of half the maximum vehicle power, the characteristic power consumption with RDE driving of passenger cars is nearly an order of magnitude lower, Fig.2. Whereas the statistically most relevant values are in the range between 4-6 %, especially underpowered SUV show higher values (> 10%) and high powered sports car lower values (< 4%). As a consequence, with increasing relative power (max engine power / vehicle weight), the impact of driving style increases and has to be considered with RDE evaluations.

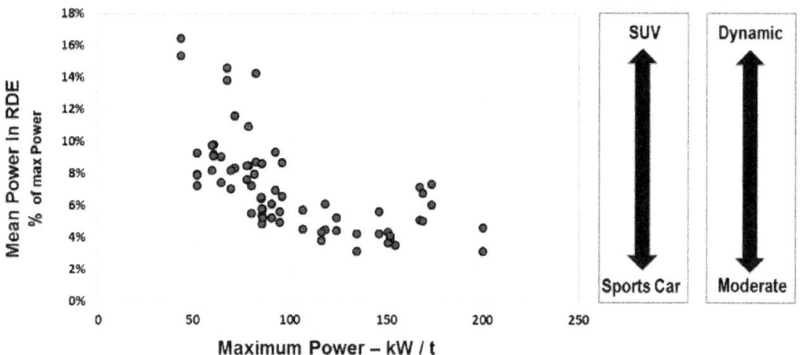

Fig. 2: Relation between mean and maximum power with RDE driving, Passenger Cars, Sports Cars, and SUV

2.1 RDE Routes

An important impact on the RDE result is originating from the topography and the velocity profile of the test route. In a simplified way, different RDE-routes can be characterized e.g. by two parameters describing firstly the vehicle speed profile („Track Dynamic Index") and secondly the power requirement (Track Severity Index). These parameters combine the velocity profile, the vehicle characteristics and the topography to enable a simple comparison of different test routes, Fig.3

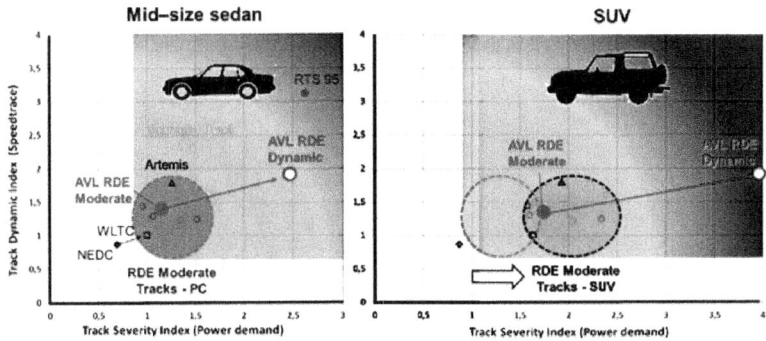

Fig. 3: Characterization of RDE test routes and velocity profiles [1]

At least for standard passenger cars and moderate driving style, the majority of the test routes are allocated within a comparable small scatter band, Fig. 3 left hand side. Such operating profiles are statistically representative for typical European customer driving profiles. The AVL RDE test route is well within the scatter band of these test routes. It has also become obvious, that aggressive driving in general has a much more significant impact on the RDE test severity than the differences between the respective test routes.

Applying the same test routes and velocity profiles for a car with less favorable power to weight ratio and higher driving resistance (power requirement for a SUV instead of a passenger car), even with moderate driving, the track severity index is shifted towards unfavorable values, Fig. 3 right hand side. For the different test tracks, this shift occurs in a quite comparable way, but with a larger severity spread compared to the passenger car. However, with a dynamic / aggressive driving style, depending on the altitude profile, the track severity index is significantly shifted towards higher values. Thus, an aggressive driving style like used in the RTS 95 does not represent a statistically relevant European customer driving profile, but rather a borderline assessment.

A more complex impact of the test route is found with Plug-In Hybrids. With such concepts, the power demand with starting the ICE can have a quite essential impact on the RDE emission.

2.2 Driving Style

As already indicated, the most significant impact on the RDE emission results from the driving style [2]. Comparing turbocharged GDI engines in different vehicles – Sports Car, Downsized SUV and Plug-In Hybrid – the driving style has quite different influence on the speed / load distribution, Fig. 4.

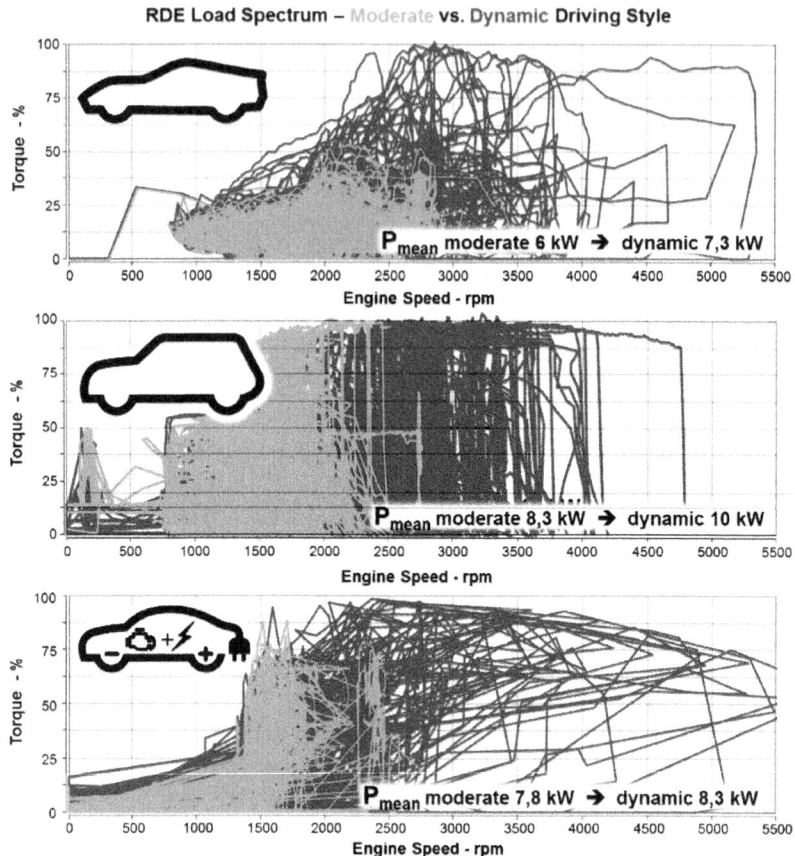

Fig. 4: Speed / load distribution with different Driving Style, Sports Car, SUV and PHEV,
all with TGDI engines

As traffic conditions and speed limits are restricting the driving on public roads, the
mean power requirements do not vary dramatically with the different driving styles.
However, there is quite a huge difference in engine dynamics resulting in spikes regarding engine torque, engine speed and consequently in maximum required engine power.
Thus, the coverage of the whole engine map is quite sensitive to the driving style.

With the Sports Car, moderate driving is concentrated just at the low part load area; neither high engine loads nor high speeds are utilized at all. However, with dynamic driving, short spikes in engine torque and speed are covering the whole engine map up to full load.

With the extremely downsized SUV (TGDI, 60 kW/t), the relation between vehicle weight and displacement in combination with a fuel economy oriented transmission setting shift the engine operation significantly towards high engine loads at low engine speeds. This is an excellent concept for best CO_2 emission in NEDC, however, already with RDE moderate driving a high share of full load at low engine speeds can be noticed. Thus, the impact of dynamic driving is more a shift of engine speed than of engine load.

With Plug-In Hybrids, the impact of driving style is strongly dependent on the Hybrid type and the respective operation strategy. In this example (P2 Hybrid with DCT, 70 kW/t), both engine speed and load are quite restricted with moderate driving style.

All these vehicles, optimized for EU6b, but not for RDE, show a dramatic impact of driving style on RDE emission. With the Sports Car and moderate driving style, the NOx and PN emissions would be even close to EU6c. However, when changing to dynamic driving style the NOx emission in RDE increase by a factor of 5 reaching the legal limit, the PN emission increases even by a factor of 14 reaching values by far outside any future conformity factor for PN, Fig. 5.

A quite significant increase of emission with dynamic driving style is generated in the uphill driving phases with high vehicle dynamics, where even with the highly powered Sports Car (175kW/t) short full load passages can be monitored.

The extremely Downsized SUV shows a quite different behavior, Fig. 6. This vehicle/engine combination is characterized by an extreme shift of engine operation towards high engine loads and low speeds.

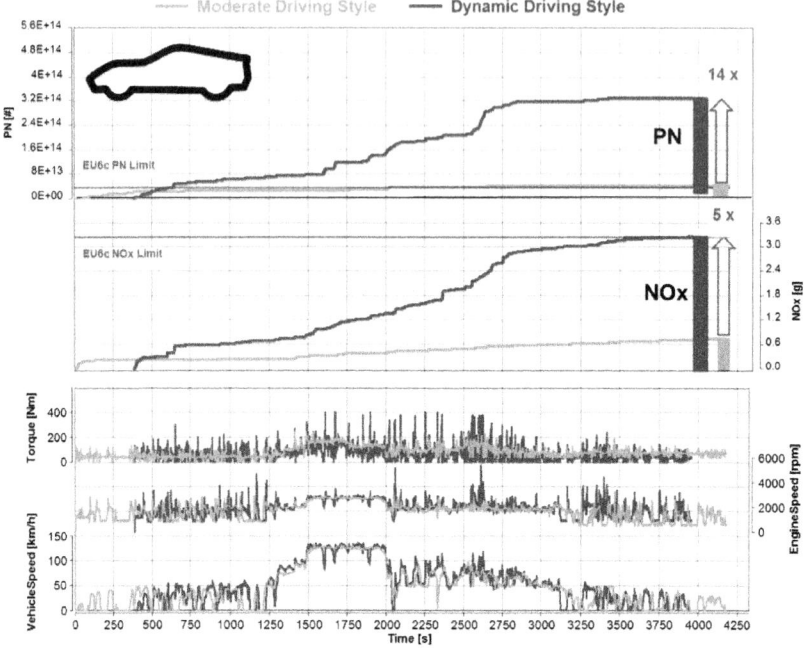

Fig. 5: Impact of Driving Style on RDE-Emission, Sports Car

To improve the drivability in this operation scheme, the TGDI is calibrated with high scavenging. While in an NEDC test this impact is within an acceptable range, with RDE driving this results in excessive NOx emission even with moderate driving style. Dynamic driving shifts the engine operation towards higher engine speed, thus the share of operation within this area is reduced partially resulting in even lower NOx.

The low specific power to weight ratio of this vehicle/engine combination results in high engine dynamics even with moderate driving. In combination with high PN emission especially at low engine speeds, PN is high even with moderate driving.

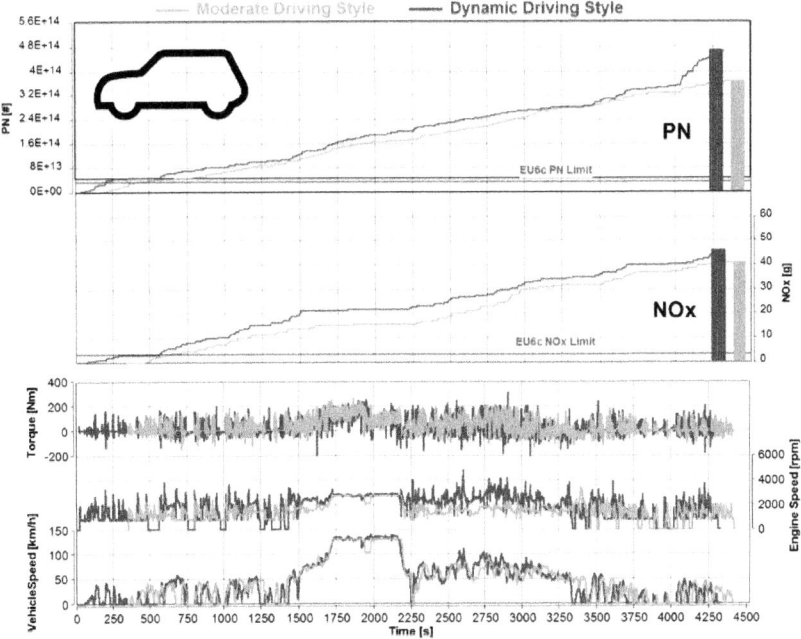

Fig. 6: Impact of driving style on RDE- Emission, extremely downsized SUV

The most complex impact of driving style on the RDE emission emerges with Plug-In Hybrids. As typically the city driving is done in pure electric operation, the actual power demand is essential for the whole RDE emission when the ICE is switched on, Fig. 7.

In this example, with moderate driving the start timing of the ICE is triggered by low battery SOC during downhill driving within relative moderate power requirements. The ICE start generates a spike in PN and NOx emission, however, in this case the impact is not critical in view of the total RDE emission.

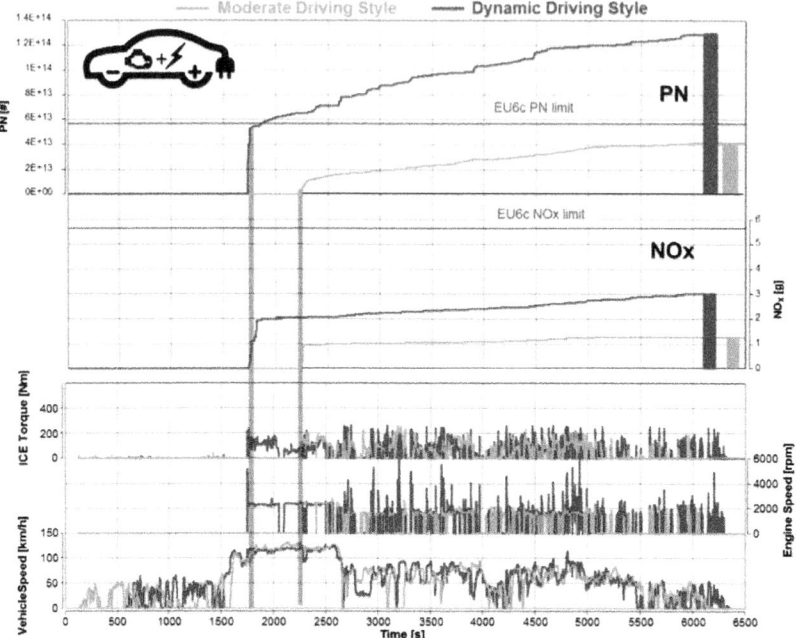

Fig. 7: Impact on Driving Style – PHEV + TGDI

With dynamic driving and resulting higher electric consumption, the ICE is switched on earlier, in this example due to power demand during an uphill high full load acceleration at the highway. Consequently, the resulting much higher emission spike has an essential impact even on the averaged total RDE result. The EU6c PN limit is reached already within this single start/acceleration phase. Thus, especially with Plug-In Hybrids, both topography and traffic condition can have a quite essential, highly stochastic impact regarding driving style. Specific, partially predictive algorithms for starting and ramping up the combustion engine are required.

Fig. 8 shows the impact of driving style on RDE-emissions for a representative sample of passenger cars including SUV´s, Sports Cars and Plug-In Hybrids.

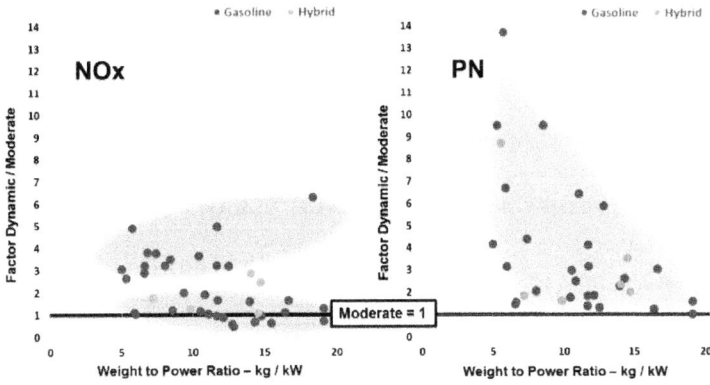

Fig. 8: Impact of driving style on RDE-Emission

Interestingly the NOx emissions on the first sight show no clear dependency on the weight to power ratio of the vehicle, Fig. 8 left hand side. A closer look reveals that for conventional powertrain solutions, two groups of vehicles can be differentiated regarding the main source of NOx emissions. Those vehicles, where the NO_x emission originates primarily from to exceeding critical catalyst space velocities (see also chapter 3), a quite distinct impact of the driving style can be found – the NOx emission with dynamic driving style are higher by a factor 3-6 than it is with moderate driving.

If an essential part of the RDE-NOx emission is generated due to scavenging strategies, the impact of the driving style on the nominal NOx is not very high compared to the previous group of vehicles. As usually dynamic driving shifts the engine operation towards higher engine speeds, the share of scavenging operation is reduced and consequently the NOx emission can be even lower than with moderate driving.

Usually the PN-emission is linked to engine dynamics. On the contrary, with a high weight to power ratio, vehicle dynamics are limited and consequently the driving style has less influence, whereas with highly powered vehicles a quite large scattering can be found, Fig. 8 right hand side.

2.3 Evaluation

From the beginning of the RDE discussions, it was a clear requirement that legal RDE regulations have to be based on statistically relevant boundaries. This is valid for ambient and engine temperature, routes incl. topography and altitude, vehicle condition, but especially velocity profile and driving style. Consequently, respective evaluation tools were designed and continuously improved. As the main focus of the initial RDE activities was centered at the Diesel engine, the effectiveness of these tools for Gasoline powered vehicles has to be evaluated separately.

Fig. 9 shows the effect of the evaluation tools EMROAD [3] and CLEAR [4] as well as the dynamic limitation via v*apos [5] with a sports car.

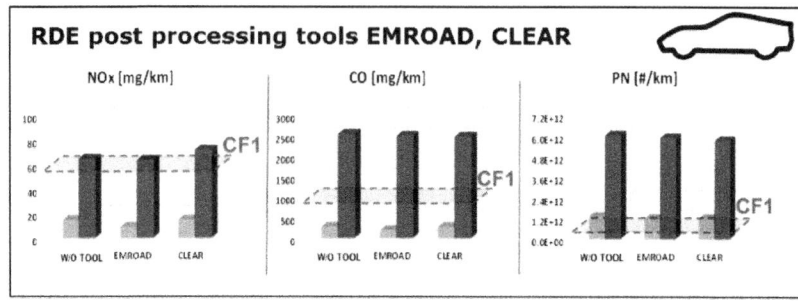

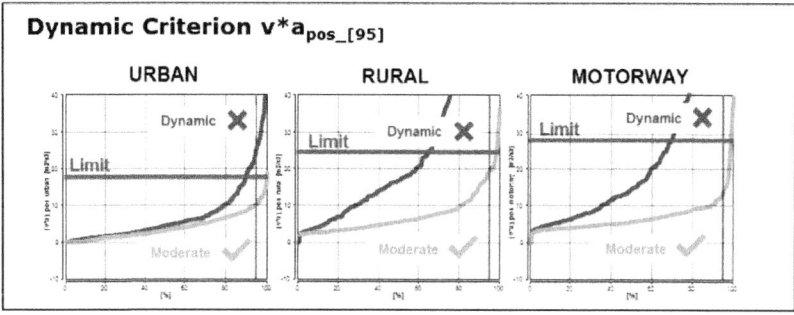

Fig. 9: Effect of Evaluation Tools and Dynamic Limitation on RDE Results – Sports Car

A highly powered Sports Car was driven both in a moderate driving style – representative for European driving style and in an extreme dynamic, even aggressive driving style just keeping the legal limits but largely utilizing the power potential (both driving styles not including unlimited highway speeds). With this Sports Car neither EMROAD nor CLEAR post processing tools show any significant impact on the NOx, CO or PN results. Even aggressive driving style would not be compensated.

The situation looks quite different regarding the dynamic limitation. The aggressive driving style was reliably detected both in city, rural and motorway operations. In this case, the extremely aggressive driving would not be a valid RDE test.

As expected, a different weighing reaction of the evaluation tools is found with the moderately powered SUV, Fig 10.

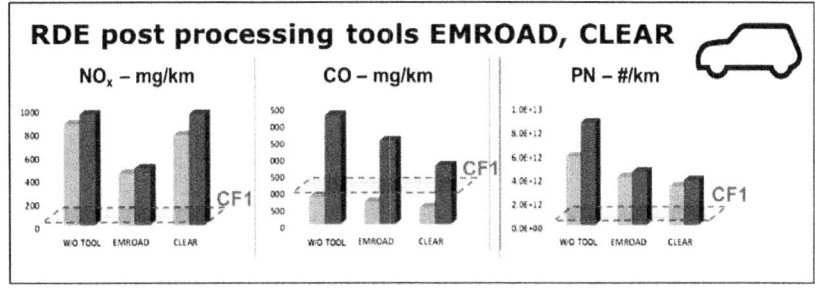

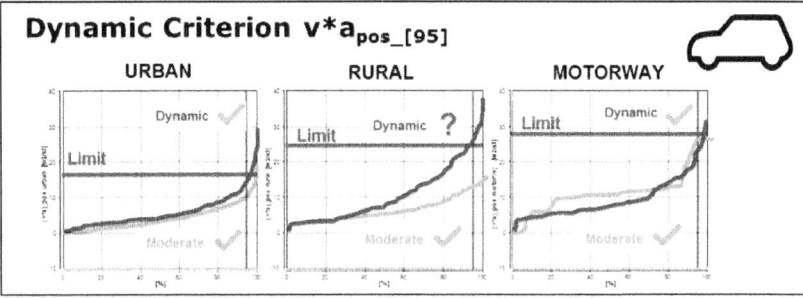

Fig. 10: Effect of Evaluation Tools and Dynamic Limitation on RDE Results –
Extremely down-sized SUV

Whereas with regard to PN both evaluation tools compensate dynamic driving style, with regard to NOx and CO the correction is to be questioned. Due the high weight to power ratio of the vehicle and the consequently limited absolute acceleration, aggressive driving style is not anymore detected as such.

Also on a statistical basis, the effectiveness of the evaluation tools with Gasoline engines seems to be limited, Fig.11.

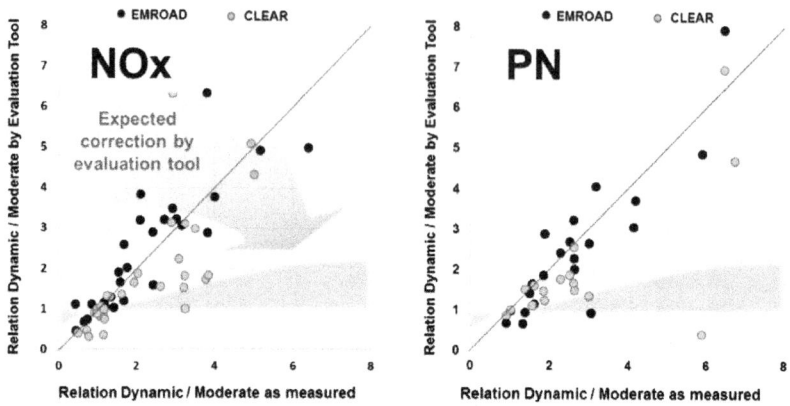

Fig. 11: Relation between dynamic and moderate Driving with and without evaluation tools

When comparing the differences between emissions in moderate and dynamic driving style by applying an evaluation tool that is supposed to compensate the impact of driving style, the differences should get smaller, in an ideal way they should move towards the green horizontal area at a relation equal 1. However, this trend can be seen just with a part of the vehicles investigated. Thus, the impact of driving style remains a major concern area for RDE development.

Fig. 12 gives an overview on PN emission with state of art Passenger Cars, SUV's, Sports Cars and PHEV's in the AVL Graz RDE-Route. Several most recent vehicles comply with low conformity numbers at least at low mileage and sufficient fuel quality. Nevertheless, especially with dedicated downsizing and high vehicle weight, compliance factors of up to 10 can be found. Thus, the relation between vehicle weight and engine displacement can be seen as a statistically relevant parameter.

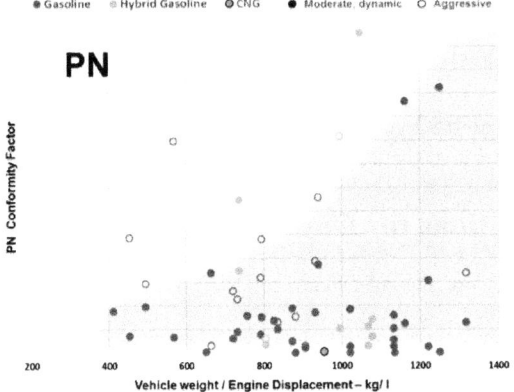

Fig. 12: PN Emission in RDE, Passenger Cars, SUV's and Sports Cars

As already stated in chapter 2.2., under specific driving conditions also Plug-In Hybrids might show quite high PN. This not only valid for PHEV's with Direct Injection Gasoline engines but also for PHEV's equipped with MPFI engines.

Whereas the PN challenge with TGDI is common knowledge, RDE-NOx issues were primarily attributed to the Diesel engines. As the dominating majority of Gasoline engines are pure stoichiometric ones, due to the excellent conversion efficiency of the 3-way catalyst, NOx emission should not be any issue, <u>Fig.13.</u>

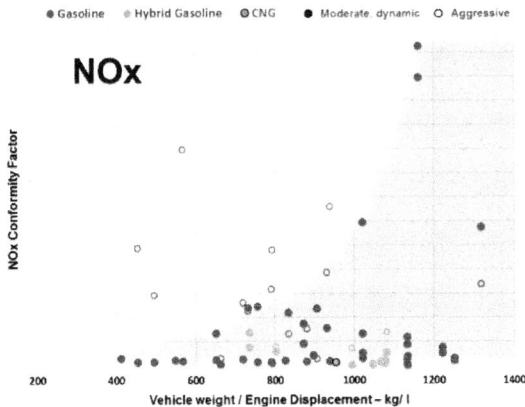

Fig. 13: NOx Emission in RDE, Passenger Cars, SUV's and Sports cars

However, similar to PN, also with regard to NOx the relation between vehicle weight and engine displacement seems to be the statistically most relevant parameter. Especially the combination of extreme downsized TGDI with high vehicle weight is most critical.

As with Hybridization the additional electric torque enables reduced scavenging, NOx is not an issue for the investigated Hybrid and Plug-In Hybrid vehicles.

3 RDE Solutions for Gasoline Engines

In a very simplified way, the most significant RDE risks can be classified in following categories:

- Scavenging at low engine speeds to improve low end torque

- Enrichment at high engine speeds / loads for component protection

- Exceeding favorable catalyst space velocities with high exhaust gas flow

- Extreme load dynamics

- In-field stability of particulate emissions

- Insufficient catalyst temperatures

- Start / Restart strategy (esp. Hybrid)

To a certain extent, these critical RDE emissions are a consequence of measures implemented for fuel economy improvement in a legal test cycle, especially as a result of extreme downsizing and fuel economy oriented powertrain concepts. Thus, measures to improve the Real Driving Emissions most often show a clear trade-off with CO_2-emissions in the NEDC test.

Fig. 14 shows characteristic RDE emissions traces for a turbocharged downsizing concept and the classification of their origin.

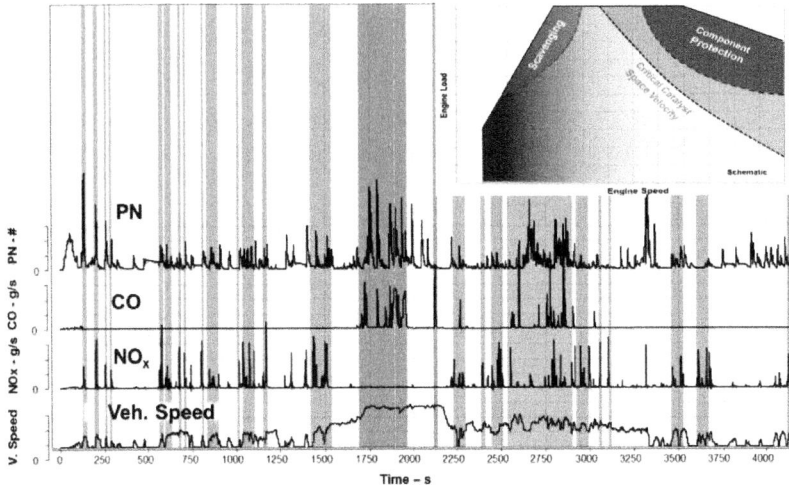

Fig. 14: Key RDE concern areas with a downsized turbocharged GDI engines

The blue bars show issues attributed to scavenging (NOx breakthrough), red bars show issues related to component protection (CO and PN breakthrough) and green bars show issues with insufficient catalyst size (breakthrough of all components).

3.1 Scavenging

Particularly in drive-away and acceleration / gear shifting events in the engine speed range between 1500 – 2500 rpm, the lean exhaust gas from scavenging exceeds the oxygen storage capacity of the catalyst and results in NOx spikes, Fig.15.

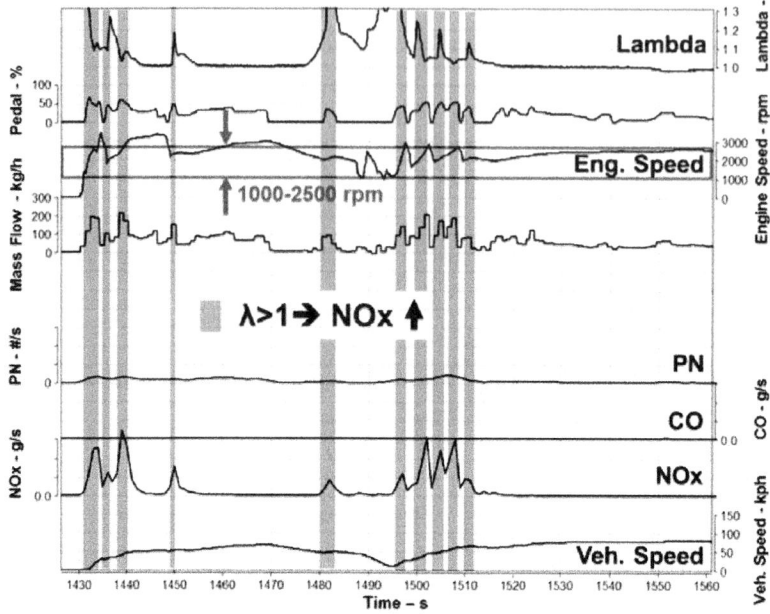

Fig. 15: Downsized turbocharged GDI: Impact of scavenging on Real Driving NOx-Emissions

Scavenging strategies were introduced as enabler for efficiency optimized downsizing and downspeeding concepts and are characteristic for fuel economy concepts with turbocharged GDI engines. Especially with vehicle concepts that include the combination of extreme downsizing, an efficiency oriented transmission strategy and unfavorable power to weight ratio, scavenging based NOx occurs already with a moderate driving style.

In general, with turbocharged engines a clear trade-off between transient response and high load /speed performance can be observed. Fig. 16 shows the trade-off between transient response – characterized by boosted torque gradient (BTG) – and maximum power. The boosted torque gradient is calculated by the torque increase above the naturally aspirated torque normalized by engine displacement.

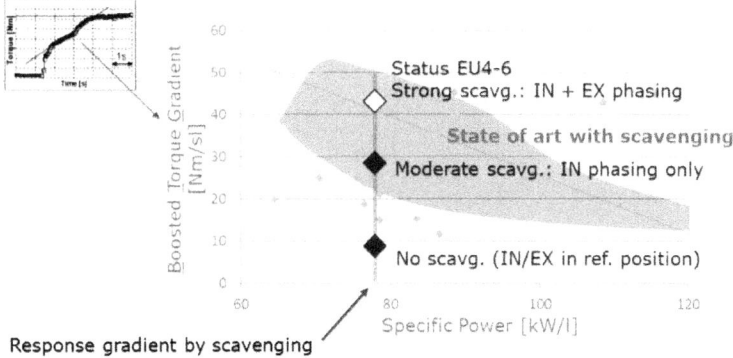

Response gradient by scavenging

Fig. 16: Tradeoff between transient turbocharger response and max. power – TGDI

Avoiding the scavenging completely by modified calibration would solve the NOx issue, however, also result in reduced boosted torque gradient, compromised engine transients and consequently drivability issues especially with fuel oriented transmission settings. The practical experience with such approaches shows both a significant deterioration of real world fuel economy due to a more frequent use of lower gears as well as compromised market acceptance of such concepts.

One alternative would be to apply scavenging within stoichiometric exhaust lambda calibration, Fig.17. However, this NOx reduction measure also causes rich combustion chamber lambda conditions and related CO_2 and PN drawbacks as well as exothermal reactions in the catalyst. To avoid catalyst degradation, a precise modelling of the exothermal reactions and the impact on local catalyst temperatures has to be applied.

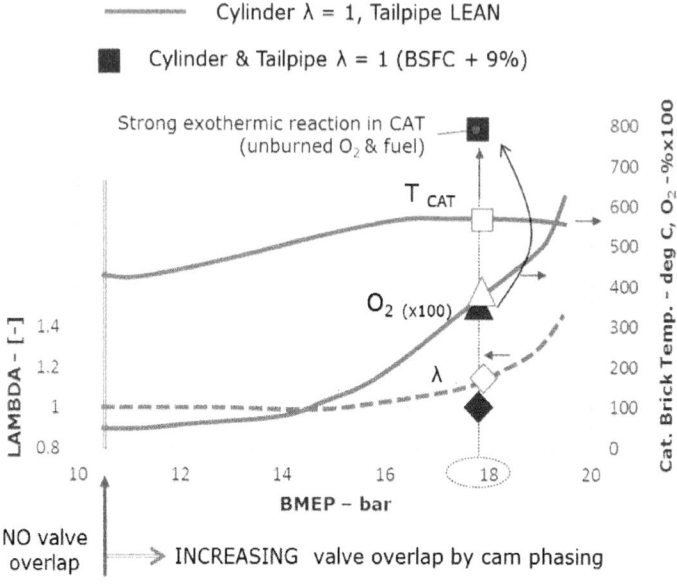

Fig. 17: Scavenging strategies with TGDI

Fuel economy neutral CO_2 solutions like electric supercharger or VTG are cost critical especially in in the lower market segments, however, increasing applications of such solutions are expected. The upcoming RDE regulation with its technological requirements is promoting such systems and might help to ramp up volumes earlier than expected in the past.

Cost neutral solutions such as larger displacements or shorter gearing, however, result in both deteriorated real world fuel economy as well as higher CO_2 certification numbers.

When determining the compression ratio, the ongoing efficiency improvement with Gasoline engines results in an increasingly complex balancing between part load and full load requirements. With most applications, the compression ratio is adjusted on the highest possible level so that within typical customer driving profiles no significant enrichment for component protection is required. However, especially with accelerations at higher vehicle speeds, rich operation can result still in increased CO and PN emissions, Fig. 18.

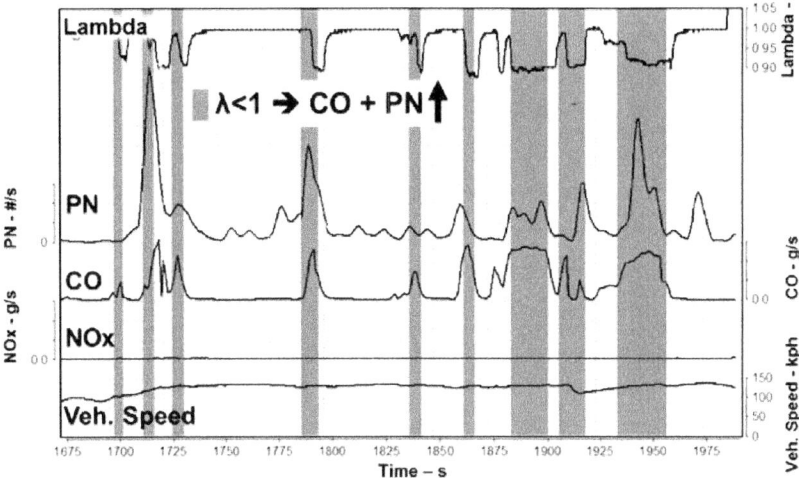

Fig. 18: Downsized turbocharged GDI: Impact of component protection enrichment on Real Driving CO und PN Emissions

Compared to various production engines, an improvement can be realized by refined combustion systems, revised gas exchange as well as an extended temperature range of the turbocharger. A further measure in this direction is an active cooling of the exhaust gases prior to the turbine by a water-cooled exhaust manifold. However, often the cooling capacity of the vehicle limits such cost effective approaches.

Charge dilution by EGR is a further potential measure to reduce or avoid enrichment. However, a complete elimination of enrichment would require a significant reduction of compression ratio and/or increase of displacement, thus resulting in a sensitive trade-off with CO_2 emissions. Both Water Injection and Variable Compression Ratio are means for further improvements of the compression ratio versus enrichment trade-off.

Also Miller and Atkinson cycle approaches result in a better compromise between compression ratio and enrichment. A high geometric compression ratio is reduced at high loads by means of valve timing strategies resulting in a reduced enrichment requirements.

3.2 Three-Way catalyst system

With Gasoline engines, the only major RDE concern area that does not immediately show a trade-off with CO_2 is exceeding the favorable catalyst critical space velocity, Fig 19.

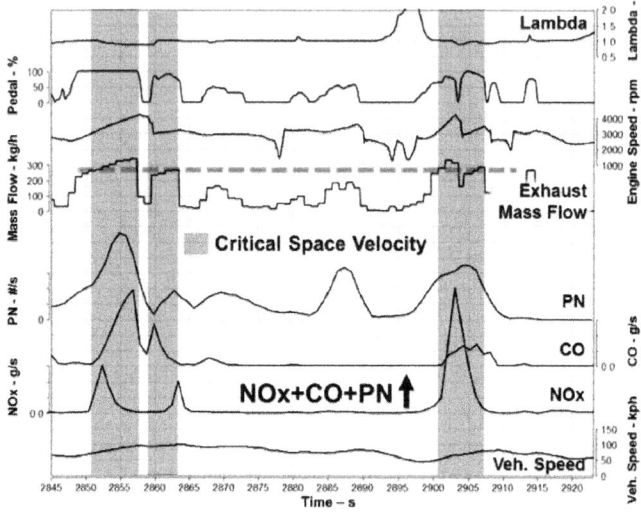

Fig. 19: Downsized turbocharged GDI: Impact of high exhaust gas flow on Real Drive Emissions

Accepting higher catalyst cost and compromises in packaging, both maximum catalyst space velocities and NOx storage capability can be improved by increasing the catalyst volume or cell density respectively the precious metal loading. However, if the increase of catalyst volume or cell density result in higher backpressure, CO_2 emission might be compromised. Furthermore, packaging boundaries of existing vehicle platforms often have geometrical restrictions regarding the most effective layout of the catalyst system.

Fig. 20 shows the RDE emission behavior of two Gasoline Turbo DI SUV vehicles in moderate and dynamic driving style. Both are equipped with an engine of same displacement without a Gasoline Particle Filter (GPF). One vehicle features a classic engine and exhaust aftertreatment specification ("Standard EU6c concept"), while the "λ=1 RDE Concept" is equipped with an integrated cooled exhaust manifold and a two-brick catalyst system that is adjusted to RDE requirements. Both vehicles have the potential to comply with EU6c emission standards in NEDC-testing.

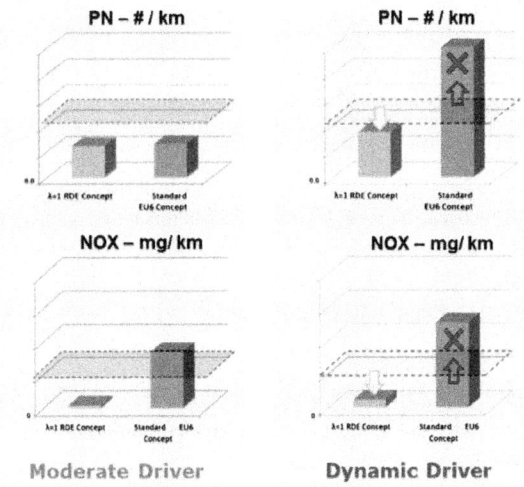

Fig. 20: RDE emission of a "Standard EU6c concept" and a "λ=1 RDE Concept", SUV

The results in Fig. 20 show once more how severe RDE-conditions are: with the SUV, the standard engine/EAS concept exceeds EU6c NOx limits even with moderate driving style. With dynamic driving style, both NOx and PN emission are significantly increased especially due to the larger number of high load phases with rich operation and catalyst raw emission breakthrough especially during acceleration phases.

The "λ=1 RDE concept" SUV features the combination of a cooled exhaust manifold, fully functional catalyst at overall λ=1conditions (increased catalyst volumes and precious metal loading of the catalyst). In contrast to the "Standard EU6 concept", the "λ=1 RDE concept" exhibits not only significantly reduced NOx emission with moderate driving, but also high robustness towards dynamic driving both with regard to NOx an PN staying below the limits even with dynamic driving.

3.3 Gasoline Particulate Filter

As Figure 12 showed, especially with dedicated downsizing and high vehicle weight, still high PN compliance factor values can be found under RDE conditions.

The need to fulfill the particulate emissions on a robust basis at critical power-to-weight vehicle ratios and under consideration of the RDE requirements up to high load dynamic operation, opens a wide field of technical challenges regarding PN optimization, as maximized robustness for real life fuel qualities and oil-born particulates must be covered as well.

The Gasoline Particle Filter (GPF) can be an attractive EAS hardware solution for PN robustness when an adequate engine hardware/ calibration upgrade is not possible.

Nevertheless, several well known technical challenges are linked to the series introduction of a GPF as well

- Layout and specification as best trade-off solution between cost, packaging, filtration efficiency demand, backpressure level and its response to minimum real life soot loading
- GPF sensor concept and reflection in the EMS
- Testbed GPF characterization and modelling
- Regeneration ability and linked EMS functional requirements
- Vehicle testing and real life worst case consideration regarding soot accumulation and regeneration.
- Durability assessment

Fig. 21 shows an overview about the GPF series introduction process and linked work packages.

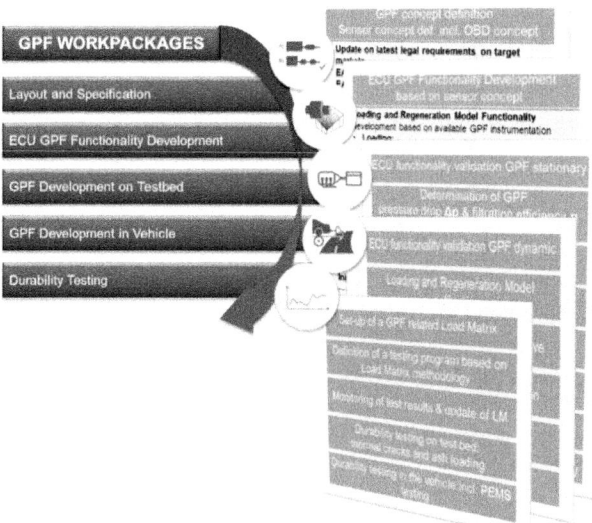

Fig. 21: GPF Work packages for series applications

Especially regarding worst case robustness and related necessity for regeneration, real world scenarios are developed and tested. One critical scenario is repeated cold start followed by a short driving distance. For example, an engine cold start, followed by 3 km driving with max. 60 km/h, stopping the engine for 5 min, "warm" start and 3 km driving and finally stop for 8h is relevant for European Real World driving conditions. Figure 22 shows the resulting soot accumulation in the particle filter at different ambient temperatures with a EU6b series production vehicle.

During the short engine operation time, the GPF is never reaching a temperature sufficient for passive regeneration in fuel cut deceleration phases, neither in close coupled, nor in underfloor GPF mounting position. As result, at -30°C within half a week a critical soot mass can be accumulated that can lead to component damaging temperatures at uncontrolled regeneration in case high load operation is followed by a long fuel cut off event.

Soot loading linked motoring control for GPF component protection as well as dedicated active regeneration strategies as support of passive potentials need to guarantee lifetime robust thermal conditions and prevention of filter clogging under RDE conditions, that show, dependent on the GPF position, a broad spectrum of thermal boundaries to be covered.

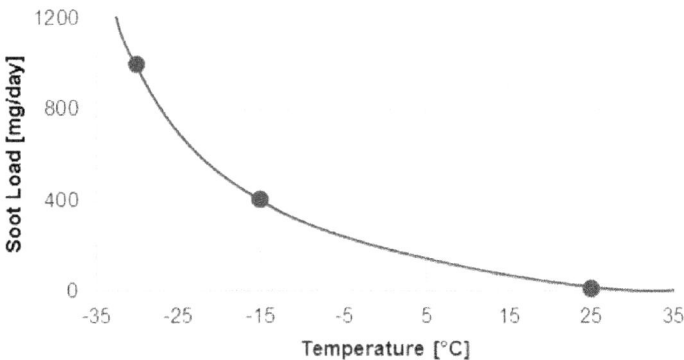

Fig. 22: Accumulated Soot load in a GPF in a real world driving scenario assumption without regeneration [6]

4   Prospects for Gasoline engines

RDE compliance will not only increase the powertrain cost in general, but will also result in quite differentiated on cost with various engine types and vehicle categories.

The most significant impact is expected with Diesel engines in small vehicles. Especially the change from LNT towards SCR aftertreatment system implies significant on cost which are considered to be crucial in this most cost sensitive vehicle category. Combined with some negative image aspects for Diesel engines resulting from the "Dieselgate", generally a reduced Diesel share especially in Europe is expected. In spite the fact that the Diesel most likely will stay the dominating power source with large passenger cars and SUV´s, the total market share in Europe will be reduced, e. g. from currently 53% towards the range of 40% in 2025 [7].

As the "Dieselgate" has also significantly pushed the trend towards electrification, some BEV´s might replace Diesel powertrains. However, especially with cost sensitive small vehicles, the Gasoline engine is in an excellent position. As the fulfillment of future CO_2 fleet values is already a challenge with the high share of efficient Diesel engines, any potential replacement requires at least comparably low CO_2 levels to future Diesel engines. In this context, the combination of 48Volt systems with the next generation of Gasoline engines will become be a highly competitive technology combination.

The most attractive approach to reduce both pollutants in RDE and CO_2 emissions at attractive cost, however, are gaseous fuels. As long as oil born particles are under control, gas engines have more or less no PN emission. Already todays CNG production vehicles show negligible PN both with moderate and aggressive driving, see also Fig, 12. Since with CNG injection into the manifold scavenging strategies are not applied, also NOx emissions in RDE are not an issue. In parallel, CNG offers a chemical potential of 25% for CO_2 reduction that can be extended by utilizing the high knock resistance for higher compression ratio. However, the emission aftertreatment, especially of methane, requires CO_2 compromises with the emission calibration. As also the vehicle weight is increased, the chemical potential of CNG reduction cannot be fully transferred into vehicle results.

A different situation can be found when looking at LPG vehicles. The low boiling points of LPG (Butane / Propane) gives the same benefits in terms of PN like CNG, the chemical potential for CO_2 reduction, however, is with about 11% significantly lower than with CNG. However, additional benefits in cold start / warm-up, homogenization and EGR tolerance can be transferred to additional CO_2 benefits so that in actual vehicle applications [8] the differences between LPG and CNG are much smaller than originating just from their chemical composition. Also regarding modification requirements, add-on weight and infrastructure, LPG is a highly attractive solution.

From a cost / CO_2 reduction perspective, both CNG and LPG are most attractive technology approaches to reduce CO_2 and pollutants simultaneously. LPG might be considered as the more simple short-term solution, CNG as the more sustainable long-term solution. With both approaches, the key challenges are not at the pure technical side, but primarily with marketing to gain sufficient customer acceptance.

Literature

1. Fraidl, G.; Kapus, P. Schöggl, P.; Striok, S.; Vidmar, K.; Weissbäck, M.: "RDE – Challenges and Solutions, 36th International Vienna Motor Symposium, May 7th – 8th 2015

2. K. Vidmar, P. Götschl, Dr. G. K. Fraidl, T. Dobes, Dr. P. E. Kapus, H. Jansen: "Real Driving Emissions –A Challenge for GDI Engines?", 8. Internationales Forum für Abgas und Partikelemission, Ludwigsburg, April 2014

3. N.N: ANNEX to the Commission Regulation amending Regulation (EC) No 692/2008 as regards emissions from light passenger and commercial vehicles (Euro 6), ANNEXES 2, TCMV, May 19th, 2015

4. N.N: ANNEX to the Commission Regulation amending Regulation (EC) No 692/2008 as regards emissions from light passenger and commercial vehicles (Euro 6), ANNEX 3, TCMV, May 19th, 2015

5. ANNEXES to the Commission Regulation amending Regulation (EC) No 692/2008 as regards emissions from light passenger and commercial vehicles (Euro 6), C4 4785157 RDE 2nd package, update TCMV October 28th, 2015

6. Reinharter, C.: "Investigation of Gasoline Particulate Filters System Requirements and their Integration in Future Passenger Car Series Applications", Master Thesis University of Graz/AVL Graz GmbH, October 2015

7. N.N., Roland Berger: "Diesel controversy –Temporary shock or paradigm shift in powertrain?", Frankfurt, October 2015

8. Aríztegui, J.; Gutiérrez, A.; Fürhapter, A.; Friedl, H.: "LPG Fuel Direct Injection for Turbocharged Gasoline Engines", MTZ 10/2015

Solutions to fulfill "Real Driving Emission (RDE)" with diesel passenger cars

Dirk Naber, A. Kufferath, M. Krüger, S. Scherer, H. Schumacher, M. Strobel

Robert Bosch GmbH

This manuscript is not available according to publishing restriction. Thank you for your understanding.

A multiscale approach to virtually render fluid dynamics on overall vehicle level

Author: Frank Hermsdorf

Co-Authors: Christoph Jahn, Prof. Dr.-Ing. Günther Prokop

TU Dresden – Institut für Automobiltechnik, Lehrstuhl Kraftfahrzeugtechnik

1 Motivation

To exploit further advancement of the modern vehicle development, it is necessary to gather information about physical as well as chemical even chains quickly and economically. In order to optimize the overall development process, it is mandatory to adjust conception of design, material and package relevant components according to the expected strains. Avoidance or prevention of corrosion on the overall vehicle validation development process is of remarkable customer interest. Whether or not a material has a tendency of corrosion is determined rather late in the experimental stage, due to complex relations of flow paths, mechanical strains, drying flow and the chemical consistency of the strained resources. Any component modification would lead to higher expenses. Apart from test rigs or test tracks, numerical load simulation is establishing in all aspects of the development process. Computer simulation to imitate and optimize technical systems is an important exponent of virtual instruments. The basis of such simulations is the physical equations of the problem in particular. In order to predict corrosive processes, the numerical method must combine various types of load simulations. On the one hand, mechanical independent variables, such as component vibrations or damage of protective coating, chemical reactions and flow paths are interacting with the determined component. On the other hand, the highly complex geometry of vehicle and component structures is challenging the calculus algorithms, especially when interlinking the calculation area to solve numerical equations. With the present approach, flowing processes in complex geometries can be displayed using computational fluid dynamics. It is a combination of various simulation methods for a diverse range of geometrical scales.

2 Approach

Complex fluid computations are subject to the following requirements:

- Import of extensive3D geometries (CAD interface)
- Simulation of single-phase and multi-phase flow
- define of inlet and outlet boundaries
- Specify boundary conditions
- Simulation of turbulent flow
- Comprehensive analysis of data (Postprocessing)
- Reproducibility of results
- Solid calculation process

As recent surveys prove, no other calculation tool can meet the overall requirements. They merely offer tool specific benefits for individual requirements.

There are different approaches to conduct multiphase or multicomponent fluid computations in a complex controlled area. They are divided in grid-based and grid-free methods of calculation (Figure 1).

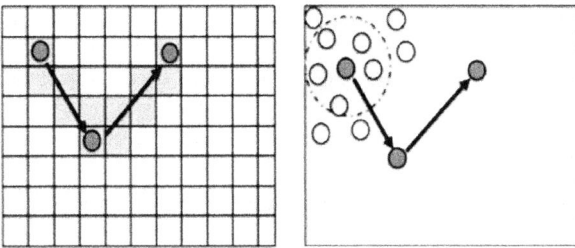

Figure 1: Mesh-based and meshless method

Discrete fluid models, as shown in Figure 1 (right hand-side), can calculate the flow parameters in the defined controlled area, without depending on the mesh. The current case applies the "Smoothed-Particle-Hydrodynamic" (SPH) as a prominent exponent of discrete approaches. In this process, the fluid phase is divided into any kind of small, discrete particles, which will then interact with adjacent particles. Using a sufficient number of particles, it provides a definition of the fluid. Among others, adhesive and cohesive forces as well as surface tensions of the interchanging particles are important parameters. Due to complex physical characteristics of discrete particles and their stochastic distribution in the fluid space, estimating the errors is much more complicated than in using the continuous calculation method. However, this method represents a fast and stable option to outline the fluid characteristics.

In contrary to the discrete calculation methods, the continuous method calculates the flow ratio depending on the nodes, as shown in figure 1 (left). Depending on the tool, it determines the general physical conditions of the fluid at the node or the middle of the cell. The results are interpolated in between two nodal points. Explicit as well as implicit approaches to calculate fluid behavior are feasible. For explicit basis, particular limiting conditions are calculated at each respective nodal point, following the previous nodal point. Approaches that follow the implicit formulation do not depend on a certain sequence. It calculates the variables of state of the nodal points under investigation from a starting point; therefore the neighboring nodes do not need to be scanned.

Furthermore, continuous as well as discrete methods can be subdivided but also combined with each other.

In consequence of the various algorithms essential advantages and disadvantages of each method can be detected. For complex geometries, the Technische Universität Dresden combines the different advantages of some discrete as well as continuous models systematically to display extensive flow paths efficiently in terms of time:

It is worth mentioning that the approximate complex geometry of an entire component can be determined using the SPH-method. For areas of special interest, local coordinates as well as the marginal conditions at the borders are selected using a script. Using these local coordinates, a component model can be extracted from the complex part in a CAD environment and is (automatically?) meshed (in CAD). The component model is impinged with the boundary conditions of fluid dynamics. Therefore a flow pattern with a refined mesh as well as continuous methods including exact wall-mounted models can be prepared. Using this iterative process, global fluid dynamics as well as exact outcome of detailed areas can be observed in a time effective manner.

3   Experimental Sample

The water container of a passenger car was used as a valid experimental sample. The real as well as virtual available geometry is particularly suitable for this analysis of different boundary conditions as it is rather complex. Gaps, material mixes, cable bushings with drip edges or different flow influences can be displayed. In addition to free flow in the form of run-off water, there is also imposed air flow through the fan of the air condition. Both have to be combined in an ideal situation. In addition, the component is easily accessible, which is a good basis for validating the calculation method. The water container is shown below.

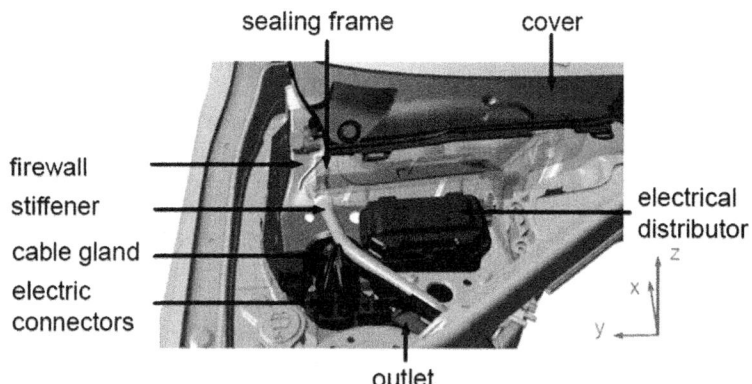

Figure 2: Test body

Based on the vehicle model, it is necessary to implement some constructive changes in CAD to use the calculation. That means that nozzles which are not within the flow area and do not exceed a nozzle limit are closed. Furthermore geometric errors are eliminated, chamfers and curves are optimized and a closed flow chamber is generated. The flow parameters are filtered at the real model and equipped with specific values from experiments or data. That way the flow parameters can be defined as an input variable. The following figure (Figure 3) shows the differentiated inlet and outlet diameters.

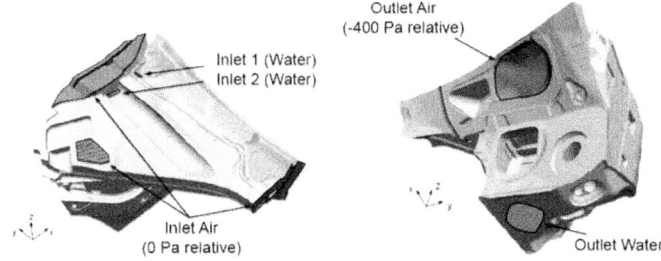

Figure 3: Boundary conditions

4 Calculation

The aforementioned experimental sample is analyzed using the following methods:

- Discrete method (DualSPHysics)
- Volume of Fluid method (Ansys Fluent)
- Euler-Euler/Euler-Lagrange method (OpenLB)

The discrete method is chosen for the global computation of the water container. DualSPHysics, which is OpenSource software of the universities of Vigo and Manchester, was used to compute the results at hand. Also, the complex flow chamber can be computed with the Lattice-Boltzmann approach. In terms of comparison, OpenSource software (OpenLB) of the University of Karlsruhe is used. The commercial tool ANSYS Fluent is applied to display the performance of all three methods. The modeling as well as the results and challenges of each toll are contrasted in the following sections.

a) DualSPHysics

DualSPHysics is favored to compute large particle flows in the field of offshore and shaft calculation. As a result, DualSPHysics is not able to handle direct input of defined inlet boundary conditions, for instance mass flows or pressure boundary conditions (date of analysis: 2015). The inlet fluid dynamics are defined by means of diffusion from idealized collecting tanks. Bernoulli's pressure equation:

$$p + \rho g h + \frac{\rho}{2} v^2 = const \tag{1}$$

Is applied when determining the flow velocity v of the tank by the fill level h for an incompressible fluid (density $\rho = const$). The result is converted into an inlet boundary condition for the test geometry.

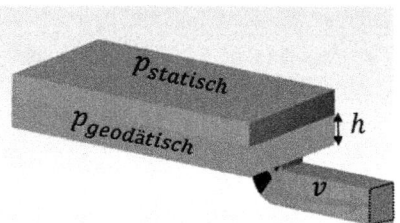

Figure 4: Inlet condition for DualSPHysics

Incoming air flow velocities have been identified iteratively and were displayed in a linear trend line. This trend line allows an exact association of tank height regarding the incoming air flow velocities that is desired.

Apart from iteratively measured incoming air flow boundary conditions, the implemented viscosity models are displayed mathematically, by which there is no direct coherence with physical laws. That said, it requires a mathematic assessment of the definition viscosity and an established relation to physical approaches. Viscosity can be defined experimentally, as well as in the computation software for the means of comparison, with a viscosity measurement instrument.

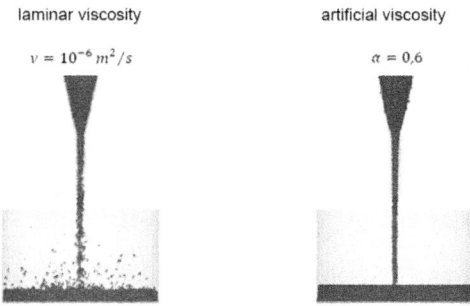

Figure 5: Real and artificial viscosity

Despite the fact that artificial and physical viscosity are not related, the test case model above shows great conformity of experimental outcome, which is why it is preferred in case of calculation situations. Another restriction is the missing interaction between water and air. The enforced air flow is ignored and the existing air is considered as ambient medium rather than flow medium. However, this restriction is taken into account for the overall evaluation. Further enhancement of polyphase interaction is being conducted by Dual SPHysics developers. As stated previously, the static height of the inlet flow medium is used to define the inlet boundary conditions. Additionally, the volume flow is defined as a product of inlet flow velocity and inlet cross section. It is essential to adjust the particle size respectively the distance between the particles if the inlet cross section is too small. The following figure shows the influence of particle distance on fluid dynamics.

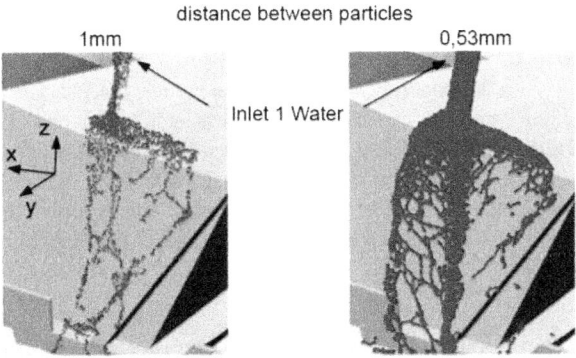

Figure 6: Particle distance

Once all boundary conditions have been defined accordingly, complex fluid dynamics simulation in the water container can be conducted stable and in a time effective manner. The following part contrasts the results.

DualSPHysics was also applied to test GPU-accelerated calculation approaches. Graphic processors have a higher performance in comparison to normal main processors and therefore offer a more efficient simulation. The higher resolutions show physical effects in great details with the same time effort of main processors. Moreover, the computing time is suitable for parametric research and optimization without costly computing clusters. As a result, GPU Computing is a vital sector of parallel and cpu-intensive applications of the future. As shown in Figure 7 it's possible to calculate the model in about the half of time.

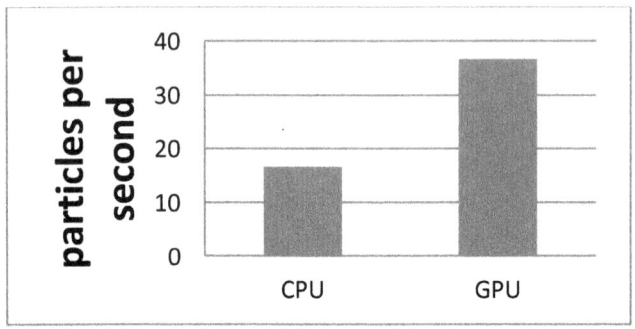

Figure 7: Particles per second on CPU and GPU

b) OpenLB

While Dual SPHysics does not rely on a grid for the particle method, OpenLB applies an automated computational grid. However, this mesh is quite different from the computational grid that is applied in Ansys Fluent which will be presented later on. OpenLB divides the geometry into voxels and partitions theses voxels into cuboids right after. While the cuboids are also used for process administration, material figures are attached. These figures define the features of every cuboid. It is differed between wall, fluid, velocity boundary conditions as well as pressure boundary conditions. Areas which do not impact the flow area are identified with a material figure 0, which reduces the complexity of the mesh. Figure 8 represents the partitioning of geometry through cuboids and voxels.

Cuboide   Voxel

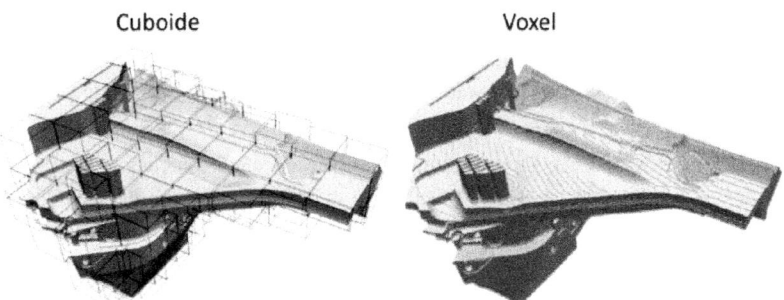

Figure 8: OpenLB segmentation

It is necessary to keep in mind that there is a conflict between the amount of voxels and the computing time or higher demand for storage capacity. On the other hand, the number of voxels is directly proportional to the accuracy of the result.

The continuous calculation requires two sets of mesh which can be linked with each other in order to define a gravity-constant.

Two different phases are analysed for the continuous-discrete method. The air phase is solved with the continuous approach in which the particle-based water phase is flowing. In comparison to DualSPHysics, velocity and pressure boundary conditions can be predetermined in OpenLB. If desired, volumetric flow can be calculated taking into account density and outlet diameter. In contrary to that inlet and outlet diameters can only be defined vertical or horizontal, which means that the inlet flow diameter must be adjusted inevitably. As a result, a channel on each inlet as well as outlet diameter has been modelled in order to guarantee this requirement. The characteristic velocities needed to be changed to display the boundary conditions in a real way. The steadiness of fluid dynamics is ensured because the predefined flow velocities are not imprinted but defined as a kind of ramp function. In contrary to that is a trade-off of transient flow computation, computation time and dividing calculation steps are at odds with each other. This is included in the quality of results.

In addition to inlet boundary conditions, further essential parameters need to be defined. They are impacting turbulence as well as dynamic response, characterize particle behaviour on the wall and the viscosity of a fluid. Once all parameters have been defined properly, stable calculations with OpenLB shall be carried out. The computing time, however, takes longer than the one of DualSPHysics.

c) Ansys Fluent

The commercial computation tool Ansys Fluent demands a complex computational grid as basis for successful calculation. Newton's approaches of the mass, energy and momentum are applied to the nodal points. This mesh is based on customized and simplified geometry in order to guarantee high mesh quality. In order to calculate fluid dynamics close to the wall it is necessary to define a high resolution boundary zone in which effects of separation can be displayed on a model basis. Besides that, correct turbulence models need to be chosen in order to display non-laminar medium flow. A definition of the scenario with regards to medium, dynamic and other variables is of importance.

Compared to DualSPHysics and OpenLB, the definition of boundary conditions is hassle-free. Pressure, velocity and mass flow can be defined at the inlets and outlets. The inlet diameters can be conferred straight from the construction geometry. As said in the first paragraph, the calculation of flow variables is based on interpolation of conservation laws between the nodes or the middle of the cell. This requires a high calculation performance and time since every cell is calculated individually. It generates even more validatable results for the part in the field of single-phase flow.

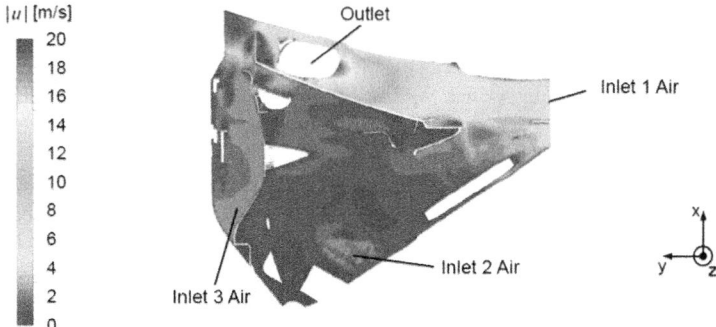

Figure 9: Ansys Fluent boundary conditions

A calculation of instationary flow with a fluid boundary layer face was impossible due to insufficient mesh quality, which could not be improved in a timely manner as the geometry was too complex for the extensive flow paths.

5   Comparison of results and validation

By means of assessing the different types of calculation methods, it is necessary to extensively validate the flow parameters. In the present case, physical boundary conditions in the dropping area as well as for the testing geometry have been determined experimentally. In order to do so, complex measurement points have been implemented.

Especially for defining the parameters, knowledge about basic relations between reality and simulation are inevitable. Dripping time that has been determined analytically was almost identically accomplished at a drop test rig. A drop of water from a water-filled pipette was exposed a defined falling distance at constant ambient temperature. Using a high speed camera, the drop time have been compared and are the basis on a microscopic level.

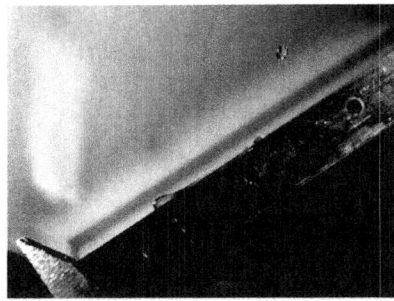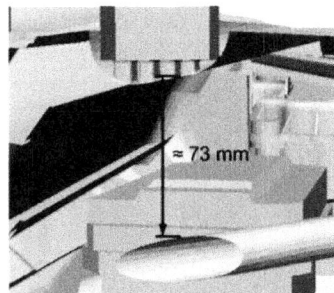

Figure 10: Measured and calculated drop time

Building upon these results, the dripping times were conducted again in a separate model structure. The dripping times calculated with Ansys Fluent and DualSPHysics, and assimilated mathematical viscosity, are slightly below the experimental data, but still in an acceptable range. The dripping times calculated with OpenLB though deviate remarkably from the experimental data, due to insufficient determinability of viscosity. It continued throughout additional, simplified tests.

Table 1: Water drop-time

drop time [sec]	
theory	0,122
experiment	0,119
DualSPHysics	0,097
OpenLB	1,01
Ansys Fluent	0,092

The mathematical simplifications for stable calculation means that the viscosity is approached insufficiently, which leads to a stiffer fluid consistency. A better solution for the test geometry can be found. In a parallel project, a test vehicle was equipped with fluid, temperature as well as condensation sensors to document the climatic influences on corrosive stressed areas in a long term test. In this vehicle the water container was equipped with fluid sensors and cameras which supply viable optical as well as metrological data for the present simulation.

Apart from a metrological comparison of the results, the optical alignment of the flow paths is significant. They show a remarkable resemblance to reality, as shown in the following figure. Especially the simulation with DualSPHysics reaches a flow pattern of decent quality. It also shows that the artificial viscosity in OpenLB, that has a stiff flow leads to a remarkable difference of surface wetting.

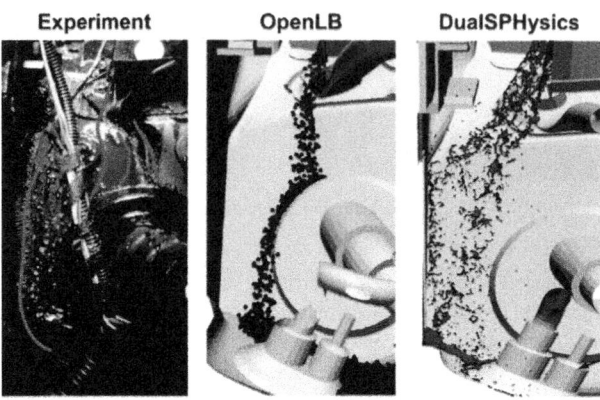

Figure 11: Waterflow real, OpenLB, DualSPHysics

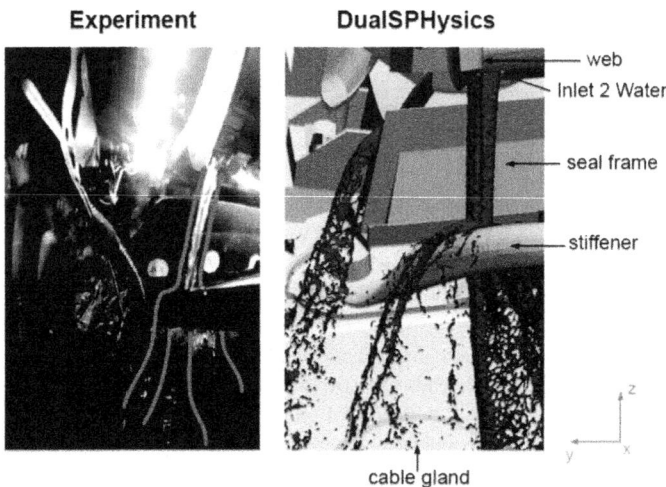

Figure 12: Result real, DualSPHysics

6 Summary and Outlook

The differences in results of fluid dynamics simulation show that there is a need to further invest into development, in order to physically record the viscosity models. However, it also turns out that Open Source tools provide feedback about possible flow paths time-efficiently in the early stage of development of products.

Simulations on complex water containers with DualSPHysics have shown specific problems related to the high amount of particles and the applied SPH method. The testing of the vehicle geometry also ascertained that quality of the results and general practicability of a specific simulation with reliable numerical approaches are of vital importance. Areas of corrosive strain can be identified when looking at the flow paths of water in transient behaviour. The commercial tool Ansys fluent was not suitable to simulate the spread of water in a temporal course. Evaluating the outcome and comparing the results of OpenLB as well as various experiments have shown that flow paths close to reality can be delivered with DualSPHysics in detail, if the implemented boundary conditions are optimized radically.

7 Bibliography

1. Wünsche, Matthias; Diploma Thesis Mehrphasige Strömungssimulation mit OpenLB am Teilfahrzeugmodell; 2015

2. Naumann, Martin; Diploma Thesis Untersuchung von mehrphasigen Strömungseigenschaften mittels einer Smoothed Particle Hydrodynamics Methode; 2015

3. Sauerzapf, Stefan; Diploma Thesis Aufbau eines Teilfahrzeugsimulationsmodells für die Strömungssimulation; 2015

8 Acknowledgment

Many Thanks for the cooperation to the Daimler AG and the BMW AG.

Investigation of visibility properties through wetted glass planes on vehicles

T. Landwehr, T. Kuthada, N. Widdecke, J. Wiedemann

IVK/FKFS, University of Stuttgart

Abstract

When driving a car, the view onto the surrounding traffic must be ensured at all times. Especially the view through the side window onto the rear view mirror and other traffic participants is very important. When driving on a wetted road, water and dirt can impair driving comfort and safety. Surface bound water droplets and rivulets on the windshield and the side glass reduce the visibility. Therefore, a new evaluation method is presented which assesses the view through pools of water on a horizontal glass plane depending on wettability of the glass and the volume of single droplets or the height of water films. It provides information about the transparency behavior of water droplets and the impact factors of a soiled side window on the view through it.

Introduction

Bad weather conditions such as rain and snow may affect driving comfort and safety. This cannot only lead to wet fading of brakes [1], but also to a limited view onto the surrounding traffic. While driving on a wetted road, water particles are whirled up by the tires. Some of the droplets are drawn into the wake and generate a fog cloud. Vehicles following behind get soiled by these small airborne droplets. However, the soiling behavior does not only depend on the driving speed and the amount of water impinging the vehicle.

Additionally, the geometry of the rear view mirror and A-pillar has an essential influence and can reduce or even prevent rivulets on the side window caused by A-pillar overflow. The wake of the mirror causes airborne and mirror housing dripping droplets to hit the side window [2]. Droplets of varying sizes and shapes arise there, being deformed due to gravity and wind driven shear forces. The vehicle surface affects the ability to roll off and the type of water accumulation and thus the degree of soiling. The resulting pools of water might impair the view through the side window and onto the rear view mirror.

Soiling tests are usually carried out in a wind tunnel. A fluorescence-medium is added to the water and the mixture is illuminated with UV light to visualize the soiled areas. Post-processing tools can evaluate the intensity and the degree of soiling [3].

This raises the question, how much water affects the view through the side window. It distracts the driver especially when it moves and does not occur continuously. In addition, an interface exists between the air and the water pool. If a beam of light impinges on the water pool, a part of the beam is reflected. Depending on the lighting situation, these reflections can lead to glaring causing visual impairment, even if one is not looking through the water pool or looking directly into the beam of light.

302

The other part of the light beam is refracted at the interface and changes its wavelength. The intensity of refraction depends on the consistence of the water pool. In some situations, the water pools are directly in the field of view and it is necessary to see trough. Hence, the refraction in the water pool on the window affects a clear view.

For this reason, it is pointed out which kind of droplet shapes are conventionally formed on vehicle windows. Based on a new method, the view through water pools is investigated on a test bench. The contact angle and the volume of droplets as well as the height of water films are the crucial parameters.

1 Contact angle and droplet sizes at vehicle windows

The soiling intensity depends on the surface condition. The wettability is the adhesive behavior of fluids in contact with solid surfaces. A measure of the wettability is the contact angle.

The contact angle can be determined by the integration of the Young's Equation (Equation 1) [4].

$$\sigma_l \cos \theta = \sigma_s - \sigma_{sl} \tag{1}$$

where σ_l is the surface tension of the liquid, σ_s the surface free energy of the solid and σ_{sl} the interfacial tension between liquid and solid.

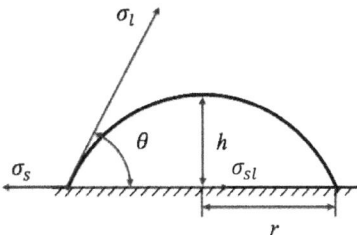

Figure 1: Contact angle of a schematic droplet shape on a solid surface

To determine the surface free energy of the solid, two droplets of known liquid are applied to the surface. Using imaging and numerical methods, the contour of the droplet and thus the contact angle is determined. The surface free energy can be calculated with the Owens, Wendt, Rabel and Kaelble (OWRK) method [5]. However, for the determination of the visibility properties through wetted glass planes on vehicles, it is sufficient to measure the contact angle with water.

Neglecting gravity, the shape of droplets can be accepted as a spherical segment. Equation 2 presents the correlation between the volume V of the droplet and its height h and the radius r.

$$V = \frac{\pi h}{6}(3r^2 + h^2)$$ (2)

The schematic contour of a droplet is shown in Figure 1. The relationship between the height, the radius and the contact angle is described by equation 3 [6].

$$\cos\theta = \frac{r^2 - h^2}{r^2 + h^2}$$ (3)

Figure 2 shows symmetrical droplets on a horizontal glass plane. Here, the influence of the contact angle and the volume of the droplet on the height and the diameter are illustrated.

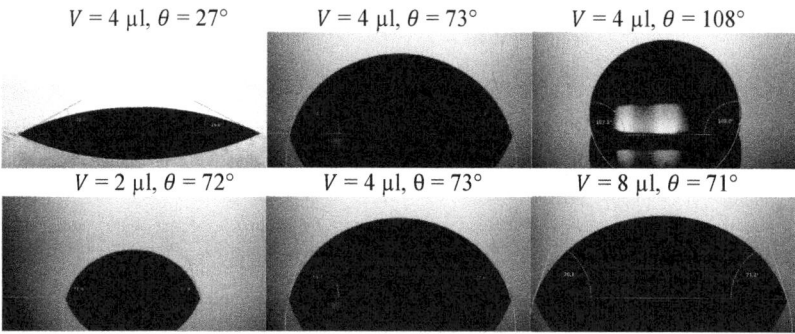

| $V = 4\ \mu l,\ \theta = 27°$ | $V = 4\ \mu l,\ \theta = 73°$ | $V = 4\ \mu l,\ \theta = 108°$ |
| $V = 2\ \mu l,\ \theta = 72°$ | $V = 4\ \mu l,\ \theta = 73°$ | $V = 8\ \mu l,\ \theta = 71°$ |

Figure 2: Droplets with different shapes

However, a regular side window is not a horizontal plane. The droplets deform and move due to the influence of gravity and wind driven shear force (Figure 3).

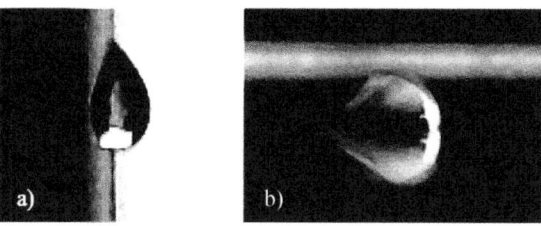

Figure 3: Moving droplet a) on a vertical glass plane due to gravity b) on a horizontal glass plane due to wind driven shear force (air flow from left)

To identify realistic contact angle of droplets on vehicle windows, 30 different vehicles of undefined surface condition were examined. The investigations showed a contact angle range of $\theta = 30°$ - $75°$ on the side window (Figure 4) and the rear window. These results confirm the study of Vollmer [7]. The contact angle on the windshield tends to be lower due to the increased abrasion by the movement of the wiper, the shear force imposed by the air flow and the consistence of the window. Its contact angle range is $\theta = 10°$ - $55°$.

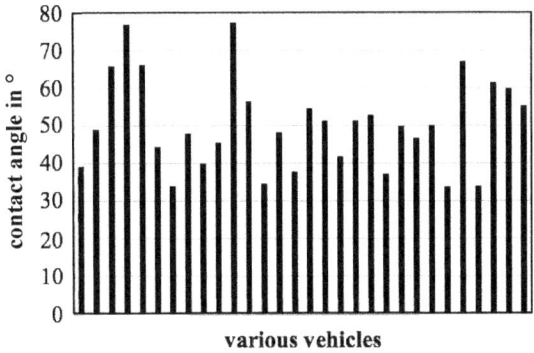

Figure 4: Contact angle on the side window of various vehicles

The composition of the glass plane has an influence on the contact angle. A polished side window features usually a contact angle below $\theta = 25°$. Treating agents are applied to the surface when washing the vehicle in the car wash. This increases the contact angle. Different wash programs from several providers produce a contact angle of $\theta = 55°$ - $70°$. Over the time, the contact angle is reduced due to deterioration of the treatment agent. Special coatings can lead to a higher contact angle which a common car wash is not able to reach.

Depending on the vehicle geometry, driving speed, water quantity and surface free energy, various forms of soiling can appear on the side window. Therefore, soiling investigations were carried out in the thermal wind tunnel. They have shown that in addition to rivulets single droplets are formed with the size of $V = 7$ µl – 2 µl up to spray with droplets much less than $V = 0,5$ µl. Thus, yield a diameter of $d = 1$ mm – 7,5 mm and heights of $h = 0,5$ mm – 2,8 mm for single droplets in contact angle range of $\theta = 40°$ - $70°$.

2   Experimental Setup

On public roads weather conditions such as sunlight, rain and wind fluctuate. In order to get reproducible results, the experiments are carried out on a test bench in a light-tight chamber with defined lighting. Figure 5 shows the setup.

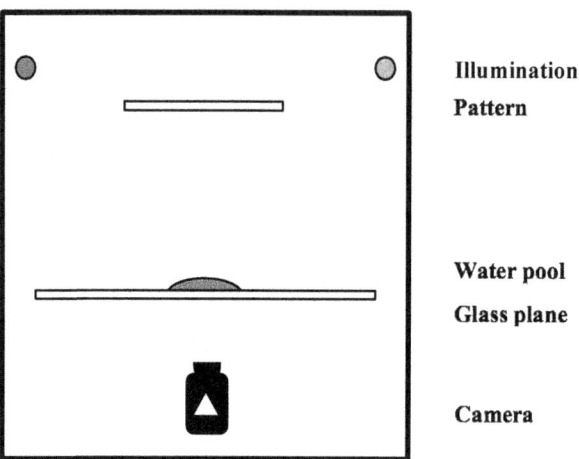

Figure 5: Schematic diagram of the test bench

The phenomena of the view through pools of water are investigated on single static droplets and on flat static water films on a horizontal glass plane. The water films have a height of $h = 0,5$ mm – 5 mm.

The glass plane is set to its initial state by polishing and treated with various treatment agents. Thereby it was possible to produce a contact angle range of $\theta = 10°$ - $110°$. A pipette is used to apply droplets with a defined volume of $V = 1$ µl - 10 µl.

The camera takes pictures of a pattern from a distance of 2 m. Depending on the water pool, a change of the pattern can be perceived.

These experiments only produce qualitative results since they depend on the camera settings and the lightning situation.

3 Experimental Results

To evaluate the phenomena of the view through pools of water, different patterns were used.

3.1 Rectangle Pattern

The rectangular pattern consists of orthogonally intersecting lines. Figure 6 shows a sector of the rectangle pattern. It should identify possible distortions and give information about the region with the worst view through the water pool depending on the contact angle and the height of the water pool. The drawn coordinate system illustrates displacements of the pattern.

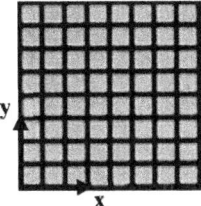

Figure 6: Rectangle pattern without water pool

The water film covers the whole pattern. In all images in Figure 7, the pattern is clearly visible. There is no distortion of the pattern. The water film generates a barely visible illumination of the resulting images and the pattern shifts the same way regardless of the film's height.

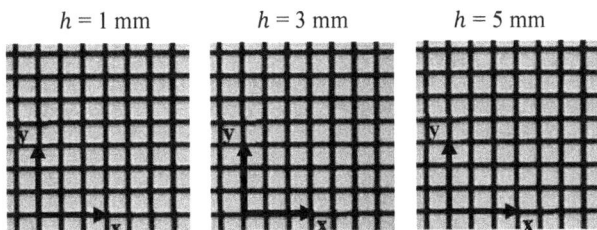

Figure 7: Rectangle pattern with different water film heights

Droplets lead to a discernible change in the pattern (Figure 8). The edge of the droplet base is delineated to clarify the position and size. There are no noticeable distortions or displacements of the rectangular pattern.

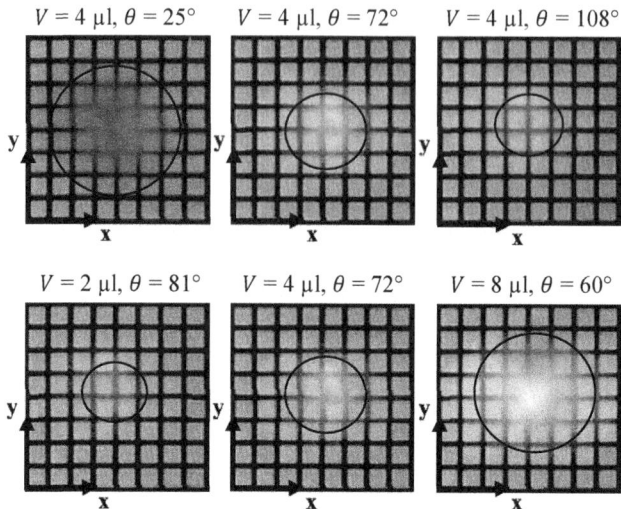

Figure 8: Rectangle pattern with different droplet shapes

The shape of the droplet has a decisive influence on the results. Compared to higher contact angels, a smaller contact angle leads at constant volume despite lower droplet height to a diminished view on the pattern. An increase in the droplet volume and therefore in the droplet height at constant contact angle reduces the transparency. The center of the base is the place with the worst view onto the pattern for all symmetrical droplets. This leads to the conclusion that the covered area by the droplet has a stronger impact than just the shape of the droplet.

3.2 Landolt C Pattern

An adequate vision is a prerequisite to be allowed to drive a vehicle. The visual acuity is the ability to distinguish two closely spaced objects from each other. Therefore, the Swiss ophthalmologist Edmund Landolt developed the optotype for visual tests in 1889 [8]. He used a ring which is open at one point. The size of the ring opening b corresponds to the thickness of the Landolt ring b, the diameter is 5 times its width (Figure 9).

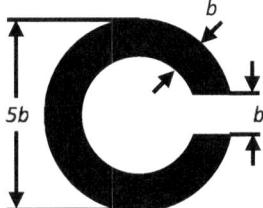

Figure 9: Landolt C

The smallest angle which can keep the objects still apart is crucial for the determination of the visual acuity and is called angular visual acuity. The Visus is the inverse of the angular visual acuity [9].

$$Visus = \frac{1'}{angular\ visual\ acuity\ [']} = \frac{effective\ distance\ [m]}{normal\ distance\ [m]} \tag{4}$$

According to the European norm EN ISO 8596 (previously DIN 58220) a driver must have a Visus of 0,7 in order to get a driver's license [8].

Therefore, a single Landolt C pattern with a Visus of 0,7 is used to detect the opening of the Landolt C and thus to assess the transparency. The Landolt C pattern can be seen in Figure 10.

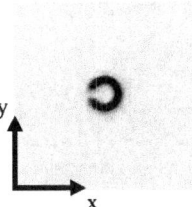

Figure 10: Landolt C pattern without water pool

Based on the investigations with the rectangular patterns for water films, no perceptible reduction of the recognizability of the Landolt C and its opening is expected. The results with the Landolt C pattern confirm the tendencies of the investigations obtained by the rectangular pattern (Figure 11). A slight brightening of the images and a displacement of the Landolt C on the images can also be perceived.

$h = 1$ mm $h = 3$ mm $h = 5$ mm

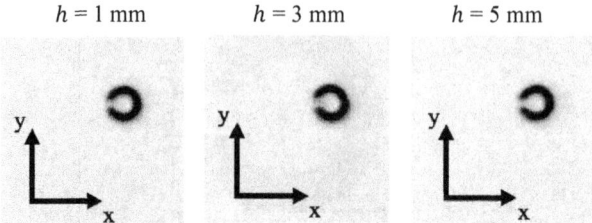

Figure 11: Landolt C with different water film heights

85 droplet shapes were investigated with a volume range of $V = 1\,\mu l$ - $10\,\mu l$ and a contact angle range of $\theta = 10°$ - $100°$. The Landolt C is located in the area with the worst view onto the pattern which is the center of the droplet. A study with 40 test persons was conducted to assess the resulting images of the Landolt C pattern. Figure 12 shows the classification in „no view", „conditional view" and „view".

$V = 4\,\mu l, \theta = 43°$ $V = 4\,\mu l, \theta = 66°$ $V = 4\,\mu l, \theta = 96°$

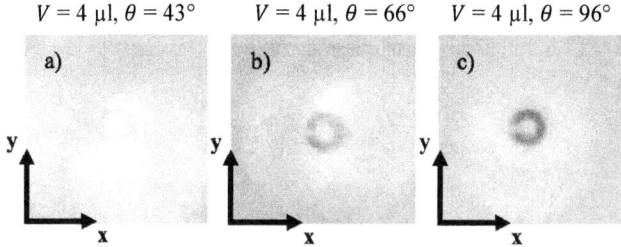

Figure 12: Landolt C pattern with different droplet shapes and classification in a) no view, b) conditional view and c) view

The results of the volunteer study are depicted in the graph shown in Figure 13. The graph can be divided into three regions. The best view through the droplet is at a small droplet volume and a high contact angle. An increase in the droplet volume and a reduction of the contact angle obstructs the view onto the pattern. If the critical region with conditional view through the droplet is exceeded, the opening of the Landolt C is no longer visible.

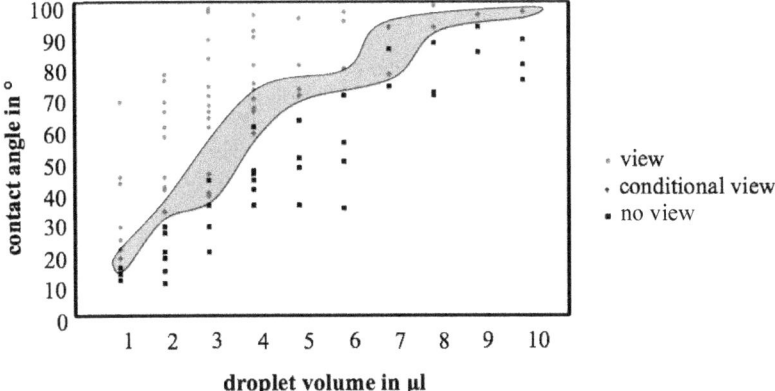

Figure 13: View through droplets depending on wettability and volume

With respect to the transparency behavior of water droplets, smaller volumes are most advantageous. A high contact angle not only favors a lower soiling of the side window because the water pools roll off faster [7] but also features an improved view through droplets. Thereby, other visual impairments like reflection or blending could result which will be part of future work.

4 Conclusion

The presented study discusses the obstructed view due to soiled side windows of passenger cars. Initially, the existing droplet shapes are discussed. Examinations on various vehicles have shown that the typical contact angle range is between 30° and 70°. Soiling tests in the wind tunnel produce surface bound water droplets on the side window with a droplet volume of $V = 7$ µl $- 2$ µl up to spray with droplets much less than $V = 0,5$ µl.

Subsequently, an evaluation method is presented to assess the view through pools of water. Therefore, droplets and water films are investigated on a test bench depending on the wettability of the glass and the volume of single droplets and the height of the water films. The results do not show a visible obstruction of the view through a water film but a displacement of the object. Droplets, however, cause a higher impairment. Large droplet volumes and small contact angles inhibit the view onto the object. A graph depicts the view through a droplet onto a pattern depending on the droplet volume and the contact angle.

This shows that the transparency through water pools depends not only on the amount of the water accumulation and thereby its height, but also on its shape. The contact angle has a decisive influence on the view.

Static droplets on a horizontal plate and water films are only simplifications of water pools occurring in reality. There are also rivulets and the effect of movement and asymmetry is not considered so far. Future studies will also examine the different lighting situation due to the time of day, other road users and various weather conditions. Additionally, it will be investigated whether many small droplets and a high contact angle could impair the view onto the surrounding traffic in other ways. Finally, these results will be used in order to improve the soiling investigations in the wind tunnel.

Bibliography

1. Spruss, I.; Schröck, D.; Kuthada, T. and Wiedemann, J.: Investigation of Braking Response Under Wet Condintion, 8th MIRA International Vehicle Aerodynamics Conference, Grove, 2010

2. Bannister, M.: Drag and Dirt Deposition Mechanisms of External Rear View Mirrors and Techniques used for Optimization. In: Society of Automotive Engineers, Inc. 2000-01-0486

3. Widdecke, N.; Kuthada, T. and Wiedemann, J.: Moderne Verfahrensweisen zur Untersuchung der Fahrzeugverschmutzung. In: Tagung „Aerodynamik des Kraftfahrzeugs", Haus der Technik e.V. München, 2001

4. DIN 55660-1: Beschichtungsstoffe – Benetzbarkeit – Teil 1: Begriffe und allgemeine Grundlagen. Berlin: Beuth Verlag 2011

5. DIN 55660-2: Beschichtungsstoffe – Benetzbarkeit – Teil 2: Bestimmung der freien Oberflächenenergie fester Oberflächen durch Messung des Kontaktwinkels. Berlin: Beuth Verlag 2011

6. Zielke, P.: Experimentelle Untersuchung der Bewegung von Tropfen auf Festkörperoberflächen mit einem Gradienten der Benetzbarkeit. „Dissertation" Universität Erlangen-Nürnberg, Erlangen 2008

7. Vollmer, H.; Gau, H.; Winkelmann, H.; Kuthada, T. and Wiedemann J.: Methode zur Bewertung der Sichtfreihaltung bei Regen. In: Tagung „Aerodynamik des Kraftfahrzeugs", Haus der Technik e.V. München, 2014

8. Dreager, V.; Harsch, V.; Kittel, R.; Blum, B. (Hrsg.): Die Bedeutung der optischen Signalgebung für die modernen Verkehrsarten, Neubrandenberg: Rethra Verlag 2006

9. Lachenmayr, B.: Sehschärfe. In: Lachenmayer, B.; Friedburg, D.; Hartmann, E.; Buser, A.(Hrsg.): Auge – Brille – Refraktion, Stuttgart: Thieme 2006

.

The new Porsche 911 Carrera – Evolution in aerodynamics, thermal management and heat protection

M. Klingbeil, J. Weissert, M. Yilmaz

Dr. Ing. h.c. F. Porsche AG

Usually in the upgrading (or face-lifting) of an existing vehicle-generation some visual changes are introduced a certain increase in engine power is implemented through engine management measures and, where appropriate, some new functionalities for increasing customer experience are added. The revision of the 991 series comes with the additional challenge of a new generation of engines.

1 Ensuring typical Porsche performance

In the new 2016 MY 911 Carrera new turbocharged engines with a cylinder capacity of 3.0 liters are used for the first time in order to realize further improvements in fuel consumption and emissions. This leads to extensive changes in overall vehicle functional level.

With the new engine concept, it was necessary to integrate a charge air cooling system into the existing rear package of the 911 Carrera to provide efficient and powerful turbocharging. The high temperature (HT) cooling system and the heat protection concept have also been adapted to ensure typical Porsche performance.

In order to realize a wide spread between both best possible performance and efficiency, the thermal management was optimized and more efficient aerodynamics, with the introduction of active radiator shutters, help to reduce the driving resistance and thus reduce the fuel consumption.

The challenges in aerodynamics, thermal management and heat protection in the light of these development requirements will be discussed in more detail below.

1.1 Integration of the charge air cooling system in the classical Carrera-Design

1.1.1 Concepts (history, differentiation of the 911 Turbo and the Carrera)

The top model of the 911 series (the "911 Turbo") has had charge air cooling for about 40 years now. The visual features determined by the functional design underline the superior performance of this car: The distinctive large rear spoiler with underlying intercooler from the first "Turbo" type 930, as well as the striking air inlets in the body side since the 911 Turbo 996 series, have both been specific elements.

With the introduction of the new turbocharged engines in the 911 Carrera and Carrera S, the challenge was to integrate a powerful charge air cooling concept in the classic, more discreet form of the Carrera and so maintain an optical differentiation from the 911 Turbo. Especially since, as is usual in a "face-lift" or product upgrade, no major adjustments in the body shell were possible and except for the front and rear bumper parts the outer skin is carried over from the predecessor.

The solution is to put the inlet openings for the charge air cooling under the extending rear spoiler to use the existing pressure field to generate a flow. In the concept phase different approaches to air ducting and for positioning the radiator and the air outlets were examined. To evaluate the feasibility of different concepts, the criteria of achievable intercooling efficiency and the impact of the flow at the rear of the vehicle on whole vehicle drag and lift were compared and balanced.

Cooling airflow for charge air coolers at the MY 2016 911 Carrera

The selected concept has the two charge air coolers placed to the sides of the engine compartment, similar to that of the actual 911 Turbo. Adequate package space for the required coolers was available in these areas due to the elimination of pre-silencers. The air outlet is also similar to that of the 911 Turbo and benefits from relatively low pressure in this area.

Although the distance of the cooler from the air intake results in a relatively long air path, this enabled the design of adequately dimensioned inlets and provides a uniform flow through the entire radiator area.

1.1.2 Measures to enhance the performance

1.1.2.1 Optimization of cooling air duct

Compared to a conventional cooler location in the front of the car an inflow at the rear of the vehicle is more challenging due to the significantly lower pressure difference between the air intake and outlet. A detailed optimization along the entire air path is indispensable to be able to achieve maximum and homogeneous airflow. The figure below shows the velocity distribution perpendicular to the radiator surface of two different concepts. It can be seen that an unfavorably optimized air supply duct, because of swirl in the incoming flow and the relatively small pressure difference across the heat exchanger, delivers inadequate flow. With relatively simple tweaks to the expansion of the air duct in front of the radiator surface and at the junction of the curved air guide to the expansion area sufficient flow can be achieved despite the low pressure difference.

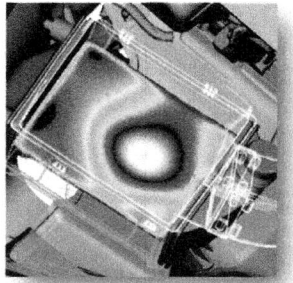 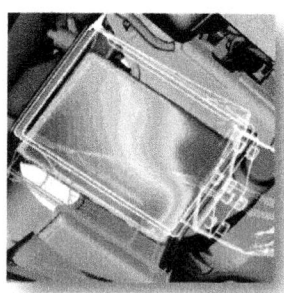

Optimization of velocity distribution through charge air cooler

1.1.2.2 Increasing the dynamic pressure at the rear spoiler

To ensure the inflow to the air intakes the rear lid of the new 911 Carrera has been completely redesigned. By lowering the lid and a new design of the fins an appropriate flow has been achieved through the intercooler even when the spoiler is retracted. The cooling air enters through a gap between the rear lid and the spoiler top surface.

The maximum cooling air flow is achieved when the spoiler is extended. In order to implement adequately dimensioned inlets and to maximize the airflow the entire spoiler module has been redeveloped. The maximum lifting height and the maximum angle of attack have been greatly increased compared to its predecessor. Thus, on the one hand the area of the air inlets is enlarged and on the other hand the ambitious lift targets of the predecessor can still be achieved despite the additional airflow through the rear area of the car.

As for the predecessor an additional seal is deployed under the spoiler blade when the spoiler is extended. This element prevents a flow under the rear spoiler and in so doing increases its aerodynamic efficiency. The rear spoiler on the 911 Carrera delivers the necessary reduction of aerodynamic lift at the rear axle simultaneously with a significant reduction in drag. With improved sealing under the spoiler the desired lift at the rear axle can be achieved with further reduced air resistance.

In the newly developed spoiler module of the 911 Carrera, the functionality of the sealing element has now been further expanded since in addition to improving the aerodynamic efficiency, an optimal inflow to the air intakes of the charge air cooling is now also required. The sealing element consists of two individual diaphragms, which when the spoiler is retracted package under the spoiler one above the other, thus requiring minimum space. As the spoiler is deployed, the two diaphragms push apart and thereby deliver the best possible seal independent of the spoiler position. Depending on the vehicle version and the driving situation, the 911 Carrera spoiler deploys to a number of different positions, for example there are discreet spoiler positions for the conditions of open sunroof, or open convertible roof. There are also specific spoiler positions for vehicles with sports suspension. With the new sealing it is now possible to achieve the best possible seal through the entire movement of the spoiler.

1.1.2.3 Adaption of rear spoiler control due to charge air cooling

By the positioning of the air intakes for charge air cooling under the spoiler blade and the use of the positive pressure in front of the spoiler to drive the cooling flow, the adaptive rear spoiler in the new 911 Carrera now has a thermodynamic function in addition to its original aerodynamic function. To take this into account, the control strategy for adaptive rear spoiler had to be extended.

Up to now the spoiler was to be extended mainly from a certain speed threshold to meet vehicle-dynamics requirements. With a new function block for the charge air cooling in the advanced control strategy the rear spoiler can now be deployed when necessary even at-slower speeds in order to cool down the engine process air.

1.2 Boosting the HT-cooling system to satisfy the demands

The Porsche vehicle cooling system is also designed for use on the racetrack. The requirement here is that the cooling system fluids remain within allowable temperature limits even when the vehicle is operated at the extremes of performance.

The following units are integrated in the HT cooling circuit:

- engine incl. oil-/coolant heat exchanger and exhaust turbocharger (ETC)

- Gearbox oil-/coolant heat exchangers (HX)

- Porsche Dynamic Chassis Control "PDCC" (oil-/coolant heat exchanger)

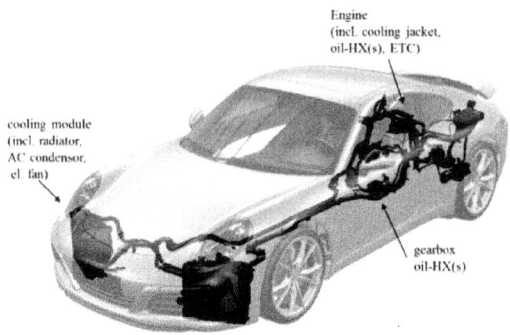

High Temperature (HT)-Cooling System of the new 911 Carrera

The newly developed turbocharged engines of the 2016 MY 911 Carrera show an efficiency advantage over a wide range of operation which has the desired positive effect concerning fuel consumption. But on the other hand at the design point of the HT-cooling system (i.e. under full load) a significantly higher waste heat has to be managed, as the "effective performance" on the racetrack is considerably higher than that of the predecessor. Because of this the high-temperature (HT) – cooling system has to be upgraded. The following actions have been taken to increase the performance of the heat sink:

Changes to the HT-cooling system between 911 Carrera MY 2012 and MY 2016:

Measure	Carrera S MY 2016 vs. MY 2012
radiator size (per side)	+ 16 %
electr. fan power (max.- per side)	+ 100 %
generator current (max.)	+ 40 %
air flowrate (radiator – max.)	+ 20 %

Enlarged, aerodynamically optimized front air intakes carry a portion of the airflow around the vehicle to the side cooling modules, mounted behind the bumper in front of the front wheels. As with the 2013 MY 911 Turbo "airblades" are attached to the outer edge of the air intakes. These air-guiding elements are increasing air flow through the air path while optimizing the flow around the front-end, thus reducing drag. The net surface area of the side radiator has risen by 16% compared to its predecessor. Thus, the effective area of the heat exchanger is increased with simultaneous reduction of the pressure loss in the air path.

The power-controlled electric fan behind the radiator has been specially developed for this application. A fan has been implemented that promotes maximum air flow for the limited amount of available power that can be permanently delivered by the electrical system. To achieve an optimum, fan blades with different number of blades and blade geometries were investigated in the vehicle environment in several iterations using CFD and finally checked as hardware in a real vehicle. By using a demand-orientated power control the maximum efficiency has been realized.

The final portion of the side air path, the ventilation outlet in the wheel housing, has been optimized to reduce the pressure loss. The ability of the outlet grille to guide the off-fan air out of the wheel housing was improved by enlarging the grill itself and re-designing the grille vanes. This was achieved with simultaneous consideration of the need for the wheel housing to protect the cooling module.

The summing of all these individual measures has resulted in an increase of the maximum cooling air flow rate by more than 20% compared to the predecessor, and this has been achieved without increasing the contribution of cooling air to the overall vehicle resistance.

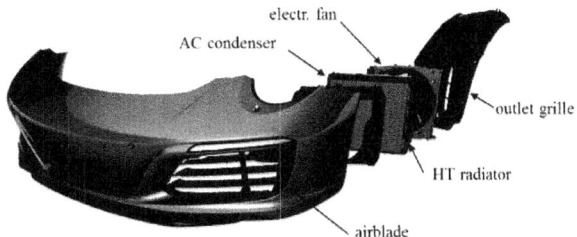

Front cooling air path of the new 911 Carrera

1.3 Heat protection for the MY2016 911 Carrera

The continuous development of the iconic 911 Carrera sports car and the increasing importance of economic and ecological aspects constantly bring new challenges for thermal protection. Furthermore, the ensuring of operational reliability becomes even more important because of increasingly densely packed engine compartments and increasing engine performance. For the customer the thermal reliability of the 911 Carrera is a matter of course, but for the vehicle development it is a challenge, which has to be ensured at all customer-relevant operating points through a suitable concept. Thus the thermal operational safety of the 911 Carrera is guaranteed, taking into account a variety of constraints, including a focus on the development of racetrack capability.

As for previous models, the heat protection concept of new 2016MY 911 Carrera has also been optimized according to function, weight and costs.

Package/routing of the exhaust system of the new 911 Carrera

1.3.1 Heat protection process

To assist in engineering the heat protection for the radical change from an aspirated engine to a turbocharged engine, the knowledge of the 911 Carrera Turbo has been used as a starting point for the development of the new 911 Carrera. The challenge was to integrate the new turbocharged engine into the engine compartment of the previous normally aspirated engine.

The engine compartment and underbody flow of the new 911 Carrera have been examined using thermal CFD simulations. By including convection, conduction and radiation effects an initial statement as to the expected component temperatures was determined. In this way it was possible to decide the size and shape of heat protection shields, the size and position of NACA ducts, the temperature class of components, the size and properties of insulation, the form of air deflectors and the performance of cooling fans at an early stage, before hardware was available. Therefore, the positioning and material selection of the new charge air cooling components for the new 911 Carrera engine took place in the early digital prototype development phase.

1.3.2   The challenges of new 2016MY 911 Carrera

The integration of the new turbocharged engines within the tight package constraints of the 2012MY 911 Carrera brought several challenges. The goal was to integrate a powerful intercooler concept in the classic design of the Carrera together with a visual differentiation from the 911 Turbo. The following points were focused upon in the development of the heat protection:

- Overcoming increased blockage in the engine compartment due to the integration of ducts and various other components such as pipes and engine control unit cover, etc.

- Optimization of the engine compartment airflow to improve convective component cooling, due to a small rear lid opening and small engine compartment fans

- Increased heat dissipation requirement due to higher power density in a tight package

- Oil sump material change to plastic, to reduce rear axle weight

1.3.3   The implemented heat protection concept

In contrast with conventional front-engined vehicles, where a powerful pressure difference between the cooling air inlet and various outlets drives a strong and constant flow through the engine compartment, the flow within the 911 Carrera engine compartment is controlled by the relatively weaker pressure difference between the upper and lower surface flows at the rear of the car, and is therefore assisted by two fans positioned under the trunk lid deck.

Along with the integration of the turbo-charged engine and its required sub-systems such as the intercooler air ducts, pipes and cover panel for the engine electronic control unit (ECU) have been introduced into the engine compartment with the help of CFD simulations, which allowed the development of a suitable engine compartment flow during an early stage of the project. As a result of this optimized air distribution in combination with a continuous demand-based regulation of the fans, all component temperatures can be kept within their specification limits, even when reducing the diameter of the fans and the corresponding engine lid opening. The fan regulation mentioned above also delivers a considerable reduction in electrical energy consumption.

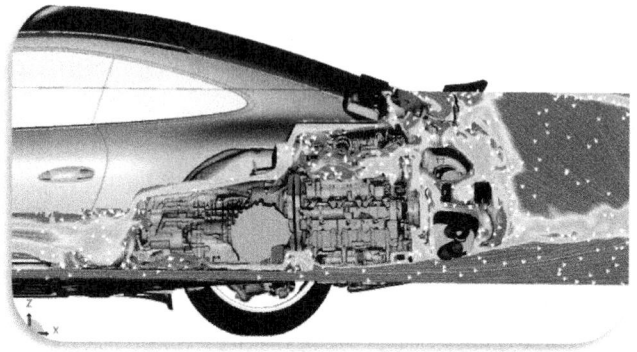

Air distribution through the engine compartment

The engine of the 911 Carrera is bounded at the sides and at the rear by the exhaust system. To protect temperature sensitive components from the high temperatures of the exhaust system in this area, the different temperature zones are separated by heat shield components. A comparison between the heat shields of the first and second generation of the 911 Carrera is shown in the figure below. The aims of separating the different temperature zones were:

- Reduction of radiation from the exhaust system to surrounded components

- Separation of different temperature zones in the engine compartment

- Convective cooling of all components in the different temperature zones+

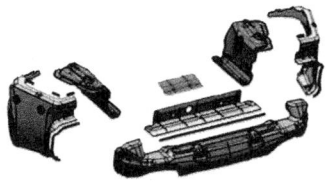
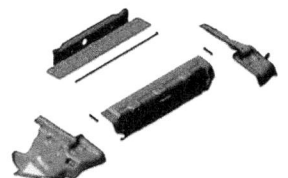

Comparison between the heat shields of the 2012 MY and 2016 MY 911 Carrera

The interaction between the heat shields around the exhaust system and the regulated engine compartment fans enables an optimal heat protection at every operating point. During driving, a great part of the waste heat of the exhaust system is discharged convectively by the underbody flow. Conversely when the vehicle is at a standstill the fans convey fresh air down into the engine compartment according to demand. An optimum time tag and rotation speed for the fans is calculated using several inputs.

The critical exhaust system waste heat at the start of an idling phase and after heat soak is blown out at the underbody and the wheel arch liner. The innovative heat protection concept enables large amounts of heat to be discharged from a smaller compartment space.

The rear weight reduction plays a very important role in relation to the vehicle balance of the 911 Carrera, for which the opportunities to reduce weight are highly restricted. The first time use of a plastic oil sump imposes another important requirement on the heat protection. The plastic oil sump is placed directly next to the manifolds because of the given package, in contrast to conventional front-engined vehicles, where the oil sump is placed far away from the hot exhaust system. Using a plastic oil sump with additional heat protection measures, the weight of the oil sump could be reduced by about 40 % in comparison to the previous one. A comparison between the oil sumps of the first and second generations of the 911 Carrera is shown in the following figure:

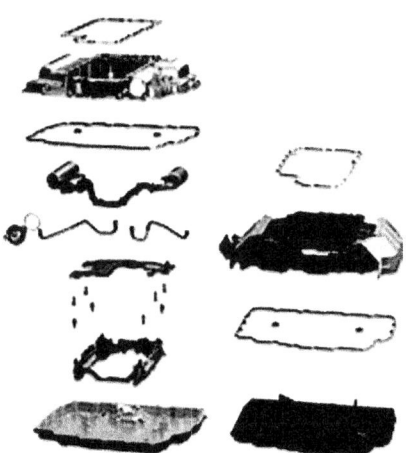

Comparison between the oil sumps of the first and second generations

2 Measures to optimize the efficiency

2.1 Optimization of the thermal management strategy during warm up

2.1.1 Comparison with predecessor

An active thermal management system was implemented in a Porsche sports car for the first time in the 2012 MY 911 Carrera. This allowed a demand orientated control of coolant flows to distribute heat in the area of the powertrain as well as the possibility to provide different temperature levels in the cooling circuit. By means of these features the warm-up time was significantly improved compared to the 2008 MY 911 Carrera (997 Model Gen. II) and a reduction in fuel consumption was achieved.

To reduce fuel consumption for the 2016 MY 911 Carrera e.g. a virtual gears feature has been introduced, resulting in a distinctly different driving profile compared to the predecessor. This leads to significantly less waste heat to be distributed during the warm-up phase.

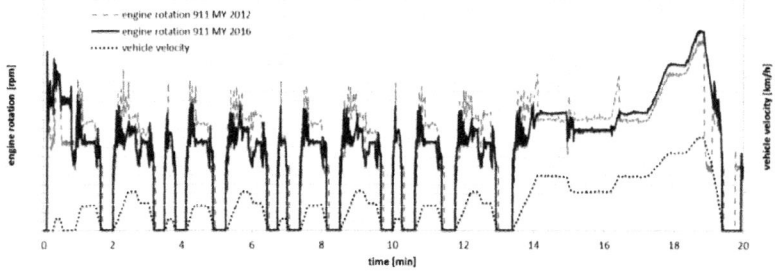

Comparison of driving profile during NEDC of the MY 2012 Carrera and 2016 MY Carrera

For further optimization of the warm-up phase, the switchable coolant pump was implemented in the MY 2016 911 Carrera, which encloses the pump wheel with a vacuum-actuated ring slider and can thus completely eliminate the flow:

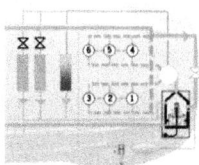 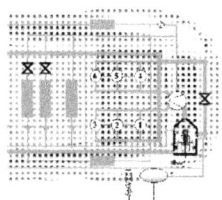

Reduced coolant-flow during warm-up
@ Carrera MY 2012 by Bypass-Valve
(only oilcooler)

Zero (coolant-) flow during warm-up
@ Carrera MY 2016 by switchable
Main-Pump

This was the first time that in a sports car "zero-flow" could be realized in the warm-up phase, whereby a very rapid heat-up of the coolant is possible.

2.1.2 Benefit/Potential

Due to the changed driving profile and the new generation of engines, the thermal management warm-up strategy had to be adapted in order to achieve a quick warm up. For this, the optimal strategy of targeted preheating of the power train was determined by both vehicle testing and through simulations.

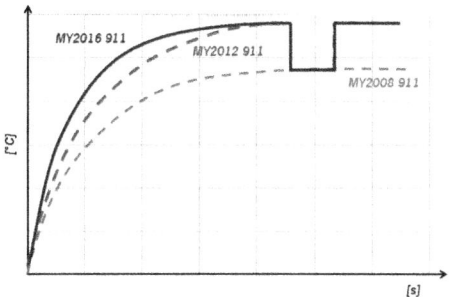

Comparision oft the warm-up phase between 3 generations of 911 Carrera

Despite the change to the driving profile of the 2016 MY 911 Carrera the warm-up time is faster than that of the 2012 MY. This corresponds to a reduction in fuel consumption in the NEDC of approx. 1 %.

2.2 Reduction in driving resistance by active radiator shutters

2.2.1 Motivation / concept

As already mentioned, the design point performance of the powertrain cooling system is required to fulfill the peak operating conditions at full load to allow operation on the race track for an extended period at high ambient temperatures. That premise ensures the claim of Porsche to develop production vehicles that can be operated on the race track without modifications. In "average usage" e.g. in normal public usage/on the highway a significantly lower cooling capacity is required, resulting in a potential to reduce the overall vehicle drag by reducing the volumetric flow rate of cooling air through the system. Therefore an adaptive cooling air control is implemented in the new 911 Carrera, which represents a unique selling point in this vehicle class. Active shutters can be closed gradually to allow through only the amount of air flow required by the two side-mounted cooling modules. By positioning the active shutters as close as possible to the outer shell of the front bumper much of the theoretical potential which is obtained by closure of the air path can be used. The closing of the radiator shutter results in a drag reduction of about 4%.

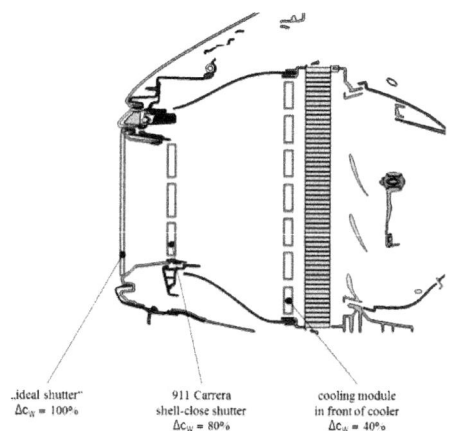

„ideal shutter"	911 Carrera	cooling module
$\Delta c_W = 100\%$	shell-close shutter	in front of cooler
	$\Delta c_W = 80\%$	$\Delta c_W = 40\%$

Aerodynamic potential of radiator shutters depending on position along the air path

Another requirement to achieve the maximum drag-reduction potential is the multi-stage adjustment of the cooling air flaps using an electric actuator. This allows only the amount of air actually needed to reach the cooler and therefore at any time to keep the air resistance as low as possible.

In addition to the air resistance the lift balance of the vehicle is naturally affected when the cooling air flow rate varies. Therefore vehicle-dynamics demands are taken into consideration by the intelligent control of the radiator shutters. In order to satisfy the core brand value for Porsche vehicles of outstanding driving dynamics, variation of the lift balance by the shutters is not allowed at very high speeds or during sporty driving.

Active radiator shutters (right picture is only for illustration: left/right synchronized during operation)

2.2.2 Intelligent Control/Application

The active radiator shutters in the new 911 Carrera allows on-demand cooling of the powertrain and AC system at any time and results in an increase in aerodynamic efficiency, which is noticeable as a fuel consumption saving especially in customer-oriented driving.

For the control of the active radiator shutters cooling requirements of the engine, transmission and air conditioning are determined and additionally the power losses of vehicle drag, electric radiator fan and the air conditioning are considered. In this way the calculation of the optimum shutter position and fan actuation for the current driving condition is calculated in real time to provide the best overall vehicle efficiency.

For maximum efficiency, the active radiator shutters can be adjusted in 5 positions (open, closed, and 3 intermediate positions).

2.2.3 Benefit/Potential

In order to get a customer relevant evaluation of the efficiency of the active radiator shutter system (hardware and application) so-called "endurance-testing" vehicles, which complete a customer-relevant driving program, were monitored and analyzed.

Because the ambient temperature has a significant influence upon the use of the air conditioning system, which has a direct influence on the control of the active radiator shutters, the percentage duration of the shutters in the open, closed and intermediate positions in an approximately10,000 km "customer-oriented endurance test" covering both winter and summer periods is shown below:

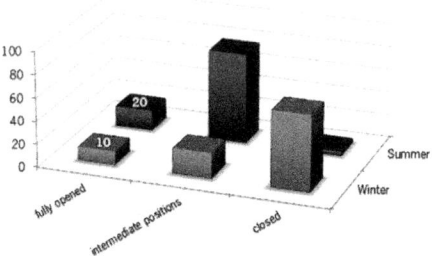

Percentage (time) of radiator shutter positions during long-term customer driving

As expected, it was found that the fully open shutter position is required in only about 10% of the driving time and 20% in the summer. For most of the time a partially or completely closed radiator shutter fulfilled the cooling air flow requirements (powertrain, air conditioning). The figure above shows that during winter approximately 80% and also during (german) summer conditions approximately 50% of the maximum potential could be realized. This leads to a seasonally-adjusted overall potential of approx. 70 % based on the maximum potential.

This confirms the effectiveness of the active radiator shutter system (hardware and application) in the new 911 Carrera.

3 Conclusion

With the integration of the new turbo engines in the new 911 Carrera a high performance charge air cooling system was installed to ensure the typical Porsche performance. This lead to a new heat protection concept which had both to deal with the tight package of the predecessor and make major weight improvements (e.g. plastic oil sump) possible. Additionally, the HT-cooling system, radiators, fans and cooling air paths were optimized to meet the demanding requirements of the racetrack.

Finally the thermal management was optimized to reduce warm-up time and more efficient aerodynamics with the introduction of active radiator shutters help to reduce the driving resistance and thus reduce the fuel consumption In this way the differentiation between vehicle operation in respectively performance and efficiency driving modes has been increased to deliver maximum suitability for daily use.

Holistic approach for the design of electrical powertrains for electric and plug-in-hybrid vehicles applying the methodology Multi-Objective Optimization

Dipl.-Ing. Adam Babik, Dipl.-Ing. Thibaut Reuschlé, Dr.-Ing. Andreas Schönknecht

Robert Bosch GmbH

1 Introduction

Reduction of CO_2[1] emissions produced by road traffic is a paramount objective which the automobile industry has to resolve. An average fleet emissions target of 95g/km CO_2 has to be achieved within the European Union until the year 2020/21. In order to comply with this demanding emissions standard, electrification of conventional combustion engine dominant drive trains is inevitable. Alternatively an increase of battery-electric-vehicles (BEV) in the OEM's fleet can contribute significantly towards the aimed CO_2 reduction. Both strategies mentioned above – the electrification of the combustion engine dominant drive train with hybrid vehicle topologies as well as the increase of BEVs within the OEM's fleet share – require new, holistic methods for the drive-train design to assure cost efficiency and end-customer's vehicle requirements. Numerical development methods, which will be introduced in this paper, can be applied to support the product development in decision making by quantifying the interdependencies between the electric, electro-mechanical and electro-chemical components on vehicle system level. For an overall minimal system cost design a multiplicity of interactions between all powertrain components has to be taken into account. The high system complexity makes it infeasible for an expert to consider all relevant correlations for a minimal system cost design with given requirements and premises on vehicle level. Therefore a simulation-based system development approach using a Multi Objective Optimization (MOO) methodology is presented in this paper, which is also investigated in further scientific research articles [1, 2]. The main focus is on electric powertrains (ePT) for battery-electric-vehicles (BEV) and plug-in-hybrid-vehicles (PHEV) to find the best cost/benefit trade-off regarding system cost while considering all defined system performance requirements. Figure 1 illustrates the technical performance and total cost of the electric powertrain (ePT) which are influenced by a multitude (more than 15) of system relevant component parameters. A conventional, component-based development method can lead to a conservative dimensioning of components and hence not into an overall cost-effective drive train system design. E.g., increasing the inverter efficiency, which results in higher inverter cost, can be overcompensated by the effects of higher system efficiency and lower energy consumption, resulting in in a lower overall system cost on vehicle level. The battery system has a major influence on the overall system cost and reducing the battery system capacity with a more efficient drive train system design reduces eventually the total system costs.

The holistic approach will be applied on two actual challenges in powertrain design:

- BEV: Investigation of one vs. two gear transmission system

---

[1] CO_2: Carbon Dioxide

- PHEV: Impact of optimization targets (CO_2 vs. system cost and vehicle acceleration vs. system cost) on system design

The simulation results carried out will be discussed in Chapter 3.

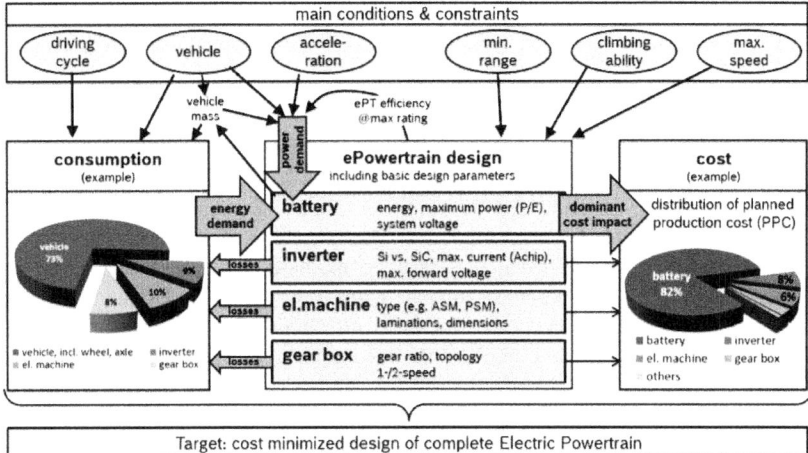

Figure 1: Requirements, design parameters and correlations between efficiency and system cost in Electric Powertrains (ePT)

2 System Optimization Environment

2.1 Overview

The vehicle system optimization environment can be subdivided into the following tasks:

- Pre-Processing
- Numerical Simulation & Optimization
- Post-Processing (Analysis & Evaluation)

The pre-process defines the optimization task, the optimization boundary conditions (e.g. vehicle driving range, driving cycle) and the vehicle class selection (e.g. compact class vehicle). This working stage requires a wide user knowledge in terms of physical relations in the vehicle's drive-train. An implausible optimization problem definition will be captured automatically by implemented plausibility checks, before an initial optimization run is performed.

During the numerical simulation and optimization process a specific drive-train composition is simulated and the main simulation results (according to the specified optimization objectives) are transferred back to the optimization algorithm. The implemented optimization algorithm, which is based on genetic evolutionary methods [3], determines the new main component parameters for the subsequent vehicle simulation run. This optimization process is running automatically in a loop until a pre-defined quantity of valid drive-train designs are evaluated.

The optimization procedure requires a fast running simulation environment. For this reason a map-based modeling for the drive-train components (E-machine & inverter, mechanical transmission system, battery cells) has been chosen. The pre-calculation of the characteristic maps (E-machine & inverter, mechanical transmission system) is based on empiric-physical in-house software tools. The technical and cost calculation will be described in Chapters 2.2 & 2.3 respectively. Subsequent to the component calculation the vehicle simulation run is performed. A detailed summary of the vehicle simulation environment is given in Chapter 2.4. Figure 2 gives an overview on the optimization workflow for an electro-mechanical drive-train.

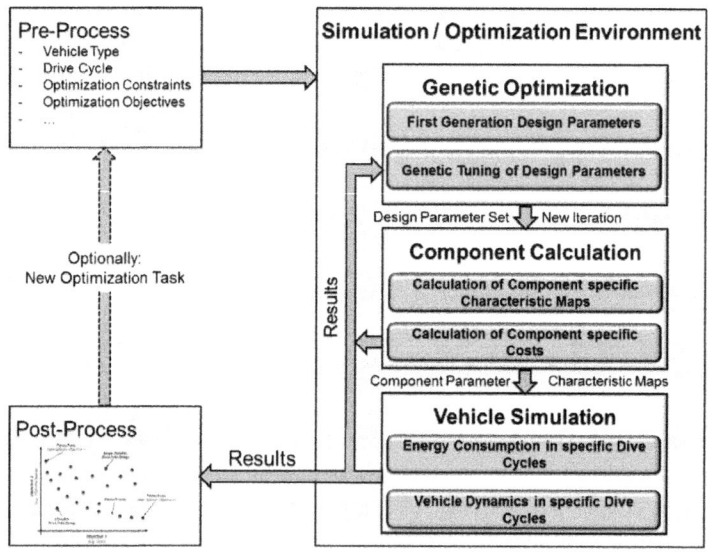

Figure 2: Workflow for the Drive-train Optimization Process;
see Figure 6 for Post-Process Details

2.2 Technical Component Calculation

The technical component calculation defines the physical behavior of a specific drive-train component. The considered components in the electro-mechanical drive-train are:

- High voltage (HV) battery system (considering high power & high energy cells)
- Power electronics (Inverter – IGBT & MOSFET technologies)
- Electric machine (induction & permanent magnet machine)
- Mechanical transmission system (1 & 2-speed transmission system)

The HV-battery system is modelled via an equivalent electric circuit, whereas the cell behavior is described via characteristic maps based on empirical data. The physical behavior of the E-machine, combined with the inverter system, is described via characteristic maps. The calculations for the power electronics as well as for the electric machines are based on Bosch's in-house calculation tools. Figure 3 shows the main parameters for a fast and vehicle system relevant characteristic map calculation for a specific E-machine.

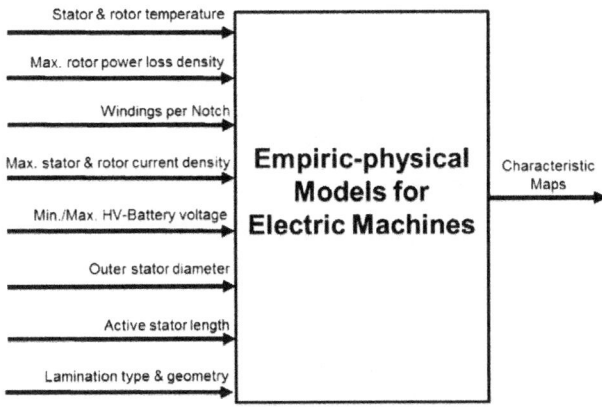

Figure 3: Main Parameters for the Calculation Characteristic Map Calculation of an E-Machine

The calculation of the characteristic maps for the mechanical transmission system are based on empirical-physical formula and take the main transmission system losses (load dependent losses (e.g. gearing losses) & speed dependent losses (e.g. churning losses) into account.

2.3 Component Cost Calculation

The component costs calculation for the electro-mechanical drive-train is based on the main component parameters which are defined in the technical component calculations. The overall costs for a specific component are calculated based on material and production costs, which make the cost calculation scalable for each individual drive-train component. The production costs take into account today's state-of-the-art production technologies. In case of the high-voltage battery systems an additional holistic approach can be selected for the cost calculation of future battery cell technologies. The drive-train costs are composed of:

- Battery system
- Power electronics (Inverter)
- Electric machine
- Mechanical transmission system

2.4 Vehicle System Simulation

Vehicle performance characteristics (e.g., energy consumption, acceleration ability, etc.) for a given vehicle system can be identified by applying numeric vehicle simulation. The vehicle simulation is performed in a dedicated tool for vehicle simulation. The realization as an object oriented model offers an efficient methodology to construct drive train topologies with standardized interfaces. These vehicle models are based on extensive component libraries (e.g., electrical machines, battery systems, etc.). Further benefits of the object oriented simulation methodology are standardized model structures as well as efficient model maintenance.

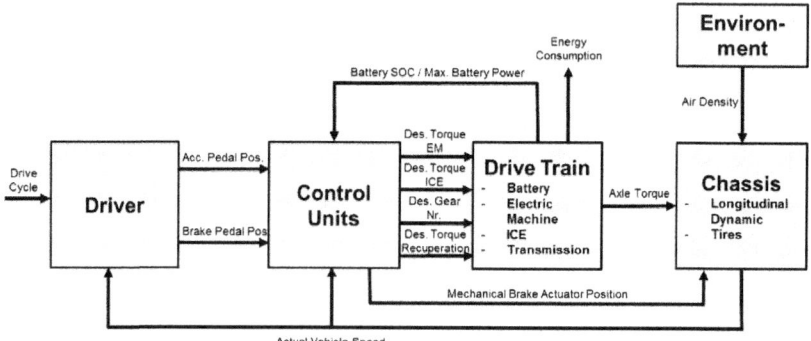

Figure 4: Simplified PHEV Model Structure

Characteristic map based models for the electric machine and mechanical transmission subsystems are used in the vehicle simulation to describe their technical characteristics. The required characteristic maps (power loss maps, generator and motor limiting curves, etc.) are derived based on pre-process calculations for each component (see also Chapter 2.2). The initial estimate assumes an optimal control strategy by the power electronics (inverter) for the electric machine. Hence, inverter losses are calculated based on the electric machine characteristic maps and provided to the vehicle simulation tool as a three dimensional (e.g. efficiency as a function of battery voltage, shaft speed and shaft torque) look-up table (see Figure 5).

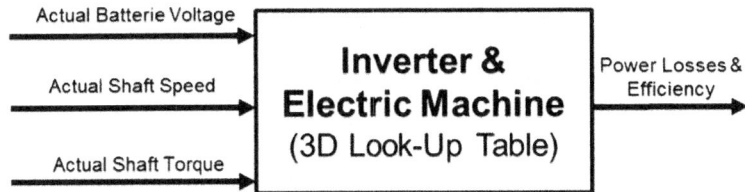

Figure 5: Power Electronics and E-Machine Model

The traction battery model is based on an electrical equivalent circuit diagram. Hence each battery cell is characterized by a resistance element, which reflects the cell's inner resistance. Characteristic maps describe both inner resistance and open circuit voltage of each cell. In this manner both state of charge (SoC) and temperature dependencies are taken into consideration. Separate maps are utilized for charging and discharging. In addition, the battery model integrates the battery management which monitors and ensures the state (current and voltage) of each cell. As a result of battery model structure based on an electrical equivalent circuit diagram, scaling with respect to number of serial and parallel cells is possible.

The physical based vehicle model takes all defined subsystem interactions into account. Interactions between drive train components and vehicle chassis (chassis and tires) and furthermore to the vehicle's environment are all taken into account. The vehicle acceleration is calculated in the simulation for each numeric integration step by opposing the driving force to the driving resistance. The vehicle speed control is realized by a generic driver, for which a feed-forward model approach is required. The presented vehicle simulation environment allows the identification of energy savings potential on vehicle system level resulting from new component control strategies.

2.5 Analysis & Evaluation

For analysis and evaluation a graphical user interface (GUI) is established for user guidance. The comprehensive post-processing environment allows the user to analyze different objectives and/or technical and cost interdependencies of the calculated drive-train designs on a two-dimensional diagram. Figure 6 shows a schematic result diagram of a multi-objective optimization with 2 optimization targets. The most suitable drive-train designs - optimized for the specific problem formulation - are plotted on the Pareto Front.

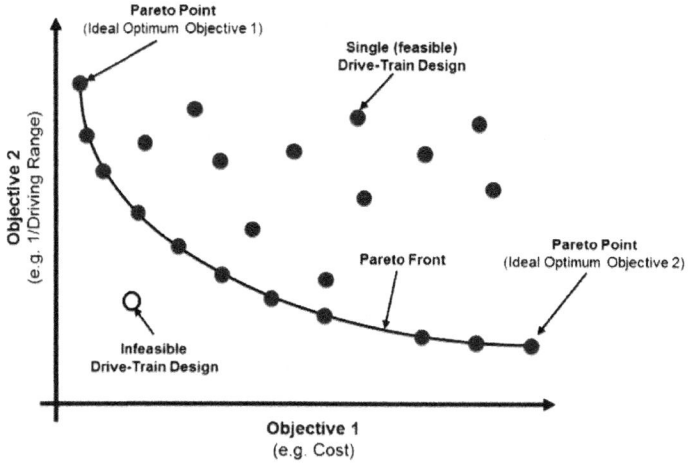

Figure 6: Schematic Representation of a Pareto Front with two Optimization Targets

3 Simulation and Evaluation

In the following Chapters (3.1 & 3.2) selected simulation results for optimized electro-mechanical drive-trains of a single compact class vehicle will be presented. Chapter 3.1 presents results for a battery electric vehicle (BEV). Chapter 3.2 provides results for a plug-in hybrid (PHEV) powertrain.

3.1 Multi-Objective Optimization of Battery-Electric-Vehicles

3.1.1 Introduction

In this paper a study comparing 1 speed versus 2 speed transmission system shall be presented. Table 1 gives an overview on the main vehicle parameters for the representative compact vehicle under investigation:

Table 1: Main Vehicle Parameters & Optimization Constraints

Compact Class Vehicle		
System Requirements	Driving Range in Drive Cycle	160 km
	Max. Vehicle Speed:	160 km/h
	Min. Driving Slope:	15% @ 50 km/h
	Acceleration Requirements: 0-100 km/h	8 or 12 seconds
HV Battery System	Cell Capacity:	34 Ah (Li-Ion)
	Serial/parallel Cells:	Variable (optimization parameter)
	Initial SoC[2]:	70% in NEDC
		50% @ acceleration maneuver
		50% @ slope drive maneuver
Electric Machine	Induction Machine	4-pole
Inverter Technology	IGBT Modules[3] or SiC MOSFETs[4]	
	Voltage Range:	650V
Mechanical Transmission	First Speed Gear Ratio:	variable (optimization parameter)
	Second Speed Gear Ratio:	variable (optimization parameter)
Drive Cycle	NEDC	
Optimization Target	Drive-Train Cost versus Vehicle Acceleration Ability	

---

[2] SoC: State of Charge
[3] Insulated-Gate Bipolar Transistor
[4] Silicon Carbide Metal-Oxide-Semiconductor Field-Effect Transistor

Transmission gear actuation losses are neglected – this simplification has an impact only during gear changes. The investigation regarding 2-speed transmission systems takes into account two different vehicle acceleration requirements (8 and 12 seconds for 0 – 100 km/h) and two different inverter technologies (IGBT and SiC MOSFET).

The optimization target was to identify the lowest drive-train costs (optimization objective 1) while considering the second optimization objective of best achievable vehicle acceleration.

3.1.2 Technical Results

Figures 7 & 8 show the energy consumption of the most energy efficient vehicles (results at the Pareto Front) in the NEDC. Both drive-train designs satisfy the corresponding acceleration and system requirements listed in Table 1. Figure 7 shows the energy consumption for designs fulfilling the 8 second acceleration requirement while Figure 8 data refers to a 12 second acceleration requirement. The relative energy savings potential is summarized in Table 2.

The energy savings potentials in Table 2 are calculated by neglecting gear shift actuation energies and by neglecting power losses caused by the load-free spur gear pairing. When taking gear shift and load-free gear pairing losses into account, these losses may outweigh the relative small energy savings in Table 2 and have a net negative effect.

The absolute energy losses relating to the drive-train components including inverter, electric machine and the transmission system are shown in Figure 9. The battery losses are neglected while the electric power requirement for auxiliaries amounts to a constant of 200 W in the system optimization.

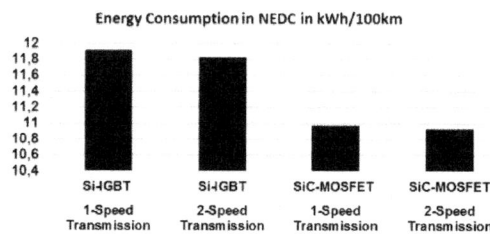

Figure 7: Energy Consumption of an optimized Compact Class Vehicle considering 8 second Acceleration Requirement (0-100 km/h)

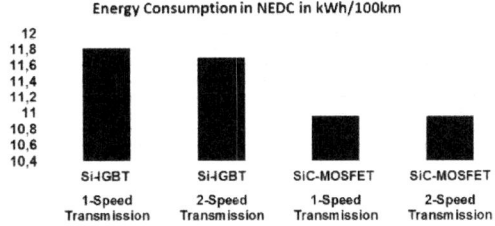

Figure 8: Energy Consumption of an optimized Compact Class Vehicle considering 12 second Acceleration Requirement (0-100 km/h)

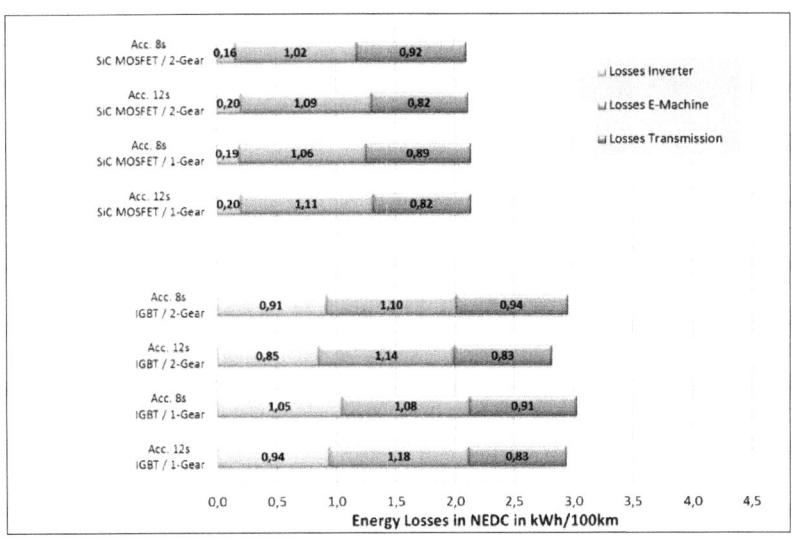

Figure 9: Absolute Drive-Train Energy Losses of an optimized Compact Class Vehicle in NEDC considering 8/12 second Acceleration Requirement (0-100 km/h) and IGBT/SiC MOSFET Inverter

341

Table 2: Energy Savings Potentials in NEDC

8 second Acceleration Requirement	
IGBT Technology:	Relative Energy Consumption Reduction due to 2nd Gear
	~0.7%
SiC MOSFET Technology:	Relative Energy Consumption Reduction due to 2nd Gear
	~0.4%
12 second Acceleration Requirement	
IGBT Technology:	Relative Energy Consumption Reduction due to 2nd Gear
	~1.1%
SiC MOSFET Technology:	Relative Energy Consumption Reduction due to 2nd Gear
	~0.2%

3.1.3 Conclusions

The comparison of a 1-speed vs. 2-speed transmission system shows a small energy savings potential (max. 1.1% for IGBT technology) for a 2-speed transmission system of a compact class vehicle under the assumed vehicle requirements presented in Table 1. For the 2-speed transmission system, the energy savings potential is higher for IGBT inverter technology in comparison to SiC MOSFET technology. This results from lower power losses in the drive train with SiC MOSFETs. Inverter energy losses for IGBT technology/2-speed transmission systems are reduced (up to 13%) in comparison to the 1-speed transmission systems resulting from lower machine torque requirements which leads to lower peak currents in the power electronics as well as lower inverter costs due to smaller chip size requirements.

The relative costs for the chosen drive-train designs are shown in Figure 10. Baseline references (100%) are defined by drive-train designs with 1-speed transmission. Figure 10 illustrates drive-train cost benefits for 1-speed transmission systems compared to the 2-speed transmission systems when same boundary conditions (inverter technology & acceleration requirement) are taken into account.

It has to be pointed out, that for larger vehicle classes (with higher vehicle weight, e.g. SUVs) and under the same boundary conditions as presented in Table 1, the energy savings potential (due to a second gear) in NEDC is higher, compared to the compact class vehicle.

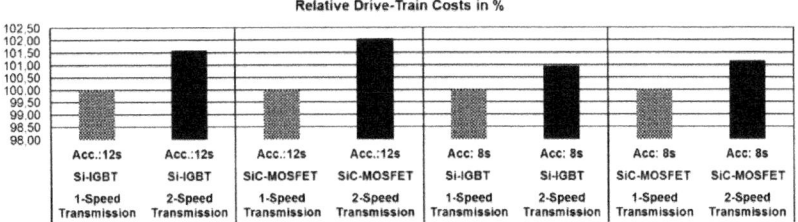

Figure 10: Relative Drive-Train Costs / Reference: 1-Speed Transmission

3.2 Multi-objective optimization of plug-in hybrid powertrains

3.2.1 Introduction

In this section, results regarding the analysis of optimization target (multi-objective) selection for plug-in hybrid powertrain are presented. The main area of interest in research of plug-in hybrid powertrains, in addition to similar questions as battery electric vehicles (BEVs), (e.g. efficiency, power, energy-content, etc.) is the influence of the internal combustion engine (ICE) and its interaction with the electrified powertrain. In order to analyze these specific aspects, several goals were defined on system-level (analogue to the goal definition on system-level for BEV-powertrain, see Section 1). These are CO_2 emissions and electric range determined during the specific plug-in hybrid electric vehicle (PHEV) certification procedure (certification procedure ECE R101 of the United Nations Economic Commission for Europe (UNECE) [5]), acceleration time from 0 to 100 km/h in electric mode and system costs. Since the considered powertrains are evaluated according to OEM-relevant criteria on system level, this work has a big practical significance. Several optimization tasks can be analyzed, e.g.: *CO_2 vs. electric powertrain costs, acceleration time from 0 to 100 km/h vs. cost, CO_2 vs. electric powertrain without battery cost, Acceleration time from 0 to 100 km/h vs. CO_2*, etc.

The costs of all components making up the electric powertrain are estimated as *ePowertrain Cost* according to the same approach as presented in Section 2.3; it is important to note that the conventional powertrain costs are not taken into account during the optimization. Indeed only components of the electric powertrain are varied during the optimization. This is consistent with an OEM point of view: given a base vehicle and powertrain combination, a plug-in hybrid variant can be established by electrification. In this approach, the ICE and gearbox systems remain constant. Acceleration performance is used as the main power requirement for all powertrains. Conversely, the certification cycle does not require significant power [5]. Hence, powertrains only designed primarily to reduce CO_2 are weak with regards to acceleration and hence less customer-friendly. However, PHEV buyers require the same dynamic driving characteristic in all-electric mode as in BEVs. In order to consider this in the evaluation of the powertrains during the optimization, a specific acceleration time is always defined, either as constraint or as an optimization goal.

The optimization task *CO₂ vs. ePowertrain Costs* has already been examined by Reuschle et al. in [6]. It is based on an OEM approach: given a vehicle class, what is the best electrified compromise between CO_2-reduction and additional cost? Indeed, in a first approach, increasing CO_2 reductions are achieved with higher and higher costs due to the need for an increasingly larger battery, and vice versa. In this paper, the result of the investigation *Acceleration time from 0 to 100 km/h vs. ePowertrain Costs* is presented. This optimization task is motivated by a completely different mind-set. It is an attempt to optimize the plug-in hybrid powertrain towards maximum performance and minimal costs without directly considering CO_2 emissions. Comparing the results obtained in this case with those obtained in the previous one is of interest. The question can be raised if it is more profitable to optimize powertrain regarding performance in order to obtain powertrain designs with a better CO_2/cost ratio which are even more customer-friendly.

The compact class vehicle segment is considered in this investigation. The P2 hybrid topology is the area of focus for this paper. The energy management strategy chosen is a causal strategy called Equivalent Consumption Minimization Strategy (ECMS) [4]. This strategy is not varied during the optimization. First of all, as the strategy is causal, it is clear that the strategy has deviations from the optimum, nevertheless it has been shown that it performs with little deviation from an optimal strategy on flat track [4]. Second, the optimization of the energy management would imply a second optimization loop during optimization and hence lead to a significant increase in computing time.

3.2.2 Definition of simulation set-ups

The simulation task and the defined parameter ranges for the multi-objective optimization are summarized in Table 3.

Table 3: Parameter Ranges for Optimization Tasks

Optimization task	1	2
	CO_2 vs. ePowertrain Cost	0→100 km/h vs. ePowertrain Cost
Battery		
Basic capacity	28 Ah	28 Ah
P/E	12 kW/kWh	12 kW/kWh
# cells serial	70..140	70..140
# cells parallel	1..2	1..2
Cell scale factor	0,6..1,5	0,6..1,5
Semi-conductor		
Type	IGBT(650V)	IGBT(650V)
Chip surface per switch	300..600 mm²	300..600 mm²
Electric Machine		
Type	PSM (IMG)	PSM (IMG)
Diameter (magnetic length)	340..400 mm	340..400 mm
Length (magnetic length)	60..130 mm	60..130 mm
# coil windings per notch	70..210	70..210
Constraints	0→100 km/h in less than 10 s	
	$V_{max} \geq 135$ km/h (electric drive modus)	
	Electric range > 50 km	

3.2.3 CO_2 vs. ePowertrain Cost and Acceleration time vs. ePowertrain Cost

The results, i.e. the Pareto-optimal designs, of the optimization tasks *CO_2 vs. ePowertrain Cost* and *Acceleration time 0 → 100 km/h vs. ePowertrain Cost* are plotted in Figure 11. Firstly, for optimization task 2, the optimization heuristic finds more designs with ePowertrain cost below 250 arbitrary unit (a.u.) than optimization task 1. These designs are optimal for acceleration time vs. cost ePowertrain but result in higher CO_2 emissions (30 g/km and upwards) than the CO_2 vs. cost ePowertrain optimized designs. Further, the design which lies at the intersection of both Pareto fronts in Figure (right) is the last design from optimization task 1 which fulfills the 10 s acceleration time requirement. This design costs nearly 250 a.u. and emits ca. 30 g/km CO_2. This overlap confirms that the heuristic during the optimization task 1 found the best cost/CO_2 trade-off design that fulfills the acceleration time requirement.

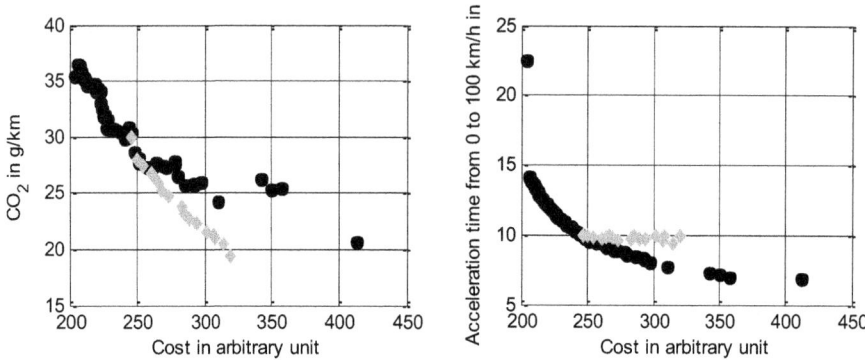

Figure 11: Pareto fronts of simulation tasks CO_2 vs. ePowertrain Cost (green/grey) and Acceleration time 0 → 100 km/h vs. ePowertrain Cost (black)

In addition, the optimization heuristic found more designs with CO_2 emissions below 30 g/km during optimization task 1 than optimization task 2. In fact, as the CO_2 emissions of the designs decrease along the Pareto front, their ePowertrain cost increases, meanwhile fulfilling almost exactly the 10 s acceleration requirement. On the other hand, designs found in optimization task 2 with CO_2 emissions below 30 g/km have better acceleration time (less than 10 s) however emit more CO_2 emissions at the same ePowertrain cost. These designs represent the minimal acceleration time reachable for this set-up.

In case of optimization task 1, the heuristic finds inverter/electric machine configurations which fulfill the acceleration time requirement during the optimization loop. As this acceleration time requirement is more demanding than the cycle requirement, this combination will varied further to optimize the CO_2/cost ratio during further iterations of the loop. In other words, more leverage is achieved on system-level to increase the battery energy-content than to make increase inverter/electric machine power to optimize the CO_2 vs. cost ratio. Hence, the performance requirement itself does not significantly influence the CO_2/cost trade-off. This can be seen in Figure 12. During optimization task 2, all three components are changed through the designs in order to improve the acceleration time as one can see with respect to each component's cost. Whereas during optimization task 1 the inverter/electric machine cost is quite constant throughout the optimization (optimal designs achieve 10 s), while only battery scaling is varied to influence the CO_2/cost ratio.

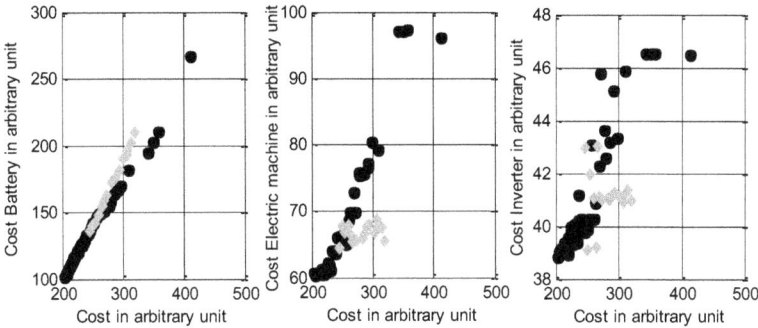

Figure 12: Costs distribution of Pareto-optimal designs of tasks 1 (green/grey) and task 2 (black)

3.2.4   Conclusion

In summary, important insights have been found via the analysis of different optimization goal pairs. The optimization task *Acceleration time 0 → 100 km/h vs. ePowertrain Cost* enables to find the absolute minimum acceleration time for a given optimization set-up. This is useful to identify the level of the performance requirement which could be set for the optimization task *CO_2 vs. ePowertrain Cost*. Furthermore, once this performance requirement has been set, the heuristic can within another optimization task *CO_2 vs. ePowertrain Costs* find the CO_2/cost optimal trade-off by varying the battery parameters. Indeed, the parameters of the electric machine and the inverter are determined by the performance requirement and are nearly negligible influence on the CO_2/cost ratio. Further optimization tasks could be investigated. First, the optimization task *Acceleration time from 0 to 100 km/h vs. CO_2* could be considered: find the best optimization objective set-up in order to obtain designs which are a good compromise between cost, efficiency and acceleration performance. In addition to that, as battery cost are the major cost factor of an electric powertrain the optimization task *CO_2 vs. ePowertrain without battery Costs* is interesting. The point thereby is to analyze how optimized powertrains differ without the strong cost impact of the battery.

4 Summary & Outlook

The presented optimization environment for electro-mechanical drive-trains supports the engineering process at an early stage of product development. Therefore, a cost-optimal drive-train configuration can be evaluated for a specific vehicle class. A modular system for electro-mechanical and electric components can be derived based on the classification of the optimization results across multiple vehicle classes and drive-train topologies. The numerical environment is flexible regarding model extensions, like new electric machine, inverter and mechanical transmission system models. Further on, new technologies (e.g. silicon carbide MOSFETs[5]) can be evaluated on vehicle system level (BEVs and PHEVs) and the corresponding interdependencies in the drive-train system can be examined. Future extensions of the optimization environment take improved optimization algorithms and the implementation of new drive-train topologies into account.

References

1. Eghtessad, M.; Meier, T.; Rinderknecht, S.; Küçükay, F.: Antriebsstrangoptimierung von Elektrofahrzeugen. ATZ-Automobiltechnische Zeitschrift, 117, 78-85, 2015.

2. Eghtessad, M.: Optimale Antriebsstrangkonfigurationen für Elektrofahrzeuge. Aachen: Shaker-Verlag, 2014.

3. Siebertz, K.; Bebber, D.; Hochkirchen, T.: Statistische Versuchsplanung: Design of Experiments (DoE). 1. Aufl., Berlin Heidelberg New York: Springer-Verlag, 2010.

4. Ambühl, D.: Energy management strategies for hybrid electric vehicles. Zürich: ETH Zürich, 2009.

5. UNECE. Regulation 101 Revision 3 CO_2 emission/fuel consumption and amendments. February 2015.

6. Reuschlé, T.; Albers, A.; Babik, A.; Schönknecht, A.: Multi-Objective Optimization of Plug-In Hybrid Powertrain - Certification Procedure Sensitivity. [Hrsg.] Heinz Schäfer. Würzburg: Expert Verlag, 2015. Elektrische Traktions- und Hilfsantriebe für die Elektrifizierung und Hybridisierung.

---

[5] MOSFET: Metal-Oxide-Semiconductor Field-Effect Transistor

Electric drivetrain modular layout based on range specifications

Dipl-Ing. Markus Orner, Dr.-Ing. Thomas Riemer, Prof. Dr.-Ing. Hans-Christian Reuss

FKFS

1 Introduction

After many years with low number of units, sales figures for electric vehicles (EVs) start rising. This leads to diversified specifications in range and drive power for EVs of the next generations. Single EV models are likely to be sold not only with different drive powers (like with combustion drivetrains) but also with different battery sizes (and thus ranges). Specific layout tools are necessary for this way of modularization, to identify optimal solutions not only for a single drivetrain but for a whole portfolio.

This paper describes a method, which includes range specifications in the drivetrain analysis. Therefore a drivetrain model is introduced and scaling parameters are explained and their impact of the drivetrain is analyzed.

After that the analysis approach is shown and constraints for the drivetrain configurations are defined.

In the last section drivetrains for two different range specifications are to be identified. Therefore a modular layout is presented and the solution space of optimal configurations is analyzed. Finally, two drivetrains are selected and the adaption of drivetrain components is discussed.

2 Drivetrain Model

Special requirements for the drivetrain model must be met, to examine the impact of range on the drivetrain layout. Thus, detailed modeling of an existing vehicle is necessary. For the analysis two drivetrain models are set up. The real drive model calculates range and consumption in real world cycles and the acceleration model determines the full load acceleration time. After that, degrees of freedom in the drivetrain are added. Especially the physical impact of the degrees of freedom must be modelled.

2.1 Basic Model

The basic drivetrain model is based on the Smart Electric Drive (Figure 1). The driving resistance [1] of the vehicle is determined from freewheeling tests and described according to equations (1)-(4), with θ representing the inertia of the drive train at the wheel.

$$F_L = \frac{1}{2} \cdot \rho \cdot c_w \cdot A \cdot v^2 \tag{1}$$

$$F_r = m_{ges} \cdot g \cdot f_r \cdot \cos \alpha \tag{2}$$

$$F_{sl} = m_{ges} \cdot g \cdot \sin \alpha \tag{3}$$

$$F_a = a \cdot \left(m_{ges} + \frac{\theta}{r_{whl}^2} \right) \tag{4}$$

The load for auxiliary consumers was determined in testing series on the Stuttgart driving circuit [2] and is set to 450W, excluding air conditioning. The efficiency model of the transmission is based on measurements of similar transmissions and differential gears.

Figure 1: Basic configuration: Smart ED

For the permanent excited synchronous machine (PSM) a modeling based on the electric and magnetic characteristics is required, to enable the use of physical degrees of freedom and to utilize the calculated power factor. The values are based on the Bosch SMG 180 [3]. It should be noted, that the original machine has a maximum torque of 200 Nm, while the machine in the Smart ED has a reduced maximum torque of 130 Nm. In the following the original torque limit is used. Operating maps for currents, voltages and power are calculated previously to the simulation, based on machine and drivetrain specifications, like the inductances L_d and L_q, the winding resistance and battery voltage. The equations for the electrical model of the machine are shown in (5)-(7) [4]. z_p describes the number of pole pairs and Ψ the flux linkage of the machine.

$$M = \frac{3}{2} z_p \cdot \left(i_q \cdot \Psi + i_q \cdot i_d \cdot (L_d - L_q) \right) \tag{5}$$

$$u_d = R \cdot i_d - L_q \cdot i_q \cdot \omega \tag{6}$$

$$u_q = R \cdot i_q + (L_d \cdot i_d + \Psi) \cdot \omega \tag{7}$$

351

Iron and friction losses of the machine are considered as well. Iron losses are described according to equations (8) and (10) as a rotational speed dependent resistance [4, p876] wired parallel to the induced voltage as shown in Figure 2. The iron resistance is very high and thus the impact on the terminal voltage u_d and u_q is neglected. Copper losses are calculated according to (9).

$$\frac{1}{R_{fe}} = \frac{1}{R_{fe0}} + \frac{1}{R_{fe1}\omega} \tag{8}$$

$$P_{v,Cu} = R \cdot (i_d^2 + i_q^2) \tag{9}$$

$$P_{v,Fe} = \frac{(u_d^2 + u_q^2)}{R_{fe}} \tag{10}$$

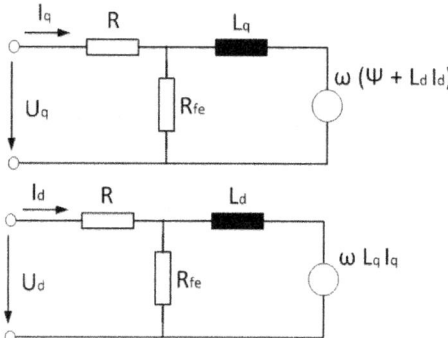

Figure 2: Equivalent circuit of the PSM

The exact values in the efficiency maps highly depend on the used machine control. Hence the maps are determined by the control methods 'maximum torque per ampere' (MTPA), 'flux weakening' and 'maximum torque per volt' (MTPV) [4]. Figure 3a and b show the operating maps in the current plane and the associated torque/speed plane. The MTPA control is used below base speed. The rotational speed in this section has no influence on the control and the current depends only on the torque. By providing a negative i_d current the reluctance of the machine is exploited.

After the maximum induced voltage is reached, the field of the permanent magnets must be weakened by an increased negative i_d current.

For high torques at high speeds the MTPV method is used, to weaken the field only as much as necessary. The maps depend also on the maximum output voltage of the inverter and the maximum current. By iterating through a rotational speed vector and torque vector, the maps are created.

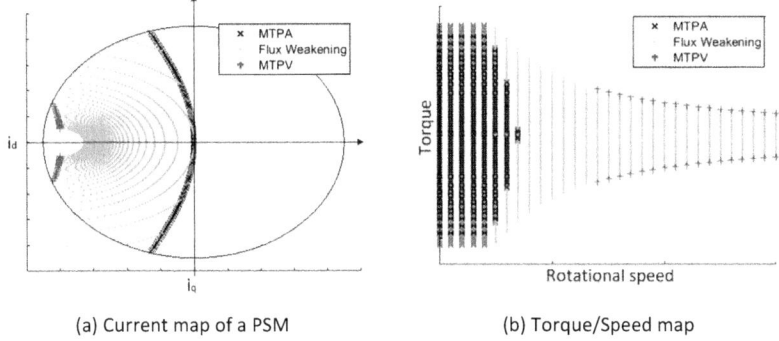

(a) Current map of a PSM   (b) Torque/Speed map

Figure 3: Control maps of a PSM

For the computation of the inverter losses conduction and switching losses are considered. Based on the battery voltage, power factor and machine current, the losses are calculated as in [5]. Therefore data sheet values are used for the inverter Semikron SKiM909GD066HD [6]. Temperature effects are taken into account by a simple thermal model with one thermal RC-circuit.

The battery behavior is derived from a model of a single cell, which is connected in serial and parallel to a battery pack. The equivalent circuit of the cell is depicted in Figure 4. The basic battery persists of 93 serially connected Li-Ion cells, with a nominal voltage of 3,65V per cell.

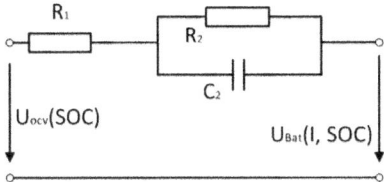

Figure 4: Equivalent circuit of a battery cell

2.2 Structure and Simplifications

The model is implemented in Matlab as a backward model. This means the values are calculated reverse to the physical causality (Fig. 5). Beginning from a given driving cycle, the driving resistances are calculated, which leads to a wheel force that would be necessary to follow the driving cycle. From there on models of the components calculate the required previous measure. All components regard losses as well.

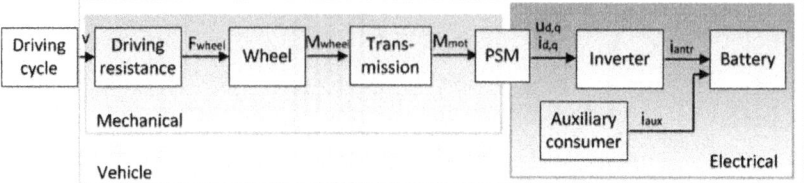

Figure 5: Schematic of the drive train model

As shown in [7] a vectorization of the problem formulation in Matlab helps massively to improve calculation time. For time efficient simulations, most equations in the drivetrain model are applied on the whole driving cycle vector. Therefore interdependencies between components and the number of state variables must be reduced, because every system state requires a time step based calculation. The model contains following simplifications: 1) The inverter model uses a constant battery voltage, equivalent of 50% SOC. 2) Temperature influence is neglected for the most parts. Only the inverter model considers temperature. 3) Machine currents are assumed to be quasi-static.

The existing states velocity, battery current, and inverter temperature only depend on parameters of their own components and can be realized with direct Matlab commands or short for-loops.

Table 1 shows a comparison between the vectorized model and a version with time step based calculation in Matlab. It can be seen, that especially for longer simulations the vector-based model is up to 200 times faster.

Table 1: Comparison model efficiency for different implementations

Time points	1180	21151
Time step based model	6,4s	102s
Vector-based model	0,22s	0,53s

2.3 Validation of the Basic Model

The validation is performed in several steps. First, the simplifications of the model in section 2.2 are validated with a complex Simulink forward model, which considers interdependencies like the impact of the battery voltage on the inverter. The two models agree very accurately with a deviation of 0,5%.

As a next step the energy consumption is compared with real driving data, calculated from the measured SOC with the original vehicle. The largest deviation in energy consumption on the Stuttgart driving circuit between the presented model and real rides is 2,7%.

At last the accurate Simulink model and the simplified Matlab model is compared for the parameter variations in section 3. Both models behave very similar and deviations are neglectable.

These investigations show that the vectorization is justifiable and the accelerated calculation time does not impair the accuracy of this investigation.

2.4 Acceleration Model

The acceleration model calculates the acceleration time of the drivetrain configuration from 0 to 60km/h. This speed range is sufficient to assess the longitudinal dynamics of the vehicle for typical urban velocities.

The model uses the maximum torque characteristic of the machine and uses the driving resistances to calculate the velocity of the next time step. This is repeated until the 60km/h limit is reached. The wheel slip is kept in the point of maximal load transmission and the dynamic axle load distribution is considered. It is assumed, that the maximum power values of the battery and inverter are sufficient.

3 Drivetrain Variation

Drivetrain analysis and layout requires parameters that can be varied in order to find an optimal solution. The scalable parameters should match to the degrees of freedom in modern modular system kits of OEMs.

3.1 Scaling Parameters

For the battery the number of cells wired in serial and parallel can be varied. A change of parallel cells changes only the battery capacity. The nominal battery voltage is not affected by the change. Varying the number of cells in serial, changes the nominal battery voltage. This leads to significant consequences for the machine, which is explained in the next section. The switching frequency of the inverter does not change with a modification in battery voltage and losses caused by harmonics in the machine are not considered. The battery mass is modeled linearly to its energy content.

The original cell size is 52 Ah. A variation of the number of parallel cells of this size would lead to a very coarse adjustability of the range. Because of that the further assumed cell size is 26 Ah and two parallel cells behave like one cell of the original size.

Also the gear ratio can be varied. By that, the operation points can be shifted to more efficient regions of the machine map and acceleration and top speed changes.

The machine has two degrees of freedom. For all variations the electric loading is fixed, to keep the electromagnetic behavior constant. The first scaling parameter is the active length of the electric machine, which increases the maximum torque at a given current. Physically the inductance of the machine increase with a longer machine. This leads to higher torques, and an increased induced voltage. The machine equations (11)-(14), (16) show that dependency [8,9]. s_{mot} describes the normalized machine length. Leakage inductances are not taken into account. The ohmic resistance is separated in an active part R_1, which scales with motor length and the passive part R_0 of the end windings. Also the number of windings can be modified. The original winding number is 13 [3]. With an altered winding number a new winding ratio $s_{wind} = windings_{new}/13$ can be calculated. A higher number of windings increases the induced voltage. The current per winding decreases in the same ratio, so that the electric loading of the machine stays constant. The iron resistance is modelled so that iron losses stay constant with a change in windings and change linearly with machine length as in eq. (16). Effects caused by the skin effect are neglected. The equations (11)-(13) and (15, 16) for the machine parameters show the modeled relationship [8, 10].

$$L' = L \cdot s_{mot} \cdot s_{wind}^2 \tag{11}$$

$$\Psi' = \Psi \cdot s_{mot} \cdot s_{wind} \tag{12}$$

$$R' = (R_0 + R_1 \cdot s_{mot}) \cdot s_{wind}^2 \tag{13}$$

$$M'_{max} = M_{max} \cdot s_{mot} \tag{14}$$

$$I'_{max} = \frac{I_{max}}{s_{wind}} \tag{15}$$

$$R'_{fe} = R_{fe} \cdot s_{wind}^2 \cdot s_{mot} \tag{16}$$

The impact of the scaling parameters on the machine characteristics is shown schematically in Figure 6.

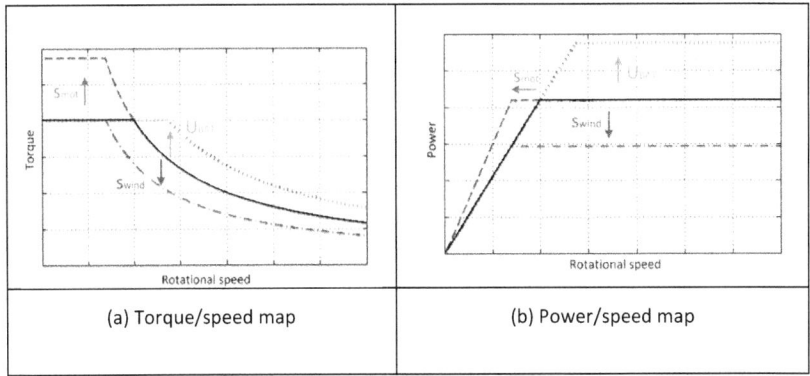

(a) Torque/speed map	(b) Power/speed map

Figure 6: Schematical influence of the scaling parameters on machine characteristics

Raising the maximum torque can only be done by scaling the machine in axial direction. For power scaling two parameters can be used: battery voltage and number of windings. Either the voltage can be increased or the number of windings can be reduced for higher power. Scaling of the machine does not change the safety behavior, because the short circuit current $i_{scc} = -\frac{\Psi}{L_d}$ varies in the same ratio as the maximum current.

Since battery voltage and number of windings change the maximum machine power, and the volume of the machine stays constant, a restraint must be considered. Otherwise the power density of the machine would become unrealistically high, which would lead to invalid machine temperatures. Therefore the maximum allowed power density in the simulation is set to $3,8 \frac{kW}{kg}$.

3.2 Single Parameter Variation

All scaling parameters can be varied in a defined range. The boundaries of this parameter space are listed in Table 2. The serially connected cells can only be varied in the range of the inverter's voltage class of 270V-450V. The minimum winding number is defined by the maximum possible inverter current.

Table 2: Scope of the variable parameters

Scaling Parameter	Basic Value	Minimum	Maximum
s_{mot}	1	0,7	1,3
Number windings	13	10	17
Gear ratio	11	6	12
Cells serial	93	85	105
Cells parallel	2	1	5

In the following section a single scaling parameter variation in the ranges of Table 2 is performed with the driving cycle explained in the next section. All other scaling parameters are fixed to the value of the basic configuration. Hereby the impact of range, acceleration time and consumption is shown in Figures 7 and 8, normalized on the basic configuration. A shorter acceleration time means faster acceleration. The arrows indicate the direction of larger scaling parameters.

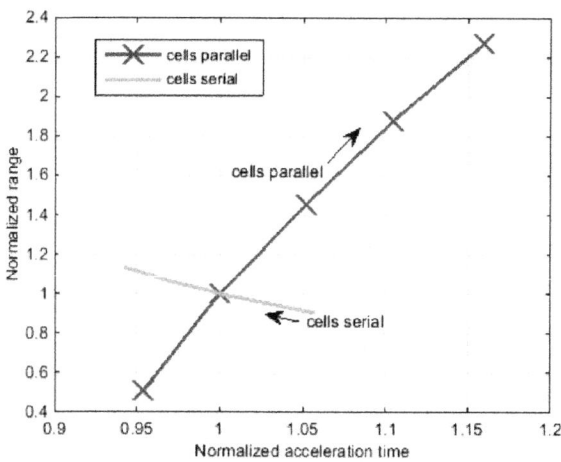

Figure 7: Effect of scaling parameters on range and acceleration time

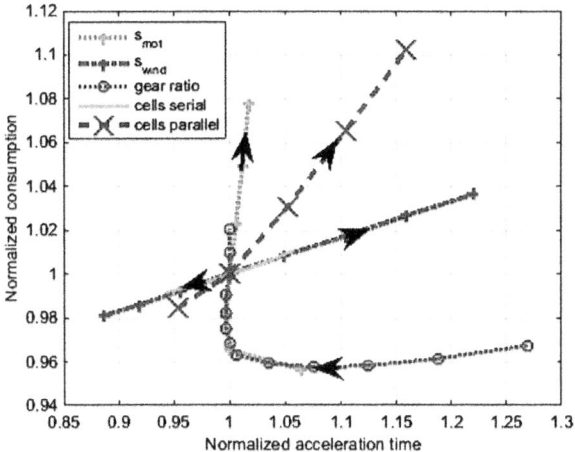

Figure 8: Effect of scaling parameters on consumption and acceleration time

The range can be influenced by the number of cells. An increase in serial cells improves vehicle acceleration up to 5,8% and range up to 14%. The effect on the range is small, because of the limited voltage scalability. A much larger increase in range can be achieved by increasing the number of parallel cells, which impairs acceleration, because of the additional mass. All other parameters have only an indirect impact on the range and are not visualized in Figure 7.

The impact on the consumption is more complex. A larger normalized machine length s_{mot} increases energy consumption, because a longer machine leads to higher copper and iron losses in the same operating points, especially in the flux weakening area. The impact on acceleration is difficult, since the basic machine has a very high starting torque and the wheel force is at the slip limit. In this case and with the limitation of equation (15) a larger machine can lead to an even slightly slower acceleration time, when the windings and battery voltage are not adjusted, because maximum output power can slightly decrease, because of higher copper and iron losses.

The same effect holds for the gear ratio. A higher gear ratio increases the maximum wheel force. Because of the slip limit, this does not lead to higher acceleration anymore. Decreasing the gear ratio leads to slower acceleration. For the energy consumption an optimal gear ratio is visible.

A variation of windings and serial and parallel cells improves or impairs both acceleration and consumption. Battery voltage and winding number can improve all evaluation parameters. They are limited by the maximum voltage and current specifications of the inverter and the maximum power density of the machine.

This analysis gives a deeper understanding for the impact of the different scaling parameters, but also shows the limitations of single parameter variations. To find optimal solutions the entire parameter space is explored in the next section.

4 Analysis Approach

A systematical approach is necessary for the following analysis of the parameter space. Also invalid drivetrain configurations must be identified and filtered.

The layout of an optimal drivetrain depends on the driving cycle which is used for the layout. In the best case it is very similar to the driving cycle of the real world drivers. Therefore a typical driving cycle of the Stuttgart driving circuit [2] is used as in Figure 9.

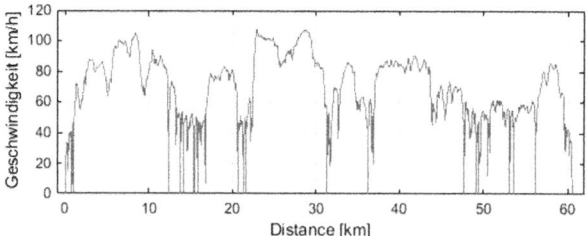

Figure 9: Representative Stuttgart driving cycle

The analysis of a drivetrain combination is controlled by the *model_supervision* function, as drafted in Fig. 10. Inputs to the function are the values of the scaling parameters (winding number, machine length ...). First the parameters for the simulation are created and the maps for the machine are generated as explained in section 2.1. After that the function for the real drive analysis is executed to determine range and consumption of a drivetrain combination. Subsequently the acceleration time from 0 to 60 km/h is calculated and the configuration is verified on minimum requirement constraints. The test for the gradability and top speed constraints is performed with modified versions of the acceleration model.

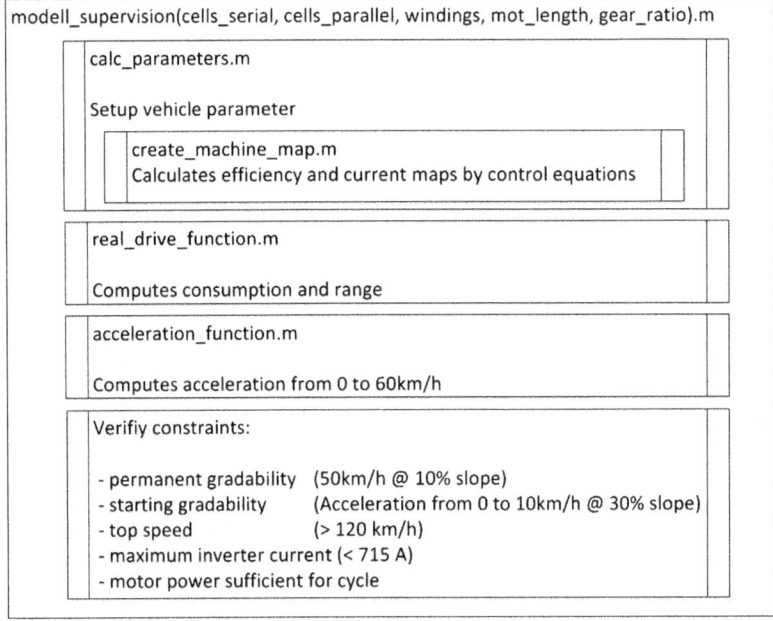

Figure 10: Software structure for the simulation of a drive train configuration

To explore the parameter space a full factorial analysis of the parameters in Table 2 is performed and 40 000 simulations are automatically executed.

The optimal configurations are detected from the solution space by the Pareto front algorithm [11]. Therefore all solutions with a range within the required range scope are extracted and the algorithm is applied.

5 Layout Example

To exemplify the method, a new drivetrain design for a Smart ED is performed. Two versions of the drivetrain, one with a range of 210 km and the other with a range of 280 km are to be found. The range tolerance is ±10km. Therefore different possibilities of realization are identified by a full factorial analysis.

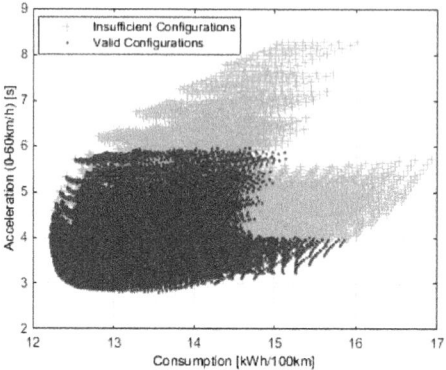

Figure 11: Result of fullfactorial analyzis with valid and insufficient solutions

The result of the full factorial analysis is shown in Figure 11, which depicts the acceleration time and energy consumption for the valid and for the insufficient solutions. The invalid solutions with a high acceleration time and high energy consumption lack motor power to accomplish the driving cycle and/or the gradability requirements.

To meet the range requirements, all solutions with the required range are extracted. All solutions for a range of 210 km ±10km and 280 km ±10km are depicted in Figure 12. As expected, configurations with a higher range have a higher consumption and a higher acceleration time.

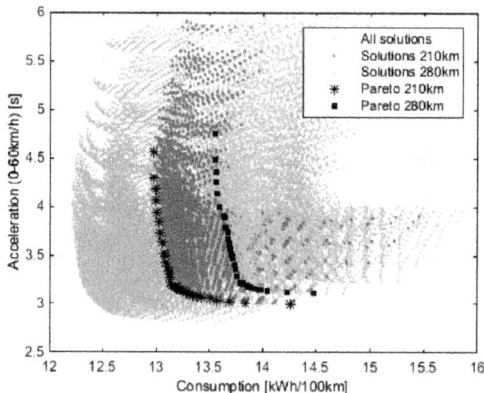

Figure 12: Pareto front with range restrictions

Table 3: Pareto-optimal solutions for both range requirements

0-60 km/h [s]	Ran-ge [km]	P/E [1/h]	Mmax [Nm]	Pmax [kW]	s_{mot}	Nr win-dings	Gear ratio	Cells serial	Cells pa-rallel
Solutions for range of 210 km:									
3,1	204	3,3	240	99	1,2	10	10,0	103	3
3,2	209	3,4	180	92	0,9	10	10,5	97	3
3,9	**203**	**2,6**	**140**	**68**	**0,7**	**13**	**11,0**	**93**	**3**
4,3	203	2,4	140	63	0,7	14	10,0	93	3
Solutions for range of 280 km:									
3,1	**279**	**2,5**	**240**	**100**	**1,2**	**10**	**11,0**	**105**	**4**
3,5	273	2,3	200	85	1,0	11	9,5	99	4
3,9	275	2,1	160	79	0,8	12	10,5	99	4
4,5	271	1,8	140	66	0,7	14	10,5	97	4

A selection of Pareto-optimal solutions, ordered by acceleration time for both versions is shown in Table 3. Higher driving dynamics are obtained by a larger machine length, lower winding numbers and a higher voltage.

Common parts for both versions should be identified to be able to save costs. If the same battery cell type is used, the so called P/E ratio should be the same. The P/E ratio describes the maximum battery power divided by the battery's energy content and is important for the aging of the battery cells. The P/E ratios for the optimal solutions of both versions are shown in Figure 13.

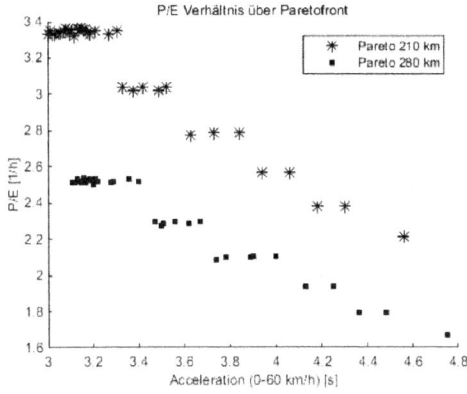

Figure 13: P/E ratio over acceleration time for Pareto optimal solutions

With a higher driving performance, the P/E ratio rises, because of a higher motor power. The ratio is also higher for the version with less range. Around a P/E ratio of 2,5 solutions for both drivetrain solutions are possible. This leads to a portfolio with the same type of cells and a model version of low range and less driving performance and a version with higher range and higher driving performance.

Two possible realizations are highlighted in Table 3. Whilst the motor length and the winding number must be varied, the gear ratio of the transmission stays constant. By that the same battery cells and the same transmission design can be used for both versions.

6 Conclusion

This paper has investigated the drivetrain design of EVs with requirements on range and modularization. Therefore a time efficient vehicle model is introduced, that takes advantage on Matlab's ability on vector operations, by reducing the number of state spaces. After that, scaling parameters in the drivetrain components are established and their impact on range, consumption and acceleration time is discussed. The parameter space is explored by a full factorial analysis. The results of this are used for an exemplary drivetrain layout with two different versions. The Pareto-optimal solutions for both versions are identified. Out of these sets of optimal solutions, realizations with many common parts are identified. A constant P/E ratio - an important requirement for the reuse of the battery cell - is met by an increased machine power for the drivetrain ratio with a higher range. Also the gear ratio stays constant. It could thus be shown that a modular drivetrain layout allows Pareto-optimal realizations, although a high number of common parts is used.

Acknowledgement

The work presented in this publication was performed in the project "*e-generation* – key technologies for the next generation of electric vehicles" which was supported by the German Federal Ministry of Education and Research (BMBF) under Project Number 13N11865.

Bibliography

1. Mitschke, M., Wallentowitz, H.: Dynamik der Kraftfahrzeuge, Springer Vieweg, 2014, ISBN 978-3-658-05068-9

2. Freuer, A.; Grimm, M., Reuss, H.-C.: Messung und statistische Analyse der Leistungsflüsse und des Energieverbrauchs bei Elektrofahrzeugen im kundenrelevanten Fahrbetrieb. 4. Deutscher Elektro-Mobil Kongress, Essen, 15.06.2012

3. Bauer, D.; Reuss, H. & Nolle, E.: Einfluss von Stromverdrängung bei elektrischen Maschinen für Hybrid- und Elektrofahrzeuge; Tag des kooperativen Promotionskolleg HYBRID, 2014

4. Schröder, Dierk: Elektrische Antriebe-Regelung von Antriebssystemen, 3. bearbeitete Auflage, Springer Verlag, 2009, ISBN 978-3-540-89612-8

5. Eckardt, B.; März, M. & Schletz, A.: Anforderungsgerechte Auslegung von Leistungselektronik im Antriebsstrang Elektrik/Elektronik in Hybrid und Elektrofahrzeugen, 2008

6. Semikron: technical data SKiM909GD066HD, 03.11.2015. URL: http://www.semikron.com/de/produkte/produktklassen/igbt-module/detail/skim909gd066hd-23930790.html

7. Matz, S.: Nutzerorientierte Fahrzeugkonzeptoptimierung in einer multimodalen Verkehrsumgebung; TU München, Verlag Dr. Hut, 2015, ISBN 9783843921404

8. T. Finken: Fahrzyklusgerechte Auslegung von permanentmagneterregten Synchronmaschinen für Hybrid- und Elektrofahrzeuge, RWTH Aachen, Shaker Verlag, 2012, ISBN 978-3-8440-0607-0

9. Bücherl, D.; Betram, C.; Tanheiser, A. & Herzog, H.-G.: Scalability as a Degree of Freedom in Electric Drive Train Simulation Vehicle Power and Propulsion Conference, 2010

10. Schoenen, Timo: Einsatz eines DC/DC Wandlers zur Spannungsanpassung zwischen Antrieb und Energiespeicher in Elektro- und Hybridfahrzeugen RWTH Aachen, 2011, Shaker-Verlag, ISBN 978-3-8440-0622-3

11. Yi Cao: Pareto Front, 27.01.2015. URL: http://www.mathworks.com/matlabcentral/fileexchange/17251-pareto-front

Combined power train and thermal management optimization using an extended dynamic programming

Mike Liebers, Robert Kloß, Prof. Bernard Bäker

Technische Universität Dresden, Chair of Vehicle Mechatronics

1 Abstract

Reducing the fuel consumption of heating and driving tasks is a major driver of innovation in today's vehicle development. Besides the design and integration of efficiency-raising auxiliaries, especially the enhancement of full electric and hybrid drivetrains are in focus. An energy management controls the interaction of all vehicle components rule based or optimization based [1].

This paper presents an approach to optimize the energy management of the electric cabin heating and traction system of articulated urban serial hybrid buses using extended dynamic programming. The traction battery and the cabin air are defining the energy storage systems. Thus the state space for dynamic programming is described by the electric charge of the traction battery and the mean cabin air temperature. Extending the dynamic programming approach by thermal comfort aspects enables the global optimal control for each energy storage system additionally to the calculation of possible energy saving potentials.

2 Introduction

Within the public funded project "Pilot Route 64 – Efficient Electromobility in Dresden" a serial hybrid bus was equipped with 117 sensors measuring inter alia temperature, pressure, flowrate, current and voltage. It is serving the bus route 64 in Dresden since 10/2014. With this framework the project aims for increasing the efficiency of this bus by combining efficient heating, ventilation, air conditioning (HVAC) components and auxiliaries with a driver support system and a superordinate energy management.

The articulated hybrid bus has a serial power train with two electric drive axles and a downsized diesel engine in combination with a generator (Genset). Compared to conventional diesel buses this combustion engine produces less power and accordingly less heat losses. Thus the hybrid bus saves up to 20 % diesel for traction tasks [3] but increases the heating oil consumption up to a sevenfold (see figure 1). The first step to improve the HVAC system in this project was the installation of two efficient heat pumps, which can reduce the total energy demand of heating tasks. Further, the reduction of thermal losses is intended using radiant heaters. The most energy-efficient usage of these two techniques combined with the vehicle powertrain needs an optimal energy management, which is the main content of this paper. Therefore this article is structured in six sections. After an overview of physical basics for modelling relevant thermal and powertrain bus components in chapter 3, the application of the principle of optimality by Richard Bellman to this problem is described in chapter 4. The paper closes with a short demonstration on a simplified example and a conclusion.

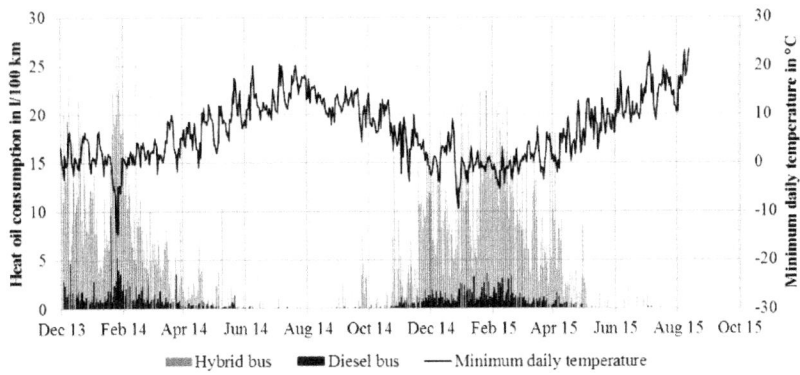

Figure 1: Daily average heat oil consumption of 10 diesel and 8 articulated hybrid buses on bus route 64 in Dresden for the period 12/2013 – 09/2015 (Temperature source: wetter.com)

3 Combined Thermal- and Power Train Model

The VDV guideline 236 determines the climate conditions in regular buses. It is significant to all German public transport companies. Depending on the ambient temperature this guideline differs between heat mode, ventilation mode and cooling mode. It sets the minimum cabin air temperature of regular buses to 18°C in heat mode. At an ambient temperature of -10°C or lower, the minimum cabin air temperature can be decreased to 13°C. The cooling mode is activated at a temperature of more than 22°C and cools the cabin at a constant temperature difference of -3 K. At an ambient temperature between 18°C and 22°C the ventilation mode is activated with no temperature difference. Regardless of cabin temperature or current mode the driver can control the workplace temperature independently. [2]

The following chapters will give an overview how the energy demand of heating, ventilation and air conditioning systems as well as the vehicle powertrain and the passenger's thermal comfort is affected by the thermal requirements of the VDV guideline 236.

3.1 Cabin model

Regarding the established VDV guideline 236, figure 2 shows the temperature limitations based on this regulation. This diagram clarifies the rising temperature difference between the heated and the unheated cabin

$$\Delta T = T_{Air\ cab} - T_{Air\ amb}. \tag{1}$$

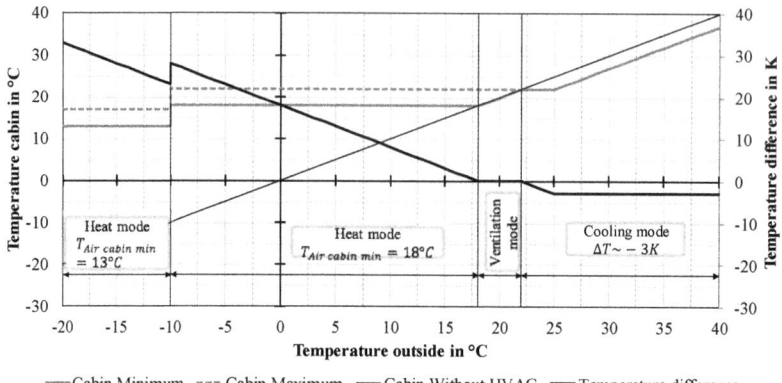

Figure 2: Temperature regulation of regular buses by [2]

The rising air temperature difference between the cabin and the ambient causes rising heat transfer losses through each element i of the steady state cabin walls

$$\dot{Q}_{Wall\ trans} = \sum_i k_i \cdot A_i \cdot \Delta T \qquad (2)$$

or through open doors $\dot{Q}_{Door}(\Delta T, t_{Door})$ (see figure 3, plot a). In (2) A_i describes the area of enveloping interior surfaces. The cabin thermal transmittance k_i combines the heat conduction through each wall element and the convection to the air inside and outside (figure 3, plot b). Thus it depends on the wall element structure as well as thermal and air flow conditions. The heat losses $\dot{Q}_{Door}(\Delta T, t_{Door})$ are caused by the air exchange between warm air inside and cold air outside during bus stops. Here, warm and low-dense air soars and leaks through open doors while cold and denser air flows from the bottom inside the cabin. Thus the mean cabin air temperature falls and the internal energy

$$\frac{E_{Air\ cab}}{dt} = m_{Air\ cab} \cdot c_{Air} \cdot \frac{dT_{Air\ cab}}{dt} \qquad (3)$$

is reduced. In (3) $m_{Air\ cab}$ is the air mass inside the cabin with a heat capacity c_{Air}.

To reach the required cabin temperature, the bus is assumed to have a full electric heating system. It combines an air/air heat pump ($\dot{Q}_{HP}$) and a warm water circuit provided by an electric water heater ($\dot{Q}_{EH}$). This system is represented by the coefficient of performance (COP) and the energy efficiency η_{EH} as each the ratio between heating and electric power P_{el}:

$$COP = \frac{\dot{Q}_{HP}}{P_{el\ HP}}, \ \eta_{EH} = \frac{\dot{Q}_{EH}}{P_{el\ EH}}. \qquad (4)$$

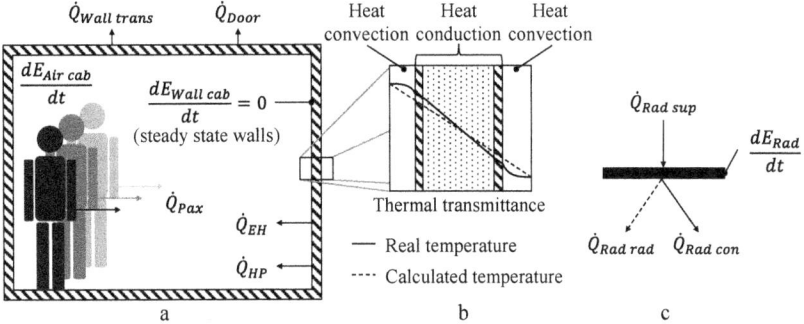

Figure 3: System boundaries of cabin air (a) and radiant heater (c) models and temperature curves through a cabin wall element (b)

In addition to the heating system, each passenger produces a convective heat flow $\dot{Q}_{Pax\ ind\ con}$, which depends on the individual body size and the difficulty of activity [4]. Therefore, the combined heat flow of all passengers inside the vehicle will be calculated with the actual number of passengers n_{Pax} to

$$\dot{Q}_{Pax} = n_{Pax} \cdot \dot{Q}_{Pax\ ind\ con}. \tag{5}$$

Disregarding the convective losses of radiant heaters $\dot{Q}_{Rad\ con}$, the thermodynamic system of the cabin air is described by

$$\frac{E_{Air\ cab}}{dt} = \dot{Q}_{EH} + \dot{Q}_{HP} + \dot{Q}_{Pax} - \dot{Q}_{Wall\ trans} - \dot{Q}_{Door}. \tag{6}$$

Plot a) in figure 3 summarizes the thermal balance of the cabin model.

3.2   Radiant heater model

At the cabin ceiling large-scale, low-temperature radiant heaters are installed to transfer heat non-substance-related. As electric heating elements, radiant heaters are supplied by electric power P_{el}, which is assumed to be completely transformed into heat

$$\dot{Q}_{Rad\ sup} = P_{el\ RH}. \tag{7}$$

This heat is transformed into internal stored energy

$$\frac{dE_{Rad}}{dt} = m_{Rad} \cdot c_{Rad} \cdot \frac{dT_{Rad}}{dt}, \tag{8}$$

radiant heat (without passengers, for radiation exchange between surfaces see [8])

$$\dot{Q}_{Rad\ rad} = \sigma \cdot \varepsilon_{Rad} \cdot A_{Rad} \cdot (T_{Rad}^4 - T_{Wall}^4) \tag{9}$$

and convection

$$\dot{Q}_{Rad\ con} = \alpha_{Rad} \cdot A_{Rad} \cdot (T_{Rad} - T_{Cab\ Air}). \tag{10}$$

In (8) – (10) m_{Rad} is the radiant heater mass with a thermal capacity c_{Rad}, a surface temperature T_{Rad} and an area A_{Rad}. T_{Wall} is the cabin wall temperature, σ is the Stefan-Boltzmann constant, ε_{Rad} is the emissivity of radiant heater surface and α_{Rad} describes the heat transfer between the heater surface and the cabin air. A radiant heater as, open system is summarized by the equations (7) – (10) and figure 3, plot c to

$$\frac{dE_{Radiant}}{dt} = \dot{Q}_{Rad\ sup} - \dot{Q}_{Rad\ rad} - \dot{Q}_{Rad\ con}. \tag{11}$$

3.3 Thermal comfort

The thermal comfort is affected by the enclosing surface temperature, air temperature, thermal stratification, air stream and the humidity [5]. This chapter shows how the two main influence factors enclosing surface temperature and air temperature can be used to enhance the HVAC system efficiency.

The operative temperature T_O describes the human thermal comfort as a combination of enclosing surface temperature $T_{Surf\ cab}$ and air temperature $T_{Air\ cab}$

$$T_O = \frac{\overline{T}_{Surf\ cab} + T_{Air\ cab}}{2}. \tag{12}$$

According to [5] the average cabin surface temperature is simplified calculated to

$$\overline{T}_{Surf\ cab} = \frac{\sum_i A_i \cdot T_i}{\sum_i A_i} = \frac{\sum A_{Rad} \cdot T_{Rad} + \sum A_{Wall} \cdot T_{Wall}}{\sum_i A_i}. \tag{13}$$

In (13) A_i represents the area of each surface element with the temperature T_i. So, low air temperatures can be compensated by higher surface temperatures. Fanger, Franke and Roedler generally confirm this statement by their thermal comfort experiments in bureau buildings as shown in figure 4.

To transfer these findings on public transportation vehicles, the comfort diagram is extended by the required air temperature in heat mode. With reference to figure 3, plot b, the mean temperature of the unheated steady state cabin walls T_{Wall} is assumed to be constant within $T_{Air\ amb} \leq T_{Wall} < T_{Air\ cab}$. Here the curves of constant operative temperature clarify the wide range of thermal comfort in an urban bus stated by VDV 236. So, radiant heaters can either be used to shift the cabin's thermal operating point along the operative temperature curve or to enhance the thermal comfort while driving. This paper discusses the efficiency-improving effect of an optimal energy management combining radiant heater and an electric air heating system at a constant level of thermal comfort.

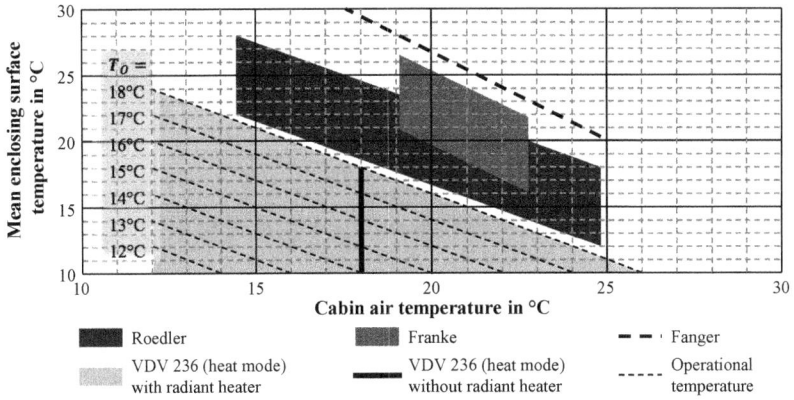

Figure 4: Extended thermal comfort diagram by [6]

3.4 Vehicle powertrain model

The intermediate circuit of the hybrid bus as central electric element connects the Genset, the traction axles, the electric auxiliaries and the high-voltage (HV-) battery as shown in figure 5. Therefore the internal combustion engine is not mechanically connected with the propulsion unit. With this powertrain topology the operating point of the Genset is independent from the power request of the traction motors P_{EM} and the auxiliaries P_{AUX}. Therefore it is possible to choose an operating point based on the best Genset efficiency.

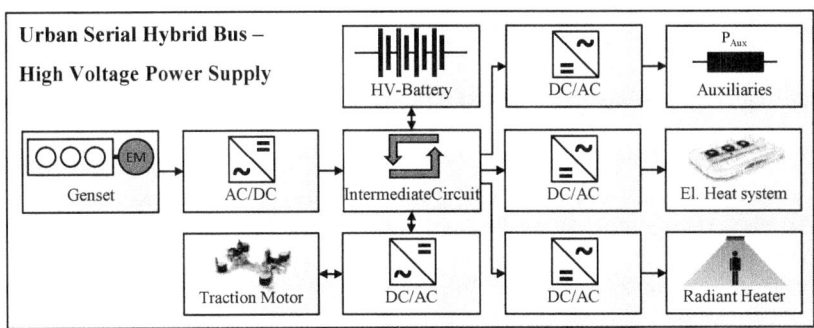

Figure 5: High voltage power supply of an urban serial hybrid bus

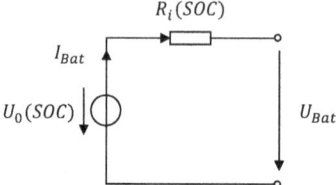

Figure 6: Two terminal network of traction battery

The battery model is map-based. It is represented by a two terminal network for what figure 6 represents the equivalent circuit diagram. The open-circuit voltage $U_0(SOC)$ is in series with the internal resistance $R_i(SOC)$. Both are dependent on the battery state of charge (SOC). Here, the electrical power balance can be derived to

$$U_0(SOC) \cdot I_{Bat} - R_i(SOC) \cdot I_{Bat}^2 = P_{Bat} = U_{Bat} \cdot I_{Bat}. \qquad (14)$$

The SOC describes the ratio of the actual battery charge Q to the battery capacity C

$$SOC = \frac{Q}{C} = \frac{\int I_{Bat} dt}{C} \Rightarrow I_{Bat} = \frac{dSOC}{dt} \cdot C. \qquad (15)$$

Additionally to the thermal comfort components electric heating system and radiant heaters, there are electric auxiliaries e.g. air compressor or 24V-auxilliaries. These are represented by chronological sequences of the electrical power demand P_{Aux} as well as demand of the traction motors P_{TM} for simplification purpose. Aiming for the required thermal conditions of VDV 236, the thermal component's power demand is calculated by models described in the previous chapters.

The Genset is used as an additional power source for the electrical intermediate circuit. Its model is based on a fuel power map and a generator efficiency map

$$P_{Gen} = \eta_{Gen}(n_{Gen}, M_{Gen}) \cdot P_{Fuel}(n_{ICE}, M_{ICE}). \qquad (16)$$

Regarding the intermediate circuit following electrical power balance can be set up

$$P_{TM} + P_{Aux} + P_{el\ HP} + P_{el\ RH} + P_{Bat} - P_{Gen} = 0. \qquad (17)$$

To reach a minimal fuel consumption of the internal combustion engine (ICE) an energy management strategy for the Genset and the thermal comfort components is needed. This strategy is calculated by a dynamic programming algorithm which is presented in the following section.

4 Using Dynamic Programming

A vehicle with serial hybrid drive has two different energy storages and converters installed. The additional degree of freedom compared to a conventional powertrain requires an energy management strategy to decide how the requested power is partitioned between the electric battery and the Genset. The resulting optimal control problem needs to be solved with consideration of the additional conditions like allowable state of charge (SOC) or engine operation limits. The supervisory control of power flow in a hybrid drivetrain can be realized with heuristic or optimization based algorithms.

In this paper the focus is on the optimization based strategies specifically on the Dynamic Programming (DP). The DP algorithm bases on Richard Bellman's principle of optimality which says that the global optimal solution of a problem consists of the optimal partial solutions. [7]

In addition to the powertrain system, the heating system of the bus is included in the optimization algorithm in this paper. To ensure a constant operative temperature T_O (cf. equation (12)), the electric heating system and the radiant heaters can be run in different operating points with different electric power inputs from the intermediate circuit. To find the optimal energy management aiming for global minimal fuel consumption a two storage dynamic programming algorithm is needed.

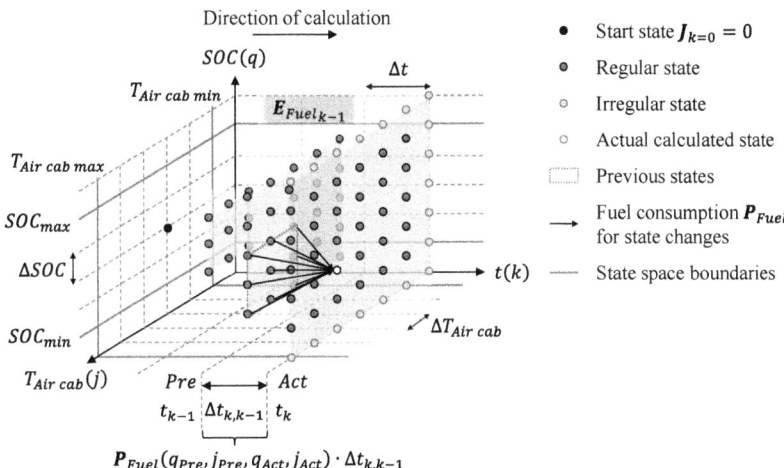

Figure 7: State space including constraints, discretization and solving equation (18) following [9]

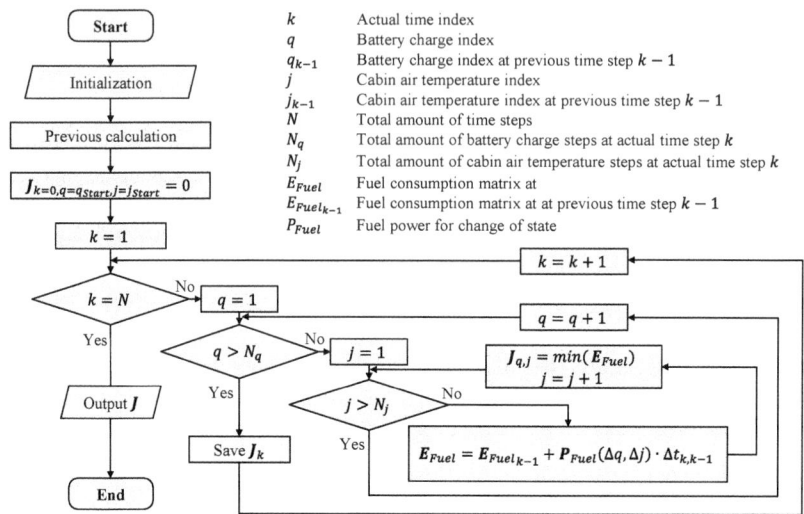

Figure 8: Two storage Dynamic Programming flow chart

For this optimization problem, the state space is defined by cycle time $t(k)$, battery state of charge $SOC(q)$ and cabin air temperature $T_{Air\ cab}(j)$ as nodes and the fuel consumption costs $E_{Fuel}(k, q, j)$ as edge weights between the nodes, see figure 7. Here k, q and j represent one state space index each. Limited by the constraints $SOC_{min/max}$ and $T_{Air\ cab\ min/max}$, the state space is discretized with constant steps of time Δt, battery charge ΔSOC as well as temperature difference $\Delta T_{Air\ cab}$. Based on Bellman's principle of optimality, the actual minimal costs J at the indices k, q and j are calculated by the minimum of the previous cumulated minimal costs E_{Fuel} at $k-1$ and the costs $P_{Fuel}(q_{Pre}, j_{Pre}, q_{Act}, j_{Act}) \cdot \Delta t_{k,k-1}$ for the change of state from previous to the actual time step:

$$J(k, q, j) = min\big(E_{Fuel_{k-1}} + P_{Fuel}(q_{Pre}, j_{Pre}, q_{Act}, j_{Act}) \cdot \Delta t_{k,k-1}\big). \qquad (18)$$

The following section will describe how the optimization is implemented in a calculation program (see figure 8). After an initialization of J at $k = 0$, $q = q_{Start}$, $j = j_{Start}$, with costs $E_{Fuel} = 0$, the algorithm starts with a forward-oriented calculation at time step $k = 1$. Till the last time step $k = N$ is reached, at every time step k the cost matrix J_k is filled with the minimum fuel consumption within the predefined state space boundaries. For a better traceability each cost matrix J_k needs to be saved.

The fuel consumption matrix E_{Fuel} results from backward calculating the changes of state using the electrical power balance (17). Here the change of cabin air temperature

$\Delta T_{Air\,cab}$ affects both the radiant heater $P_{el\,RH}$ and the electric heat system $P_{el\,HP}$, $P_{el\,EH}$ as described in (4), (6) and (12). Varying the SOC changes the battery power P_{Bat} (see equation (14) and (15)). Due to this, the electrical Genset power P_{Gen} remains the only variable to fulfill the balance (17). So, using (16) the fuel power P_{Fuel} can be calculated.

Finally the cost matrix J contains the cumulated global minimal costs E_{Fuel} within the defined state space boundaries $SOC_{min/max}$ and $T_{Air\,cab\,min/max}$ and the chosen discretization Δt, ΔSOC and $\Delta T_{Air\,cab}$ for the given power demands P_{Aux} and P_{TM}. Using a backward calculation, the optimal path from final state at t_{end} to the initial state at t_0 can be found. It is appropriate to set the final state on the same level as the initial state.

5 Results and Discussion

The following simplified example will show how the described Dynamic Programming approach works on measured data from bus route 64 in Dresden. Here the heat pump and the electrical heater are combined with the radiant heaters to demonstrate under which conditions the usage of radiant heaters is reasonable.

To heat the cabin air, both the heat pump and the electric heater produce a heat flow of $\dot{Q}_{EH} = \dot{Q}_{HP} = 32\,kW$. Due to restrictions of the electric system, 4 radiant heaters with a total electric power of $P_{el\,RH} = 2{,}4\,kW$, an area of $A_{Surf\,rad} = 2{,}88\,m^2$ and a surface temperature range of $T_{Air\,cab} \leq T_{Surf\,rad} \leq 100\,°C$ are mounted at the cabin ceiling above the passengers (see the bus pattern in figure 9). The unheated cabin surface temperature and the ambient temperature is set to $T_{Wall} = T_{Air\,amb} = 10\,°C$.

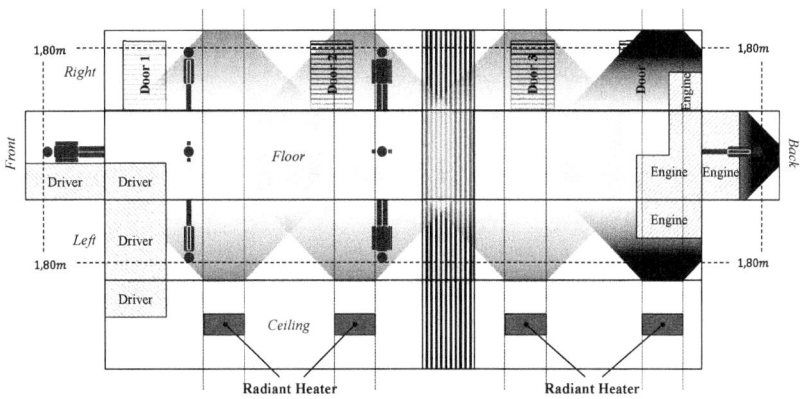

Figure 9: Pattern of an articulated bus with 4 exemplary installed infrared heaters in the ceiling

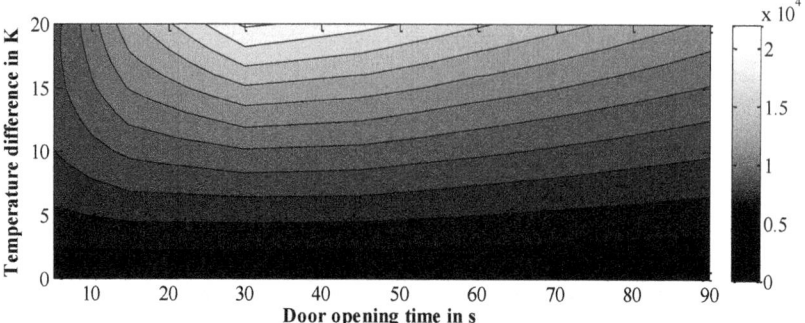

Figure 10: Heat losses $\dot{Q}_{Door}(\Delta T, t_{Door})$ through all 4 open doors on an articulated hybrid bus

The cabin has a total interior area $A_{Surf\,cab} = 181\,m^2$ enclosing an air mass $m_{Air\,cab} = 132{,}7\,kg$ with a constant mean thermal transmittance $k_{Wall\,cab} = 5\,W/m^2 \cdot K$. In this short example passenger's heat flow $\dot{Q}_{Pax}$ is neglected. The heat losses through open doors $\dot{Q}_{Door}$ were determined empirically in winter 2014/15. Generally they depend on air temperature difference ΔT and the door opening time t_{Door}. The results of these experiments with 4 open doors are visualized in the heat loss map in figure 10. To identify these heat losses a signal representing the door status is needed. Figure 12 shows an excerpt of the measured door status (0: closed, 1: opened) for the discussed example.

The state space is defined by the cycle time t with a discretization of $\Delta t = 1\,s$ and the cabin air temperature $T_{Air\,cab}$ with a discretization of $\Delta T_{Air\,cab} = 0.01\,K$. The initial conditions at t_0 are $T_{Air\,cab} = 18\,°C$ and $T_{Rad\,Surf} = T_{Wall} = 10\,°C$. Due to the relatively small active radiant heater area compared to the total interior surface, the cabin air temperature is limited to $14\,°C \leq T_{Air\,cab} \leq 20\,°C$. As a first step, the battery state of charge is neglected. So the optimization algorithm aims for minimizing the energy demand at the intermediate circuit.

Figure 11 presents the average electrical power of the radiant heaters with variant electrical heater efficiency η_{EH} (cf. equation (4)) for this example. Below an efficiency of $\eta_{EH} \leq 50\,\%$ the radiant heaters are activated operating continually with $P_{el\,RH} = 2{,}4\,kW$. Their calculated load drops with higher energy efficiency of the air heating system. It is obvious that a heating system with efficiency of 70 % or higher does not require electrical radiant heaters in this example. Regarding to the energy efficiency it can be concluded that a heat pump with a $COP > 1$ does not need radiant heater support under these assumptions. Therefore, the following discussion is about the shown transitional area of $50\,\% < \eta_{EH} < 70\,\%$ of the electrical air heater.

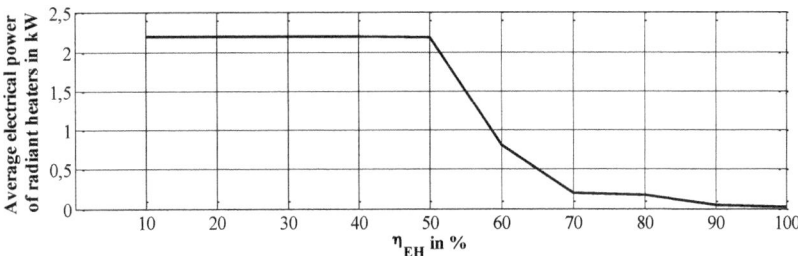

Figure 11: Mean electrical power of radiant heaters depending on electric air heaters efficiency

Figure 12 shows the optimal energy management of electric air heaters and radiant heaters resulting from the presented DP algorithm calculated with $\eta_{EH} = 60\,\%$. It can be seen that the algorithm distinguishes between the two door states opened and closed. Open doors cause the maximum radiant heater power $P_{el\,RH} = 2,4\,kW$ before the bus stops to increase the average enveloping surface temperature. Hence the cabin air temperature is reduced simultaneously to lower the heat losses $\dot{Q}_{Doors}(\Delta T, t_{Door})$ (see figure 10). While driving, the radiant heater surface temperature is decreased between $22\,°C \leq T_{Rad} \leq 32\,°C$. This behavior (see grey-shaded driving situation in figure (12)) is now more closely analyzed.

To hold the cabin air temperature $(dE_{Air\,cab}/dt = 0)$ the heat demand $\dot{Q}_{EH}$ is only affected by the heat transfer losses through the cabin wall (see equation (6)). To decrease the heat transfer losses energy-optimal, the radiant heaters are operated with an electric power of $P_{el\,Rad} = 0,26\,kW$ to reduce the cabin air temperature to $T_{Air\,cab} = 17,8\,°C$. According to the equations (4) and (6), the total electrical power demand difference between with and without DP is $\Delta P_{el} = -0,06\,kW$ (see table (1)).

It is obvious, that the DP algorithm aims for reducing the electrical power demand at the intermediate circuit $E_{el\,IC}$. In figure 12 the energy difference between driving cycle without and with DP $\Delta E_{el\,IC}$ is illustrated. Despite the extra electric load of radiant heaters, the total energy demand can be reduced in this simplified example using the Dynamic Programming algorithm.

Table (1): Comparing the electrical power demand with and without DP while driving

	k	A	ΔT	$\dot{Q}_{Wall\,trans}$	$P_{el\,Rad}$	$P_{el\,EH}$
	$W/m^2 \cdot K$	m^2	K	kW	kW	kW
Without DP	5	181	8	7,24	0	12,07
With DP	5	181	7,8	7,06	0,24	11,77

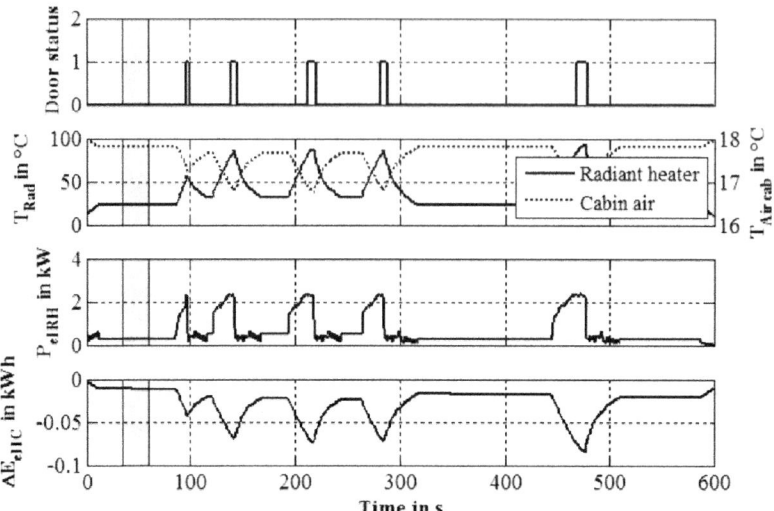

Figure 12: Optimal energy management of electric air heaters ($\eta_{EH} = 60\,\%$) and radiant heaters

6 Conclusion

This paper presented an approach for an optimization based energy management using an electrical air heating system supported by electrical radiant heaters to increase the energy efficiency of an urban serial hybrid bus. Combining the official thermal requirements by VDV 236 with thermal comfort aspects regarding both air and enclosing surface temperature enables an extra state of freedom. The developed algorithm uses this additional state of freedom with Richard Bellman's principle of optimality. It considers the vehicle powertrain system, the electrical auxiliaries as well as two types of electric air heaters finding the global minimal fuel consumption. With a simplified example, the algorithm was successfully tested for plausibility. The reduction of energy consumption of the electric air heating components supported by electric radiant heaters was verified.

Future work is to enhance and validate the created models with measured data to check and improve their accuracy. To the steady state wall model, the dynamic heat transfer model as well as the enhanced radiation model should be paid special attention. The latter could be based on the open source simulation program "Dynamisches Raummodell zur wärmetechnischen und wärmephysiologischen Bewertung" by Prof. Bernd Glück. Furthermore the algorithm will be applied for different realistic use cases on bus route 64 in Dresden. So it is possible to calculate comprehensive optimal based control trajectories and derive rules for a heuristic online energy management. Since the VDV guideline 236 is also relevant for the whole public transport the presented approach can also be transferred to other vehicles e.g. trains.

Literature

1. Farzad Rajaei Salmasi, "Control Strategies for Hybrid Electric Vehicles: Evolution, Classification, Comparison, and Future Trends", IEEE transaction on vehicular technology, vol. 56, no. 5, 2007

2. Verband Deutscher Verkehrsunternehmen (VDV), „Klimatisierung von Linienbussen", VDV Schriften 236, 1996

3. Michael Faltenbacher, Annekristin Rock, Olga Vetter, „Abschlussbericht Plattform Innovative Antriebe Bus", 2011

4. Eckehard Specht, „Der Mensch als wärmetechnisches System", Otto-von-Guericke-Universität Magdeburg, 2005

5. Peter Kosack, „Beispielhafte Vergleichsmessung zwischen Infrarotstrahlungsheizung und Gasheizung im Altbaubereich", Research report, Version 1, 2009

6. Arno Dentel, Udo Dietrich, „Dokumentation Primero-Komfort, Thermische Behaglichkeit – Komfort in Gebäuden", HafenCity Universität Hamburg, 2008

7. Richard E. Bellman, "Dynamic Programming", Dover Publications, Reprint no. 6 (1957, 2003)

8. Bernd Glück, „Dynamisches Raummodell zur wärmetechnischen und wärmephysiologischen Bewertung", Rud. Otto Meyer-Umwelt-Stiftung, 2004

9. Steffen Kutter, Bernard Bäker, "Predictive Online Control for Hybrids – Resolving the conflict between global optimality, robustness and real-time capability", Vehicle Power and Propulsion Conference (VPPC), Lille, 2010

Analysis of the effects of high coil temperatures on performance and drivability of electric sports cars

Tobias Engelhardt, Dr.-Ing. Axel Heitmann, Stefan Oechslen
Dr. Ing. h.c. F. Porsche AG

Prof. Dr.-Ing. Hans-Christian Reuss
Institute for Internal Combustion Engines and Automotive Engineering,
University of Stuttgart

1 Abstract

Electric motors feature temporary high power densities. To avoid overheating, the allowable torque is derated dependent on temperature signals from thermal sensors. The investigation of thermal sensors in the coil ends with a special test device shows absolute and dynamic deviations. In this paper, the influence of deviations of the thermal sensor on performance and drivability of the vehicle is investigated.

The investigation is executed using simulation models of the vehicle and the electric motor. The model of the electric motor is a lumped-parameter thermal network which includes the thermal sensor. The absolute deviations of the thermal sensors are simulated by temperature offsets. Slow sensors which cannot follow dynamic temperature profiles result in overshooting temperatures. Thus the slowness must be compensated by a temperature offset.

The offsets due to absolute and dynamic deviations have to be combined to avoid overheating. The used thermal sensor of the examined electric motor causes a considerable loss of performance. The drivability is mostly influenced by the dynamic of the sensor. A slower sensor improves drivability but at the same time degrades the performance significantly.

2 Introduction

Electric powertrains for sports cars gain attention due to their instant torque delivery and high power density. However, this high power density is temporary since it leads to high temperature rises. Thus the continuous power density is lower than the short-time power density. If driven at high loads, e.g. on a race track, the temperature of the critical components reach limiting temperatures. The load must be reduced to avoid overheating which would otherwise damage the powertrain. In this paper, the load reduction is called derating.

Nowadays derating is mostly implemented in the controller of the power electronics. It protects itself and the electric motor from overheating. This paper focuses on derating due to the electric motor. The power electronics are assumed to be non-critical. Within the electric motor there are mostly two critical components: The coil ends and the rotor. The examined electric motor is a permanent magnet synchronous machine. The critical parts of the rotor are therefore the magnets. However, in this paper the rotor is assumed to be non-critical. Thus derating only occurs due to the protection of the coil ends. The windings in the coil ends consist of copper with an insulation layer. Phases are also separated by insulation foil. In the researched electric motor, the windings are cast with epoxy resin. The insulation system has class H (DIN EN 60085) which allows 180 °C. This limiting temperature is strictly held for comparability in this paper.

A simple implementation of derating is to reduce the maximum torque in dependency on the temperature signal. The temperature signal of the coil ends is mostly measured by a thermal sensor. It is challenging to measure the highest copper temperature since the heat distribution in the coil ends is heterogeneous and the positioning is subject to tolerances. Furthermore, commonly used sensors are too slow to display the actual copper temperature in dynamic load cycles.

Among others electric sports cars should have two features: High performance and drivability. As the meaning of performance is obvious, drivability has to be defined. In this paper, drivability is defined as the ability to have the driver trust in the response of the powertrain to his torque request. Hence, fluctuations due to derating affect the drivability negatively.

The aim of this paper is to investigate the influence of deviations of the thermal sensors on performance and drivability of the vehicle on a race track.

3 Accuracy of Temperature Signals

The temperature signals of the coil ends in the researched electric motor are measured by Pt100 sensors. There are four sensors within the coil ends. The hottest one triggers derating. However, measurements on the test bench show high deviations between the sensors. The investigated electric motor is a hand built prototype with the sensors also placed manually. The heterogeneous temperature distribution within the coil end combined with the imprecise manual positioning of the sensors make measuring the hottest copper temperature difficult. To obtain an idea of the margin of error, a special test device has been developed (figure 1). It features a copper winding in a slot with several sensors positioned accurately within the copper. A sketch of the slot is displayed in figure 2. Different types of sensors are used and distributed over the cross-section area. The copper winding is cast with epoxy resin. The slot cross-section area corresponds to the coil ends of the electric motor. Underneath the slot is a cooling jacket.

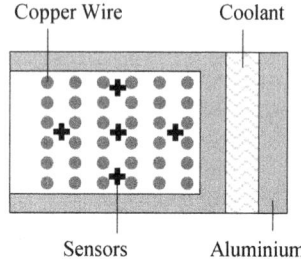

Copper Wire   Coolant

Sensors   Aluminium

Figure 1: Test device on test bench

Figure 2: Sketch of slot cross-section area of the test device

DC currents are induced into the copper winding to simulate copper-losses on a test track load cycle. The measurements result in two findings:

Total Temperature Deviation

Figure 3 shows all measured temperatures within the copper windings of the test device. The highest temperature deviation in steady-state conditions was 85 K. The lowest temperatures were measured close to the cooling-jacket and the axial walls of the slot. The highest temperatures were measured by special high-voltage Typ-K sensors in the middle of the cross-section area. The congestion around 160 °C was measured in the middle of the cross-section by regular Typ-K and Pt100 sensors. These sensors also showed deviations of around 20 K. This leads to the conclusion that even with little positioning tolerances absolute temperature deviations are to be expected.

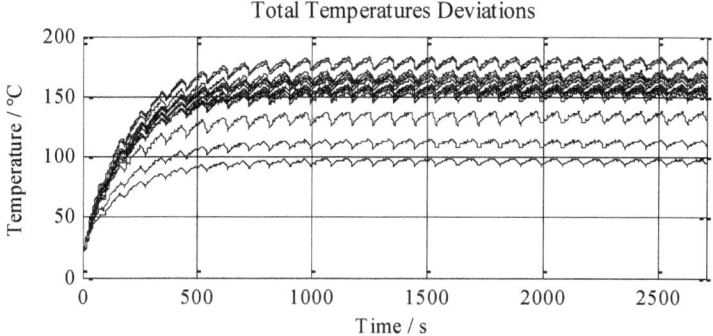

Figure 3: Overview of all measured temperatures over several laps of a PG Weissach load cycle

Dynamic of Temperature Signals

The time constant of the used Pt100 sensor is high compared to 0.5 mm Typ-K thermocouples due to its higher heat capacity and thermal resistance. The thermal resistance is higher because of the insulation shrink hose and the thermal connection to the copper wire. Equation (1) shows the differential equation of the sensor [1]. In analogy to mechanics, the heat capacity C works as dampening constant and the thermal resistance as an inverted spring constant. ΔT is the temperature difference between the sensor and the copper wire. The measured temperature of the sensor is therefore damped by its heat capacity. The thermal resistance reduces the driving force to change temperature. In the following the behavior of the sensor is referred to as "slow" sensor.

$$C \cdot \frac{dT}{dt} + \frac{1}{R} \Delta T = 0 \tag{1}$$

As results, the slow Pt100 sensors cannot follow the actual copper temperature at the dynamic test track load cycle. Figure 4 shows the measured temperatures of a Pt100 and a Typ-K sensor in comparable position. As can be seen, the Typ-K delivers a much more dynamic signal. The absolute deviation must be evaluated carefully since position sensitivity is high.

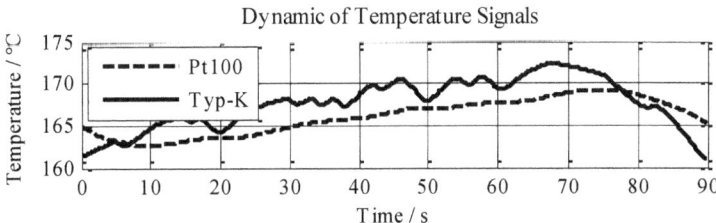

Figure 4: Comparison of a Pt100 and a Typ-K sensor over one lap of a test track load cycle

The investigation with the test device showed a high sensitivity of the position and the sensor type on the measured temperature. The temperature signal is likely to have significant deviations due to absolute temperature offsets and slow sensors.

4   Simulation Models

The investigations of this paper are carried out using simulation models. To simulate lap times and thermal effects a vehicle model and a thermal model of the electric motor are required.

4.1 Vehicle Model

The vehicle model uses a combined forward/backward algorithm similar as described in [2]. The advantage is a faster and more robust calculation compared to pure forward simulations based on driver controllers. The vehicle model is implemented in Matlab® and calculates longitudinal dynamics only. To be able to calculate lap times, a desired speed curve has been simulated by a full vehicle dynamics simulation model featuring a much more powerful vehicle [3]. As can be seen in figure 5, the longitudinal vehicle model always tries to reach the desired speed which is only possible in breaking zones and corners. Thus changes in motor power result in changes in lap times. The vehicle model also includes a battery model which calculates the actual DC voltage and the energy consumption.

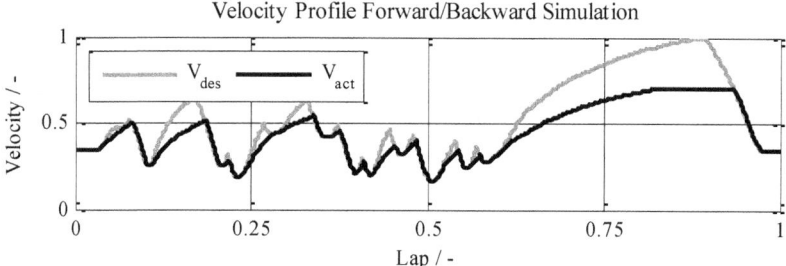

Figure 5: Forward/Backward Simulation with normalized desired and actual velocity

4.2 Thermal Model of the Electric Motor

The thermal model of the electric motor is essential for this research since derating originates at the coil ends only.

4.2.1 Thermal Network

The thermal model is implemented as lumped-parameter thermal network which is common for similar applications [4], [5]. Figure 6 shows the thermal network of the electric motor with five heat capacities and five thermal resistances. It consists of heat capacities for the coolant (C1), both coil ends combined (C2), the stator iron including the copper windings in the slots (C3), the rotor iron including magnets (C4) and the thermal sensor (C5). The sensor model is necessary to investigate the dynamic of the sensor.

The only heat sink is the coolant which is modelled with infinite heat capacity to keep its temperature constant. There are three heat sources: Copper losses in the coil end, copper and iron losses in the stator, iron and magnet losses in the rotor [6]. The losses are derived from efficiency measurements. Copper losses are calculated by the measured currents and the winding resistances. Iron and magnet losses are calculated by subtracting the copper losses and the calculated drag losses from the total losses. The distribution of copper losses between coil end and stator correlates to the copper masses. The distribution of iron losses between stator and rotor is difficult and described in the following section. The linear equation system of the thermal network is solved as matrix operation similar to [7].

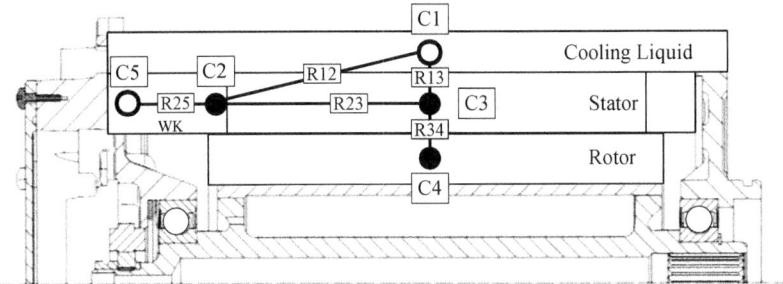

Figure 6: Thermal network of the electric motor with sensor

4.2.2 Parameter-Fitting

Predicting the values of the heat capacities and thermal resistances by analytical or numerical calculations is challenging. Also, as Dajaku [4] points out, a physical mistake is made by considering the stator and the rotor as one heat capacity due to their heterogeneous heat distribution. Another source of error is the distribution of iron losses between stator and rotor iron. In order to avoid those difficulties, most parameters of the thermal model, including an iron loss distribution factor, are found by fitting the simulated temperatures to measurements with a genetic algorithm (NSGAII, [8]). Errors in physical modeling are compensated by this process.

The coil end temperature and the rotor temperature are crucial for simulating derating behavior. Furthermore, not only the actual copper temperature but also the sensor temperature is important, since this paper analyzes influence of the sensor. Pt100 sensors have been used for measuring the coil end temperature on the test bench. In chapter 3 Pt100 sensors were described as slow. In order to achieve a copper temperature dynamic close to reality, the heat capacity of the coil end is fixed. It is calculated by the copper mass and the specific heat capacity.

The command variables are the sensor temperature and the rotor temperature. The highest measured coil end and rotor temperature signal from the test bench are used. A load cycle of the Nürburgring is used for optimization, which is long and diversified. The parameters of the thermal network are varied by the genetic algorithm until the temperature curves of the simulated sensor and rotor temperatures converge with the signals of the test bench.

Equations (2) – (4) define the optimization problem. There are ten optimization parameters, two command variables and no additional conditions. The thermal resistance of the airgap is split in two optimization variables $R_{34,l}$ (at 0 rpm) and $R_{34,h}$ (at max. rpm) which yields a linear speed dependent thermal resistance. The iron loss distribution factor is labeled as f_{fe}. F_{sen} and F_{rt} are squared deviations of sensor and rotor temperature.

$$\underline{y}^* = \min_{\underline{p}}\left\{y = f\left(\underline{p}\right)\right\} \tag{2}$$

$$\underline{p} = \left(R_{12}, R_{13}, R_{23}, R_{34,l}, R_{34,h}, R_{25}, C_3, C_4, C_5, f_{fe}\right) \tag{3}$$

$$\underline{y} = (F_{sen}, F_{rt}) \tag{4}$$

4.3 Validation

The vehicle model is not validated because the vehicle only exists virtually. However, it has been verified by comparing lap time and fuel consumption with an existing vehicle of similar power-to-weight ratio.

The thermal model of the electric motor has to be validated properly, since its behavior is key to this investigation. Thus the simulated temperature curves are compared to temperature signals from the test bench. The load cycle is not the Nürburgring but the PG Weissach.

Figure 7 displays the temperature curves of the validation load cycle over time. The graphical examination shows high congruency. Table 1 contains the temperature deviations at maximum temperature and the mean temperature deviation over time. All values are well below 5 K which is the tolerated deviation. The thermal model is therefore declared valid.

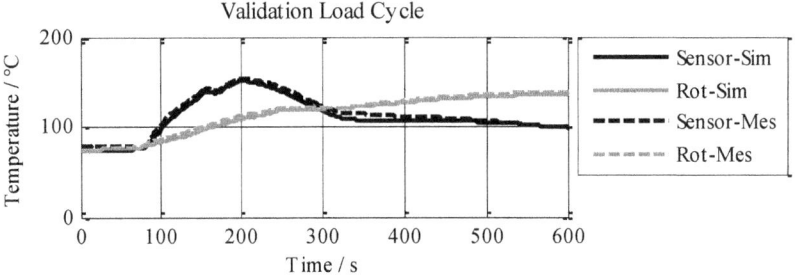

Figure 7: Validation of the electric motor on the test track load cycle (at 250 s rotor-derating has reduced torque to 0 Nm)

Table 1: Validation values electric motor on PG Weissach load cycle

Deviation max. Temp. Sensor	Deviation max. Temp. Rotor	mean Deviation Sensor	mean Deviation Rotor
2.0 K	2.4 K	3.4 K	2.1 K

4.4 Derating-Strategy

The implementation of derating is shown in Figure 8. The normalized torque is a function of the temperature signal, which is provided by the sensor in the coil ends. The normalized torque decreases linearly after exceeding the derating temperature. The normalized torque multiplied by the current maximum torque yields the limited allowable torque. There is no distinction between positive or negative torque.

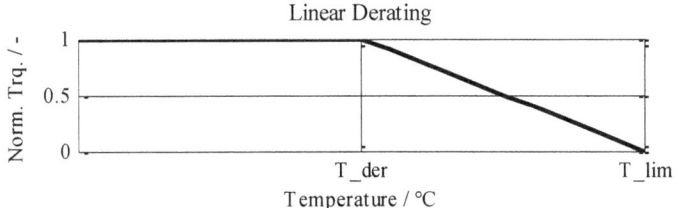

Figure 8: Linear temperature dependent derating curve

5 Performance and Drivability

Derating influences the continuous performance and the drivability of the vehicle. High performance on race tracks is obtained by exploiting the maximum thermal potential of the electric motor. Thus the coil end temperature should remain as closely as possible to the limiting temperature. The derating-strategy controls the coil end temperature dynamically by rapidly changing the allowable torque. This however, degrades drivability since the driver notices this as fluctuations. Optimizing performance and drivability at once therefore results in a conflict of interest.

The performance is measured in lap times. To measure the drivability is more complex. The dominant heat source in the coil ends are copper losses. At maximum load of the electric motor, copper losses decrease with increasing speed [9]. The temperature dependent derating of figure 8 will thus allow higher normalized torque at higher speeds than at lower speeds. This leads to fluctuations of the normalized torque. If these fluctuations are repeatable and thus the normalized torque is a function of the motor speed, the driver does not notice them negatively. Only fluctuations referred to the same motor speed degrade drivability. In order to measure speed dependent fluctuations, the fluctuation number is introduced. Every time step of the simulation the normalized torque is provided by the derating-strategy and referred to the actual motor speed.

As can be seen in equation (5), the actual normalized torque m_f is subtracted from the normalized torque when the actual motor speed n_{em} has been reached before. The differences in normalized torque are accumulated for all time steps i.

$$f_s = \sum_{i=1}^{z} \big| m_f(n_{em}, i-1) - m_f(n_{em}, i) \big| \qquad f_s = [0, \infty) \tag{5}$$

The fluctuation number is a phenomenological unit which makes evaluating difficult. Thus fluctuation numbers are used relatively for comparison.

6 Results

The investigation is carried out using simulation models. The virtual vehicle has rear wheel drive with one electric motor, inverter and a one-speed gearbox. The electric motor produces 200 kW at 800 V DC voltage and 250 Nm of torque. Its maximum speed is 15000 rpm. The vehicle mass is 1250 kg and the maximum velocity is 200 km/h.

The test cycle is the PG Weissach test track. The initial temperatures of the electric motor are 70 °C. Seven laps are simulated to reach steady-state derating with all settings. The lap time of the last lap is considered for performance and the fluctuation number of the last lap for drivability.

The supply temperature of the coolant before the inverter remains constant at 65 °C. Changing the supply temperature has a similar influence on performance and drivability as changing the offset temperature of the sensor.

6.1 Absolute Temperature Deviations

The temperature signal in this investigation is the sensor signal of the fitted thermal model (chapter 4). The temperature deviations are simulated by adding an offset to the sensor temperature of the thermal model. The applied offsets are -20, -10, 0, 10 and 20 K.

However, negative offsets lead to overheating of the coil ends. The influence on performance and drivability of overheating the coil ends is presented but not evaluated in this investigation. In reality sensors are even more likely to measure too low, thus have negative offsets (negative tolerance). To make sure temperatures will not exceed 180 °C, this negative tolerance must be considered in the derating-strategy. The limiting temperature must be reduced by the negative tolerance of the sensor. This has the same effect on the performance as positive offsets of the sensor temperature. The results of the positive offsets represent both real positive temperature deviations and adaptions of the limiting temperatures due to negative tolerances.

Figure 9 shows the simulation results for -20, 0 and 20 K offset. The velocity curves look similar. The velocity of the 20 K offset is slightly lower than the -20 K offset on the long straights. The sensor temperatures (not displayed) all converge to 180 °C. The actual coil temperatures in the second diagram differ. With -20 K offset the temperature goes clearly over 180 °C. Even without offset the coil ends overheat by 7 K due to the slow sensor. With +20 K offset the coil temperature remains clearly below 180 °C and therefore wastes thermal potential.

The curves of the normalized torque look similar, but the offset in temperature is passed through. With +20 K offset, the normalized torque drops earlier and lower than with 0 or -20 K offset. This has an impact on both performance and drivability.

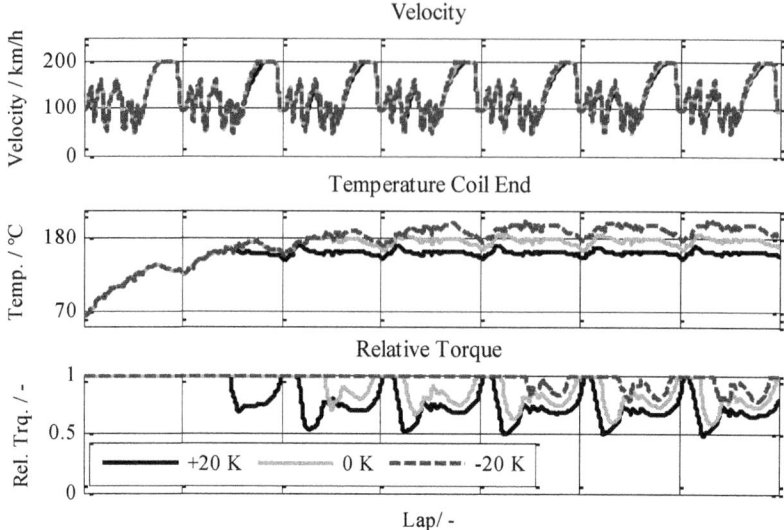

Figure 9: Simulated Velocity, coil end temperature and normalized torque over 7 laps
PG Weissach with sensor offset temperatures

In figure 10 lap times and fluctuation numbers are displayed over the temperature offset. As expected, lap times decrease with increasing offset value.

The fluctuation number increases with the offset value but has a peak at 10 K. The lower fluctuation number at negative offsets is due to the high resulting coil end temperatures. At these high temperatures high losses can be endured. Derating occurs therefore only at sections with very high iron losses. High iron losses develop at lower and medium velocities. Thus, on long straights the coil temperature decreases (figure 9). At the next slower section, the normalized torque is not reduced until the temperature has reached derating-temperature again. With increasing offset value the fluctuation number increases because the coil end temperature is lower and thus derating occurs more often.

At very low coil end temperatures (+20 k offset) another effect leads to less fluctuation. To maintain the low coil end temperature little iron losses are necessary. Hence the derating-strategy is able to keep the sensor temperature almost constantly at 180 °C. Iron losses are therefore a function of velocity and thus motor speed. As described in chapter 5, this decreases the fluctuation number.

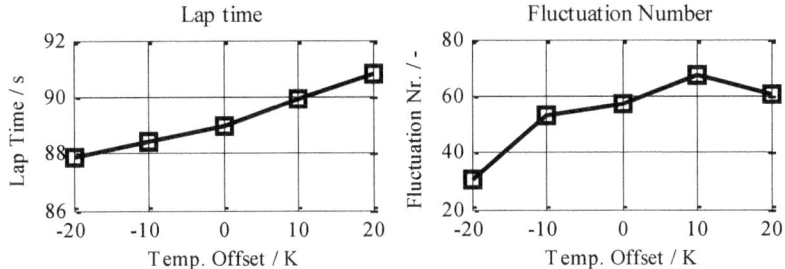

Figure 10: Lap time and fluctuation number over limiting temperature

Negative offsets are not evaluated because of the resulting overheating of the coil ends. The difference in lap time from 0 to 10 K is 0.9 s and 0 to 20 K is 1.8 s. The difference in fluctuation number from 0 to 10 K offset is 10.7 units and 3.4 units from 0 to 20 K.

6.2 Dynamic of the Temperature Sensor

The dynamic of the sensor is simulated by varying the heat capacity of the sensor and the thermal resistance between sensor and coil end in the thermal model. The zero line is the result of the optimization of chapter 4 (normal sensor). This yields a slow temperature signal since the used Pt100 has a comparably high heat capacity. To simulate a very dynamic sensor, the heat capacity of the sensor converges to zero (dynamic sensor). The sensor temperature equals the coil end temperature. In the examined electric motor windings are cast with epoxy resin. This reduces the thermal resistance considerably. To simulate an electric motor without epoxy resin, the thermal resistance is multiplied by five (5R sensor).

Figure 11 shows the sensor temperatures over five laps. The black and grey curves represent the coil end temperature and the temperature of the normal sensor (Pt100) and are similar to the measured curves of the test device in figure 4. The black curve represents actual coil end temperature. The grey and especially the black dashed line are more dynamic and also lower than the black line in the first two laps. If derating is activated by a slow sensor, coil ends overheat. The second diagram displays the actual coil end temperatures. With the normal sensor the coil ends overheat by 6.9 K with the 5R sensor by 17.2 K. This happens because the sensor temperature rises too slowly which leads to an overshoot of temperature.

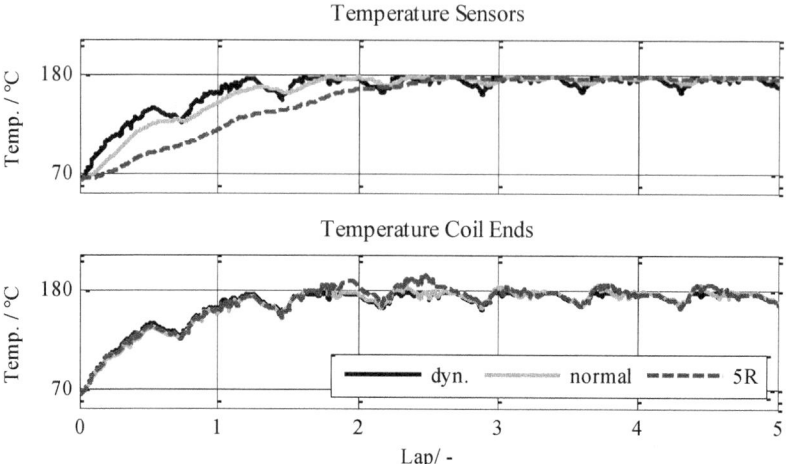

Figure 11: Simulated sensor and coil end temperatures for dynamic, normal and 5R sensor

For further investigations, the slow sensors are adjusted to remain below 180 °C. Thus an offset equal to chapter 6.1 is applied to compensate the temperature overshoot. The normal sensor requires an offset of 9 K and the 5R sensor of 27 K. Those values have been derived iteratively. Overheating may still occur in even more challenging load cycles.

In figure 12 the effect is displayed. The second diagram shows the sensor temperatures which converge to 180 °C as expected. The coil temperatures of the slow sensors normally rise at first until the sensor temperature reaches derating-temperature and the normalized torque is reduced. At this point, sensor temperature and coil temperature are similar due to the offset. The offset compensates the slow sensors. In steady-state condition, the sensor offset is passed through to the coil end temperature which remains clearly below 180 °C. The lower coil end temperatures have a negative effect on the performance since the thermal potential is not fully exploited.

The normalized torque fluctuates less with slow sensors. Whereas the normalized torque curve of the dynamic sensor has high frequency fluctuations, the 5R sensor has not.

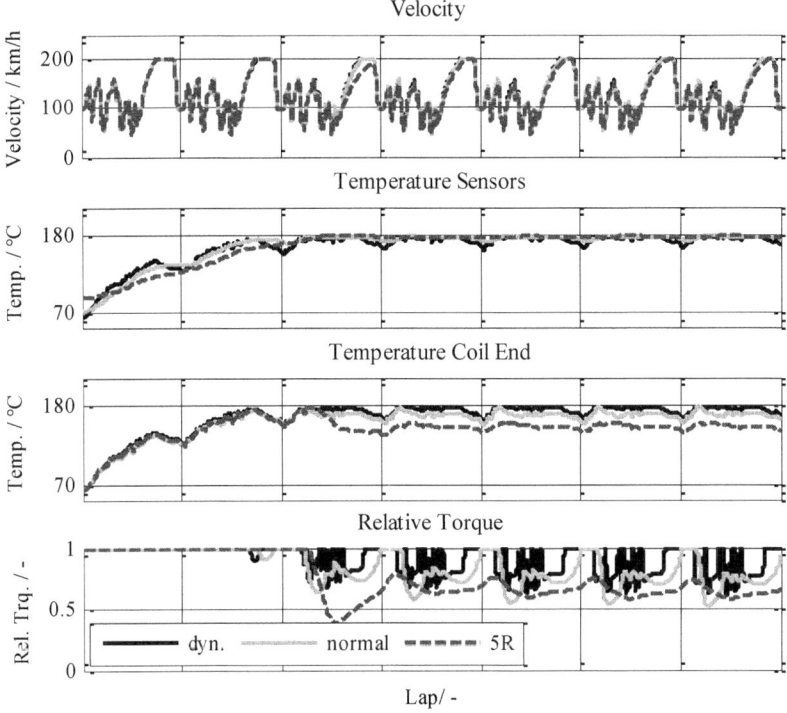

Figure 12: Simulated Velocity, sensor and coil end temperature, normalized torque over 7 laps PG Weissach for dynamic, normal and 5R sensor

Figure 13 shows lap time and fluctuation number over the sensor type. The lap time increases with sensor slowness, mainly due to the temperature offset. As expected, the fluctuation number decreases with sensor slowness. The difference in fluctuation number is higher between the normal and the 5R sensor than it is between the dynamic and the normal sensor.

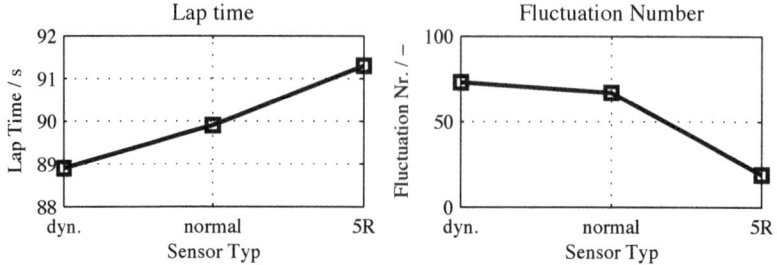

Figure 13: Lap time and fluctuation number over sensor type

The lap time increases by 1.0 s from the dynamical to the normal sensor and 2.4 s from the dynamical to the sensor with 5 times the thermal resistance. Those differences are considerable.

The difference in fluctuation number from the dynamical to the normal sensor is 6.2 units only, which is barely noticeable. The difference between the normal and the 5R sensor is 54.2 units. This improvement is clearly noticeable for the driver.

7   Conclusion

In this paper the influence of deviations of the thermal sensor on performance and drivability of the vehicle with active derating is investigated. Absolute and dynamic deviations of thermal sensors have been displayed by a special test device. To carry out the investigation, simulation models are required. Besides the vehicle model the thermal model is essential. The lumped-parameter thermal network features a heat capacity for the thermal sensor to make this investigation possible.

The absolute deviations of the thermal sensors are simulated by temperature offsets. Sensors are more likely to have negative offsets due to positioning tolerances. To ensure not exceeding 180 °C, the limiting temperature in the derating-strategy has to be adapted. This has the same effect as adding a positive offset to the temperature signal. A realistic positive or negative offset of 10 K increases the lap time by 0.9 s.

Slow sensors result in overshooting temperatures. Thus the slowness must be compensated by a temperature offset. In steady-state conditions and active derating, the offset of the sensor temperature is passed through to the actual coil end temperature and thus degrades performance. The used Pt100 sensor causes a lap time increase of 1.0 s. However, slow sensors improve drivability. Since its improvement degrades the performance considerably, drivability should rather be improved by a derating-strategy.

The offsets due to absolute and dynamic deviations have to be combined to avoid overheating. Hence a commonly used Pt100 sensor requires 19 K offset which results in a lap time increase of about 1.8 s already. This is a considerable loss of performance which should not be tolerated. In context, a reduction of motor power from 200 to 145 kW results in a 1.8 s higher lap time (without derating).

Potentially, the insulation system allows raising the limiting temperature for the short period of race track operation. However, this should be used for knowingly improving the performance through the derating-strategy, not for compensating inaccuracies of the temperature signal.

References

1. Baehr, H. D.; Stephan, K.: *Wärme- und Stoffübertragung. 6. neu bearbeitete Auflage*, Springer-Verlag Berlin, Heidelberg, 2009.

2. Wipke, K. B.; Cuddy, M. R. und Burch, S. D., *ADVISOR 2.1: A User-Friendly Advanced Powertrain Simulation Using a Combined Backward/Forward Approach*, in IEEE Transactions on Vehicular Technology, Vol. 48, No. 6, 1999.

3. Rüger, S., *Vollhybridantriebsstrang für ein sportliches Hybridfahrzeugkonzept*, Braunschweig: Shaker, 2014.

4. Dajaku, G., *Electromagnetic and Thermal Modeling of Highly Utilized PM Machines*, Aachen: Shaker Verlag, 2006.

5. Hak, J.: *Einfluß der Unsicherheit der Berechnung von einzelnen Wärmewiderständen auf die Genauigkeit des Wärmequellen-Netzes.* Archiv für Elektrotechnik, Band 47, 6. Heft, S. 370 – 383, 1963.

6. Finken, T.: *Fahrzyklusgerechte Auslegung von permanenterregten Synchronmaschinen für Hybrid- und Elektrofahrzeuge.* PhD Thesis, RWTH Aachen, Shaker Verlag, 2011.

7. Kipp, B., *Analytische Berechnung thermischer Vorgänge in permanentmagneterregten Synchronmaschinen*, Hamburg, 2008

8. Deb, K.; Pratap, A.; Agarwal, S. und Meyarivan, T., *A Fast and Elitist Multiobjective Genetic Algorithm: NSGA-II*, in IEEE Transactions on Evolutionary Computation, Vol. 6, No. 2, 2002.

9. Soong, W. L., *Design and Modelling of Axially-Laminated Interior Permanent Magnet Motor Drives for Field-Weakening Applications*, Glasgow, 1993.

Experimental validation of the Maxwell model for description of transient tyre forces

Andreas Hackl, Wolfgang Hirschberg, Cornelia Lex
Graz University of Technology, Austria

Georg Rill
OTH Regensburg University of Applied Science, Germany

1 Introduction

Modelling and simulation of safety relevant Driver Assistance Systems (DAS) and Vehicle Dynamics Controllers (VDC) which act in standard and limit situations lead to increasing accuracy demands in the description of dynamic reactions of tyre contact forces, e.g. [4], [7]. For that purpose, first-order approaches are widely applied in this field of vehicle dynamics and handling, which originate from Schlippe & Dietrich [13], were modified by Pacejka [10] and later on refined by Rill [11], [12]. This approach is typically characterised by the first-order differential equation

$$\tau_{x,y}\,\dot{F}^{D}_{x,y} + F^{D}_{x,y} = F^{S}_{x,y}\,,\tag{1}$$

where the superscripts D and S distinguish between dynamic and static tyre forces and the subscripts x and y indicate the longitudinal and lateral directions of the tyre contact forces F. Anyway, the coefficient τ which corresponds to a dynamic relaxation length is not constant, but depends on the wheel load F_z and the tyre slips s_x and s_y respectively.

The line of modelling of the visco-elastic mechanism of tyre deformation x is the key of a proper description of τ. This can roughly be done by a simple spring-damper element of type Voigt-Kelvin, having two parameters c and d, see Figure 1. However, fixed values of them can only cover a limited range of amplitudes and frequencies, as it is well known for elastomer materials, cf. [1], [2]. Therefore, the inclusion of the more complicated Maxwell element is proposed, which has got two additional parameters c_M and d_M and one internal variable x_M.

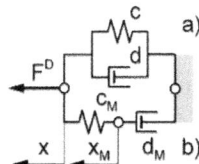

Figure 1: Maxwell model, which is a combination of the Voigt-Kelvin model a) and an additional Maxwell element b)

Based on previous researches, e.g. [1] [2], the scope of the present paper is to investigate the qualification of the Maxwell model for an appropriate dynamics description. The four model parameters have to be identified under the condition of practical applicability from the engineering point of view. This is based on measurement data from an extensive laboratory testing programme. However, prior to this, preliminary investigations were carried out in order to check the principal suitability of the method to cover the specific problems resulting from the rotating tyre. With the aim to run vehicle dynamics models on uneven, but not rough roadways, a frequency range of at least 3 Hz was considered. Due to the research project's current focus on lateral vehicle dynamics, the evaluation of the correspondent relations in longitudinal direction will be dealt with on a later occasion.

2 Enhanced Tyre Dynamics

The TMeasy tyre model describes the dynamics of the tyre forces and torques by taking the tyre compliance into account, [11]. In a first and quite simple approach the compliance of the tyre is modeled by a linear spring in parallel to a linear damper, Figure 2 a).

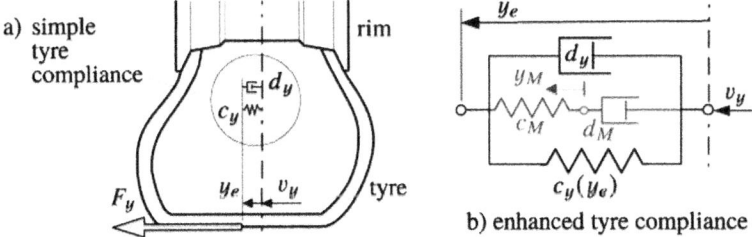

Figure 2: Tyre dynamics in lateral direction

This Kelvin-Voigt model is defined in lateral direction by a Taylor expansion of the dynamic tyre force

$$F_y^D = F_y\left(v_y + \dot{y}_e\right) \approx F_y\left(v_y\right) + \frac{\partial F_y}{\partial v_y}\dot{y}_e = F_y^{st} + \frac{\partial F_y}{\partial v_y}\dot{y}_e, \tag{2}$$

and accompanied by the simple force law

$$F_y^D = c_y y_e + d_y \dot{y}_e, \tag{3}$$

where F_y^{st} denotes the steady state lateral tyre force, v_y is the lateral component of the contact point velocity, y_e denotes the lateral tyre deflection, and c_y, d_y are the stiffness and damping constants approximating the tyre compliance in the lateral direction. Combining the relations in (2) and (3) results in a first-order differential equation for lateral tyre deflection. Making advantage of the TMeasy tyre modelling concept one finally gets

$$\left(v_{Ty}^* d_y + f_G\right)\dot{y}_e = -v_{Ty}^* c_y y_e + f_G v_{Ty}^* s_y, \tag{4}$$

where s_y denotes the lateral slip, $f_G = F_G / s_G$ represents the global slip derivative of the generalized tyre characteristic, and a modified transport velocity v_{Ty}^* through the patch is used to simplify the expressions.

This simple approach is quite effective, but curve fits to measurements show, that the stiffness and damping parameters depend on the excitation frequency [2]. That is why an enhanced compliancy approach is investigated in this paper. At first, the linear spring is extended to a non-linear one by simply adding a term proportional to the tyre deflection

$$c_y = c_y(y_e) = c_y\left(1 + p_y|y_e|\right), \tag{5}$$

where the parameter p_y defines both, the character and intensity of the non-linearity. Secondly, a Maxwell model is placed in parallel to the Kelvin-Voigt model, Figure 2b). Now, the simple force law given in (3) has to be replaced by

$$F_y^D = c_y y_e + c_M(y_e - y_M) + d_y \dot{y}_e. \tag{6}$$

Combining this equation with (2) finally results in

$$\left(v_{Ty}^* d_y + f_G\right)\dot{y}_e = -v_{Ty}^* c_y y_e - v_{Ty}^* c_M(y_e - y_M) + f_G v_{Ty}^* s_y. \tag{7}$$

As extension to (4) the time derivative of the lateral tyre deflection y_e depends now additionally on the stiffness c_M, the damping d_M and the internal displacement y_M of the Maxwell element. The force balance applied to Maxwell element delivers a second first order differential equation with respect to y_M

$$d_M \dot{y}_M = c_M(y_e - y_M) \quad \text{or} \quad T_M \dot{y}_M = -y_M + y_e, \text{ where } T_M = d_M / c_M \tag{8}$$

driven by the lateral tyre deflection y_e and characterised by the time constant T_M that is defined by the stiffness and damping parameters c_M and d_M of the Maxwell element.

3 Optimisation strategy and parameter study

At first, an optimisation strategy validation and a parameter study were done. Therefore, a simulation model with fixed parameters was used to generate a modelled lateral force $F_{y,m}$, which was then superposed by random noise to consider a realistic measurement behaviour. With this created force, two different optimisation strategies were investigated, including the performance under presence of measurement noise and also the convergence of the parameters. To keep the optimisation time in a reasonable extent, a parameter study was done to find adequate manoeuvres to identify each parameter, and also an effective sequence for optimising the different parameters and thus increase the convergence probability of the parameters. Afterwards, the simulation inputs (i.e. manoeuvres) were defined which are suitable to parametrise the parameters of the Maxwell model, furthermore an appropriate sequence for the optimisation strategy of the different required parameters was developed.

3.1 Modell parameter and test manoeuvre

To test the optimisation strategy and check the convergence of the parameters, a parameter set with reference to [11] was used to model the lateral force $F_{y,m}$, see in Table 1. In prior investigations, a sine steer input with varying slip angle α at a constant normal force F_z has proven to be a suitable test manoeuvre to investigate the frequency response and parametrise the model values, see [1] and [2]. Thus, a sinusoidal slip angle input α with constant amplitude and three different frequencies was chosen, see Table 2.

Table 1: Parameters to describe the lateral steady state tyre characteristics at $F_{z,Nom} =$ 3600 N which were used for optimisation validation and parameter study. For a parameter definition see [11], the dynamic model parameters are described in Section 2.

Initial slope	$dfy0$	60000	N / -
Maximum force	fym	4000	N
Slip s_y where $f_y = f_m$	sym	0.15	-
Sliding force	fys	3900	N
Slip s_y where $f_y = f_{ys}$	sys	0.25	-
Fictitious velocity	v_N	0.01	m / s
Tyre stiffness	c_y	120000	N / m
Tyre damping	d_y	600	Ns / m
Nonlinear stiffness	p_y	0	1 / m
Stiffness Maxwell element	c_M	$c_y / 2$	N / m
Damping Maxwell element	d_M	$d_y \cdot 5$	Ns / m

Table 2: Overview of the manoeuvre conditions to validate the optimisation strategy and investigate the convergence of the model parameters, listed in Table 1.

Tyre load	F_z	3600	N
Sine slip angle amplitude	α_{Amp}	1	deg
Sine slip angle frequency	f	0.5, 1, 3	Hz
Longitudinal speed	v_x	60	km/h
Time step	t_s	2e-3	sec
Number of sine periods	N	10	-

3.2 Optimisation strategies

Two different optimisation strategies were investigated. The first one, a nonlinear least-square algorithm (NLSQ) [8], can be used to find parameters of nonlinear characteristic at a short optimisation time. The second one, a hybrid optimisation strategy which uses Particle Swarm optimisation in combination with the Nelder-Mead Simplex method (HPSO) [9], was found as a suitable algorithm to find the global minimum of non-linear optimisation problems. The normalized root-mean square deviation

$$NRMSD = \frac{RMSD}{\hat{y}_{max} - \hat{y}_{min}}, \quad \text{with} \quad RMSD = \sqrt{\frac{\sum_{t=1}^{n}(\hat{y}-y)^2}{n}}, \quad (9)$$

was used as cost function of the optimisation problem for different manoeuvres. Therein, $\hat{y}$ describes the simulated measurement signal superposed by random noise and y the simulated signal, both with the length of n time steps. After about 25 optimization runs with these two methods, it was seen that both of the strategies were not able to find all four parameters with a high accuracy. Especially the NLSQ had problems to find a satisfying result at higher noise to signal ratios. In addition, the results strongly depend on the starting values. The results of the HPSO are more adequate but with a higher cost of optimisation time. Overall it was seen that the HPSO algorithm was able to find more accurate parameter values, especially with higher added noise.

3.3 Parameter study

With both optimisation algorithms it was recognized that the value of the spring parameter c_y was the parameter to be most easy identified compared the others. When setting the overall boundary conditions from 20% up to 500% of the correct value, nearly 100% of the HPSO and around 60% of the NLSQ optimisation runs found the correct value with less than ±5% deviation.

Regarding the Maxwell element parameters c_M and d_M, the results of the two algorithms differed strongly. In the majority of cases, the NLSQ algorithm was not able to find acceptable values. In contrast, the HPSO procedure found results with a deviation of $\pm 10\%$, in around 75% of the optimisation runs. In addition, the convergence ratio increased with higher frequency of the sinusoidal input. Another advantage of the HPSO result was that the average values of the resulting Maxwell elements deviated less than 5% over the investigated frequency range.

The largest deviation during the parameter optimisation was seen for the damper values d_y. Especially for manoeuvres with a small frequency, both algorithms were not able to find acceptable values. Convergence increased with frequency, but was still worse than for the other parameters.

With this investigation, it was shown that the stochastic optimisation method HPSO is suitable to handle the quite sensitive nonlinear optimisation problem. Thus, this algorithm is chosen to parameterise the Maxwell model with measurement manoeuvres. However, to identify all parameters with sufficient accuracy, the respective manoeuvres require further detailing.

3.4 Manoeuvre definition

In addition to choosing a suitable optimisation method, the presented investigations show that a more detailed manoeuvre setup is needed to ensure that all parameters can be identified with sufficient accuracy.

To achieve this, two additional manoeuvres were applied. The first one is a quasi-stationary manoeuvre to define the nonlinear behaviour of the lateral spring characteristics c_y. In this way, it is possible to reduce the number of optimisation parameters to three. For the damper parameter d_y, another manoeuvre was required. In [2] it was shown, that a slip angle step is a suitable manoeuvre to validate the Voigt Kelvin damper characteristics d_y such it was also chosen for this application. To consider the parameter influence between the Voigt Kelvin damping d_y and the two Maxwell element parameters c_M and d_M, two more iterative processes with the sine- and step manoeuvre were carried out as shown in Table 4.

Another effect, seen at the simulation with different frequencies, was an increasing value of the spring characteristic when just using a Voigt-Kelvin model instead of a combination with a Maxwell element. This reflects the dynamic behaviour of the Voigt-Kelvin model in dependence on the manoeuvre frequency, which was previously investigated in [1] and [2].

In Table 3, an overview of the parameters and the manoeuvres to identify them is given including weighting factors for the final estimates. For example, the spring rate d_y is mainly identified with step steer manoeuvres and optimised under narrow boundaries with sine steer manoeuvres.

Table 3: Measurement manoeuvres and their respective weighting factors to identify the parameters of the Maxwell model.

Parameters		Quasi - Stationary [%]	Sine [%]	Step [%]
Tyre stiffness	c_y	100	0	0
Tyre damping	d_y	0	20	80
Stiffness Maxwell element	c_M	0	80	20
Damping Maxwell element	d_M	0	80	20

4 Test bench setups and test procedures

Experimental data were used to obtain all required model parameters based on the manoeuvres and optimisation strategy chosen in Section 3. The used data were measured on a suspension test rig using specific test setups and procedures, which are presented in this section.

4.1 Test bench

A brake and suspension test rig designed for investigation of durability and fatigue of components of quarter vehicle suspensions was used to conduct the measurements described in Section 4.2, [3]. A wheel assembly consisting of tyre, rim and wheel carrier where mounted to the test bench using a rigid suspension (no spring and damper elements). The test bench has a drum with a standardised outer diameter of 1.219 m. The drum speed can be between 0 and 1300 rpm. It can also be pivoted around the vertical axle to generate a slip angle α of $\pm$ 15 deg. The vertical tyre load is controlled using a hydraulic cylinder with a maximum cylinder force up to 25 kN as well as a maximum pulse frequency of 35 Hz and a maximum actuator acceleration of 10 m/s² in dependence of the travel range.

Forces and torques were measured in the wheel hub using a high-precision wheel force transducer [6]. Forces, torques, drum roll angle α, drum roll speed Ω_d, vertical travel z_c, of the wheel carrier, and the ambient temperature at the test bench are measured with a sampling rate of 1 kHz.

4.2 Test setups and procedures

A radial tyre size of 205/55 R16 was used for the presented investigations. Two different test setups were used. At first, the lateral tyre stiffness was measured using stationary measurements with a non-rolling tyre, where the tyre deflection was varied. The required test bench setup is shown in Figure 3, the test procedure is described in Section 4.2.1. In the second test setup as shown in Figure 4 and described in Section 4.2.2, dynamic tests with varying side-slip angle were performed on a rolling tyre at a longitudinal speed v_x of 60 km/h.

The measurement procedure has included several parameter variations. However, all measurements used for the investigations presented here were conducted at tyre normal force of 3600 N, a tyre inflation pressure of 2.75 bar and an ambient temperature of 20°C.

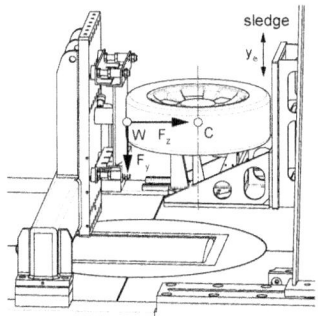

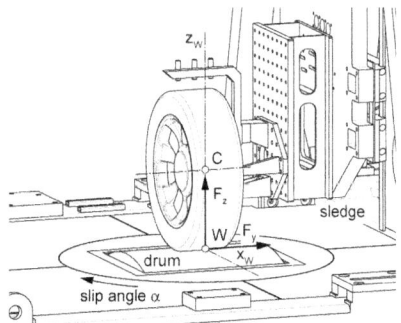

Figure 3: Test bench setup for lateral spring characteristics (W – system; C – system; F_y – lateral tyre force; F_z – tyre normal force; y_e – lateral tyre deflection)

Figure 4: Test bench setup for dynamic measurements (step steer and sinus steer inputs) with varying drum roll angle α

4.2.1 Measurements to determine lateral tyre stiffness

The linear and non-linear lateral tyre stiffness characteristics have been identified using stationary measurements of the non-rolling tyre. The tyre was deflected in the lateral direction y_e, and the resulting lateral and vertical tyre force were measured. Since the sledge of the test rig can only move in vertical direction, the assembly of wheel carrier, measurement rim and tyre was mounted in a rotated position, see Figure 3.

So for this setup, the vertical sledge movement of the test rig corresponds to a lateral tyre deflection y_e. Also, the normal tyre force F_z cannot be controlled in this setup, such it has to be adjusted with a hydraulic pump before the test procedure. During the measurement procedure, the normal force decreases with increasing lateral deflection. To compensate for this effect, the lateral tyre force F_y is normalised with respect to the measured normal force F_z by

$$F_y{}^N = F_y \, \frac{F_{z,0}}{F_z}, \tag{10}$$

where $F_y{}^N$ is the normalized lateral force and $F_{z,0}$ indicates the nominal tyre load.

4.2.2 Dynamic measurements

Using the test bench setup shown in Figure 4, different step steer and sinus steer manoeuvres were conducted. To parameterize the steady-state tyre characteristics, step steer manoeuvres with steps of drum roll angle α between 1, 2 and 4 deg within a range of ± 12 deg were conducted.

To identify the damping characteristics, step steer inputs from α = -1 to 1° were performed. To validate the dynamic tyre models, sine steer inputs with an amplitude of α = 1° were performed with frequencies of 0.25, 0.5, 1, 1.5, 2, 2.5 and 3 Hz were conducted. A detailed description of an enhanced measurement program is given in [1].

5 Parameterisation of the lateral tyre parameters

After defining the optimisation strategy (see Section 3) and having suitable measurements from a test bench (see Section 4), the results of the parameterisation of the Maxwell model under the selected conditions are presented in this section.

5.1 Steady state tyre characteristics

To parametrise and validate the dynamic behaviour of the tyre, a steady state tyre model is required. Therefore, the tyre model TMeasy is used and the parameterisation is supported by the utility software *TFView* as described in [5].

As described in Section 4, the steady state characteristics are parametrised with step steer inputs during constant drum speed of 60 km/h. The identified characteristics shown in Figure 5 were parametrised with slip angle steps of 1, 2 and 4 deg. The circles symbolise the measurement points and the dashed line the parametrised model.

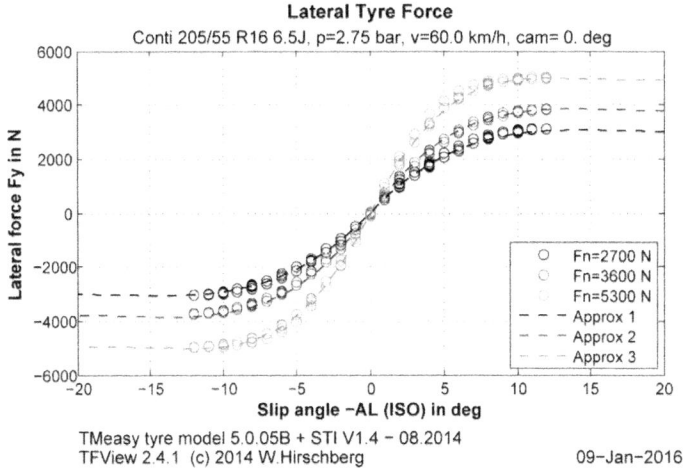

Figure 5: Parametrisation of the steady state tyre model TMeasy using the utility *TFView*

5.2 Nonlinear spring characteristics

For the parametrisation of the nonlinear characteristics, a lateral displacement y_e was produced and the lateral Force F_y^N measured as described in Section 4.

Because of the very slow lateral movement, a quasi-stationary behaviour was assumed. With (5) and (6), the related force equation comes to

$$F_y^N = c_y \cdot y_e + c_y \cdot p_y \cdot |y_e| \cdot y_e, \tag{11}$$

where the parameters c_y and p_y were determined with the least square method.

Figure 6 shows the results of the modelled nonlinear lateral spring in comparison to the measurement data as well as the deviation between these two.

5.3 Lateral damper and Maxwell element

After determining the stationary tyre characteristics and the nonlinear spring characteristics, an iterative identification process started using sine and step manoeuvres of the slip angle α. As described in Section 3, the same procedures were chosen for the sine manoeuvre as for the parameter study, with exception of the frequency sampling. This is necessary because of the dependency of the Maxwell element parameters on the frequencies.

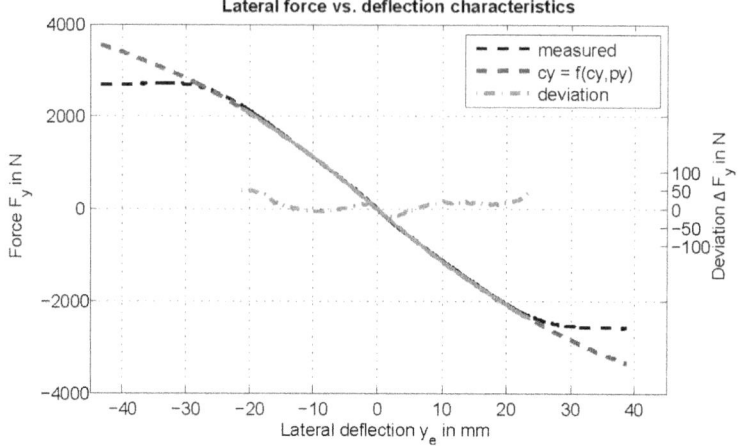

Figure 6: Comparison of the identified nonlinear lateral spring characteristics and the measurement data. Estimated parameters: $c_y = 121819$ N/m and $p_y = -7.5466$ 1/m

Thus, the identified parameters were weighted with its frequencies and in the case of the Maxwell spring, this finally calculates to

$$c_{M,avg} = \frac{1}{\sum f_i} \cdot \sum_{i=1}^{N} c_{M,i} \cdot f_i \,, \tag{12}$$

where $c_{M,i}$ denotes the estimated parameter per frequency f_i. The Maxwell damper d_M was calculated in the same way. For the slip angle step with the sine amplitude of $\alpha_{Amp} = 1$ deg, a step from $\alpha = 1$ deg to $\alpha = -1$ deg was chosen to parametrise the Voigt Kelvin damping d_y. Starting from the quasi-stationary manoeuvres and the slip angle step manoeuvres, the base values of the spring c_y and damper d_y were identified. With these two values, the iterative estimation process was started and the average results of ten independently executed optimisations processes are determined as listed in Table 4.

From the resulting Maxwell parameters shown in Table 4, it can be seen that with increasing iteration steps, the value of Maxwell spring element c_M decreases and the value of the damper element d_M increases.

Since the parameter values do not change significantly after the third iteration, the iteration is stopped here.

Table 4: Identified parameters of the Maxwell model, determined using three different measurement manoeuvres and an iterative optimisation procedure.

M. Nr.	c_y in N/m	d_y in Ns/m	c_M in N/m	d_M in Ns/m
Opt. boundary	[lower boundary	upper boundary] in % of start value M.Nr.-1		
Quasi- stat. 1	121819	-	-	-
	[0 ∞]	-	-	-
Step Nr. 1	121819	332	-	-
	[100 100]	[0 ∞]	-	-
Sine Nr. 1	121819	315.4	32580	5891
	[100 100]	[95 105]	[20 500] · c_y	[5 2000]·d_y
Step Nr. 2	121819	159.1	26065	7068
	[100 100]	[50 200]	[80 120]	[80 120]
Sine Nr. 2	121819	151.1	14380	8951
	[100 100]	[95 105]	[50 200]	[50 200]
Step Nr. 3	121819	282.5	15099	9243
	[100 100]	[50 200]	[75 125]	[95 105]
Sine Nr. 3	**121819**	**282.5**	**12039**	**9243**
	[100 100]	[100 100]	[75 125]	[100 100]

6 Validation of the estimated model parameters with measurement data

Finally, this section deals with the validation of the parametrised non-linear tyre dynamic model. Therefore, the simulation results of the parametrised Maxwell model are being compared to both measurement data and simulation data of two other models.

The first model for comparison is the state of the art Voigt Kelvin model, which is parametrised with a linear spring and damper value. The two required parameters are estimated from the quasi-stationary and first step manoeuvre; cf. Table 4, Step Nr.1. The second setup represents a Voigt Kelvin model with a nonlinear spring characteristics. The parameter c_y now depends on the frequency f and is given by $c_y = c_y(f)$. In addition, this adapted Voigt Kelvin model requires two constant parameters, namely parameter p_y for the nonlinear spring behaviour and d_y for the damper characteristics. The value for the damper parameter is the same as in the first Voigt Kelvin model.

The nonlinear parameter p_y is used as shown in Figure 6 and the optimised characteristics for the spring value $c_y(f)$ is given in Figure 7.

Figure 7 shows the results of the validation of the sine manoeuvre with different frequencies. On the left subfigure, the characteristic of the frequency-dependent spring (grey asterisks) is presented, compared to the two other models with a constant spring (grey circle – state of the art Voigt Kelvin model and black asterisks – Maxwell model). These characteristics show the similar behaviour to previous investigations with a different tyre [2]. On the right subfigure, the results of the NRMSD are presented. It can be seen that the errors of the Maxwell model and the optimized Voigt Kelvin model are smaller for frequencies higher than 1Hz. For frequencies which are smaller or equal to 1Hz, the error of the Maxwell model is slightly higher. From the evaluation of smaller frequency ranges, it seems that the added Maxwell element causes a small time delay which is responsible for the increased error value. Further investigation will be held on these smaller frequency ranges.

In Figure 8, the characteristics of the lateral force during a sine manoeuvre with a frequency of $f = 2.5$ Hz is shown. On the left subfigure, a time period is presented which shows that the deviation of the time delay of the approximate TMeasy and Maxwell model is smaller than that of the state of the art Voigt Kelvin model, which is named *TMeasy fixed parameters* in Figure 8. On the right subfigure, it is shown that the behaviour of the Maxwell element as well as the model with the variable spring characteristic (*TMeasy approximation*) do have higher stiffness than that of the state of the art model.

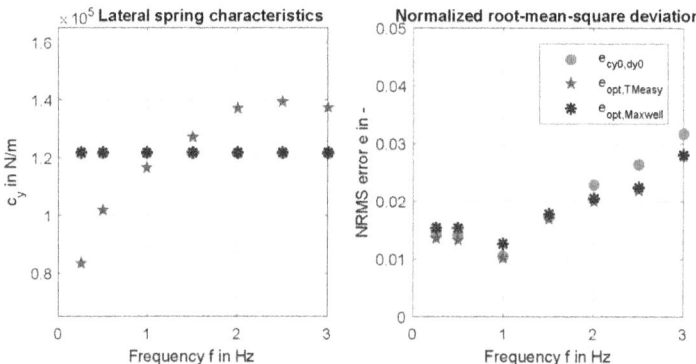

Figure 7: Result of the influence of different sine slip angle frequencies on the spring characteristic of the Voigt Kelvin model (left), and the results of the comparison of the three model setups on the NRMSD (right).

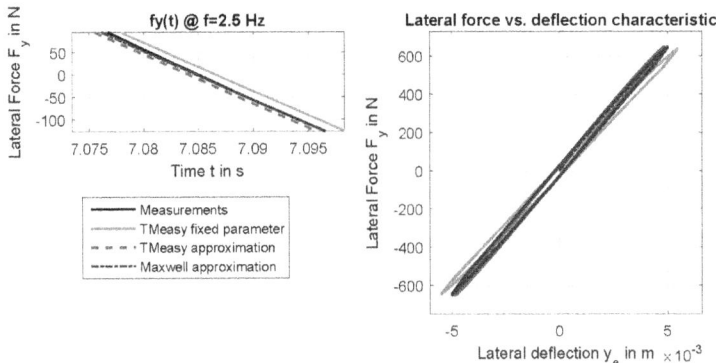

Figure 8: Comparison of the parametrised models with a measured sine manoeuvre with a frequency of f = 2.5 Hz and slip angle amplitude of 1deg.

The comparison presented in Figure 9 shows a similar behaviour for measured step steer input manoeuvres to that of the sine inputs with higher frequencies. The time delay between measurement and model increases in the case of fixed parameters. This means that the accuracy is lower than for both of the other approximations.

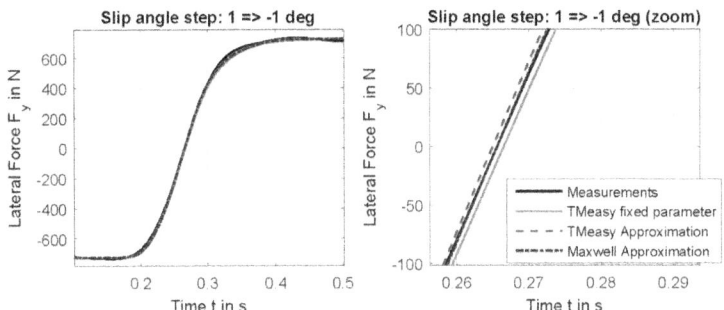

Figure 9: Comparison of the parametrised models with a measured step steer input from slip angle α = 1 deg to α = - 1deg

In general, it was shown that the Maxwell element improves the simulation accuracy of the transient behaviour of the tyre dynamics, especially for higher frequencies.

415

It was also shown that the parametrisation of this model requires different measurement manoeuvres and thoroughly chosen optimisation methods. Further investigation on alternative optimisation algorithm, including their cost functions, will be done to improve the parametrisation process.

In a next step, a more detailed investigation of smaller frequency ranges is needed to improve the behaviour on the overall frequency range. Another possibility to investigate the parametrisation process and validate the behaviour of the Maxwell model is to transform the model in the frequency domain.

In summary, further investigations have to be done to reduce the complexity of the parametrisation an increase the parameter convergence. However, the Maxwell element proofed to be a suitable element to model the dynamic hardening of the viscoelastic materials of the tyre under rotation.

Conclusions

The present paper deals with the modelling extension of the first order tyre dynamics approach for vehicle dynamics and handling simulation. The consideration of tyre dynamics is an essential condition for reliable computation of transient driving manoeuvres, such as they are typical while interventions of advanced driver assistance systems (ADAS). Since conventional first order models simply consider spring-damper elements of type Voigt-Kelvin for the description of the tyre's compliance in longitudinal and lateral direction respectively, their operational validity is limited to a certain amplitude and frequency. Thus, the treatment of elastomer components requires enhanced compliance modelling, which is done here by means of a Maxwell element. Furthermore, the hyper-elasticity seems to be worth to be taken into consideration.

The Maxwell model includes four instead of two parameters so far, and it is the key issue whether for a certain tyre these can be identified from usual dynamic bench testing. Otherwise, the approach would theoretically be nice but without practical relevance. Firstly, the extended modelling principle is briefly described. In order to principally check the identifiability of the model parameters, pre-computations based on ideal, "artificial" testing data having random disturbance are carried out. This preliminary studies show the principal suitability to identify the particular parameter of the Maxwell model by means of an optimisation strategy. In particular, Particle Swarm optimisation in combination with the Nelder-Mead Simplex method was found as a suitable algorithm to find the global minimum of the non-linear optimisation problem. Furthermore, recommendations for suitable testing manoeuvres can be derived from the preliminary studies.

In the following, an extended measurement programme on a test bench under laboratory conditions was carried out to acquire data to validate the above mentioned modelling approach. The testing conditions are briefly documented, where particularly sine inputs on the slip angle with different amplitudes and frequencies under different tyre speeds were applied. Advantageously the results shown in the paper are restricted to a nominal tyre load of 3600 N.

From the investigations it was shown that the Maxwell element combined with a nonlinear static stiffness is qualified to improve the simulation accuracy of the transient tyre behaviour, especially for higher frequencies. Furthermore it was shown that the parametrisation process of this model requires different measurement manoeuvres and suitable optimisation methods. Further investigation on alternative optimisation algorithm, including their cost functions, will be done to improve the parametrisation process. It can be pointed out, that the identification of the four parameters of the Maxwell model c_y, d_y, c_{yM}, d_{yM} plus the nonlinear stiffness parameter p_y require different effort in the sequence from easy to difficult: c_y, p_y, c_{yM}, d_{yM} and d_y.

Although the Maxwell element has proofed to be a suitable to model the dynamic hardening of the elastomer materials of the tyre under rotation, further investigations have to be done to reduce the complexity of the parametrisation process and to increase the parameter convergence. For that, the higher testing amplitudes and frequencies can be utilised in order to reduce the sensibility of the parameter identification.

Bibliography

1. Hackl, A., Hirschberg, W., Lex, C., Rill, G. (2015) *Experimental validation of a non-linear first-order tyre dynamics approach*. Proc. 24th International Symposium on Dynamics of Vehicles on Roads and Tracks, Aug 17-21, 2015 Graz.

2. Hackl, A., Hirschberg, W., Lex, C., Rill, G. (2015) *Tyre dynamics: Model validation and parameter identification*. Proc. EAEC-ESFA Congress Bucharest 2015, Springer Int. Publishing, Switzerland.

3. Harrich, A., Tonchev A., Hirschberg, W. 2006: *Der neue dynamische Bremsen- und Radaufhängungsprüfstand an der TU Graz*. Proceedings of brake.tech, Munich, December 7-8, 2006. Munich: TÜV SÜD.

4. Hirschberg, W., Weinfurter, H., Jung, C. (2000) *Ermittlung der Potenziale zur LKW-Stabilisierung durch Fahrdynamiksimulation*, VDI-Berichte 1559, Düsseldorf, p. 167-188.

5. Hirschberg, W., Rill, G., Weinfurter, H. (2007) *Tire model TMeasy*. Vehicle System Dynamics, 45 (Suppl. 1) p. 101-119.

6. Kistler, 2009: *RoaDyn S635 System 2000 – Radkraftsensor für schwere Pkw und Hochleistungsfahrzeuge, Typ 9267A1*, URL: http://www.kistler.com/?type=669&fid=39214&model=document, accessed on May 28, 2015.

7. Lex, C., Eichberger, A. (2011) *Der Reifen als Einflussgröße für Fahrerassistenzsysteme und Fahrdynamikregelungen*. OEAMTC Symposium Reifen und Fahrwerk, Vienna.

8. MATLAB version (8.6.0.267246) R2015b, Natick, Massachusetts: *The MathWorks Inc.* 2015, URL: http://de.mathworks.com/help/optim/ug/lsqnonlin.html, accessed on Jan 11, 2015

9. MATLAB version (8.6.0.267246) R2015b, Natick, Massachusetts: *The MathWorks Inc.* 2015, URL: http://de.mathworks.com/help/gads/particleswarm.html, accessed on Jan 11, 2015

10. Pacejka, H. B. (2006) *Tire and Vehicle Dynamics*, Butterworth-Heinemann, Oxford.

11. Rill, G. (2012) *Road Vehicle Dynamics, Fundamentals and Modeling*. CRC Press, Taylor & Francis Group, Boca Raton.

12. Rill, G. (2006) *First order tire dynamics*. III European Conference on Computational Mechanics: Solids, Structures and Coupled Problems in Engineering, Lisbon.

13. Schlippe, B. von, Dietrich, R. (1942) *Zur Mechanik des Luftreifens. Zentrale für wissenschaftliches Berichtswesen der Luftfahrtforschung* (ZWB), Berlin-Adlershof.

14. URL: http://www.ftg.tugraz.at, accessed on Oct 01, 2015.

Measuring a reference friction potential by anti-lock braking tests

Thorsten Lajewski, MSc, Dr. Jochen Rauh
Daimler AG

Prof. Dr. Steffen Müller
TU Berlin

1 Motivation

Physical limits of a road vehicle are defined by the maximal transferrable forces between vehicle and tires. These maximal forces are called friction potential. Knowledge of this potential is important to predict stopping distance and save cornering speeds. Today's state of the art driver assistance systems and stability control either assume a high or a low friction coefficient. They then adapt the assumed value if they reach the friction limits. Advanced driver assistance systems can be improved by the knowledge of the correct friction potential in advance. For example, a collision mitigation system could dynamically adapt its intervention distance.

Only a small part of the friction potential is required for regular driving. In these driving situations the potential itself is not measureable. It depends on a multitude of influence factors. These factors can be attributed to the three interacting components tire, road and an intermediate matter, which can be a layer of water, snow, ice or other impurities.

The tire properties, which influence the friction potential, are the tire geometry, rubber type, profile height, pressure, temperature and load. The road properties are macro and micro texture, wear and surface temperature. In case of water based intermediate layer, the aggregate state and height are the defining factors.

Some of these factors, for example tire pressure, are continuously measured in series-production vehicles while others such as the road micro texture can only be determined in a laboratory setting.

The advantages of knowing the friction potential have sparked the interest of many research projects and have led to many different friction estimation methods.

1.1 Friction Estimation

The friction estimation methods can be grouped into the two categories called effect based and cause based methods (Müller, Uchanski, & Hedrick, 2004). Effect based methods use changes in the vehicle dynamics to detect the friction potential.

These methods use the lateral vehicle dynamics (Solmaz & Başlamışlı, 2012), the aligning moment (Hsu, Laws, & Gerdes, 2010), the longitudinal tire stiffness (Rajamani, Piyabongkarn, Lew, Yi, & Phanomchoeng, 2010) or detect the non-linear region of the tire model (Villagra, D'Andréa-Novel, Fliess, & Mounier, 2011). The disadvantages of these methods are that they require a tire model and a sufficient use of the available friction potential. Thus, continuous friction estimation through these methods is not possible. Their main advantage is that they can estimate the correct friction potential between the road and the tire. In addition, most of these methods do not require any additional vehicle sensors.

Cause-based methods detect some of the influencing factors and use empirical friction information or a model based approach to estimate friction changes. These methods often require additional sensors like modified radar systems (Viikari, Varpula, & Kantanen, 2009), special weather sensors or cameras (Varpula, et al., 2009). Since cause-based methods only detect some of the influencing factors, the real friction potential is never known but the methods are able to estimate the friction potential continuously.

In order to overcome the disadvantages of both approaches, recent research projects like the EU friction project (Varpula, et al., 2009) combine several methods using data fusion algorithms.

1.2 Friction Potential Definition

The friction potential estimation methods aim for different use cases. Therefore, the methods use different definitions of the friction potential and require different estimation accuracies. For example in (Rajamani, Phanomchoeng, Piyabongkarn, & Lew, 2012) the authors want to use friction estimation to improve the performance of traction and stability control systems. Therefore, they require very accurate friction estimates for each individual tire. In this case, the friction potential is the maximum of the tire slip-force curve. This friction potential can be calculated by equation 1.

$$\mu = \frac{F_x}{F_z} \tag{1}$$

In (Hartmann, Amthor, & Jarisa, 2015) the authors use friction estimation in advanced driver assistance systems like collision mitigation systems. In this case, a friction potential definition using the maximal accelerations the vehicle can achieve is sufficient. This can be calculated by equation 2.

$$\mu = \frac{|a_{max}|}{g} \tag{2}$$

Here, the friction potential is defined by the combination of the four tire operation conditions and the performance of the vehicle assistance systems.

In (Bruzelius, et al., 2010) the metrics precision, availability, response time, robustness, correctness and adaptation have been developed to compare different friction estimation methods. Several of these metrics require a known reference friction potential. The authors recommend different methods to measure this friction potential ranging from using special test vehicles to using the acceleration during ABS intervention as shown in equation 2.

Since advanced driver assistance systems like collision mitigation systems can be improved - given they predict the correct braking performance -, the latter method is evaluated in this paper.

2 Measurements

A large number of anti-lock braking tests are conducted to measure the correct reference friction value and gain insight into the repeatability of friction measurements.

All braking tests have been measured on a test track with the same vehicle. The measurements have been performed on the same part of the track, which is a two lane concrete highway section. The starting speeds of all braking tests have been chosen in the range of 100kph and 130kph resembling normal highway speeds.

The conducted measurements are split into five groups, which are shown in Table 1. The tires used in group one and two are the same, while the road condition has been changed from dry to wet. Here, the wet condition means that the complete surface is wet, but there is no standing water.

In group three the environmental conditions have been the same as in group one but instead of summer tires a set of winter tires has been used. In group four the tire set used in group three was fitted in the opposite running direction. The difference between group three and five is that another set of the same winter tires was used.

Table 1 Overview Measurement Groups

Measurement Group	Tire	Road Condition	Number of Measurements
1	Summer Tire	Dry	25
2	Summer Tire	Wet	23
3	Winter Tire	Dry	51
4	Winter Tire	Dry	27
5	Winter Tire 2	Dry	50

During the tests all influencing factors deemed important have either been kept constant or their influence has been checked by varying these factors on purpose. The tire properties geometry and rubber type are constant in each measurement group. Before the tests all tire sets were driven for approximately 1000km, so that the starting profile height was comparable to new tires but production residues were removed. The profile height has changed slightly during the measurements due to the repeated braking tests conducted with the same set of tires. The tire pressure has been set to the recommended pressure before the start of the tests. The tire temperature has not been controlled. Besides the unavoidable small change of the profile height, repeated braking tests also increase the temperature and thus the tire pressure. In order to account for these effects, each measurement group was split into several days of testing to give the tires time to cool down. The tests have been paused after 10-15 measurements. The measurements showed no significant change after the break, showing that the temperature influence is in the range of the general measurement noise.

The road properties have been kept constant by using the same part of the road for all measurements. In order to eliminate any development of the road surface some measurements were done on the second lane, while the majority took place on the first lane. The measurements on the second lane showed no significant difference to the measurements on the first lane, indicating that the road influence was not changed due to the repeated tests.

During the measurements the vehicle was equipped with several additional sensors to confirm the sensor performance regarding the acceleration measurements. The additional sensors used are a six degree of freedom inertial measurement unit (IMU) and a VBox GPS. In order to account for the vehicle pitch while braking, the suspension travel has been measured with potentiometers.

3 Analyses

3.1 Evaluation Intervals

The acceleration signals of all brake tests have been averaged over three different intervals. The first evaluation interval is the first 0.5 seconds starting at the point when all four wheels are controlled by the anti-lock braking system. This interval is similar to the interval used in (Müller & Müller, 2015) for measurements on public roads, where large velocity changes are not possible.

The second interval defined in (United Nations, 2014) and used for the EU wet label specification is the interval, in which the vehicle speed is between 80kph and 20kph.

The last interval is defined by the speed interval from 100kph down to 0kph. This interval is directly related to the stopping distance and thus an important measure for collision mitigation systems. The stopping point of the vehicle has been defined as the point, where the longitudinal acceleration passed through zero after the ABS control stopped.

Figure 1 shows the filtered ESP acceleration sensor output for one of the measurements. All three evaluation intervals and the start velocity of the test are marked in the figure. The ABS controller is active on all four wheels in the area with the white background. It can be seen that after the start of the brake manoeuvre it takes a while until all wheels are controlled. At low speeds the ABS control stops because the wheel rotation cannot be resolved accurately enough for the ABS controller.

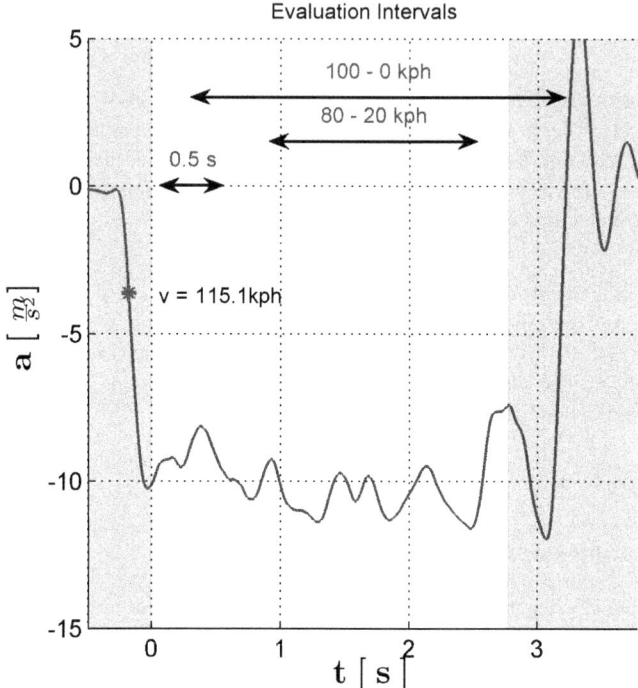

Figure 1: Evaluation Interval Description

3.2 Measurement errors

The acceleration defined as friction potential is the horizontal acceleration of the centre of gravity in an inertial coordinate system, while the vehicle ESP acceleration sensor measures the acceleration in the body coordinate system. This leads to several systematic error components, which need to be considered.

The first error component is due to the gravitational acceleration component the acceleration sensor measures. This error is down to misalignment during the fitting of the sensor and the vehicle pitch motion during braking. The misalignment errors can be corrected by averaging the acceleration measurements while the vehicle stands still. The vehicle pitch angle can be calculated using the suspension travel (z_s). Before the measurements the average suspension travel on the front and rear is measured $\left(z_{s_{f0}}, z_{s_{r0}}\right)$. Using the current suspension travel (z_{s_r}, z_{s_f}) and the wheelbase (wb) the pitch angle (θ) during braking can be calculated.

$$\theta = \tan^{-1}\frac{(z_{s_r}-z_{s_{r0}})-(z_{s_f}-z_{s_{f0}})}{wb} \tag{3}$$

It has to be noted that the pitch due to tire deflection is unknown and thus not included in the correction. The additional pitch due to the tire deflection can be estimated by assuming a maximal pitch angle of around 3° and a tire compression of 10%. Hence, the error is around -0.05 m/s².

The corrected horizontal acceleration can be calculated by

$$a_{corrected} = \frac{(a_{measured}-a_{offset}-g\ \sin\theta)}{\cos\theta} \tag{4}$$

On top of the gravitational component the acceleration sensor measures two additional components due to pitch velocity (ω) and pitch acceleration ($\dot{\omega}$) described by equation 4.

$$a_{error} = -\omega^2\Delta x + \dot{\omega}\Delta z \tag{5}$$

In this equation Δx and Δz are the distances between the pitch axis and the sensor position in x and z direction in the body coordinate system. Since the pitch axis is unknown, these systematic errors are not corrected.

Due to the pitch velocity the GPS sensor measures an additional velocity component, which can be calculated by

$$v_{error} = \omega\Delta z \tag{6}$$

This error has not been corrected.

In addition to the errors caused by vehicle body motion, measurement errors due to scale errors and non-linearity in the sensor measurement range exist.

3.3   Sensor comparison

Figure 2 shows a comparison of the standard ESP acceleration sensor, the IMU and the GPS based acceleration of group one in the evaluation interval 80kph - 20kph. The derivative of the GPS-velocity has been calculated using the regression line of the velocity measurements in the relevant interval. The average absolute error between the inertial measurement unit and the ESP acceleration sensor in the 80kph – 20kph interval is 0.11047 m/s², the maximum relative error between the two calculated sensor measurements is 2.52%. This value is below the specified maximum sensor error of the ESP acceleration sensor, pointing to a good measurement quality.

The average error between the GPS based acceleration and the ESP sensor in the same interval is 0.06 m/s², thus even better than the comparison of the other two sensors but the maximum relative error is 8.75% indicating a larger variance between the sensor measurements.

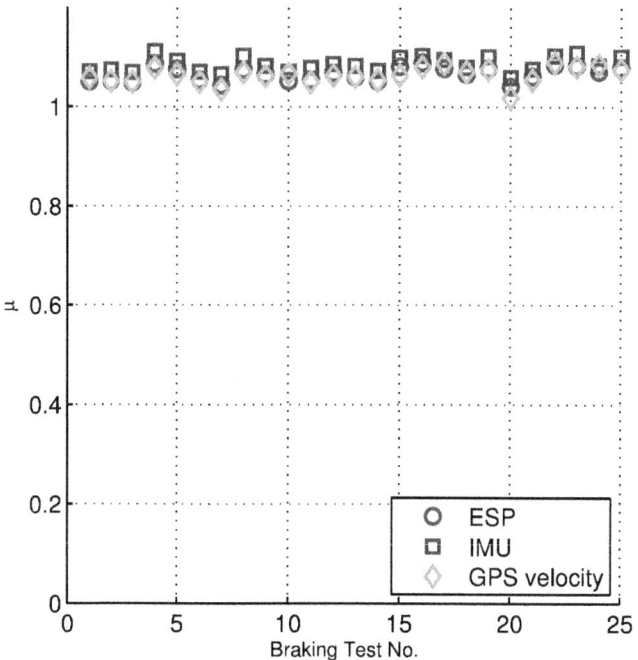

Figure 2: Sensor Comparison 80kph – 20kph interval

In the interval 100 kph – 0 kph the measured differences between the sensors are larger. The average difference between the intertial measurement units and the ESP sensor is 0.14 m/s² and between the GPS sensor and the ESP sensor is 0.09 m/s². The maximal relative error is 3% and 3.2% respectivly. The larger errors can be explained by the vehicle body motion at the start and end of the braking manoeuvre. Due to different placement of the sensors, they are affected differently by the body motion (see equation 4 and 5).

While in the 100kph – 0kph interval only the end of the braking manoeuvre is included, the 0.5s interval concentrates the evaluation on the start of the braking event. This explains the even larger differences. The average error between the IMU and the ESP sensor is 0.20 m/s² and the maximal relative error 6.47%. The GPS sensor provides wrong and unusable results. Here, the average error is 1.02 m/s² and the maximal relative error amounts to 39.31%. This error is partly explained by the positioning of the GPS antenna on top of the vehicle roof leading to a large distance between the sensor and the pitch axis. Also the number of discrete measurement points in the 0.5s interval is limited by the used measurement frequency of 100Hz resulting in 50 data points. In some cases these 50 points are not enough to average out measurement noise. A few measurements also seem to be influenced by poor GPS quality.

In the following analysis the ESP sensor measurements have been used because they are available in every modern car and showed a good correlation to the IMU measurements in all three intervals.

3.4 Interval comparison

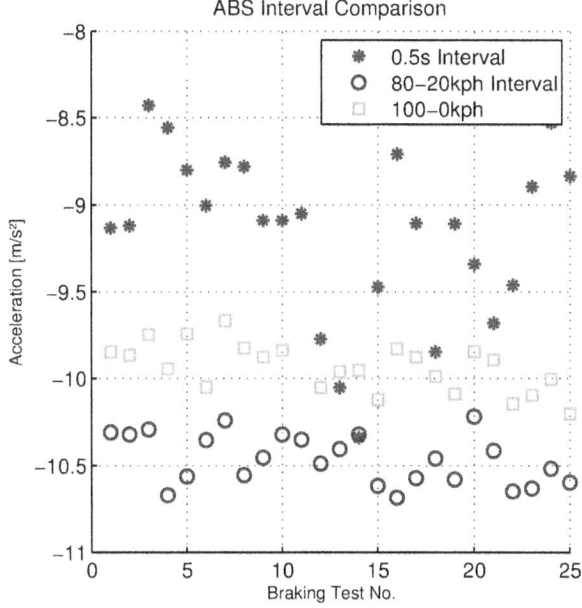

Figure 3: Interval Comparison with Corrected ESP Acceleration sensor

Averaging the acceleration measurements in the three intervals leads to different results, which are shown in Figure 3 for the measurements of group one. It can be seen that the variance of the 80kph – 20kph interval and the 100kph – 0kph interval is significantly smaller than the variance in the first 0.5s interval. The standard deviation (S) of the friction value in the first interval is +/-0.0147 and +/-0.0142 in the second interval. Thus, they both are in the same order of magnitude. Taking into account the limited number of samples, the reference friction interval can be calculated using the 99% confidence interval of a two sided t-distribution (see equation 7).

$$\mu_{ref} = \bar{\mu} \pm 2.797 \, \frac{S}{\sqrt{24}} \tag{7}$$

This results in the interval [1.058, 1.075] for the 80kph – 20kph interval and [1.004, 1.021] for the 100kph-0kph interval. Hence, both measurement intervals can determine the reference friction potential up to the first decimal place. It has to be noted that the measured mean ($\bar{\mu}$) is different for both intervals and both intervals have no intersection.

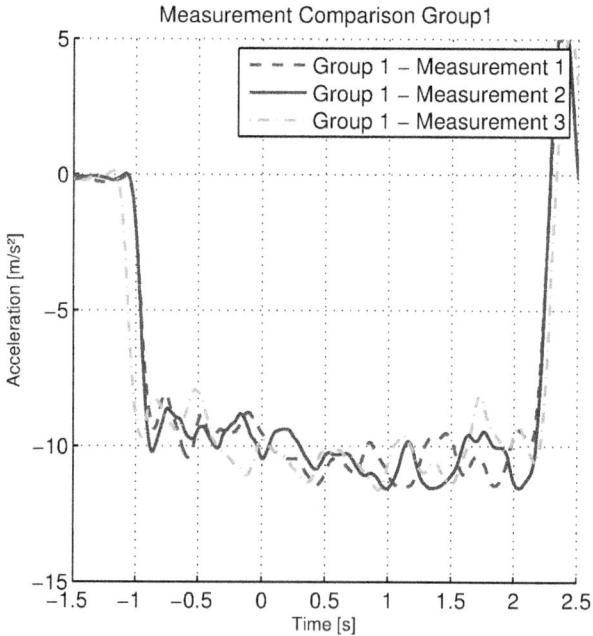

Figure 4: Raw Measurement Data Comparison

Figure 4 shows the measured accelerations for the first three measurements of group 1. The time t = 0 indicates the start of the 80kph-20kph interval. The figures visualises, the small variances in the interval 80kph – 20kph well. This can be seen by the time it takes the vehicle to stop from the start of the interval, which is almost the same for all three plotted measurements. The figure shows also, that at the start of the braking manoeuvre the reached deceleration is significantly lower (above -10m/s²) than in the 80kph-20kph interval. Almost all measurement points of the interval are below -10m/s².

Despite of the larger variance, the 0.5s interval allows us to specify the reference friction with the same accuracy in the interval of [0.905, 0.962]. This interval is lower than the other two. A possible explanation for this behaviour is that the ESP controller requires some time until it finds the optimum operating point.

Over all measurements the values of the 0.5s interval have been around 12% lower than for the 80kph – 10kph interval. The 100kph – 0kph interval has only been 3% lower.

3.5 Measurement Groups Comparison

The comparison of the different measurement groups gives insight into the influence of the factors changed between them. In group one, the average measured friction value is 1.07. Compared to this value, group two is 24% lower with an average measured friction of 0.813. The results of all measurements of the two groups can be seen in Figure 5. The standard deviation of group two, reaching a value of 0.0311 for the 80kph – 20kph interval, is two times larger than the standard deviation of group one. This is caused by different water heights during the braking tests. The calculated reference friction interval for group two is [0.79398, 0.83137].

Up to this point the major part of the analysis has been conducted with the data of group one. The reason for this is that the measurements with winter tires (group 3, 4, 5) showed a significant development over the driven tests. Thus, statistical properties like variance or mean do not apply to these measurements. Nevertheless, the results are very important to consider, when using winter tires to estimate a reference road friction using ABS braking tests. The results of all groups with winter tires can be seen in Figure 6. First, it has to be noted that the first few brake measurements show a friction value around 0.8. This is significantly lower than the grip value of the summer tire. Secondly, all three measurement groups show an increase in grip level the more braking tests are performed with them. The measurements of group three and group five show that this behaviour is reproducible. The increase of measured friction potential is very similar for the two sets of the same winter tire type. After around forty brake tests the measured friction potential is above 1.0 and thus in a very similar range to the summer tires. This constitutes a change of friction potential of over 20%.

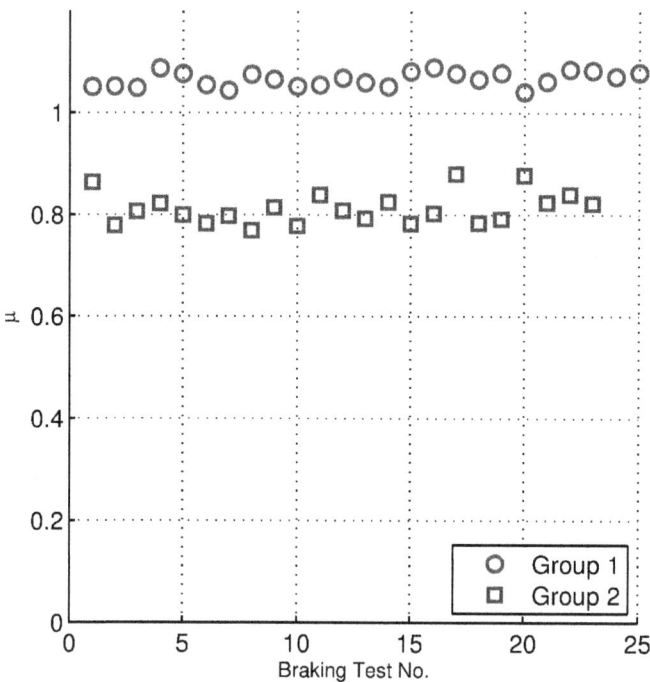

Figure 5: Measurement Results of Group 1 and Group 2 80kph-20kph Interval

The change of profile height during the 50 performed tests was minimal but the changes to the profile geometry are significant. The sharp edges of the profile blocks became smooth radii and the siped treads are bent upwards. The assumption is that these changes lead to a more homogenous pressure distribution in the contact area of the tire and hence to a better friction potential during braking. In order to prove this, the measurements of group four have been conducted with the tires used in group three after those tests had been performed. The results of group four show that the friction potential of the first few measurements is below 0.8 but starts rising once again after a few tests. Since the only change between the two groups has been the running direction of the tires, this proves that the increase in friction potential has to be related to the shape of the profile blocks.

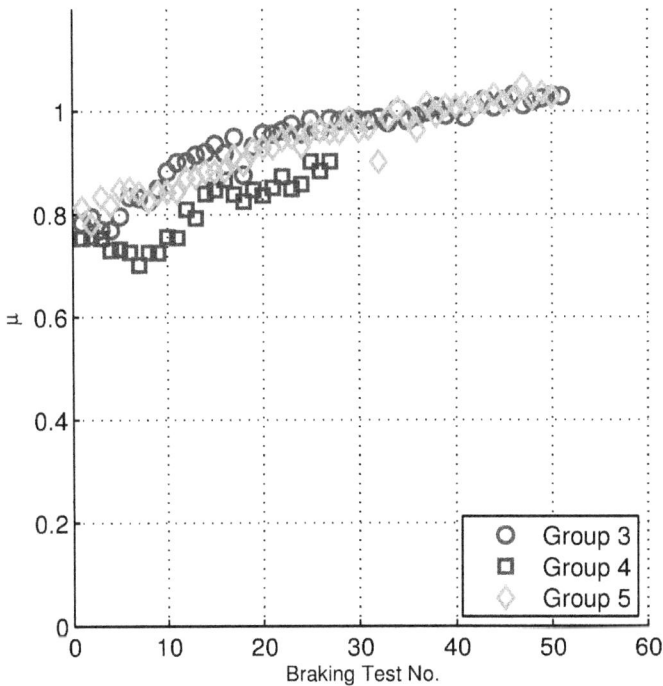

Figure 6: Measurement Results of Group 3, Group 4 and Group 5 80kph – 20kph Interval

For the winter tires only the first few measurements provide usable results, since repeated ABS braking manoeuvres and thus conditioning of the tires for an optimal braking performance does not take place in common driving situations. Since these tests do not provide enough data for statistical analysis, the same uncertainties as calculated for the summer tires are used. As mean value the average of the first two measurements of group three and group four was taken. This way the friction potential of the first measurements of groups four and five can be assumed to be in the interval [0.784, 0.809].

4   Conclusions and further work

The knowledge of the real friction potential is an important measure to validate friction estimators. The conducted tests have shown that different evaluations of friction measurements using braking tests can lead to different friction potential values. A correct definition of the reference friction depending on the use case is required.

Repeated ABS braking tests may lead to changes in the tire properties. These have to be considered. Common statistical parameters like mean and standard deviation assume a normal distribution of the measurements, which is not given for the winter tires used in the test. Thus, these tires should not be used in repeated tests or the measurement results are lowered by a significant amount: over 20% for the tires used in this paper.

Using the measurement results, the reference friction potential can only be determined up to the first decimal point but averaging over longer intervals reduces the variance. Especially the beginning of a braking event shows large variation. The test vehicle showed that in this phase the ABS does not yet operate at the optimal point and thus the predicted friction coefficient is too low.

The results presented in this paper are only valid for the test scenario, using the same vehicle, the same track, the same tires and the same test process. Thus, they are not transferrable to other settings. Additional influence factors need to be tested in order to get a general overview. These factors include testing other vehicles and roads. Also only one type of summer tire and one type of winter tire have been tested. In further tests all these parameters should be varied in order to gain insight into their influence on the measurement results.

Bibliography

1. Bruzelius, F., Svendenius, J., Yngve, S., Olsson, G., Casselgren, J., Andersson, M., et al. (2010). Evaluation of tyre to road friction estimators, test methods and metrics. *Int. J. Vehicle Systems Modelling and Testing, 5*, pp. 213-236.

2. Hartmann, B., Amthor, M., & Jarisa, W. (2015). Fahrbahnzustandserkennung als grundlegender Baustein für das Umfeldmodell. *VDI-Tagung Reifen-Fahrwerk-Fahrbahn und Innovative Bremsentechnik.* Hannover: VDI.

3. Hsu, Y.-H. J., Laws, S. M., & Gerdes, J. C. (2010, July). Estimation of Tire Slip Angle and Friction Limits. *IEEE Transactions on Control Systems technology, 18*(4).

4. Müller, G., & Müller, S. (2015). Messungen von Reibwerten unter Realbedingungen zur Erhöhung der Fahrzeugsicherheit. *VDI Tagung Fahrzeugsicherheit.* Berlin.

5. Müller, S., Uchanski, M., & Hedrick, K. (2004, January 29). Estimation of the Maximum Tire-Road Friction Coefficient. *Journal of Dynamic Systems, Measurement and Control, 125*(4), pp. 607-617.

6. Rajamani, R., Phanomchoeng, G., Piyabongkarn, D., & Lew, J. Y. (2012, 12 6). Algorithms for Real-Time Estimation of Individual Wheel Tire-Road Friction Coefficients. *Transactions on Mechatronics, 17*(6), pp. 1183-1195.

7. Rajamani, R., Piyabongkarn, D., Lew, J. Y., Yi, K., & Phanomchoeng, G. (2010, August). Tire-Road Friction-Coefficient Estimation. *IEEE Control System Magazin, 30*(4), pp. 54-69.

8. Solmaz, S., & Başlamışlı, S. Ç. (2012, July). Simultaneous estimation of road friction and sideslip angle based on switched multiple non-linear observers. *IET Control Theory and Applications, 6*(14), pp. 2235-2247.

9. United Nations. (2014). *Regulation No. 117 Uniform provisions concerning the approval of tyres with regard to rolling sound emissions and/or adhesion on wet surfaces and/or rolling resistance.* United Nations.

10. Varpula, T., Kutila, M., Pesce, M., Köhler, M., Hüsemann, T., Hartweg, C., et al. (2009). *Friction Final Report.*

11. Viikari, V. V., Varpula, T., & Kantanen, M. (2009, December). Road-Condition Recognition Using 24-GHz Automotive Radar. *IEEE Transactions on Intelligent Transportation Systems, 10*(4), pp. 639-648.

12. Villagra, J., D'Andréa-Novel, B., Fliess, M., & Mounier, H. (2011). A diagnosis-based approach for tire-rorad forces and maximum friction estimation. *Control Engineering Practice, 19*(2), pp. 174-184.

Parametrical approach for modeling of tire forces and torques in TMeasy 5

Dipl.-Ing. Ronnie Dessort, Dr.-Ing. Cornelius Chucholowski
TESIS DYNAware GmbH

Prof. Dr.-Ing. Georg Rill
OTH Regensburg

1 TMeasy in a Nutshell

1.1 Introduction

For the dynamic simulation of on-road vehicles, the model-element "tire/road" is of special importance, according to its influence on the achievable results. Sufficient description of the interaction between tire and road is one of the most challenging tasks of vehicle modeling. Two groups of tire models can be classified: handling models and structural or high-frequency models. Usually, various assumptions are made in modeling vehicles as multibody systems. Therefore, in the interest of balanced modeling, the precision of the complete vehicle model should stand in reasonable relation to the performance of the applied tire model. Handling tire models are characterized by a useful compromise between user friendliness, model complexity, and efficiency in computation time on the one hand, and precision in representation on the other hand.

TMeasy represents a handling tire model based on a semi-physical model. It includes a massless force element acting between the road and the wheel. The unevenness of roads is approximated by small local planes in the contact region of the tire. TMeasy generates all components of the contact force vector and contact torque vector including a first order tire dynamics. The wheel modeled by a rigid body must incorporate mass and inertia properties of the rim and the tire. TMeasy is available as part of several commercial vehicle simulation packages like TESIS DYNAware products and is very successful used in handling applications – offline and in realtime.

TMeasy parameters are easy to guess: even with a crude knowledge of size, payload as well as friction property of the tire-road combination a first guess gives feasible results - good enough for simulation of extraordinary tires, [1]. Of course, the parameters can be adjusted by curve fits to meet given tire measurements or vehicle dynamic results more precise, [2]. Another advantage lies in the ease with which tire properties can be scaled to represent different road and tire conditions.

1.2 Wheel Position and Orientation, Axis Systems, Forces and Torques

The position vector r_{0C} and the unit vector e_{yR} represent the location and orientation of the wheel with respect to the earth-fixed axis system, Figure 1.

In normal driving situations, the contact patch between tire and road forms a coherent area and the effect of the pressure and tension distribution can be fully described by a resulting force vector applied at a specific point P of the contact patch and a torque vector. The vectors are described in a wheel-fixed axis system, which coincides with the W-axis (Wheel-axis) system defined in the ISO-Directive 8855. The z_W-axis is normal to the local road plane. The x_W-axis is mutually perpendicular to the z_W-axis and to the wheel rotation axis e_{yR} and fixes also the third element of right-handed W-axis-system y_W. The components F_x, F_y, F_z of the contact force vector are named longitudinal force, lateral force, and normal force or wheel load according to the direction of the axes in the W-system. A cambered tire generates a tilting torque T_x around the x_W-axis. The non-symmetric distribution of the normal forces in the contact patch causes the torque T_y around the y_W-axis which is responsible for the rolling resistance. The irregular distribution of sheer stress caused by friction in the road plane induce a torque T_z around the z_W-axis, which in particular is important in vehicle dynamics and is induced by two main effects $T_z = T_B + T_S$. The bore or turn torque T_B is generated by drilling the tire around the z_W-axis perpendicular to the footprint like in a parking maneuver. The self-aligning torque T_S is mainly induced by lateral force, because the center of the friction tension does not coincide with the geometrically defined contact point P.

TMeasy provides the resulting tire force and torque vectors F_C and T_C applied at the wheel center C as an interface to multibody systems.

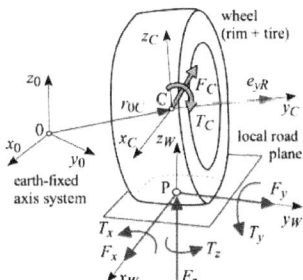

Figure 1: Axis systems, forces and torques

1.3   Contact Geometry

The calculation of the geometric contact point is an important part of TMeasy since it takes the shape of the tire and the road plane into account.

The track may have irregularities described by an arbitrary function of two spatial coordinates, $z = z(x;y)$. The current position of the wheel center C and the unit vector e_{yR} of the wheel rotation axis are known, Figure 2. On an uneven track, the contact point P cannot be calculated directly. Four points P_1 and P_2 as well as P_3 and P_4 with longitudinal and lateral distance are used to define a bent area representing the local road surface and calculate the final track normal vector e_n. The size depends on the tire dimension. The intersection of the rim center plane with the local track plane will now determine the longitudinal and lateral directions e_x and e_y, as well as the geometric contact point P.

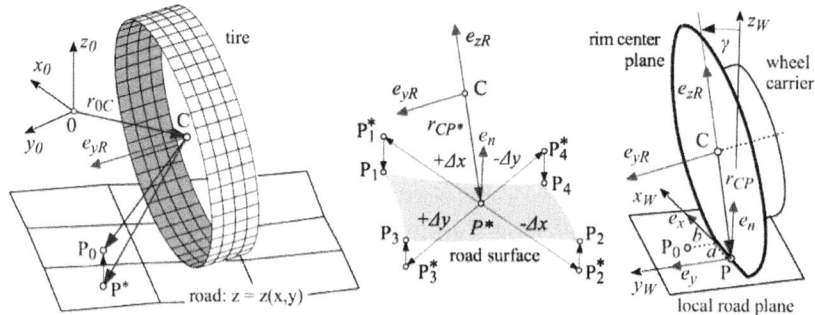

Figure 2: Contact geometry

On a cambered tire, however, the static contact point Q will represent the contact patch more appropriately than the geometric contact point P, Figure 3. Assuming that the pressure distribution on a cambered tire corresponds with the shape of the deflected tire area, the acting point of the resulting vertical tire force F_z is shifted from the geometric contact point P to the static contact point Q. The size of the deflected area corresponds with a generalized vertical tire deflection Δz and its center determines the lateral deviation y_Q of the contact point. The static contact point Q described by the vector $r_{0Q} = r_{0P} + y_Q \, e_y$ will represent the contact patch very well in any situation, because it is always placed inside the contact area. In contrast, the geometric contact point P as indicated in Figure 3 may even be located outside the contact area in situations where the tire is close to liftoff.

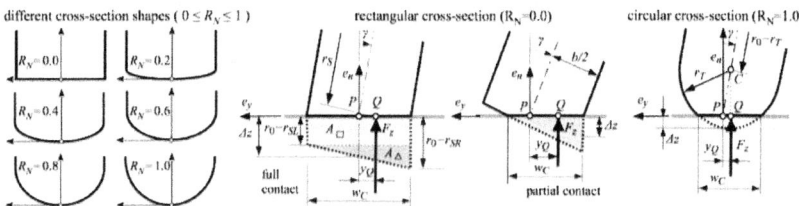

Figure 3: Different tire cross-section shapes and influence on the location of the contact points

The shape of the deflected area strongly depends on the shape of the cross section of the unloaded tire and on the camber angle in addition. Whereas passenger car or truck tires usually have a nearly rectangular cross-section, motor-cycle tires often do have a circular or a more rounded cross-section. Within TMeasy complex shapes of cross-sections are approximated by a simple roundness parameter. The plots on the left side of Figure 3 illustrate how increasing values of the roundness parameter R_N will morph a rectangular cross-section continuously into a circular one.

1.4 Steady-State Forces and Torques

1.4.1 Wheel Load and Tipping Torque

The vertical tire force F_z is calculated as a function of the tire deflection Δz and its time derivative $\Delta \dot{z}$. In a first approximation, the wheel load is separated into a static and a dynamic part,

$$F_z = F_z^{st} + F_z^D = a_1 \Delta z + a_2 (\Delta z)^2 + d_z \Delta \dot{z} \qquad (1)$$

where the static part is described as a nonlinear function of the tire deflection and the dynamic part is roughly approximated by a linear damper element. Because the tire can only apply pressure forces to the road, the normal force will be restricted to $F_z \geq 0$. TMeasy replaces the parameter a_1, a_2 of the parabola by the values of the tire radial stiffness at the payload and double the payload. The tipping torque is taken into account by applying the wheel load F_z at the static contact point Q.

1.4.2 Slips and Forces into Longitudinal and Lateral Directions

The brush model approach used in TMeasy delivers the longitudinal and lateral slips as

$$s_x = \frac{-(v_x - r_D \Omega)}{r_D |\Omega| + v_n} \quad \text{and} \quad s_y = \frac{-v_y}{r_D |\Omega| + v_n} \qquad (2)$$

where v_x and v_y are the components of the contact point velocity, r_D is the dynamic rolling radius, and Ω denotes the angular velocity of the wheel about its rotation axis. A small fictitious velocity $v_N > 0$ added to the denominator avoids numerical problems on a locked wheel, where $r_D |\Omega| = 0$ will hold. Please note, that this slip definition is used internal in TMeasy only and may differ from various definitions in the literature.

The longitudinal and the lateral forces are described as functions of the longitudinal and the lateral slips $F_x = F_x (s_x)$ and $F_y = F_y (s_y)$, Figure 4. During general driving situations, e.g., acceleration or deceleration in curves, the longitudinal slip s_x and the lateral slip s_y will appear simultaneously. In order to generate an appropriate combined force, the longitudinal and lateral slips are normalized, slightly modified, and composed to the combined slip s_C, Figure 4.

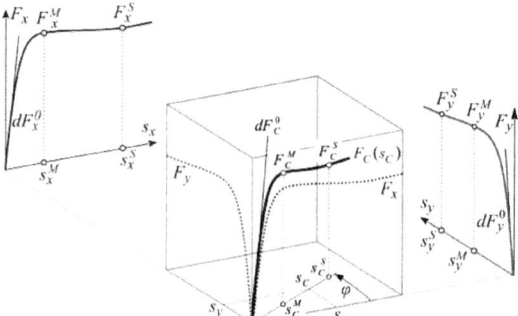

Figure 4: Combined tire forces

The combined tire force characteristic $F_C = F(s_C)$ is defined in TMeasy by characteristic parameters: the initial inclination, the location and the magnitude of the maximum, as well as the sliding limit and the sliding force, Figure 4. These parameters are appropriately derived from the corresponding values of the longitudinal and lateral force characteristics, [3].

1.4.3 Self-Aligning and Bore Torque

The dynamic tire offset or the pneumatic trail mainly depends on the lateral slip, $n = n(s_y)$. Acting as a lever to the lateral force F_y it generates the self-aligning torque.

$$T_s = -n(s_y) \cdot F_y \qquad (3)$$

In particular during steering motions, the angular velocity of the wheel has a component perpendicular to the contact patch which is defined as bore motion of the tire. If the wheel moves in the longitudinal and lateral direction too, then a very complicated deflection profile of the tread particles in the contact patch will occur. Within TMeasy the contact patch is substituted for that purpose by a thin ring with the equivalent bore radius R_B. The corresponding bore slip and the bore torque will then be determined by

$$s_B = \frac{-R_B \omega_n}{r_D |\Omega| + v_n} \quad \text{and} \quad T_B = R_B F_C(s_B) \qquad (4)$$

where the equivalent bore radius R_B serves as lever arm, $R_B \omega_n$ describes the circumferential sliding velocity and F_C is determined by the combined force characteristic. However, the simple steady-state bore torque model will serve as a rough approximation only. In particular, it is less accurate at slow bore motions ($s_B \approx 0$) that will occur during parking maneuvers. However, a straightforward extension to dynamic tire forces and a dynamic bore torque will generate realistic parking torques, [5].

1.4.4 Three-Dimensional Slip

In particular during steering maneuvers at standstill, a longitudinal, a lateral, and a bore slip will occur simultaneously. By extending the combined slip s_C defined in Figure 4 with the bore slip s_B to a more generalized and three-dimensional slip

$$s_G = \sqrt{s_C^2 + s_B^2} \tag{5}$$

the effects of the bore motion on the combined tire forces and vice versa can be taken into account. The generalized force characteristic $F_G = F_C (s_G)$ will now provide, by

$$F_C^* = F_G \frac{s_C}{s_G} \quad \text{and} \quad T_B = R_B F_G \frac{s_B}{s_G} \tag{6}$$

a modified combined force and the bore torque in the corresponding parts of the generalized force characteristic. A similar decomposition of the modified combined force finally results in the longitudinal and lateral forces.

1.4.5 First Order Tire Dynamics

The tire forces F_x and F_y acting in the contact patch deflect the tire in the longitudinal and lateral direction, as depicted in Figure 5. In a first-order approximation, the dynamic tire forces follow from

$$F_\Diamond^D = F_\Diamond\left(v_\Diamond + \dot\Diamond_e\right) \approx F_\Diamond(v_\Diamond) + \frac{\partial F_\Diamond}{\partial v_\Diamond}\dot\Diamond_e \tag{7}$$

where $\Diamond \in \{x; y\}$ denotes the longitudinal and lateral direction as well as $\dot\Diamond_e$ names the corresponding tire deflection, respectively.

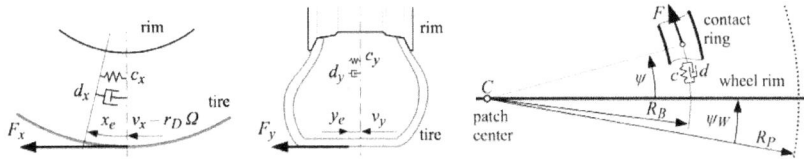

Figure 5: Tire deflection in the longitudinal, lateral and circumferential direction

In steady state the longitudinal and lateral tire forces will be provided as functions of the slips. On the other hand, the dynamic tire forces can be derived from

$$F_x^D = c_x x_e + d_x \dot{x}_e \quad \text{and} \quad F_y^D = c_y y_e + d_y \dot{y}_e \tag{8}$$

where c_x, c_y and d_x, d_y denote stiffness and damping properties of the tire in the longitudinal and lateral direction. Combining the relations in (7) and (8) finally results in first-order differential equations for the longitudinal and lateral tire deflection,

$$(v_{Tx}^* d_x + f_G)\dot{x}_e = -v_{Tx}^* c_x x_e - f_G(v_x - r_D\Omega) \tag{9.1}$$

$$(v_{Ty}^* d_y + f_G)\dot{y}_e = -v_{Ty}^* c_y y_e - f_G v_y \tag{9.2}$$

where $f_G = F_G/s_G$ defines the global derivative of the generalized tire characteristic and modified transport velocities v_{Tx}^* and v_{Ty}^* are used to simplify the expressions.

In a similar approach the dynamic bore torque is modeled by

$$T_B^D = cR_B^2\psi + dR_B^2\dot{\psi} \quad \text{with} \tag{10.1}$$

$$[d(r_D|\Omega| + v_n) + f_G]\dot{\psi} = -c\psi(r_D|\Omega| + v_N) - f_G\Omega_n \tag{10.2}$$

where c_t, d_t approximate the circumferential stiffness and damping properties of the tire, ψ describes the torsional deflection of the tire, and Ω_n names the bore angular velocity defined by the component of the wheel angular velocity normal to the road.

By neglecting possible dynamics of the tire offset, the dynamic self-aligning torque can be approximated by

$$T_S^D = -nF_y^D \tag{11}$$

as a product of the steady-state tire offset and the dynamic tire force. In this approach the dynamics of the self-aligning torque is controlled by the dynamics of the lateral tire force only.

This first-order dynamic tire model used in TMeasy is completely characterized by the generalized steady-state tire characteristics f_G, and the stiffness c_x, c_y, c and damping d_x, d_y, d properties of the tire. Via the steady-state tire characteristics, the dynamics of the tire deflections and hence the dynamics of the tire forces and the bore torque will automatically depend on the wheel load F_z and the generalized slip s_G.

2 Parameter Fitting

In this chapter the quantities named with capitals, which, e.g., have to be measured from any (real or virtual) tire test rig, are defined by the TYDEX-format, [4]. At present the tire model TMeasy requires 52 model-parameters in total. Some of them depend on the geometric properties of the tire others may be set automatically. This paper focuses on those TMeasy parameters that have a strong influence on the handling performance of a vehicle. In [5] one can find a detailed explanation of how to get a valid parameter set even if no tire measurements are available. In fact the manageable amount of physically based parameters is one of the significant advantages of TMeasy in contrast to other (semi-)empirical tire models. The easy parameter handling enables engineers to perform pre-application, even if precise tire information is missing. With more effort, model parameters can also be adjusted to fit tire measurements with high accuracy. A general fitting procedure and, based on this, exemplary results are presented in the following. The actually fitting of dynamic parameters and its validation was performed with the Car Professional package of the simulation framework DYNA4, [6].

2.1 General Procedure

Some exemplary standard test cases to derive the respective model parameters are shown in Table 1. Using an individual characteristic curve, the parameters can either be calculated from this or be adjusted for best possible fit.

First of all, the basic geometry information (No. 1) of the unloaded tire has to be established. Almost all other parameters except the tire roundness factor are wheel load dependent and therefore the subsequent test cases must carried out at different wheel loads.

The longitudinal, lateral and vertical stiffness (No. 2) can be directly calculated from the respective force-excitation diagrams. In reality the damping properties of a tire are nonlinear especially for high frequencies and are very difficult to measure. Since the influence on low frequency handling maneuvers is not significant, TMeasy contains a linear damping behavior. Because of numerical stability the value of the damping coefficient may not be zero and may not be too big. If loss angle measurement is not available, an initial guess can be derived from the attenuation factor of an oszillation. More tips can be found in [5].

The dynamic rolling radius is an important value, because it is part of the slip definition. It is achieved by measuring the wheel turns and distance covered of a free rolling wheel on a tire test rig or in a car (No. 3), without any inclination or side slip. In TMeasy it is modeled by a linear blending function between the unloaded and loaded tire radius. The weighting coefficient used in this function can be adjusted to fit the characteristic curve properly.

The longitudinal and lateral as well as the self-aligning properties (No. 4 and 5) can be fitted via quasi-static tests based on linear sweeps of the longitudinal slip and the side slip angle, respectively. The latter test case is based on test No. 5.1 and serves as adjustment of the parameters describing the behavior of the pneumatic trail, Figure 6.

Table 1: Overview of the TMeasy fitting procedure

#	Test case	Characteristic curve	Model Parameters
1	*Tire dimension*	*NOMWIDTH, ASPRATIO, RIMDIAME*	UNLOADED_RADIUS, WIDTH, RIM_RADIUS
2.1	Stiffness & damping	FX vs. LONGDISP	CLONG, DLONG
2.2		FYW vs. LATDISPW	CLAT, DLAT
2.3		FZW vs. DSTWOWHC	CVERT, DVERT
3	Dynamic rolling radius	KROLRAD vs. FZW	RDYNC0
4	Pure longitudinal slip	FX vs. LONGSLIP	DFX0, FXMAX, SXMAX, FXSLD, SXSLD, AMPLFX_<POS,NEG>
5.1	Pure lateral slip	FYW vs. SLIPANGL	DFY0, FYMAX, SYMAX, FYSLD, SYSLD
5.2		MZW vs. SLIPANGL	PT_NORM, SY_CHSI, SY_ZERO
6	Pure bore slip	MZW vs. SLIPANGL	CTORS, DTORS, RB_ADJUST
7.1	Combined camber and bore slip	FYW vs. INCLANGL	CAMF
7.2		MXW vs. INCLANGL	ROUNDNESS
8	Pure long. dynamics	FX vs. RUNTIME	CXDYNC
9	Pure lat. dynamics	FYW vs. RUNTIME	CYDYNC

In the next step, the parameters of pure bore motion should be determined in a dynamic steering sine-sweep at standstill (No. 6). If no test bench is available, you will find hints in [5] to estimate torsional compliance as well as the bore radius.

In a real car steering and thus cornering as well as tire inclination lead to bore angular velocity and bore slip, respectively. This is why a proper value of RB_ADJUST is needed to determine an influence factor acting on the estimated camber force (No. 7.1). The inclination angle shifts the geometric contact point and thus produces a tipping torque (No. 7.2). Due to a simultaneous influence of the tire roundness on both the camber force and the tipping torque, one could not perform a serial parameter fitting but have to find a pareto optimality instead.

Finally in the dynamic equations (9) the longitudinal and lateral relaxation length can be slightly modified by introducing scaling factors acting on the stiffness quantities, Figure 7. The identification of these parameters in test No. 8 and No. 9 could be done by validation of full vehicle or tire component simulation with measured step response, for example pulsed braking or steering step. Many other parameters influence the dynamic behavior and the simple approximation is not able to simulate mass oscillation. Nevertheless it is precise enough for typical handling maneuvers and the development of driving stability controllers.

The rolling resistance is not important for a handling tire model. Usually it is expressed as the ratio of drag force to wheel load. In the handling model of TMeasy it is calculated as torque $T_y = -\text{sign}(\Omega) f_R\, r_0\, F_z$, where f_R (RRCOEFF) denotes the dimensionless rolling resistance coefficient. More sophisticated equations can be used to overwrite the internal value for high fidelity consumption analysis.

TMeasy specifies the tire characteristics for the nominal or payload FZ_NOM and its double. In normal driving situations the wheel load is mostly smaller than the double payload. Extreme dynamic effects may produce peaks in the wheel load with values exceeding the normal range. To ensure realistic parameter values at all wheel loads, the parameter FZ_MAX is used to limit the wheel load dependent tire parameter interpolation.

2.2 Results

The described fitting procedure was used to generate a TMeasy dataset for a common tire of the automotive mid-sized class. At this point only the lateral tire properties are presented and discussed.

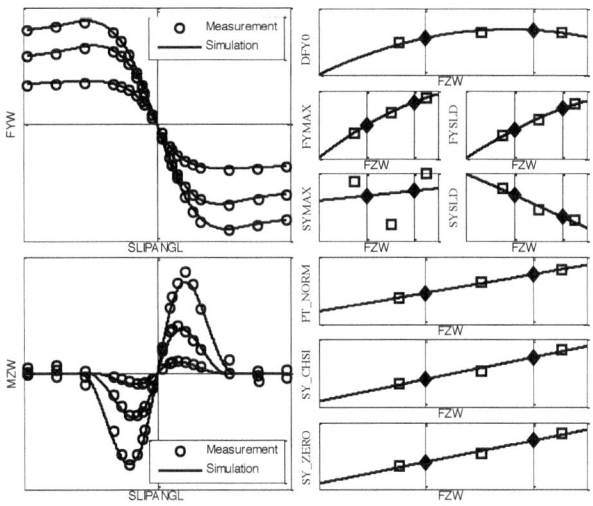

Figure 6: Fitting of lateral force and self-aligning torque parameters (quasi-static)

445

On the left side in Figure 6 the resulting characteristic curves of self-aligning torque MZW and lateral tire force FYW are shown after parameter fitting No. 5. The preceding step in each case was to identify an optimal parameter set at each wheel load. These values are shown on the right side of Figure 6 by square markers for every model parameter describing the stationary lateral tire properties. In the next step a least square approximation with linear or parabolic trial functions was used to calculate the model parameters at the payload and its double. These final parameters are marked with filled diamond.

In general all parameters related to forces are calculated by a parabolic function (e.g. FYMAX). Here one needs to ensure that the following conditions are fulfilled for each wheel load in the range of $0 \leq F_z \leq FZ\_MAX$: FYMAX $\geq$ FYSLD, SYMAX $\geq$ SYSLD, SY_ZERO $\geq$ SY_CHSI. To avoid non-physical behavior FZ_MAX has to be set in such a way, that FYMAX and FYSLD are always increasing in the relevant wheel load range. As one can see in Figure 6, the linear and parabolic trial functions approximate the wheel load individual parameter fitting results quite well. Based on this TMeasy depicts the measured lateral force and self-aligning torque with high accuracy.

An enhanced formulation of the tire dynamics is implemented in DYNA4, [6]. The scaling factor CYDYNC affects the stationary lateral stiffness and therefore the so called lateral relaxation length. Here this quantity is identified on the basis of side slip angle steps at different wheel loads. The steering gradient is approximately 60°/s. On the left side of Figure 7 the time-based course of the lateral force at a wheel load nearby payload is shown. As depicted on the right side of Figure 7, a linear curve fit of the wheel load specific optimal parameter values is suitable. The dashed horizontal line indicates scale 1, which means that here the dynamic has to be slightly faster than originally calculated with eq. (9). This leads to an accurate dynamic response both in the step up and step down phase. Simulations at other loads yield similar results.

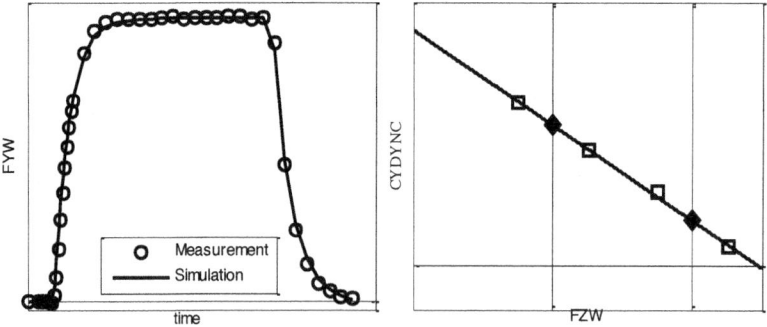

Figure 7: Lateral step response and parameter fitting (dynamic)

3 Optimization and Validation

In the presented particular example the tire test rig measurement procedure No. 7 from Table 1 was not available, whereby for the parameters describing the bore motion behavior only default (RB_ADJUST = 1) or approximated (CTORS ≈ 2500 N/m and thus DTORS ≈ 3 Ns/m) values could be set initially. Especially the latter parameters are based on simple assumptions and will serve as a very first guess, if no measurements or additional information are available, [3].

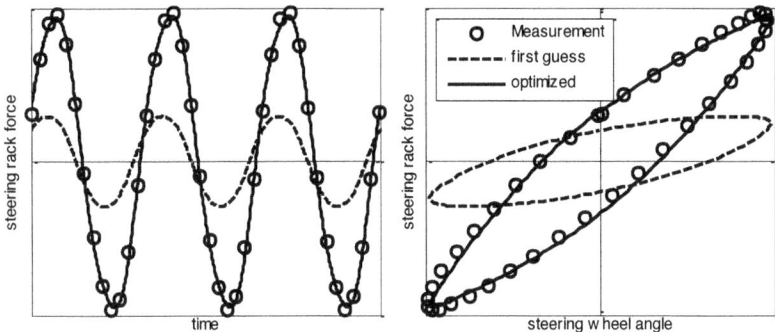

Figure 8: Sinusoidal steering input at standstill compared to full vehicle measurement

For this specific tire, a matching concerning the steering rack force extracted from a full vehicle measurement opens the opportunity to adjust the initial guess of the mentioned parameters. In Figure 8 the results of a sinusoidal steering with constant amplitude and frequency at standstill are depicted. Regarding its magnitudes, this might be especially interesting for rear steering scenarios. Here one can see a smaller amplification and a phase shift in the simulation results of the initial parameter set (dashed lines) compared to the measurement. In this test case the influence on these characteristics of both the bore radius adjustment factor and the torsional stiffness is shown in Figure 9. The diagrams contain also both the target amplification and target phase calculated from the measurement and thus all valid parameter sets {RB_ADJUST; CTORS} leading to these targets via the level curves. Due to its minimal influence on the bore dynamics, the parameter DTORS is calculated as solely depending on CTORS. Now one gets a realistic parameter set by the intersection of those level curves, what finally leads to RB_ADJUST ≈ 0.95 and CTORS ≈ 7000 N/m. Therefore the bore radius adjustment factor seems to be very well estimated by the model itself, whereas the torsional stiffness needs to be significantly greater. In general, higher stiffness produces faster first order dynamics, thus a higher corner frequency and less damping of the input amplitude, respectively. As depicted in Figure 8 by the solid line, simulating with this adjusted parameter set results in a high congruence regarding the measured steering rack force.

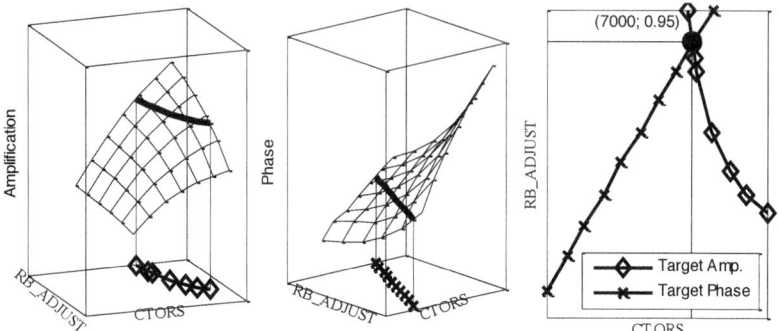

Figure 9: Parameter adjustments to improve bore motion behavior

For validation a comparison of the torque around the z_W-axis MZW in a lateral step response test on the tire test rig is shown in Figure 10. One can see, that the bore torque T_B dominates the sum of torques (solid line) most of the steering phase (i.e. where $T_B \neq 0$ holds) and generates a characteristic peak. Simulating with the just derived parameter set depicts this behavior very well. In the dynamic phase the self-aligning torque T_S increases and reaches its stationary value, which draws to the measured end value.

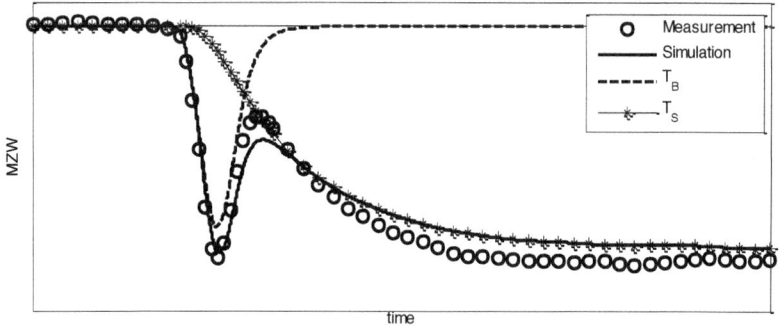

Figure 10: Self-aligning torque at a step response test on the tire test rig

4 Conclusion

The present paper describes the general approach of the semi-physical tire model TMeasy for vehicle dynamics and handling simulation and its enhancement for bore torque simulation in Version TMeasy 5. A parameter fitting process realized by TESIS DYNAware and the validation of real tire behavior by simulation with DYNA4 is presented.

Even with first guess parameters, the TMeasy tire model behaves in a realistic and plausible manner. Parameter estimation is intuitive and datasets from previous model versions can be easily migrated. After parameter fitting, the simulation results correlate well with both the tire test rig and full vehicle measurements. The enhancement of a three-dimensional slip calculation in the latest version does not modify the model behavior for high slip conditions, but improves the results not only for highly dynamic situations but also for low speed maneuvers such as parking.

References

1. G. Rill, An engineer's guess on tyre parameter made possible with TMeasy, in: Proceedings of the 4th International Tyre Colloquium (P. Gruber and R. S. Sharp, eds.), (University of Surrey, GB), 2015.

2. W. Hirschberg, F. Palacek, G. Rill and J. Sotnik, Reliable Vehicle Dynamics Simulation in Spite of Uncertain Input Data, Proceedings of 12th EAEC European Automotive Congress, Bratislava, 2009.

3. G. Rill, Road Vehicle Dynamics, Fundamentals and Modeling, Taylor & Francis, Boca Raton, 2011

4. H.-J. Unrau, J. Zamow, TYDEX-Format, release 1.3, TYDEX Workshop, 1997, retrieved January 13, 2016, https://www.fast.kit.edu/download/ DownloadsFahrzeugtechnik/TY100531_TYDEX_V1_3.pdf

5. Georg Rill. An engineer's guess on tyre parameter made possible with TMeasy. In P. Gruber and R. S. Sharp, editors, Proceedings of the 4th International Tyre Colloquium, University of Surrey, GB, 2015.

6. http://www.tesis-dynaware.com (retrieved January 13, 2016)

Software-in-the-Loop at the junction of software development and drivability calibration

René Linssen, Frank Uphaus
Daimler AG

Jakob Mauss, QTronic

1 Current model usage in drivability calibration

Different causes lead to a reduction of testing vehicles in absolute numbers or in relation to a broadening product portfolio. At the same time development cycles get shorter or have to tackle more complexity in the same time frame. In drivability calibration, the usage of powertrain test benches is a solution to this dilemma. This strategy, commonly referred to as "road-2-rig", can be thought one step further: Transferring all components to a simulation-only environment and extending the approach to a "road-2-rig-2-simulation" strategy. The key question is availability and quality of models of the key components.

Most test benches cannot be operated without some kind of plant model. Vehicle simulation and tire simulation are the essential models when using a powertrain test bench. Other test bench types such as "Engine-in-the-Loop"-concepts use even more simulated parts such as gearbox and all components related to hybridization of a powertrain. Thus there is a set of models available at different test benches. The models are capable of bringing real road behavior to the test bench and thus enabling calibration of drivability behavior in a stable and reproducible environment.

Taking a look at powertrain software development it can be seen, that here too simulation is being used. In this domain plant models of the relevant powertrain components are a commonly used tool. A very well established type are the models used at Hardware-in-the-Loop (HiL) test benches. With respect to plant models, all relevant types (Engine, gearbox or powertrain and vehicle) are available.

With respect to the Electronic Control Units (ECU) used for engine and powertrain control, it is noteworthy that several ECUs of vehicles of Mercedes-Benz use control unit software developed in house. Having all rights on the source code, the possibility of creating virtual control units for simulation is widely used. At the core of this concept is a second build process for a simulation module integration platform. However, the weak link in this concept showed to be the engine control unit.

The introduction showed the current usage of models in the software development and calibration domain. This article shows why it is necessary and beneficial to integrate all models into one simulation platform, how it has been achieved and what kind of results can be expected if this approach has reached a mature state.

2 Why Software-in-the-Loop simulation?

The fundamental difference between simulation environments for design purposes such as fluid simulation be it one dimensional or of higher special accuracy, and the software-in-the-loop (SiL) approach is the sheer number of modules. A short example from drivability calibration will show how many modules are required to address a simple task in a virtual environment.

Picture 1 shows the initial phase of a full load acceleration of a vehicle with manual gearbox and diesel engine. The maneuver starts after the vehicle has been brought to a halt after phase 1. The transition between phase 1 and 2 is determined by the change in accelerator pedal value as can be seen by the dashed line jumping from 0% to 100%. Caused by the increase in injected fuel and resulting effective torque, the engine revs up. The speed increase is limited by the amount of fuel injected due to smoke limitation constraints. This is the first example of coupling plant model and control logic. This closed loop requires a model of the air path and the information of the pressure and temperature condition at the inlet valve for the function calculating the maximum allowed amount of fuel.

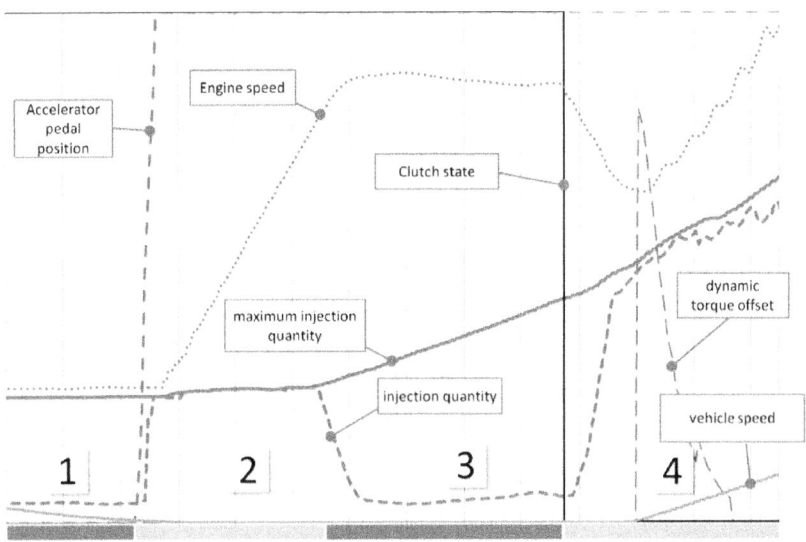

Picture 1: Initial phase full load acceleration

During the 3rd phase, the pedal value remains at 100%, the amount of fuel being injected is reduced however. This is another closed loop between plant model and control logic. In case of the clutch being open (clutch state at 100%, gearbox not in neutral) the engine speed is limited to a lower value than its usual maximum allowed rotational speed. In the end, this enforces a constraint on the kinetic energy of the engine in case of a sudden, probably unintended closing of the clutch. Detecting this

453

state requires knowledge of the state of the pedals and manual gearbox as well as the engine speed.

In the 4th phase, the injected fuel quantity is being influenced by two additional coupled systems. As a powertrain has the characteristics of a spring-mass-damper-system, a sudden increase in torque at the engine side will result in powertrain oscillations. These oscillations are in a frequency range that is perceptible by the vehicle occupant. Their amplitude can be reduced by a dynamic torque offset (see "dynamic torque offset", dashed line in phase 4). By adjusting the injected fuel quantity ("actor") based on measured engine and wheel speeds ("sensor") the engine torque can be influenced dynamically using closed loop control. Simulating this behavior requires parts of the control logic as well as an elastic powertrain model, ideally including backlash.

Another limitation of the fuel quantity can be active in phase 4. In this measurement, the injected quantity is close to, but not reaching, the maximum allowed quantity. Calculation of this limit again requires coupling of a plant model and control logic using sensors (boost pressure and additional information) and actuators (primarily injected fuel quantity in addition to various actuator positions of the air path).

In the 4th phase closing the clutch causes acceleration of the vehicle. To gain a realistic trend of the vehicle's speed the parameters of the clutch and wheel slip as well as the vehicle model need to be determined with high accuracy. Closed-loop interaction of plant model and control logic as used for electronic stability control or anti-slip control requires additional control units which will not be included in the work presented here.

A short maneuver with a duration of less than 3 seconds has clearly shown the need for a detailed coupling of actuators to sensors (plant model) and coupling by control logic from sensor to actuator (control logic model). Phenomenological completeness can only be achieved using an integrated system simulation. This chapter has shown the reasons for pursuing this goal. Description of the actual components and the assessment whether phenomenological completeness can be achieved will follow in the next chapters.

3 Structure of the integrated system simulation

Tools and process description for creation and integration of various components will be described next. Plant models and control unit logic models will both be considered, with an emphasis on new aspects of simulation of control logic.

3.1 Virtualized control units

Two control units were required for the integrated system simulation: a superordinate powertrain control unit and the actual diesel engine control unit. There are substantial differences concerning input / output characteristics, sources of the control unit's software and supported bus communication.

- Functions running on the **powertrain control unit** usually are independent of the actual engine (diesel, gasoline, electric motor). If bus capacity is not sufficient to achieve this goal exceptions from this philosophy can be made. The powertrain control unit supports several CAN busses as well as a FlexRay bus. There are several analog ports like the one for acquiring the accelerator and the gear stick position as well as digital ports like SENT actuators (Single Edge Nibble Transmission, SAE J2716 SENT). The control unit's software is developed at Daimler and available in graphical representation as MATLAB/Simulink models and as C Code.

- The **engine control unit** can be seen as remote controlled by the powertrain control when it comes to the actual torque demand values. The engine control unit's responsibility is to convert torque demands into fuel quantities and ensure actual injection into the combustion chamber as well as to perform closed-loop control on gas properties on the engine inlet and outlet side to meet emission requirements and perform necessary component protection. Due to this control unit being specific for an engine type, the vast majority of sensor–actuator couplings can be found here. There are several CAN busses. Parts of the control logic are developed at Daimler, other parts supplied by the ECU manufacturer. By linking together both parts, one binary is being created and used subsequently.

Source code, in this case C Code, of all user software tasks running on the RTOS (real-time operating system) of the control unit is the basis for virtualization of the powertrain control unit. Basic software components, such as drivers for receiving and sending CAN-messages or for communication with sensors as well as the RTOS itself, are provided by the simulation environment (QTronic Silver, [1]). This way, the compiled C Code can be used on a PC without making any changes on the code-base. The emulator code required can partly be generated automatically using available information from existing files (address to label file "a2l" or can bus files "dbc") – this works for CAN, sensors and actuators – and partly requires manual configuration. Names and execution times of the tasks that need to be run by the emulated RTOS need to be configured manually as an example. The build process in principle is a script that compiles existing C Code for usage on a computer using Windows instead of compiling C Code for usage on a processor of the powertrain control unit.

The initial effort for virtualization of the powertrain control unit took two weeks. The effort required for upgrading to a new software version depends on the amount of changes between the two versions but is much less than the initial effort.

Virtualization of the engine control unit is substantially different, as the software in part is provided by the control unit supplier without us, Daimler, having access to the source code. In addition, the close interaction between in-house and supplier software makes it very hard to separate both parts. To still be able to provide the integrated system simulation with a virtual engine control unit, processor emulation (c.f. [2]) was used, even for those parts that stem from in-house source code. This can be achieved using information from three sources: The binary file of the control unit software ("hex" file), starting addresses of the tasks ("map" file) as being generated in conjunction with the binary by the linker and the address-2-label file ("a2l") as known from engine calibration work. The real control unit can thus be turned into a virtual control unit the same way as described for the powertrain control unit: The simulator provides drivers for CAN, sensors and actuators as well as an RTOS for scheduling the tasks initially, time based or crank-angle based. In contrast to the previous implementation, in this case the tasks are being executed by interpreting the program code from the binary ("hex"-file) using a chip simulator. Thus, the speed of execution of the tasks drops by a factor of 10 compared to compiled C Code. The virtual engine control unit runs with half real time on a typical PC. Picture 2 shows the process for generating a model of a single task ("function").

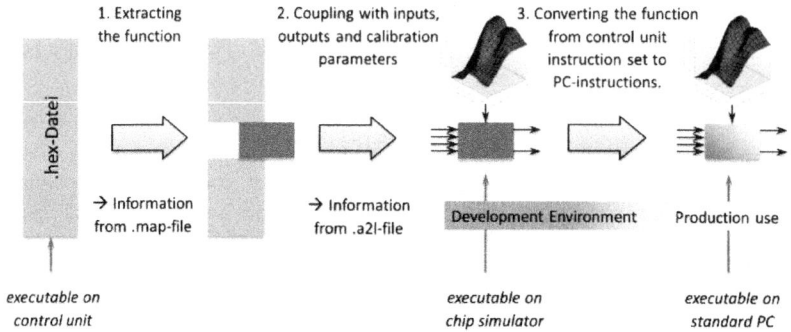

Picture 2: Generating a model of non-source code control unit function

The initial effort for engine control unit virtualization was much larger than the one for the powertrain control device as there was no proven and mature process model for control unit virtualization. Instead a process model for control unit virtualization was developed simultaneously. Again, the effort for upgrading to new software depends on the amount of changes and is a fraction of the initial effort.

Picture 3 shows the structure of the control unit's logic and separates the in-house from the supplier software. Coupling elements and functions performing hardware access are marked in addition. Functions performing hardware access cannot be run on the chip simulator, as they implement bus and protocol communication and access analog-digital converters. Their number however is much smaller than the number of regular ECU functions. In addition, changes to functions performing hardware access are seldom. This explains the large difference between initial effort of virtualization when all the functions were identified and solutions for these special functions were developed and the subsequent upgrades to new software versions using the process model developed in conjunction.

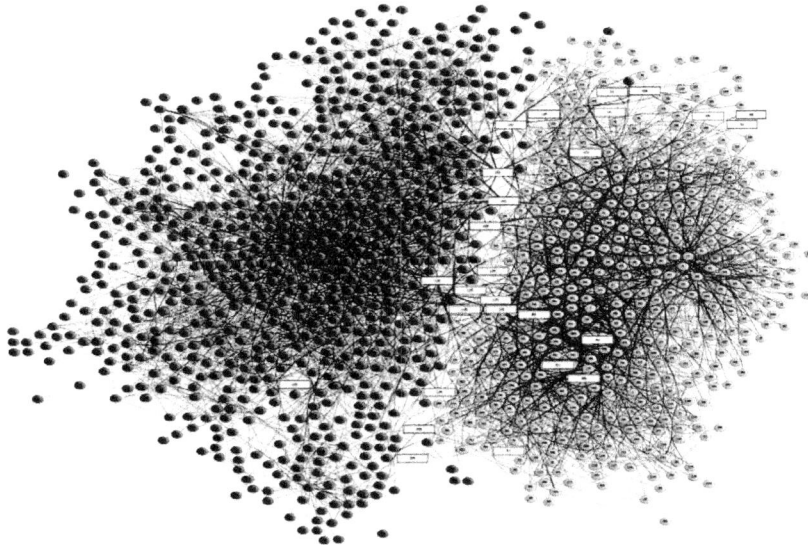

Picture 3: Structure of the engine control logic with in-house software (right) and supplier software (left) as well as coupling and hardware-related software (white rectangles)

The user of the integrated system simulation has access to all calibration parameters of the virtualized control units. The common calibration tools such as ETAS Inca and Vector CANape are supported. Table 1 shows the number of functions and calibration parameters of both control units as an indicator of their size and complexity. This also serves as a reference point when assessing the virtualized control unit's performance, as done in chapter 4.2.

Table 1: Complexity of the virtualized control units (approximate values)

Control unit	# functions	# maps	# curves	# value
Engine	1500	2500	3000	35000
Powertrain	400	600	1000	12000

3.2 Plant models

For the sake of simplicity the powertrain chosen for demonstrating the integrated system simulation was a manual gearbox. The plant models required were an engine model, a gearbox model including differential for a front wheel drive vehicle, side shaft models, wheel models (driven axle and non-driven axle) and a vehicle model.

Out of the possibilities of engine models at Daimler, an already existing and proven diesel engine model was chosen. This model had been in use at a Hardware-in-the-Loop test bench ("HiL"). The HiL-model had the advantage of complying with the signal interface of the real and thus also with the virtualized control unit. In most cases, removal of the HiL-hardware-specific input and output blocks of the engine model and direct use of the open signals was sufficient to make it usable in the integrated System simulation. The engine model closes the loop from sensors to actuators. Ultimately, approximately 40 signal connections between engine model and virtualized engine control unit have to be considered.

The powertrain was modeled in MATLAB/Simulink using SimDriveline by means of a physics based approach instead of signal based approach. Inputs of the powertrain are the effective torque of the engine and the states of the 3 remaining powertrain actuators (clutch, brake pedal and gear chosen) as well as 2 boundary conditions (road incline and head wind velocity). Outputs are different rotational speeds, with 2 (engine speed and wheel speed) having importance for feedback into the control units and the maneuver shown in picture 1.

All plant models were turned into QTronic Silver modules using code generation features of MATLAB/Simulink and an up-to-date Microsoft Visual Studio compiler using a build procedure that is an option of QTronic Silver.

3.3 Integration

Silver integrates all 4 components and ensures signal name based interconnection. Using a naming und unit convention as interface standard, this allows for flexible exchange of modules. By means of a simple user interface the simulation can be started and operated interactively.

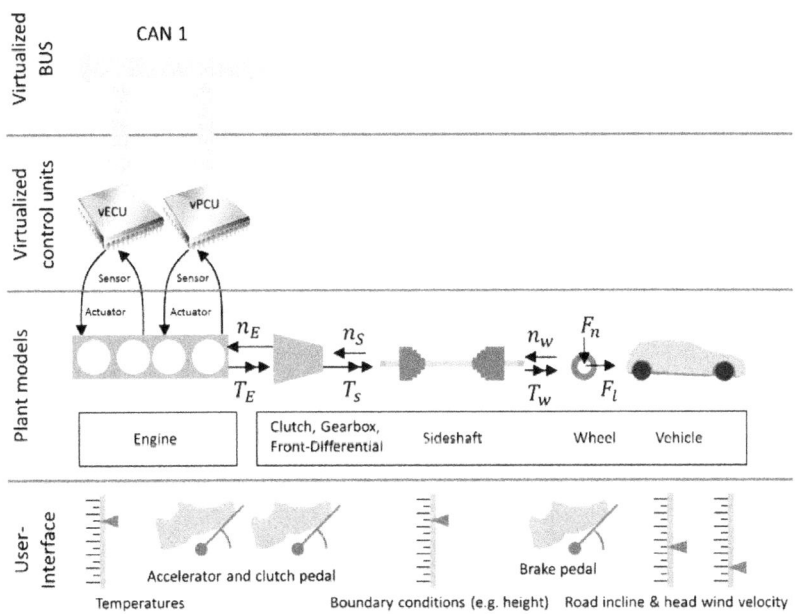

Picture 4: Integrated system simulation, software-in-the-loop simulation, „SiL"

Picture 4 shows the structure with the two virtualized control units and the two plant models. Subdividing the powertrain into components is possible but not necessary as the focus of the simulation is not on the plant side. Instead, the component being tested mostly in this integrated system simulation is the software itself, thus the term software-in-the-loop abbreviated "SiL".

4 Properties of the SiL-simulation

Two properties of interest will be investigated in this chapter: The quality of the simulation and especially the question whether phenomenological completeness can be achieved and the performance of the simulation in terms of required calculation time.

4.1 Simulation accuracy

Validation is being performed using characteristic maneuvers for drivability calibration. The maneuver used here is the same as was used for deriving the necessity of an integrated system simulation. The maneuver will be conducted in the SiL-environment by operating the simulation based on a measurement ("Replay").

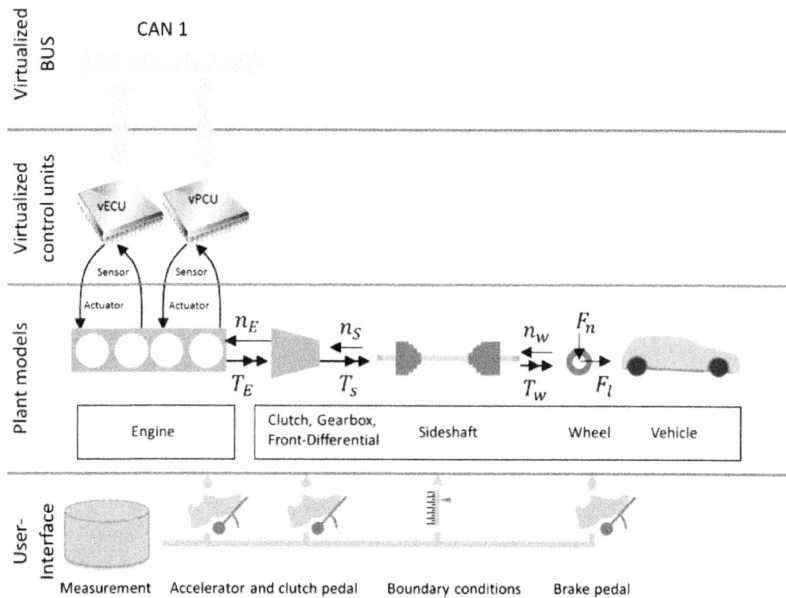

Picture 5: Validation using „Replay" of the maneuver

This kind of operation is essential for the usefulness of the SiL-simulation for use cases from drivability calibration. The first step in the development of the SiL-simulation was ensuring phenomenological completeness.

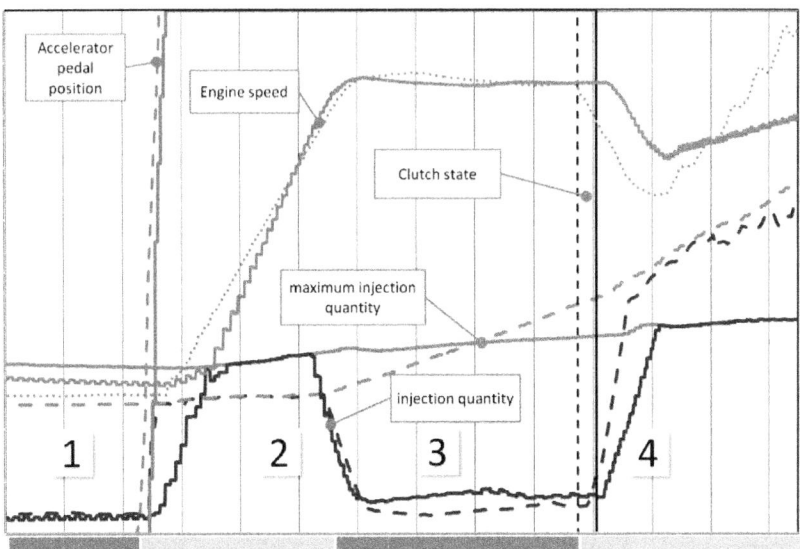

Picture 6: Validation of the maneuver (dashed lines are measurements)

Picture 6 shows both measured and predicted signals for the maneuver using the measured values of gear, accelerator pedal position and clutch. There is good agreement between measurement and simulation. The speed gradient in phase 2 is slightly larger compared to the reference measurement. This is a result of a higher maximum injection quantity which in turn is a result from differences in starting conditions between simulation (engine was just started before the maneuver) and measurement (just had completed the previous maneuver).

In phase 3 and 4 the maximum injection quantity rises much faster in reality than the simulated quantity. Investigations showed that a sluggish increase in boost pressure, caused by a turbocharger inertia chosen much larger than actually present caused this behavior. As a result, limitation of the actually injected quantity to the maximum allowed quantity occurs in simulation in contrast to the reference measurement. By updating the parameters of the HiL-model or switching to a different in-house engine model, it is possible to position the simulation on the right spot on the tradeoff between accuracy and calculation speed.

In summary: The goal of phenomenological completeness can be achieved. The simulation quality of the plant models is sufficient to see activation and results of individual control logic functions.

4.2   Simulation performance

QTronic Silver provides information about the computation time requirements of the individual modules for targeted improvement measures. The step size used for the simulation is 1ms base, with tasks in the control units being executed only if required by periodicity or crank angle (as in real control unit) and the engine plant model using 1ms and the powertrain plant model using 2ms step size respectively. Picture 7 and 8 show the distribution of computation times among the individual modules of the SiL-simulation (setup as shown in picture 4, control units as in table 1):

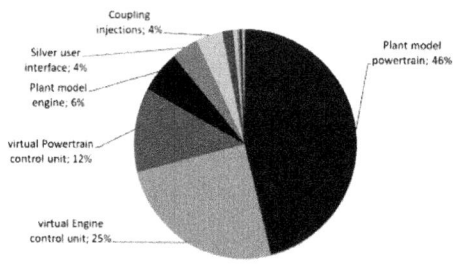

Module	Percentage
Plant model powertrain	46,2%
virtual Engine control unit	24,9%
virtual Powertrain control unit	11,9%
Plant model engine	6,0%
Silver user interface	4,0%
Coupling injections	3,9%
Coupling ign. Switch	1,5%
Coupling engine control unit	0,7%
Coupling general	0,3%
Coupling powertrain control unit	0,3%
Coupling pedals	0,2%
Automation	0,2%

Picture 7: Distribution of computation time   Picture 8: Computation time details

It can be clearly seen, that the plant model of the powertrain has highest computation time requirements. This is a result of interaction between two models: The powertrain plant model provides engine speed and position values requiring torque values whereas the engine plant model provides torque information requiring engine speed and position. This currently forces the powertrain plant model to be executed more often than actually required by the dynamics of the powertrain plant. By continuous optimization of specific SiL properties, maturity and performance are being improved. However, it is of importance to emphasize the low computation requirements of the engine model that was used previously at a HiL test bench in real-time.

The analysis was done on a laptop with Core i7 processor 2620M (2nd generation Core i7, released early 2011) having 4,0 GB of memory and running Windows Enterprise 64 Bit.

To leverage the advantages of parallel execution in QTronic Silver, all larger modules should require a comparable computation time. Using a setup with less frequent execution of the powertrain plant model and thus less overall computational demands significant improvements with respect to the real-time factor could be achieved. This

462

strategy however leads to limitations in operating range of the simulation. A strategy enclosing all plant models is favored instead.

5 Use cases

The usage of the SiL-simulation can be explained best using the common sequence of calibration activities in the development cycles. Following the classical V-cycle of software development, the "A"-cycle (as in Applikation, the German translation of calibration) covers all calibration activities (cf. picture 9). It starts with early calibration work, having specific calibration tasks under extreme conditions at the top and is the followed by testing and validation activities, which have a high importance when nearing start of production. During development of a vehicle, several such cycles (usually two per year) are scheduled.

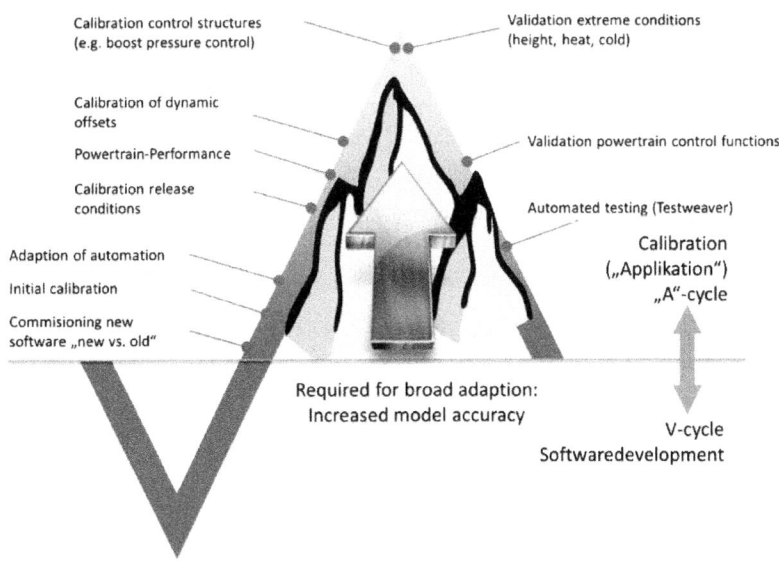

Picture 9: Usage of Software-in-the-Loop simulation along the calibration cycle

The beginning of the A-cycle is marked by commissioning the new software. Performing an initial calibration of new parameters and adapting automation (such as Inca-Flow) are common tasks in this phase. The SiL-simulation has a supporting role. This phase has just moderate requirements with respect to plant model accuracy.

When comparing the new software with a previous version the SiL-simulation is of use, too. Only in this environment everything can be kept the same except for the control unit's software. These use cases are already within reach of the SiL-simulation.

The initial phase is followed by early calibration work such as defining release conditions. With these tasks, not the precise value is of interest, rather if and when a physical phenomenon occurs. Model accuracy required increases slightly. Calibration tasks covering powertrain aspects are characterized by functional complexity and interaction between control units. Powertrain behavior itself can be modeled with reasonable accuracy. Due to this, a whole new class of sophisticated use cases can be addressed. A further use case is performance predictions of the overall vehicle.

At the top of the A-cycle there are calibration tasks like boost pressure or exhaust gas recirculation control, or even tasks concerning emissions and fuel consumption. These fields are usually dubbed air path and combustion calibration. Applications from these fields, as well as getting valid simulation results under extreme ambient conditions, require excellent plant models. Exhaust aftertreatment calibration is a similar field.

In a later phase of the A-cycle emphasis is put on validation activities. Their importance rises with start of production getting closer. Validation of air path, combustion and after treatment calibration requires excellent model quality.

The requirements for validating powertrain control logic are less demanding. Automatic testing at the end of the A-cycle leverages the advantage of having a fully reproducible environment. In addition, operating conditions can be reached which are difficult to reach in real driving conditions.

6 Summary and Outlook

Software-in-the-Loop simulation with multiple xCUs and sophisticated plant models has reached a new level at Daimler. Phenomenological completeness for drivability calibration has been achieved. In addition, the Software-in-the-Loop simulation connects to all established calibration tools and exhibits excellent automation capabilities, adding to its value for numerous use-cases.

In the future, the complete range of powertrain variants will be covered and quality and computation speed will be improved further. There will be more cooperation with other departments and the exchange of modules between departments will be simplified using a rights management.

Chip Simulation will be further developed together with QTronic, especially with focus on implementing AUTOSAR interfaces in QTronic Silver. This approach has shown its potential with the implementation of the virtual CAN bus. Further elements of the base software will follow. As a result, Chip Simulation will depend less on specific engineering and will become a standard tool for powertrain development.

7 Literature

1. J. Mauss: Virtuelle Steuergeräte für die Antriebsentwicklung. 17. MTZ-Fachtagung VPC – Simulation und Test 2015, Hanau bei Frankfurt am Main, 30.09 – 01.10.2015

2. J. Mauss, M. Simons: Chip simulation of automotive ECUs. 9. Symposium Steuerungssysteme für automobile Antriebe, 20.-21.09.2012, Berlin. In: Nietschke und Predelli (Hrsg.): Steuerungssysteme für Automobile Antriebe, expert Verlag Renningen 2012.

Continuous delivery for simulation-model development

Marius Feilhauer, Dr. Jürgen Häring, Dr. Jens Buchner

ETAS GmbH
Borsigstraße 14
70469 Stuttgart
Germany

Abstract

The usage of simulation models is popular within the development in Automotive Industry. Simulation models are used to early evaluate new functionality in Model-in-the-Loop setups and to test Electronic Control Units in Hardware-in-the-Loop arrangements. Along with the increasing functionality in current vehicles, the necessity of suitable simulation models growths. Autonomous driving is one area where functional complexity growths rapidly and a complete virtual vehicle model is necessary to enable testing within a simulation environment.

We present an approach to handle the complexity and variety of simulation model development efficiently. Therefore the principles of Continuous Delivery, known from software development, are applied to increase productivity and reliability of simulation model development.

1 Validation of Autonomous Driving Functions

The path to autonomous cars is not only related to available technology but mainly to the possibilities of testing and validating autonomous driving functions (see [1]). To cover the huge variability of conditions autonomously driving cars will be exposed to in the real world, most test and validation activities will depend on appropriate simulations of driving behavior, vehicle dynamics, and vehicle interactions with its environment. The complexity of the simulation models depends on the complexity of the systems implementing autonomous driving functions, their interactions, and the actual scenario. It can be characterized, e.g., in terms of the number of sensors and traffic participants, or the amount of communication with other vehicles. The sketch in Figure 1 visualizes different influencing factors to a driving scenario which have to be covered by simulation models.

Figure 1: Schematic visualization of different influencing factors to a driving scenario.

To enable tests in an early stage of the development process within the Automotive Industry (AI), simulation models are extensively used in Model-in-the-Loop (MiL), Software-in-the-Loop (SiL), or Hardware-in-the-Loop (HiL) setups. Autonomous driving is just one example where these methods are applied. Here a complete vehicle model might be necessary including its powertrain, dynamics, and a virtual driver. These models are built with tools such as ETAS ASCET or MATLAB®/Simulink®. Typically, the test teams have to test many similar variants of the unit under test e.g., various control units including an emergency break assist for different types of vehicles, or various control units for different combustion engines. With each variant to be tested, also appropriate variants of the simulation models have to be provided. Hence, each variant of the simulation model has to underlay a well-defined development process to ensure that it is error free. As an example, various test stages have to be executed and the documentation has to be adapted.

If a change in the simulation model affects several of its variants, each of it has to be tested again and its documentation has to be updated. Each manual step in this process has to be re-done for each model variant, which increases effort and slows down development and test.

Handling Test-Complexity by Using Automation

Continuous Integration (CI) defines an automatically triggered process, where steps like code generation, compilation and unit tests or code validation get executed. A high execution-frequency of this process pipeline ensures a short feedback cycle for involved developers. In addition to CI, the idea of Continuous Delivery (CD) is to constantly produce software that is ready to be released to customers (see [2]). Therefore all relevant aspects from code commitment to product release need to be automated as far as possible. This enables short iteration cycles which result in improved product quality, shorter times to market, and enhanced productivity. Developers no longer have to bother with manual testing and releasing but can make use of a deployment infrastructure to automate formerly manual steps. While continuous delivery is still quite new but widely accepted within the community of software developers, it is not used comprehensively in the area of model based development at all.

The key to improve development efficiency while creating reliable simulation models lies in applying the principles of continuous delivery to the development process of simulation models as well. Although many tools from classical software development for infrastructure automation can be re-used, some additional components to test and build simulation models are necessary. To seamlessly build, test and deploy simulation models, this gap has to be closed to create a pipeline which enables complete automation. This pipeline is built on a concept which combines model architecture, generic interfacing and model modularization. It enables a highly flexible and automated integration of different model configurations.

With this approach, the ETAS simulation model development department created a process which enables rapidly creating customer specific variants of simulation models that are extensively tested. Additionally, the development of new product features can rely on extensive tests already in an early stage of development.

2 Continuous Delivery Principles

One key challenge in modern software development is managing complex projects and delivering valuable software, including the functionalities customers expect, as fast as possible. Besides functional implementation a release process is required involving different stages of building and testing the product up to the final software release. When trying to manage the release-process manually several different steps are necessary during each release. Implementing these repetitive tasks manually involves a high probability of introducing errors. This is why *releasing software* has been seen as a risky step after a development phase. The idea of continuous delivery is to automate the complete release process as much as possible to increase the release frequency and reduce the number of errors introduced by human. This process is described in detail in [3] and the deployment pipeline has been visualized schematically as shown in Figure 2:

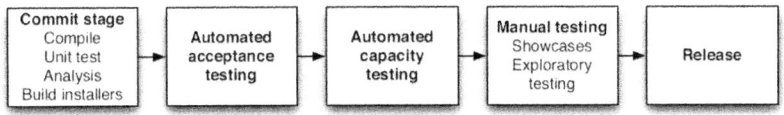

Figure 2: The deployment pipeline [3].

When having such a highly automated deployment pipeline every change made to the applications source code, its configuration or environment can trigger the creation of a new pipeline instance [3]. The consequence of this statement is, that every change to the software leads to a new release. This results in short release cycles compared to "classical" software release processes. Having short release cycles results in small differences between each release and therefor the risk of failing the deployment pipeline gets minimized and failures can easily be tracked down to their root cause. Additionally, the time-to-market of new product features is reduced since they can be shipped as soon as they have passed the deployment pipeline. This implies that the pipeline must fail as soon as one of its stages fails. If for example the software cannot be compiled, unit tests cannot be applied. Therefore a feedback-loop to the involved developers has to be guaranteed and must enable a reaction to the error. Figure 3 shows a sequence diagram including different stages of the deployment process and its feedback-loops.

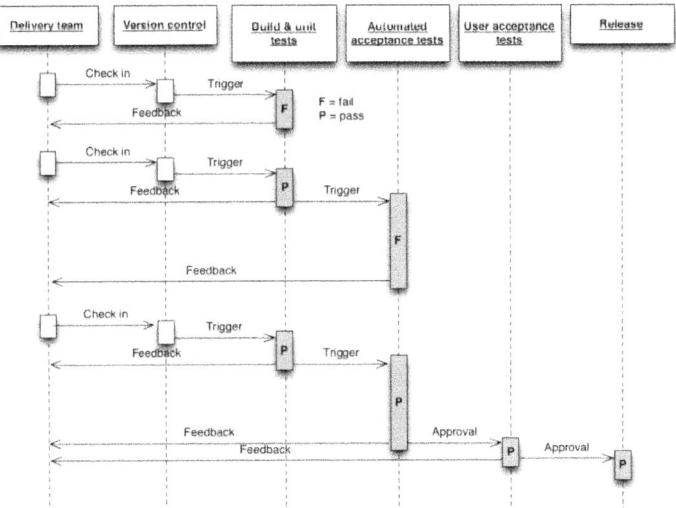

Figure 3: Changes moving through the deployment pipeline [3].

To enable the deployment pipeline to be triggered often, its stages should be kept as fast as possible. If tests for a single component take hours to complete, a short feedback cycle of the overall process cannot be achieved. The problem even increases over time as additional tests and components become part of the software.

From a technical perspective there are different tools necessary to support the deployment process. To make every change to the software system traceable, a Version Control System (VCS) has to be used for all involved parts. Besides the actual source code this includes third-party libraries as well as build and environment configurations such as the used compiler-version or the operating system. Only when having all components traceable each release can be tracked down to its sources and involved systems. When bugs appear on the customer side, the sources of each release have to be clearly identifiable to find the cause of the error. The most widespread VCS are Apache Subversion [4] or Git [5]. Additionally, a continuous integration server can support the process over all involved stages. One open source example is Jenkins [6]: It acts as a central server-based orchestrator to trigger process steps in a defined order. Additionally, it enables all involved persons to easily get an overview of the current project status via a web-interface. By having a huge and active open-source community, many plugins for different use-cases are available to customize Jenkins to project specific requirements. Other tool components of the CD process are necessary to enable scripting for each process stage: The build process including all necessary source and library files has to be described, the tests to be automatically executed have to be provided, and deployments to different execution environments have to be described. Depending on the project the involved tools might be very specific as Java-based software can be built using e.g. ant, whereas C-, JavaScript- or MATLAB®-based code require their own build approach.

3 Application of Continuous Delivery to Simulation-model Development

To give a common understanding of how an ordinary simulation model development might have been done, an exemplary development process is described. The presented process includes many steps which should no longer be applied when trying to automate the development process, the so called "Antipatterns" [3]. The exemplary ordinary model development process is visualized in Figure 4.

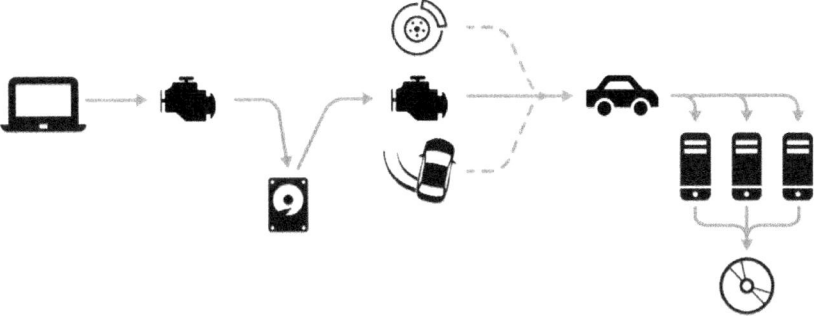

Figure 4: Exemplary ordinary model development process.

As a starting point, an engineer may develop a simulation model, such as the model of a 4-cylinder, 2-liter diesel-engine, using tools like ETAS ASCET or MATLAB®/Simulink®. When the developer finalizes the model based on certain acceptance criteria it may get stored on a local hard disk. At a later point in time the engine model has to be combined with other simulation models to create a higher level vehicle system. The combined vehicle model has to be built for different execution environments and the vehicle model might be deployed to other departments or customers.

Using this ordinary process as an example, many stumbling blocks can be identified which waste time and restrict the development of reliable and tested simulation models. When performing the process steps manually, the most noticeable drawback is the little amount of time spent on the actual model development in relation to the overall process. Process steps such as model integration, building or deploying are monotone and prevent the engineer from model development. Nevertheless, they are necessary to implement a full vehicle model. Therefor these steps must be automated: This reduces the possibility of human mistakes during the process steps and can increase the process' execution frequency. The only part developers should spend time on is the development of simulation models and associated test cases.

Another necessity is not to store simulation models on local drives only. They have to be integrated in a common, centralized version control system. This enables to retrace every change applied to the simulation models. By applying automated backups on VCS servers the risk of data loss gets minimized. To be able to trace product releases back to the appropriate model source code, the whole development process must be under version control. This can easily be achieved by consequently applying a VCS and all its functionality, like branching and tagging, to the model development process.

Principle benefits of the use of a VCS in combination with automatic builds are 1. That developers are motivated to commit even small improvements, since the check-in is the de-facto step to save results of work. 2. The commits go along with an enforced manual documentation as well as the possibility to use text-compare tools to clearly track back changes. 3. The combination of the previous two facts with automatic testing states can be seen as an analogy to the application of scientific methods to software development: The hypothesis is stated by assuming that the committed changes are going to work well, while making small changes is sharpening the meaning of the hypothesis. The application of test-routines provide the empirical basis if the hypothesis is falsified or not. This means, clear facts are generated linking errors with causes. This, in-turn, is a reliable basis for the developers to learn about the products feature status and its quality. 4. Applying the procedure automatically, fast, and with each commit accelerates the learning frequency and hence the products improvement cycles.

To realize an automated model build, which combines various simulation components, clearly defined interfaces have to be available. This enables compatibility among different simulation models and variants as well as an automated combination can be realized.

As common in modern software development, simulation models have to be tested comprehensively. Small model parts have to be tested using unit tests and more complex high-level simulation models can be tested within integration and system tests. In the style of software tests such different test-levels have to be used in model development, too. Each single component, each model and each combined system has to pass its tests. Otherwise it is not possible to ensure reliable and verified simulation models. With regard to CD existing test cases of simulation models can easily be used in an automated process as regression tests: This ensures the models functionality and enables a short iteration cycle if problems arise. If for example the simulation model for vehicle dynamics may change, already available tests for vehicle dynamic models can be applied and the full vehicle model can subsequently be tested automatically. Due to the automated process every adaption to existing simulation models will trigger the whole process to be executed, available test cases to be performed and verified system models to be build.

Succeeding process steps such as compiling for various platforms, integrating the software into an integration platform, and performing tests for specific systems can be automated in a similar way. Figure 5 visualizes a CD process for model development where automated process steps are placed on a gray background.

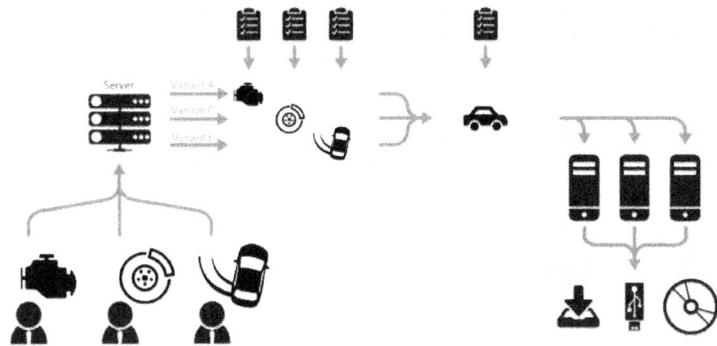

Figure 5: Continuous delivery applied on model development.

To conclude, we state that all steps of the development process after the implementation of the actual simulation model should be automated as far as possible. This mindset is well established in classical software development, where different tools supporting the CD process were developed. When applying CD on model development, such tools can and should be reused whenever possible. However, as simulation model development has its own tools and methods, additional tools need to be exploited. In the following paragraph we present the tool-chain used by the ETAS model development team.

Tool-chain used by the ETAS Model Development Team

We describe the development for MATLAB®/Simulink® based simulation model development. As a central repository, an Apache Subversion (SVN) server is used. This enables a reuse of an existing and robust VCS solution.

Test-case creation and execution is covered by ETAS RT2: Test-modeling, assessments and extensive test-reporting is available to ensure a short feedback-cycle to the development process. Furthermore, the execution of all involved tests can be automated. ETAS RT2 creates HTML based test reports which can easily be accessed by every developer to quickly identify problems.

The combination of various simulation models to, e.g., a full vehicle model is realized using a MATLAB®-based build tool. Based on an XML specification multiple MATLAB®/Simulink® models taken from a library can be integrated automatically to an overall model. Therefore, different model variants can be built and subsequently tested. This enables to cover many combinations within an automated CD process. The combined systems get tested with predefined ETAS RT2 test cases during the CD pipeline.

In addition to model build tools, a Jenkins server is used as a central orchestrator to schedule the different stages of the deployment pipeline. This includes recognizing modified files in the VCS, triggering the MATLAB®-based build scripts, executing unit tests, combining multiple models and executing consecutive tests. Finally, built and tested software can automatically be deployed on an ETAS RTPC to execute the resulting model in a real-time environment and release the tests for HiL applications. Figure 6 shows a schematic overview of involved software components as used by the ETAS model development team.

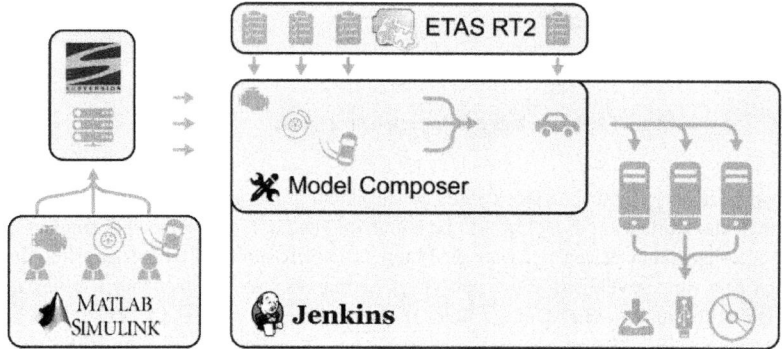

Figure 6: Tools involved in the CD process of the ETAS model development team.

4   Benefits and Challenges of Continuous Delivery Applied to Model Development

The implementation of a CD tool chain for model development is challenging since the highly automated process differs from the conventional software development process and specialized tools supporting the area of model development are rare. Nevertheless, applying CD to model development implies many advantages over the conventional, i.e., "manual" approach.

One of the main advantages is to have a reliable, fast and repeatable process which ensures a **high quality of the product** because of many involved test stages. Since every commit to the VCS triggers the deployment pipeline, a high execution frequency of the process is guaranteed and the differences between each release get minimized. This also **reduces the uncertainty during a release process** and if the development pipeline fails, the root cause of the error can be identified easily. By "automating almost everything" [3] **process errors caused by humans get reduced**. This additionally enables engineers to concentrate on current development tasks while having a **short feedback cycle**.

When adding new tests over time while keeping the old tests, the number of tests continuously increases which results in an **increasing test coverage** of the involved models. This process therefore improves the overall product reliability as every developer constantly knows the current quality of the product.

A great challenge when trying to apply CD to an existing development team is to motivate everyone to support the new approach and improve it by identifying bottlenecks. Especially when setting up the whole infrastructure for the CD process many decisions have to be made: Which repository structure to choose, how to develop single testcases, which additional tooling to support the CD process should be used? There is not always a clear answer and the whole team has to find a suitable way which obviously introduces additional benefit to everyone by increasing productivity and reliability.

Another challenge is to minimize necessary overhead and to keep additional tooling for all developers as simple as possible. Having to acquaint oneself with new software which does not obviously increase productivity will be experienced negatively and obstruct a successful long-term establishment of the CD approach.

Bibliography

1. H. Winner and W. Wachenfeld, *„Absicherung automatischen Fahrens"* in 6. Tagung Fahrerassistenz: Der Weg zum automatischen Fahren, Lehrstuhl für Fahrzeugtechnik, TU München, Ed, 2013.

2. B. Fitzgerald and K.-J. Stol, *"Continuous Software Engineering and Beyond: Trends and Challenges"* in 1st International Workshop on Rapid Continuous Software Engineering: Proceedings: June 3, 2014, Hyderabad, India.

3. J. Humble and D. Farley, *Continuous delivery: Reliable Software Releases through build, test, and deployment automation*, 6th ed. Upper Saddle River, NJ: Addison-Wesley, 2012.

4. The Apache Software Foundation, *Apache Subversion: Enterprise-class centralized version control for the masses.* Available: https://subversion.apache.org/ (2015, Dec. 14).

5. Git, Git: --fast-version-control. Available: https://git-scm.com/ (2015, Dec. 14).

6. Jenkins CI, Jenkins CI: Welcome to Jenkins CI! Available: https://jenkins-ci.org/ (2015, Dec. 14).

Generic development of software components and reuse for projects and variants

Walter Nagler, ZF Friedrichshafen AG

Abbreviations

ECU   Electronic control unit

HiL   Hardware in the loop

SA   System architecture

SD   System design

SWA   Software requirement

SWD   Software design

UML   Unified Modelling Language

XMI   eXtensible Markup Language Metadata Interchange

XML   eXtensible Markup Language

1   Software as Construction-Kit: Benefits and Motivation for Reuse

Reuse of sourcecode has existed since the very first days of programming. This was even the case in assembler: Writing a subroutine and calling it from more than one location is already reuse.

This usually works quite well as long as <u>one</u> developer is working on <u>one</u> program.

Automotive suppliers nowadays have a growing number of software developers spread over several projects, teams, departments, divisions or companies. This makes reuse a lot harder. And on top of that, it becomes even more difficult if development is spread over different locations, countries, languages and time zones.

But this is (for several reasons) the usual setup in many companies today, so why should we reuse although it is hard under these conditions?

Because reuse brings the following benefits:

- – Productivity (Do not re-invent the wheel several times)
- – Quality (Reviews, metrics, …)
- – Reliability (Proven in use)
- – Functional safety
- – Efficiency
- – Decrease time to market

These benefits usually outweigh the following disadvantages:

- Portability

- Flexibility

- Extensibility

Many approaches to the reuse of source code are already familiar on the market.

ZF has developed a solution to reuse even more than just the sourcecode to improve the ratio of benefits and disadvantages.

2 Reuse Along the Software Development Process Following Automotive Spice at ZF

In addition to the source code, several further processes are involved in the engineering process. The solution developed at ZF makes reuse of artifacts possible from System Requirements Analysis to System Test.

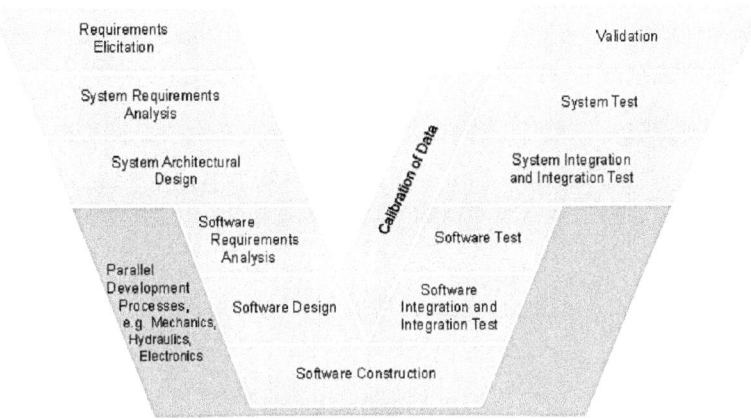

Figure 1: System engineering process

2.1 Software Requirements Analysis ENG.4

Software requirements usually arise from system requirements via system architecture. Because those system requirements usually correlate directly with customer requirements, they are completely project-specific to avoid interference between different projects.

If a system requirement can be fulfilled by a generic software requirement, a "linked copy" of the generic software requirement can be simply inherited to the project context. This "copy" enables baselining within the project while keeping clean the generic part and creates more flexibility on the project side for linkage and variant handling.

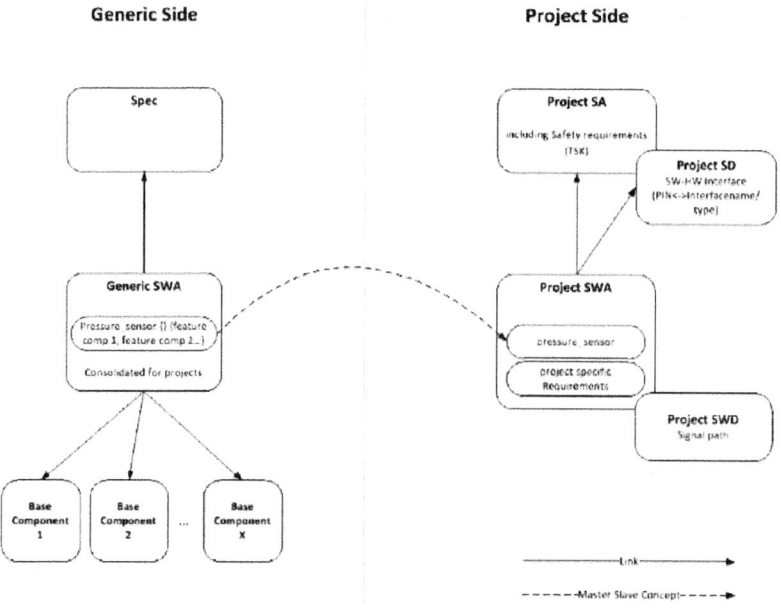

Figure 2: Relations between requirements (see "Abbreviations")

In case of changes to existing generic SWA, the impact on certain projects has to be considered for not creating contradictions. For Change Management at both sides, special queries can be used to identify SWA that has changed since the last baselining process and to automatically apply changes from one side to the other, if applicable.

2.2 Software Design ENG.5

Sparx Enterprise Architect is used for the software design with UML.

Each generic component pretends its own function-based interface, consisting of "providing" and "requesting" ports.

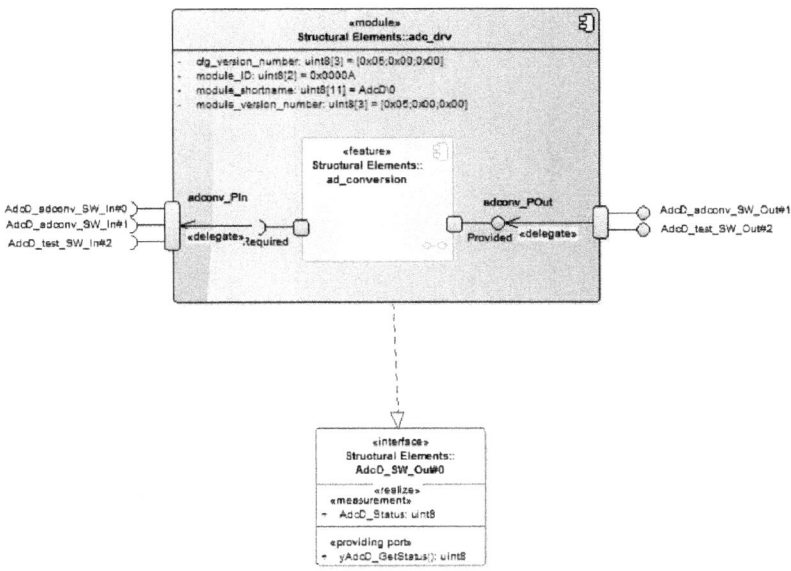

Figure 3: Component interface

The effort required for project integration is higher with this design than it would be if the components were directly linked to each other, because the project has to connect all these interfaces to create the first working software.

But on the other hand, this design reduces dependencies between all components to create the maximum flexibility in project design. This makes it very easy to decide if components shall be implemented generically or within the project scope and enables easy substitution of components, working with simple wrappers if interfaces do not fit.

In order for the software designers of the projects to have best tool assistance, the software designs of generic components are offered as separate XMI files. The projects include these files directly into their models:

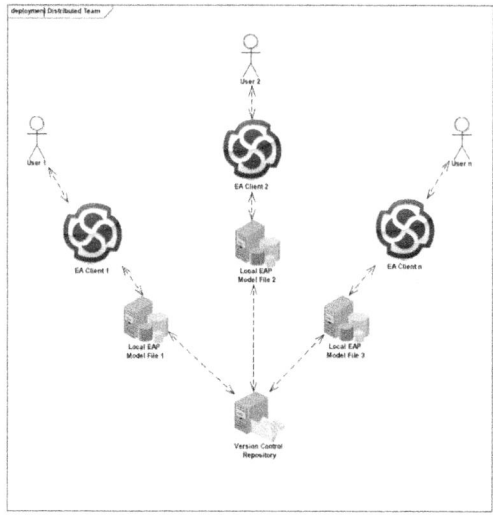

Figure 4: Sparx distributed version control repository

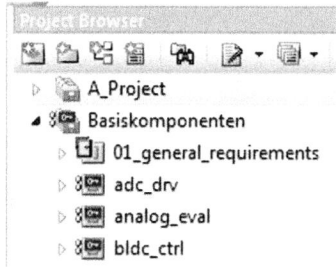

Figure 5: Project view of the model including generic components

Thus, the software designers of the projects can directly access the software designs of the generic components in all detail levels and even link their designs directly to the interfaces of the components.

As suggested by AUTOSAR, each generic component has a "configuration" area separated from the generic source code (e.g. for configuring the number of channels a component shall have). The components deliver sample configurations which the projects overwrite with their own needs. The baselining of these configurations is handled on the project side.

For continuous control in change management, a configuration includes version information that the component checks and refuses configuration at compile time if the found version information in configuration does not meet minimum demands of the component.

2.3 Software Implementation ENG.6

All components bring two makefiles: a generic one which just has to be included in the makefile of the project plus a configurable part which must be overwritten in the project context.

In contrast to components pre-integrated into libraries, the code is reused on source code level to have full possibilities of variant creation at compile time. This tends to produce long runtimes of the software build. We came across this issue through a highly parallelized build process combined with modern multicore computer systems.

A central storage for each component and project provides the possibility to create a fully working private workspace in very few steps. This area reflects the current development branch and is therefore fully under version control.

This storage is also the single source of truth for the **Continuous Integration System**, which executes automated builds, module testing, static code analysis and inquiry of all metrics. In addition to these (archived) artifacts, engineers have read access to the sandbox of the Continuous Integration System and can immediately pick up any detailed information they need without the need of setting up a complete workspace on their own and waiting for build results.

2.4 Software Integration and Integration Tests ENG.7

As test automation platform, "EXAM" (by Micronova GmbH) has been established for testing at ZF. This model-based solution provides several abstraction levels for high reuse.

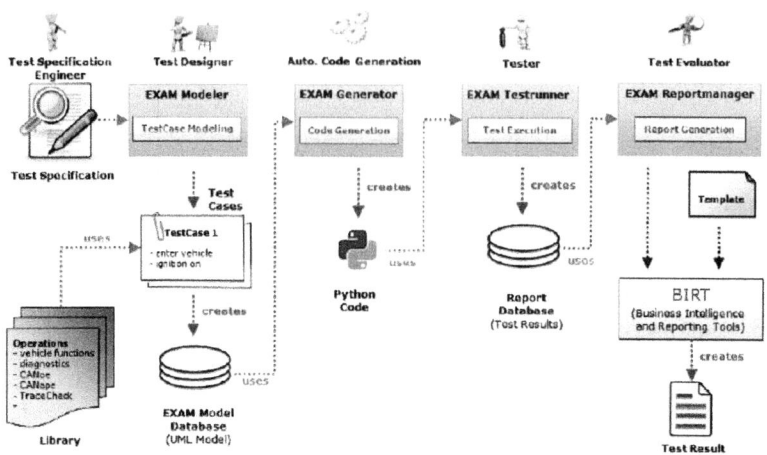

Figure 6: EXAM process

Operations implement the technical details and are organized in different libraries. All generic libraries are available to all projects. Test cases combine these operations into working code according to the purpose of the test suites.

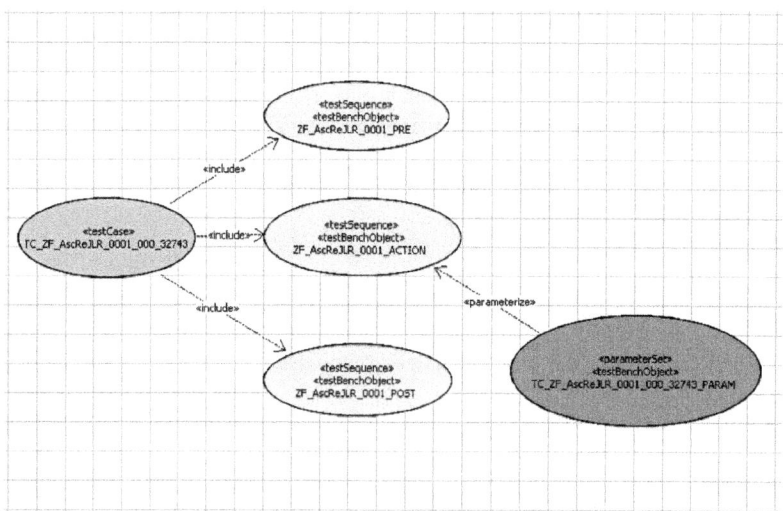

Figure 7: Usage of parameter sets

For efficient handling of variants in projects, test cases can run sequences with different parameter sets.

2.5 System Tests ENG.8

Using the same software platform for automated system tests as for integration tests allows high reuse even in testing.

For this aim, all test routines do not directly reference a specific target. Instead, they use "interfaces". This indirection makes it possible to use the same procedures for integration testing on the HiL and system testing on the target. This is achieved with different interfaces (e.g. "HiL", "Target", …).

Creating an interface "software" makes it even possible to develop the body of test procedures in advance, without accessing a hardware test bench. This approach enables efficient utilization of expensive test bench hardware.

3 Conclusion and Lessons Learned

The software construction kit has already been used to create the hardware-related software in several ECU projects (transmission, power electronics and other similar products) in several variants.

In the planning phase of this construction kit, significant difficulties have been expected in the technical realization.

In retrospect, most of them were easier to fix than expected.

It was much harder to deal with unforeseen details. But all of the technical (tool-related) issues just came along with initial effort required to find and implement the solution.

It turned out to be more difficult to find generic requirements that meet all the requirements of the different projects as some requirements were even conflicting. This task does not only produce initial effort, instead it will continue over the whole lifetime of the platform. In order to be able to meet these requirements, it is essential to provide flexibility for project-specific realization in all disciplines over the entire process. However, a generic solution shall be preferred wherever possible.

More lessons learned:

Problem research with a generic approach produces much more communication work than project-centered development.

The relevance of bugs often depends on the constellation in the project software.

There is a constant risk of losing project focus when concentrating too much on the component level.

The impact of bugs in reused software is much higher because it likely affects multiple projects. Because of this, generic software needs to be reviewed and tested more intensely before releasing it for the projects.

The benefit of reuse drastically increases when it is applied across all disciplines of software engineering not only to the source code.

Agile processes in automotive industry – Efficiency and quality in software development

Dr. Axel Schloßer, Jürgen Schnitzler, Thomas Sentis, Dr. Johannes Richenhagen

FEV GmbH, Aachen

Introduction

Future innovations in cars are predominantly driven by software. This covers all areas of customers experience from powertrain over driveline and infotainment systems up to connectivity and even the automation of driving tasks. Within this broad variety of software applications different software suppliers with sometimes completely different histories and objectives have elaborated appropriate development processes and organizations suitable to their respective business case.

Looking to extremes we find mainly sequential V-Models to develop e.g. engine control software with millions of lines of code controlling highly complex physical systems while satisfying tremendous demands on quality and safety requirements. These development cycles usually span several years. On the other hand we encounter highly iterative agile models e.g. to develop Apps which are highly dedicated to specific usually simple tasks or Calibration Methods, with innovations cycles of weeks or even days in a very flexible manner. Both approaches are contributing to the Agile Manifesto with his shift in prioritization in four areas (Figure 1).

Quality, safety and even security demands are much lower as for other software and is often covered by releasing new versions instead of testing the software thoroughly beforehand. The V-Model approach is widely spread and accepted within the automotive industry while agile processes are very common within the IT industry. These processes are accompanied by respective organizational structures and controlling instruments.

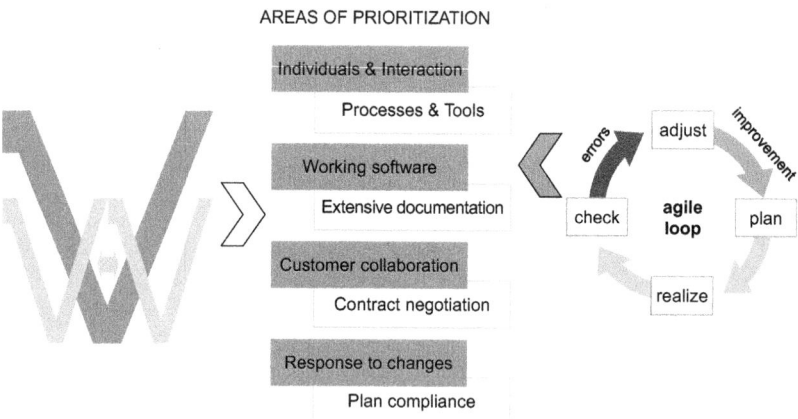

Figure 1: V-Model model according to ISO 26262 vs prioritization of the agile manifesto

Over the last years a broad discussion about introducing agile methods into the automotive industry has been led in the community. Looking at different project challenges, the prioritizations of the agile manifesto and V-Model approaches are not compatible or even contradictory to each other. (Figure 2). The most obvious differences are the mainly sequential procedure of the V-Model versus the cyclic procedure of agile models one hand. One the other hand planning horizons of the approaches are completely diverging. While the V-model approaches focuses on a dedicated long-term plan the agile approaches are focusing on continuous short-term planning. Throughout this paper we will refer to the key challenges Collaboration Type (CT), Collaboration Control (CC), Deliverable Maturity (DM) and Planning Horizon (PH) depicted in Figure 2.

Nevertheless all approaches have their advantages in their respective field and nowadays these contradictory approaches merge within mobility systems or even within a car. Customers do not accept any longer to own a car over years without getting updates on their software features as they are used to from personal devices. On the other hand no one will accept to fix errors or bugs due lack of testing or quality while driving a car.

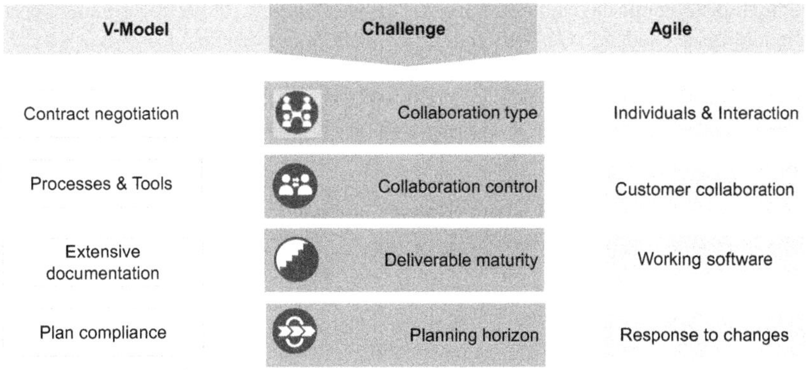

V-Model	Challenge	Agile
Contract negotiation	Collaboration type	Individuals & Interaction
Processes & Tools	Collaboration control	Customer collaboration
Extensive documentation	Deliverable maturity	Working software
Plan compliance	Planning horizon	Response to changes

Figure 2: Contradictory prioritization of objectives

This paper will present three different examples of combining advantages of both approaches to benefit from both worlds. The first example deals with the development of calibration methods, a desktop software which started extremely agile (Extreme Programming) and became more rigid over time (e.g. Scrum) due to several drawbacks. The second example covers the development of automation systems for test benches shifting from a V-Model approach to agile methods. The last third example outlines the introduction of agile principles for automotive embedded software. Hence this paper will cover 3 different areas with three different starting points.

Calibration methods

With a growing complexity and scope of work in vehicle calibration a strong need for suitable methods evolved to increase efficiency and assure quality in calibration work. These methods were meant to support calibration engineers with the evaluation of measurement data and optimization of calibration data sets. To ease the application of methods throughout the wide range of calibration engineers, the developed solutions were incorporated into intuitive software solutions under the brand TOPEXPERT and such distributed.

Specialists in the field of algorithm and software development were grouped together to foster the interdisciplinary skills bridging the gap between calibration and software engineering. Calibration engineers contributed with their domain expertise and thus were responsible for the requirements definition.

In the first years a large amount of TOPEXPERT tools was created with quick gains and high benefits for the calibration work. This was possible through a very close collaboration (see challenge CT) between the methods department and the respective calibration engineer. Daily meetings between developer and calibration engineer were common, so that the development progress could be tracked closely allowing a planning horizon of only few weeks. Questions could be settled soon and the creation of requirement documents almost seemed to be obsolete. The collaboration type between customer and developer was extremely relying on individuals and their interaction. In fact the software development was as agile as it could be and happened in a way of extreme programming where the customer can interfere where and when he wants to [2].

The planning (see challenge PH) of the development projects however was quite lean. There was a roadmap containing the topics to be realized in a year based on a prioritization by the customers and only rough effort estimates, because of missing detailed requirement definitions. The lack of detailed requirement documents at the begin of the software implementation in some cases even led to rework of solutions, because customer requirements changed fluently in the course of the development, which was a result of the agile collaboration type.

With the growing number of software solutions further drawbacks of the extend of development agility appeared. In first instance these drawbacks concerned deliverable maturity (see challenge DM) causing huge maintenance efforts. Mostly single developers were responsible for one application and a new project only was started in case another project was finished. So the software developers experienced a high degree of freedom. They were even able to choose a software language, which seemed to be most suitable for them. By this freedom of course every developer could work very efficiently and make use of his individual strengths. On the other hand there was a close connection between an application and its developer. Software changes were most likely to be performed by the developer of an application which could become resource critical.

Especially maintenance efforts for common functionalities like import routines were high, because necessary adaptations had to be performed in all applications separately and even in different software languages.

Single development projects decided for a SCRUM-based development practice to avoid requirement changes during their monthly development iterations by a less agile type of collaboration (see challenge CT). The step to a department wide SCRUM-approach however was driven by the aspects of cost and scalability that should be improved by a larger planning horizon (see challenge PH). Next to the mentioned efforts for maintenance this enfolds also the efforts for new developments aiming at the reuse of software components and in general the possibility of a wide planning horizon, allowing a more elaborated project planning including developments for new customers and a separate product driven development .

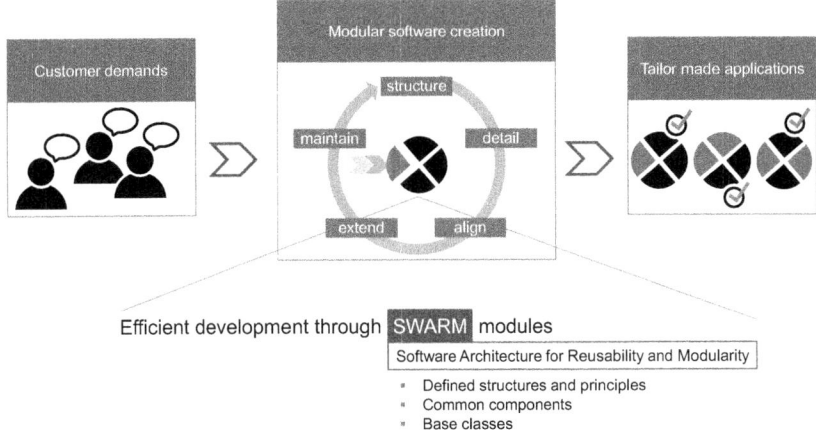

Figure 3: An efficient way from customer demands to tailor made applications

As technical prerequisite for reusability of code C# was introduced as common programming language for all new TOPEXPERT software applications. At the same time the organizational structure of the methods department changed. Next to the existing support team three development teams were founded for the application platforms:

- FEVcal with responsibility for all applications in the field of "Modeling and DoE",

- FACE for all "Offline Data Evaluation" applications and

- VTA for solutions concerning "Vehicle Test Automation".

With the creation of new code emphasize was put on common interfaces to ensure reusability of common code parts. At the same time the various customer demands were structured to achieve modular contents to be used in several applications (Figure 3). An additional development team, the "Core" team supported the software architectural preconditions to achieve reusable code parts ranging from basic class definitions, e.g. for data maps, data services, plugin functionalities up to entire tool parts, which we call activities. This established "SWARM" concept, standing for "Software Architecture for Reusability and Modularity" is the basic enabler for minimized maintenance efforts and efficient software development. The SWARM architecture could only be applied successfully with a more strict collaboration type (see challenge CT), especially concerning requirement definitions, and an even wider planning horizon (see challenge PH). A fix roadmap plan of at least one year and a rough three year plan allow the design of software modules to supply tailored solutions for various application domains. This has been achieved by concretizing the "Vision" of Scrum concepts with the sustainable Swarm architecture.

Each developer team is led by a product manager, which is in charge of requirement definition, prioritization within the team and the design of an entire application platform for the application field of his team. The design plan of all tool platforms results in a roadmap plan for at least one year.

Now the main challenge in the planning of calibration tool development concerns the alignment of roadmap based product development together with customer driven software development services as well as requests for short notice tool adaptions. Considering this background only a very flexible process of software development can be successful, which is based on clear requirement definitions and a stable roadmap. For this reason a two-stage SCRUM based development process was established going into the direction of a SCRUM of SCRUMS [3].

The first stage deals with the assignment of development requests from different customers to the available developer teams and the second stage about the distribution of tasks amongst the developers within the developer teams (Figure 4). To have a stable basis for this process a thorough definition of the requirement specifications from external customers and the product managers of the roadmap based tool development is crucial (see challenge CT). Often developers themselves assist in the creation of the requirements specification to gain all necessary information. The finished requirements are analyzed by a technical specialist who then creates a work break down structure containing the necessary development tasks with effort estimations. The tasks are transferred into a product backlog for the developers and the estimates are base for corresponding project plans.

However a project plan is not something that can be seen separately. To achieve a balance in workload and still be able to allow the insertion of high prioritized tasks, the development capacity on one hand and the tasks to be realized on the other hand are assigned to different developer teams on a monthly base. Therefor the managers of the development teams sit together and distribute the tasks moderated by the department manager. For urgent tasks even team comprehensive planning can be done, for example to realize pair programming by technical specialists from different teams, which will then separate for a certain time from their own teams to have full concentration for their tasks.

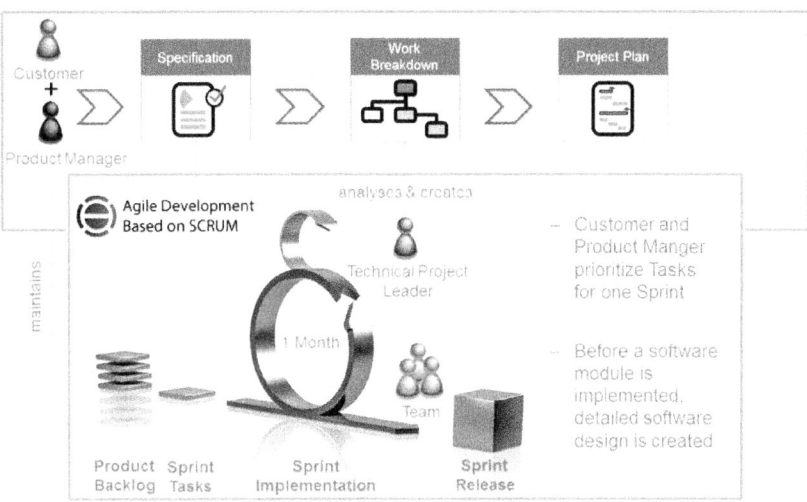

Figure 4: SCRUM process for calibration tool development

In general the assigned workload for the developer teams coming from the product backlogs of different project requests will be put into a sprint backlog suited for each team. Then, developers distribute the tasks amongst each other. This leads to an efficient development, because the developers can pick tasks, which are most suitable for them. At the same time it is achieved that not only a single developer is occupied with one application. A wide group of developers is involved, which allows quick responses in case of urgent code adaptions, e.g. when a bug occurs (see challenge DM). In daily stand up meetings, which in some teams only take place every second day, the status of implementation is communicated and tasks can also be switched to another team member.

The effort estimates from the work breakdown structures enable a progress tracking ranging from single development tasks up to entire development projects and allow a reliable planning for the entire department. By this a very high efficiency is achieved through the modular based development of applications with minimized maintenance efforts. Still the monthly iteration cycles offer enough flexibility to react on urgent development requests.

Coming from an extreme way of agile programming the shift to a more rigid way of development brought several noticeable benefits, especially in the field of Collaboration type (see challenge CT) where a shift to more detailed specifications could avoid iteration loops in development and at the same time was the enabler for a longer Planning Horizon (see challenge PH). Together with the SWARM Software Architecture the more complex planning of the development work yielded in a high efficiency increase by reusability of code. Last but not least the Deliverable Maturity (DM) could be improved leaving the way of extreme programming, when a larger group of developers became involved in the development of single code modules.

Testbench automation

Industry is facing fundamental changes in propulsion system development trends. In addition to the development of conventional internal combustion engine concepts, the trend to utilize electrified or combined hybrid systems is increasing. This trend is triggered by future global emission regulations, increasing costs of fuel and changes in customer behavior. Extended development activities resulting from these trends create specific demands towards the test environments and the test cases. At the same time, this trend emphasizes the need for highly efficient testing facilities in order to fulfill requirements towards shorter time-to-market.

Intelligent test bench automation software is one of the key factors to support faster development of new products from the concept over calibration and validation up to the end of line testing in production facilities. An enormous wide range of different test methods at each state of the development process of e.g. engines, transmissions or even the whole powertrain needs to be supported by this test bench automation software. In order to support future needs the test bench automation software itself is underlying continuous development [5].

The real challenge of this development is the huge number of change requests concerning features. The frequency of change requests increased heavily at shorter implementation cycles. Therefore agile development methods needed to be introduced in the development process of our test bench automation products. These support fast reaction on changes and esp. accepting the idea that requirements will evolve throughout a project (see challenge PH). Therefor the collaboration with the customer needs to be intensified during each step of the development process. In addition to that each change increases the risk of compatibility and reliability of the product. Increasing the number of changes at the same time can even exponentially increase these risks on the maturity of the deliverable (see challenge DM).

So an introduction of agile methods with reaction times between hours and weeks need to be aligned with an exponential increase of testing effort before releasing the product at the test benches. Hence moving to agile methods in 2009, FEV also introduced continuous integration processes including automated testing methods executed each night (see challenge CC).

Further on the deployment of the software was also integrated in this process. Production monitors were positioned in the development centers in order to indicate the result of the automated tests in real time. What we observed by increasing heavily the automated testing of our software code was also the great support for software developers themselves.

Even during their development of a new functionality they used the automated testing methods to see already at the beginning of their changes if they run into quality issues. A very good key index was the average number of commits of a developer, which increased from one per week in 2009 to nowadays 5 per day. For a full test of the test automation product today 4000 different test cases are executed and with the further increasing functionality and complexity the number of test cases needs to be increased even more. Due to limited resources a new approach of generation of test cases is needed to ensure the compatibility and quality of the test bench automation product.

Therefore FEV developed a model based testing framework MBT which is highly configurable and real-time capable. The special features of this system are the generation of test agents and test adapter out of formally describable models. With the modeling languages of the MBT framework, it is now possible to specify instances of the system. The system configuration models are automatically tested on conformity to the description of the configuration of the system model. An additional integrated error model allows identifying possible risky errors in the implementation in order to prioritize the way of testing.

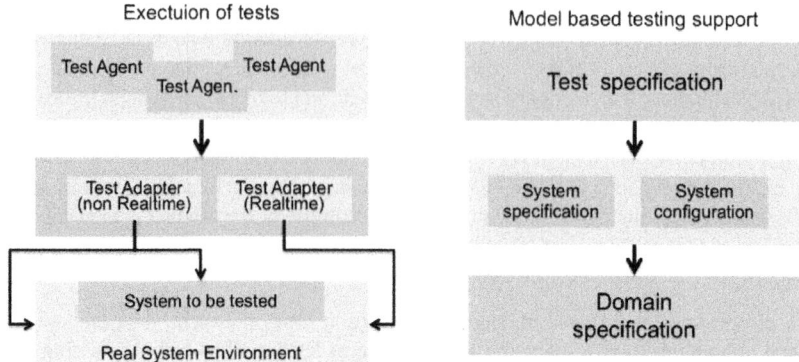

Figure 5: Workflow of testing execution and model based testing support

An integration of agile methods in our development process of test bench automation products brought us first of all in the position to develop faster and deliver more releases per year (see challenge PH). But the customer needs to accept a much deeper collaboration even up to the definition of specification (see challenge CT). Software tools are mandatory to e.g. speed up the process of negotiation and approval of requirements and specifications. Automation is absolutely mandatory in order to ensure the maturity of the deliverables (DM). Without the integration of automated testing methods within a continuous integration process it is impossible to keep or even improve the quality level of our products. As a next step the model based testing MBT approach will allow us to automate our development processes even further.

Embedded SW

Development of embedded control software is today strongly driven by early and long term planning, elaborated agreements and contracts with a huge number of sub-projects and suppliers.

When embedded software was introduced in modern vehicles, it was a component part provided by the supplier - e.g. injection control was provided by the injection system supplier. Later, control units provided by different parties could be combined via generic communication protocols such as Controller Area Network (CAN). Today, control units from different suppliers actuate multiple components at the same time, contain software from multiple parties and communicate via diverse communication protocols to an even higher number of additional controllers. For series vehicle solutions, all development activities and development parties must be aligned in a reliable and timely manner. Hence, the development cost share of software extends continuously (Figure 6). The same applies for the extend of SW code which already reaches a level of 100 Mio. Lines of codes for modern cars and will increase by a factor of 2 or 3 in the future [4].

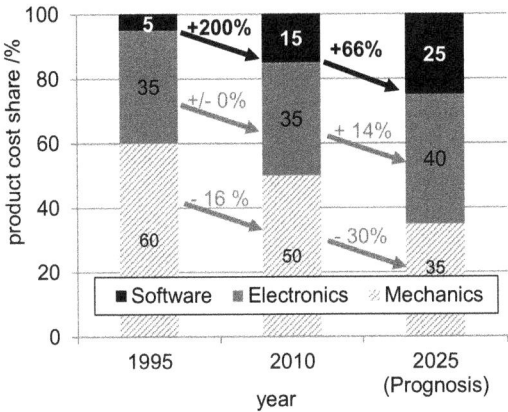

Figure 6: Increase of SW activities for vehicle product development on the cost level

In order to establish corresponding working processes within the V model approach, different norms and standards have evolved over the past. Process maturity models such as the domain-specific Automotive Software Process Improvement and Capability Determination (ASPICE) or the generic Capability Maturity Model Integration (CMMI) serve for the assessment of processes, process compliance and enable continuous improvement. ISO26262 specifies software development tasks on system, hardware and software level to define, realize and verify requirements introduced by safety-critical functions. In fact, it has become a general reference to best practices to ensure embedded software quality. Thus, high software quality could be achieved for different features: electronic gas pedal, Adaptive Cruise Control (ACC) or high voltage Battery Management Systems.

However, these and future software challenges cannot be handled anymore at required short development cycles, with multiple variants and under high cost pressure. In order to cut-down development time, agile development principles are becoming more popular in embedded control software development. However, a simple take-over of any agile methods is not applicable. Instead, technologies and organization must be setup in an agile way considering domain-specific requirements (Figure 7).

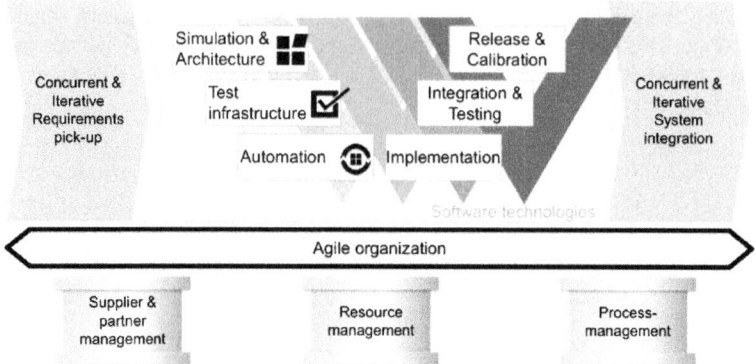

Figure 7: Technical and organizational aspects of Agile Automotive Software development

The overall acceptance criterion for agile embedded software development is the concurrent and iterative processing of requirements (see challenge CT) and their realization as high quality code in a working system. From technical perspective, different challenges must be mastered: simulation environments must are needed to enable early and automated verification and validation of executable specifications. System and software architecture must be tailored in a ways that quality assurance is possible on a suitable unit level. A test infrastructure must exist to define appropriate tests at the right time and to monitor test results. All aspects of software development must be facilitated by a maximum degree of automation, especially for redundant and regressive tasks with a low level of complexity. With these prerequisites, implementation, integration and testing can happen at the required delivery milestone defined by the agile process. Here, it is very important to make use of the provided automation facilities for short deployment cycles.

From organization side, additional measures need to be realized. Resources and processes need to be aligned, e.g. releases and system integration must be coordinated in a way, that activities still requiring human resources are not dominating the workload making the developer team inefficient. Additionally, suppliers in the project network need to be aligned, especially if they still work with less agile processes.

The application of Continuous Integration (CI) within the tool framework ASSIST is one example to demonstrate the capabilities of agile methods in modern embedded control software development. Continuous Integration tools established in other software domains cannot be applied directly. Hence, an own framework was developed for model-based software development (Figure 8). Control code is generated automatically from executable models which must fulfill extended quality requirements [1].

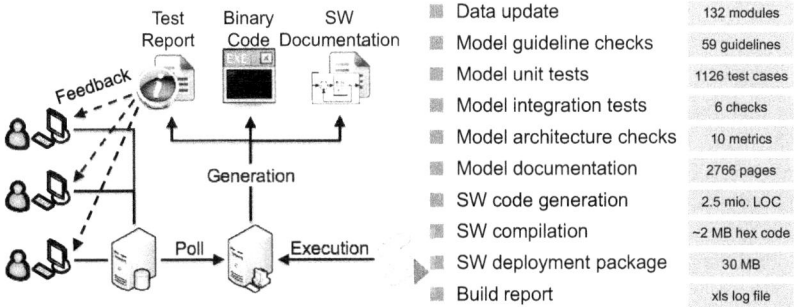

Data update	132 modules
Model guideline checks	59 guidelines
Model unit tests	1126 test cases
Model integration tests	6 checks
Model architecture checks	10 metrics
Model documentation	2766 pages
SW code generation	2.5 mio. LOC
SW compilation	~2 MB hex code
SW deployment package	30 MB
Build report	xls log file

Figure 8: Continuous Integration for Automotive series software development (2009)

All developers contribute their work products on a common data repository. An integration server polls the work products regularly and runs a build script automating various development steps: This includes diverse aspects of quality checks on model and code level: best practice guidelines, architecture and integration plausibility checks, dynamic unit tests with simulation on model and text code and binary code level. Additionally, so far manually created work products such as textual code, software documentation or compiled binary code as well as a comprehensive build report are generated automatically.

This ASSIST approach unifies various advantages: through automation, developers can concentrate on tasks which require human intelligence and experience. Project managers are able to identify the quality status and hence project risks with fast feedback from the project start. Executable code can be deployed to the customer in short frequencies expanding the degrees of freedom for project planning (see challenge PH, DM). Referring to the agile trade-off outlined in the introduction, CI shifts embedded software development towards agile approaches: collaboration control is enabled by a dashboard of diverse status reports while automation supports a defined SW process. Deliverable maturity is still kept on a high level without giving up advantages of agile processes (DM).

This example shows the benefits of agile approaches – but even further measures are applicable: SCRUM as an agile development process is appealing for the alignment of various parties like Original Equipment Manufacturers (OEM) and suppliers. As SW requirements are not completely known at project start, with SCRUM, they can be elaborated over time without bypassing a waterfall process. This shifts the process trade-off in a more agile direction since the collaboration type is rather dominated by bilateral alignment than a fixed contract at project start (see challenge CT, CC).

Summary and Outlook

Presenting the evolution of development frameworks for three different automotive software areas this paper has shown that both agile and V-Model approaches have their merits and drawbacks. There is not one or the other fitting for all project types. In fact both can benefit from certain concepts of the other. V-Models are more appropriate when fixed milestones and deliverables, a fixed price, documentation (e.g. testing and quality) or coordination of dependencies with other development groups are major priorities. When it comes to uncertain requirements, frequent changes, faster feedback loops or self-organizing teams agile approaches are first choice.

In practice a mixed approach combining elements of both approaches according to the underlying business case and organizational structure has proven to be successful. This becomes evident since all three examples presented started from different origins and ended up combining elements to their specific needs.

In case of the calibration methods the introduction of agile approaches has led to a more stable environment for developers. To avoid the well-known scope creep in agile environments a mid-term roadmap is imposed to ease the burden of the product manager in maintaining the product backlog.

The most significant changes in development of calibration methods were the steps from a very agile collaboration type and planning horizon into a more rigid but still agile direction. Together with the SWARM-software architecture these changes led to a very efficient but still flexible way of development.

Looking at the test bench automation the introduction of agile methods helped to cope with changing requirements and to provide faster and more stable products designed to customer needs.

Last but not least the development of embedded controls benefitted significantly from the introduction of agile methods, namely Continuous Integration and SCRUM. Keeping quality on a high level, agility was especially realized in terms of deliverable maturity and planning horizon: high quality software can be delivered in shorter time. At the same time, the project schedule can be adapted over the lifecycle if desired by all parties.

In general as of today it seems that the most benefit for the discussed areas can be generated by embedding agile methods into more rigid V-Models. This can be easily depicted by inverting the priorities of the agile manifest (Figure 9).

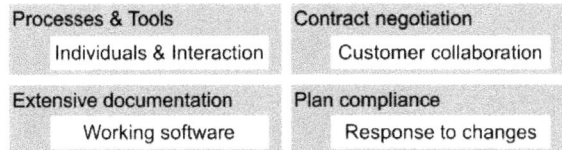

Figure 9: Embedding agile methods into long-term V-Models

Looking into the future software development will face further demands in speeding up and becoming more flexible while at least keeping or even improving software quality and fulfilling safety as well as security requirements. This will lead to further merging of V-Model and agile approaches and has to accompanied by standardization and automation of development tasks to be able to cope with ever more software in ever more complex environments.

This requires clear production rules relying on standardization and automation but also encouraging of individuals to deliver happily products with a limited scope and higher responsibility. Basis to this will remain the overall system design as a highly complex and innovative engineering task.

In the end the automotive industry will generate software rather in a production than in an engineering manner.

Bibliography

1. Richenhagen, J., Orth, Ph., Schloßer, A.: Continuous Integration for automotive model-based control software development, VDI Verlag, Düsseldorf, 2011, AUTOREG 2011, ISBN: 978-3-18-092135-8, S. 235-246

2. Goll, J., Hommel, D.: Mit Scrum zum gewünschten System, Springer Vieweg, Wiesbaden, 2015, ISBN 978-3-658-10720-8, S.115

3. Sutherland, J.: Agile can Scale: Inventing and Reinventing SCRUM in Five Companies, Cutter IT journal, Vol. 14, No.12, 2001, p. 10

4. Robert N. Charette, IEEE Spectrum, This car Runs on Code, 02/2009

5. Modellansatz hochkonfigurabler Softwarelösungen mit Echtzeitanforderungen, Abschlussbericht Forschungsprojekt der IB Sachsen-Anhalt, ZWB-Nr.: 1304/00082, 2014.

ARENA2036 – DigitPro: Development of a virtual process chain

P. Middendorf; D. Michaelis, P. Böhler, J. Dittmann, F. Heieck

Institut für Flugzeugbau, Universität Stuttgart

1 Introduction

The demand for a reduction of weight in automotive application is increasing in several research and development projects. The reason for that are the decline of resources, strict laws for emissions and an increased awareness of natural environment. Fiber reinforced plastics (FRP) can be used to decrease the weight of automotive structures as load path directed material distribution is possible.

As FRP structures are specifically manufactured for each application the integration of additional functions as for example sensors or heating elements implemented directly into the structure is possible to save secondary weight. Using a virtual process chain the development of complex structures can be optimized.

These topics are focused in the research project DigitPro and LeiFu as part of ARENA2036.

2 Content of Research: ARENA 2036

The use of composite materials in automotive structures needs innovative methods of construction and design. On the one hand a thorough understanding of the material behavior is important and on the other hand the development processes have to be investigated in order to consider the composite material.

The definition of possible structures is done in LeiFu and the virtual process chain built in DigiPro is applied on those structures. Different technology demonstrators are used to validate the different simulation approaches.

The main goal is to build up a complete automotive floor structure out of different manufacturing technologies. The virtual process chain is used on braided parts considered in the floor structure. To investigate the needed simulation approaches a generic geometry is developed to validate numerical and experimental results. Finally the optimized simulations will be applied to the final braiding structures.

3 Virtual Process Chain

The final structure includes some challenges for the braiding process which are identified at the beginning. This complexity is transferred to the generic geometry to produce the same effects appearing during the braiding process.

The generic geometry is used to investigate the braiding and infiltration simulation approaches. The limits of existing simulations can be identified and the approaches have to be improved in these areas to show the potential of numerical prediction of the used processes.

The generic geometry is built of an s-shaped curvature with constant triangular cross-section with defined radii in the corner (cf. Figure 1). The critical geometrical areas are included in the generic geometry although a simplification is used. The challenges for the braiding process are even increased.

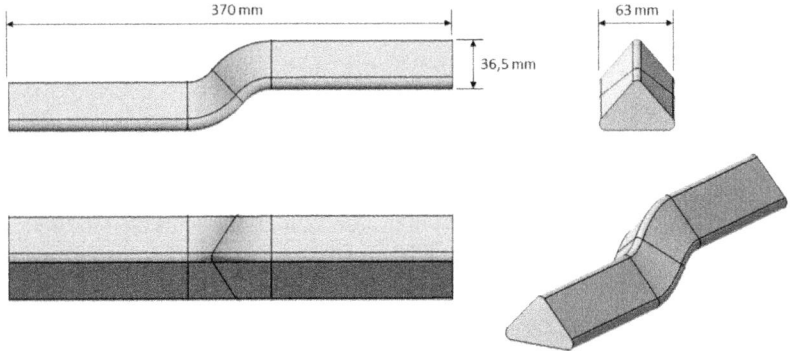

Figure 1: CAD-Modell of the generic braiding geometry

The virtual process chain considers different steps which will be approved using the generic geometry:

- Braiding Process Simulation:

Procedure: explicit FE simulation of braiding process

Goal: Prediction of deflection, gapping and fiber angles of the final braid considering the physical effects such as friction and material properties and process boundary conditions

- Infiltration Simulation:

Procedure: explicit FE and FV simulation of infiltration process

Goal: Prediction of locally varying permeability and flow front expansion as result of the locally varying fiber architecture and process boundary conditions

- CAM-Interface:

Procedure: Linkage of virtual braiding process simulation and real manufacturing process

Goal: Definition of manufacturing boundary conditions in reality based on results of the braiding simulation in order to achieve best braiding results without expensive experimental testing

- Optical Validation and Structure Simulation:

Procedure: Return of real fiber architecture in virtual simulation environment using an optical sensor

Goal: Validation of braiding process simulation and feedback of fiber architecture for following structure simulation

3.1 Data Transfer: HDF5

For design and development of technical structures in automotive applications several different software tools are used for the different development steps in order to get best results in the shortest time. The transfer of the results from one simulation software to the next is very important while not losing any necessary information. A closed virtual process chain is needed.

In the project DigitPro the HDF5 (Hierarchical Data Format) data format is used and specified for composite application. This data format is only a container where the organization of data within can be defined. For this definition the identification of useful and important input and output parameters of each simulation step is to be collected. An example for that is the fiber architecture which is created in the braiding process simulation and has to be transferred to the infiltration and to the structural simulation.

In the data container almost every data from mesh information to results of each simulation is stored. For the different simulation steps different meshes have to be used and the locally varying results are transformed into a global coordinate system.

Using the "HDF-View" the data format can be read and a mapping software called "DYNAmap" is investigated by DYNAmore for storing the results of each simulation into the data format. [1]

3.2 Braiding Process Simulation

The virtual process chain for this generic geometry starts with the braiding process simulation.

Despite most publications using analytical or pure kinematic approaches for predicting the fiber architecture [2] an explicit finite element approach is used in this case. Whereas only mandrel and yarn movement is considered in state of the art approaches the FE simulation is able to take into account the material behavior as well as interactions between yarns and machinery such as friction. Thereby it is possible to show appearing effects on the final fiber architecture which is important for following simulation steps.

The software tool used for the braiding simulation is ESI Virtual Performance Solution 2014.0. The simulation is shown in Figure 2.

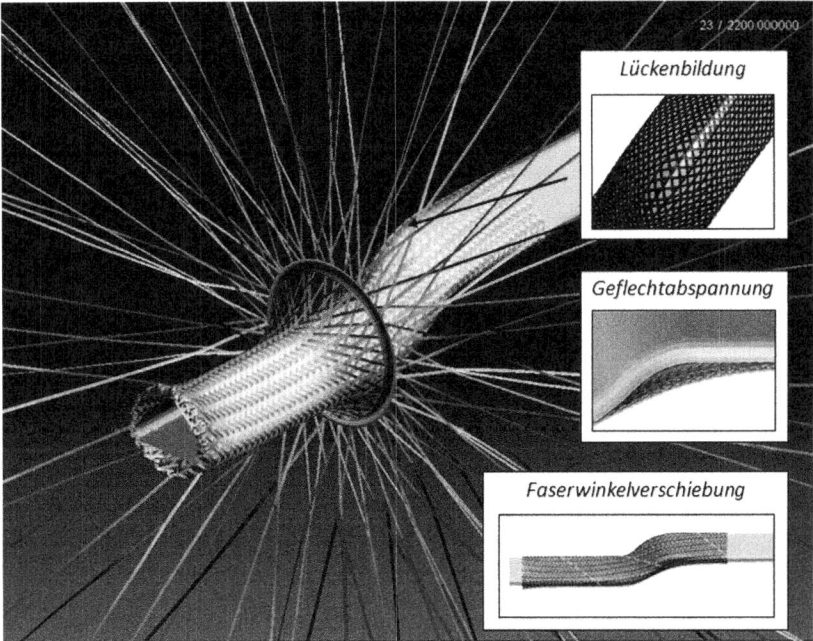

Figure 2: Braiding simulation using ESI VP Solution 2014.0

The braiding machine is modelled in the same configuration as in reality using 64 braiding yarns, moving on sinusoidal paths, and 32 standing yarns. The speed of the mandrel is calculated to achieve a braiding angle of 45°. Spring elements are used as bobbins to apply a constant tension force on each yarn. The mandrel is moved on a predefined path in a way that every section of the yarns is always perpendicular to the braiding plane in the center of the machine.

The modelling of the yarns is very important as it defines what effects can be predicted. In this project a bar and a shell approach is used with focus on the shell approach. The width of the shells is adjusted to the same width as 12k yarns have in reality whereas the thickness of the yarns is defined by the contact thickness between the different yarns. To have separated high tension and very low bending stiffness the material model MAT140 is used which was created for forming of flat textiles.

The results are transferred to following simulations as the infiltration simulation. Therefore a mapping into the defined HDF5 format is necessary.

3.3 Infiltration Simulation

For the infiltration simulation the permeabilities of the used textile are needed. Therefor a planar test device with an optical sensor is used and the tests are performed for braided textiles. This permeability test can also be done numerically and the different yarn approaches for the braiding process simulation can be compared.

The experimental tests are done with an infiltration pressure of 2 bar, a fiber volume content of 52% obtained by four layers of triaxially braid with an areal weight of 589 g/m^2. The infiltration media is Glycerol 85% with a dynamic viscosity of 109mPas at room temperature. In the results shown in Figure 3 can be seen that the flow front is almost homogeneous and the permeabilities can be determined as K_1=1.63e-11m² and K_2=1.59e-11m².

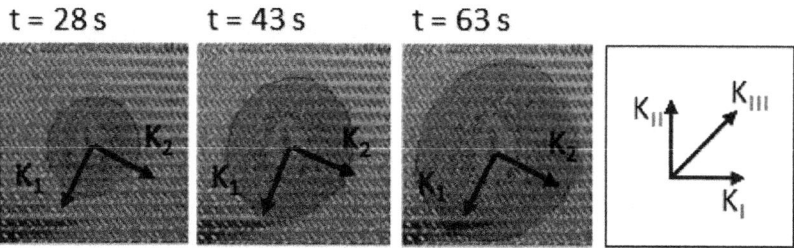

Figure 3: Results of permeability determination for braids

As these determined values are calculated for the main flow front directions and are not aligned to the fiber directions they have to be converted and unfold to K_I = 1.60e-11m² (standing yarns 0°) and K_{III} = 1.62e-11m² (braiding yarns ±45°).

The virtual infiltration is done by using ESI PAM-RTM considering the fiber directions out of the braiding simulation and the permeabilities calculated above.

The finite element simulations show almost no difference between the two yarn approaches (c.f. Figure 4). The reason behind is the locally averaged fiber orientation.

By mapping the bar element directions on the infiltration mesh each element without direction is interpolated from the neighbored elements which leads to no elements without orientation. This mapping or the shell approach is not necessary as the covering range of the braiding results is much higher and closer to reality. But by averaging the results the difference is negligible.

The flow front is influenced mainly by the permeabilities in the different directions called by the different fiber orientations.

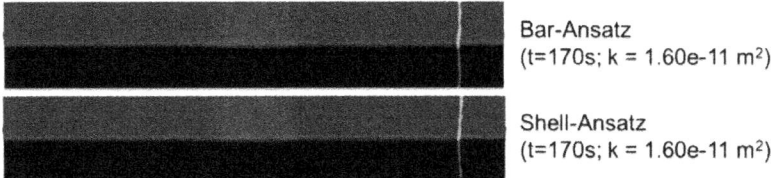

Bar-Ansatz
(t=170s; k = 1.60e-11 m²)

Shell-Ansatz
(t=170s; k = 1.60e-11 m²)

Figure 4: Infiltration simulation for different yarn approaches considering the results of the braiding process simulation

3.4 CAM-Interface

To have comparable results in simulation and reality it is important to have identical boundary conditions. Next to the machine and the material the movement of the mandrel through the braiding center is important as the fiber architecture depends on that movement. To achieve identical motion a tool for the definition of the movement

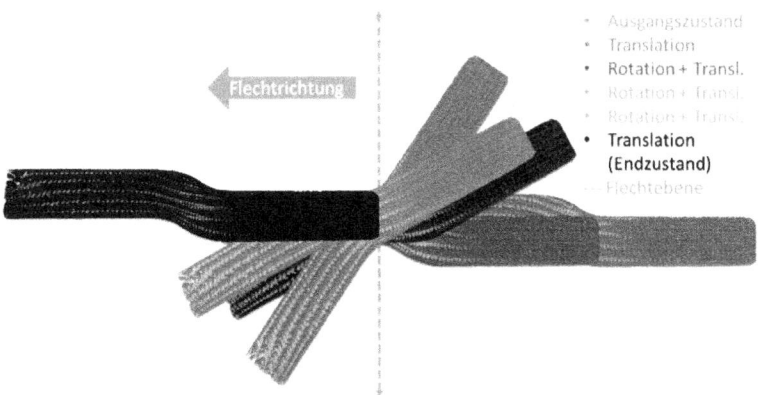

Figure 5: Perpendicular movement of the mandrel to the braiding center

is developed based on the geometry of the mandrel. As simplification the goal is to have a perpendicular direction of each mandrel section to the braiding plane (cf. Figure 5).

For each section a specified velocity of the mandrel is defined which will influence the fiber orientation on the mandrel.

As shown in Figure 5 the movement is a superposition of a translation and a rotation of the mandrel. As in reality the motion is defined by the movement of the so called tool center point defined by the robot, the movement in the simulation is defined by a fixed point as well. The mandrel is separated in different sections and for each section the velocity is calculated. In Figure 6 the mandrel movement is shown for the same time-step and it can be seen that the results fits adequately.

Figure 6: Comparison of numerical and experimental mandrel movement

3.5 Optical Validation

To use the braiding process simulation as prediction method for the final fiber architecture a validation for each investigation step is necessary. Therefor an optical measurement system developed at the institute of aircraft design is used. It consists of an objective, a CCD-sensor and an illumination device (c.f. Figure 7). For positioning of this system a CAD-based path algorithm is used which calculates each position of the sensor regarding the curvature of the structure. These positions are reached by a robot as the commands are translated directly in KUKA code.

Each picture is then analyzed and the fiber directions are separated by using an algorithm which is developed by FIBRE Bremen. It is based on a grey-scale analysis

The orientations are then mapped on a finite element mesh and can be compared to the results out of the braiding process simulation (cf. Figure 8).

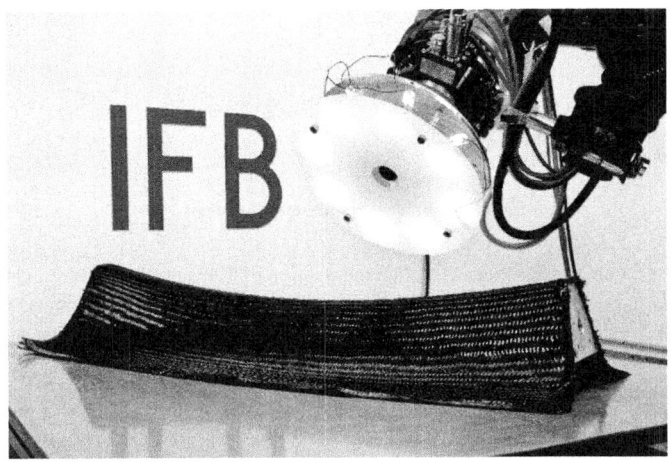

Figure 7: Optical measurement system

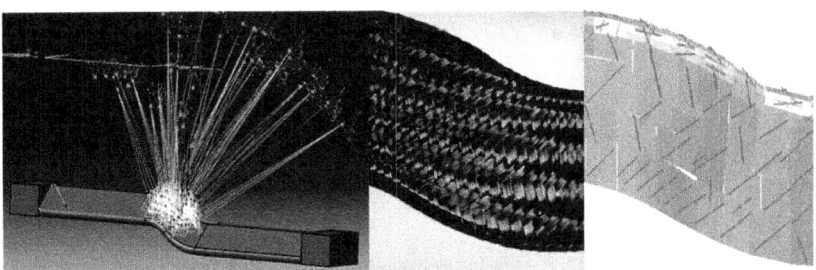

Figure 8: Positioning of the sensor and mapping of fiber orientation on FE-mesh

Three different areas of the generic geometry are analyzed (sections 4 to 6, c.f. Figure 9) and it can be shown that a good correlation of braiding process simulation and experiment is achieved.

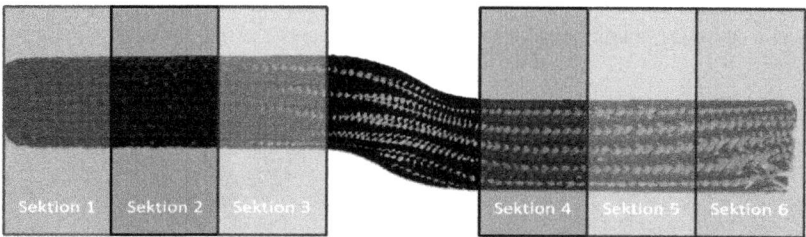

Figure 9: Sections for optical anaysis

In Figure 10 can be seen that an averaged deviation of 1.4° appears between simulation (ShS) and automated analysis (AV) which shows that the simulation can be used for prediction of the braided fiber orientation for the generic structure. In Figure 10 also the manual determination (RB) of fiber directions is shown which gives an averaged deviation of 1.3°.

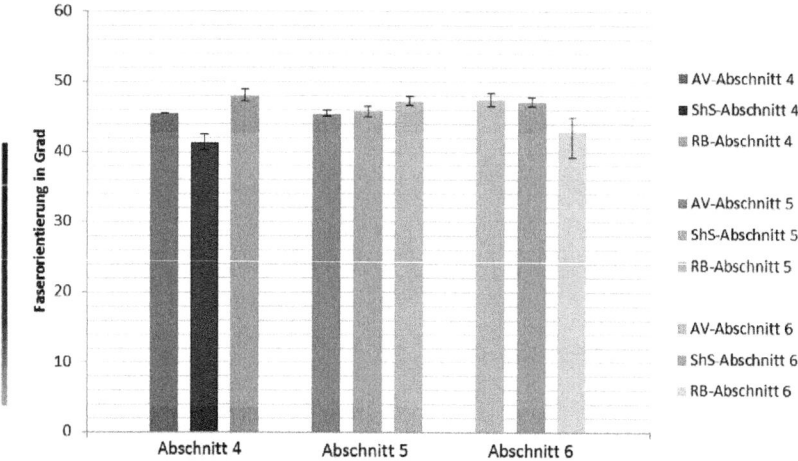

Figure 10: Determination of fiber orientation using automated detection, simulation and manual detection

The detected fiber orientation can be used as validation method for the simulation on the one hand and on the other hand to build a suitable model for the structure simulation to predict the behavior of the final structure considering real fiber architecture. In state of the art approaches the target value for the fiber directions is used and local variation cannot be considered. In the project DigitPro the real fiber orientations are used to improve predictability.

4 Summary and Outlook

In the project DigitPro which is a spare project of ARENA 2036 the complete virtual process chain is investigated.

A real automotive structure is chosen and a generic geometry is developed in order to get a simplified structure with almost every challenge of the target structure. The structure is designed by using numerical approaches starting with a braiding process simulation to virtually get the fiber architecture, predicting effects like gapping or deflections and defining the machine setup. This is followed by an infiltration simulation for defining the permeabilities and predicting the flow front evolution. This will lead to better and faster definition of infiltration setup and save time and costs.

A CAM-interface is investigated to transfer the virtually defined setup to the machinery and to have comparable results in simulation and in reality. Only with this interface a validation of the simulation approaches is possible.

This validation is done by using an optical sensor which is able to detect fiber orientations automatically.

A following structure simulation can show the potential of saving weight for the target structure using numerical optimization within the closed virtual process chain.

Acknowledgement

The authors thank the „Bundesministerium für Bildung und Forschung (BMBF)" which funds the research work in this project.

Literature

[1]   J. Dittmann, P. Böhler, D. Michaelis, M. Vinot, C. Liebold, F. Fritz, H. Finckh und P. Middendorf, „DigitPro – Digital Prototype Build-up Using the Example of a Braided Structure," in *IMTC*, Chemnitz, 2015.

[2]   A. Pickett, J. Sirtautas und A. Erber, „Braiding Simulation and Prediction of Mechanical Properties," *Applied Composite Materials,* 2009.

[3]   A. Miene, M. Göttinger und A. S. Herrmann, „Quality assurance by digital image analysis for the preforming and draping process of dry carbon fiber material," in *SAMPE EUROPE International Conference and Forum*, Paris, 2008.

Highly versatile plug&produce assembly systems

Dr. Stefan Junker, Marian Vorderer

Robert Bosch GmbH, Corporate Sector Research and Advance Engineering

1   Introduction

Due to decreasing innovation cycles, new technologies are implemented into new product ideas continuously faster. At the same time, well established manufacturers of goods are facing highly agile small and medium sized enterprises bringing new and innovative products to the market very fast. While this competition allows customers to select products fitting best to their demand and budget, producing companies are challenged with decreasing product life cycles, a higher product variety, fluctuating demands as well as strong cost pressure. As this so called mass-customization effect is more and more affecting also classical high volume-low variant markets, production systems which provide both, high productivity and adaptability, are required.

An example for this development can be found in car production, where the number of different car models has approximately quadruplet in the past 60 years (Figure 1). In addition, customers can choose from a continuously growing number of features, ranging from new propulsion concepts and body variants to an increasing number of comfort and safety features. As a result, the amount of different components delivered by automotive suppliers is increasing exponentially while cost pressure is increasing.

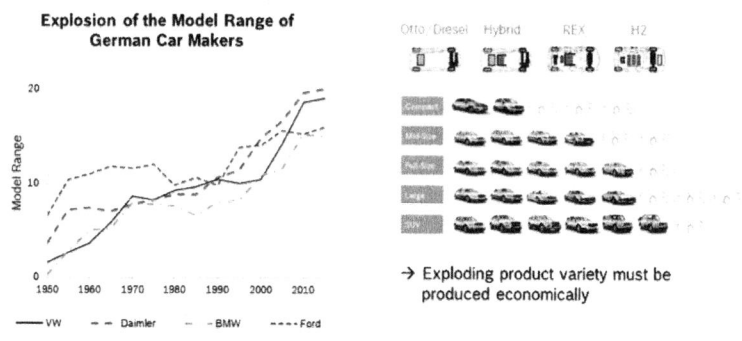

Fig. 1: Increasing product variety leads to more complex product portfolios which have to be manufactured in a cost efficient way

These highly volatile markets also reflect in the requirements for competitive production equipment. While at low volume, producer's high flexibility could be economically provided by human operators, manual production of high volume parts is not an economical feasible option. In order to keep up competitiveness, future automated assembly systems therefore must be able to comply with changing products and their variants quickly and cost efficient at minimized downtimes. This paper introduces a new concept for the design of highly adaptable assembly systems. Besides a highly modular structure, the presented approach is supported by innovative IT- and control technologies.

2 Structure Analysis of Current Machinery Concepts

For the economical production and assembly of products and components two different machinery concepts have been evolved. Special machinery is optimized towards throughput and homogeneous products, whereas modular machinery is targeted towards more efficient processing of changing demands.

2.1 Special Machinery

The automation of assembly processes in current production facilities is mainly realized with special machines (Figure 2). The expected product throughput and variants are determined at the beginning of the engineering cycle during planning stage resulting in requirements regarding production technique. A unique design can be found in all sub areas of the special machine. This includes mechanical structure, control technology, electrical installation and program of the machine.

The mechanical structure of the machine, its structure and its connections are rigid and optimized for the assembly process. Therefore, subsequent changes to the layout lead to additional effort in engineering and implementation. The architecture of the control concept is centralized requiring all information and signals routed to a control cabinet. Bus systems reduce the cabling effort, however, fixed wiring is necessary in order to integrate the components, such as actuators and sensors [1].

Experts are prerequisite for the creation of machine programs to implement the assembly specific sequence most commonly as a fixed written program. Application of industrial control technique requires utilization of proprietary engineering tools leading to additional configuration and administrative efforts.

The realized special machines are suited to handle the previously during planning stage defined product range including variants based on software alternatives, but they are not designed for later adaptation to new products or processes.

The mechanical, electrical and software engineering of special machinery causes significant expenses due to their custom design. Current approaches [2] tackle these expenses by re-utilization of construction templates such as assemblies (mechanical construction), macros (electrical construction) and libraries (software development). However, practical application leads to minor savings effects. The integration of the machine is associated with high expenses for construction and wiring, whereas disturbances like incompatibilities, design faults or unknown dependencies are detected lately during machine assembly. This applies as well to commissioning.

Due to the highly specialized machine design an adaption to not foreseeable products and variants is not provided. The adaption to new products and/or processes requires a modification of the machine, which leads to significant downtime in practice.

Fig. 2: Special purpose assembly machines consist of custom designed mechanics, electronics and controls limiting the adaptability to varying products and processes

2.2 Modular Machinery

The costs caused by the custom design of special machines during planning, engineering and adaption are more predictable for machines based on a modular concept. The mechanical structure consists of a base frame (Figure 3) with a standardized grid to house several modules. These modules are interchangeable and represent specifically adapted process steps. The automation of an assembly task is customized by compiling the necessary modules.

A main controller is assigned to the base frame to handle the modules. The main controller coordinates and synchronizes each local module controllers which control the individual process of the corresponding module. The connection to the modules uses a standardized interface for energy supply, media and data. Thus, customized wiring is done inside the module only. Outside the module, a standardized wiring can be found within the base frame.

A standard can be found as well in the software implementation. The standard supports software developers by defining interfaces and libraries in order to simplify the integration of the module in the machine's base frame. Following the standard, the developer can focus on the implementation of the process related code, because general tasks like sequence control of the machine, communication and data documentation is provided already. The software engineering for the machine's modules as well as their integration into the base frame of the modular machine requires software experts like the special machinery.

Fig. 3: Modular assembly machines consist of standardized modules mounted into a base frame. Each module can be replaced in order to react to varying products and processes

The process modules necessary for the specific assembly task have to be provided permanently. While engineering and commissioning of new process modules can be done in parallel to production. The integration of approved modules into the system requires a short downtime of the machine only.

The key advantage of modular machines in relation to special machinery consists of limiting engineering efforts for the module's design. The interlinking of the modules to an operational machine is predefined by the standard. Changes of assembly requirements can be addressed easier by adapting the automation through the development of new modules. This also applies in the case of faulty modules, where modules can be exchanged completely to reduce the downtime.

The disadvantage of the modular concept is the module's mechanical limitation narrowing down the designer. In some cases, multiple processes are integrated into one module due to space limitations. This might distinguish the advantage of the modular approach.

2.3 Trends and Examples at Academia

Many examples for simplifying the engineering process for assembly machinery can be found in the academic field. The following three projects are a selection of visionary concepts for the evolution of assembly systems.

2.3.1 Plug&Produce: Holonic Assembly System (University of Tokyo)

The Holonic Assembly System (HAS) was developed at the end of the 20th century at the University of Tokyo [3]. The concept of plug&play from the personal computer domain was extended by automatic network setup and interaction between devices within the system. The application to industrial utilization coined the term plug&produce.

The control system's architecture followed the holon principle, which is characterized by a contrary behavior: self-autonomy and cooperation with other holons. The realized demonstrator consists of three robots with the ability to perform handling tasks (stacking of discs). Each device was associated with an execution holon in order to control itself (self-autonomy) and simultaneously negotiated with the dynamic execution of work orders of the overall system (cooperation) (Figure 4).

The origin coordinates of each device were determined by the user manually and stored within the corresponding data model. An automated commissioning was enabled by adding information about adjacent devices and product data.

Although the basic approach plug&produce, automatic analysis of workbench and assembly precedence graphs enhanced the relevance of the project for a potential real utilization, the application was limited to simple pick-and-place applications. More complex operations like the combination of equipment (robots with intelligent tool changer) could not be realized using this approach.

The demonstrated approach permitted a much better adaption of assembly systems to changing boundary conditions. Thus, in case of changes within the assembled product or errors, the efficiency of the assembly system could be maintained.

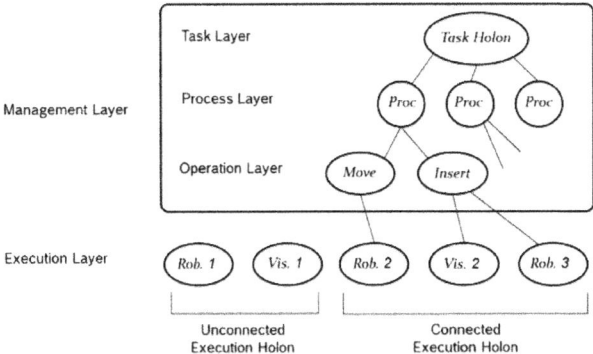

Fig. 4: Layers within the Holonic Assembly System (HAS) in order to enable plug&produce. Connected execution holons autonomously connect to the task holon. [3]

2.3.2 Skill based Programming: SMErobot (Fraunhofer IPA)

Based on the fact that robots are mostly used in mass production, characterized by many repetitions at low variance of the parts to be produced, the publicly funded project SMErobot^{TM} targeted the application of robots for the production of small batches in small and medium enterprises (SME). Applications within SMEs are characterized by rapid adaption to new products and processes instead of short cycle times. The users have distinct expertise in the manufacturing process. However, little knowledge of robot programming. Therefore, it was important to carry out commissioning, configuration and operation of the robot cell without special programming knowledge [4].

For demonstration purposes, an exemplary cell was chosen with an application from woodworking: a robot with tool changer able to transport wood panels using vacuum grippers and to perform drilling and milling operations. The selection of the machining processes and their parameterization was carried out by the user on a screen at the station (process-oriented programming) without any manual programming effort via plug&produce technologies. [5]

The core idea of plug&produce-based system integration consists of describing the functionality of the devices (intelligent tools and robots) by means of skills. Related attributes parameterize them and a combination to higher-value functions is possible. The process "DrillHole" is an example for combining skills: a robot with movable flange (skill "Move Programmable"), with different tools (drill skill "CanRotate") connected to the robot flange (skill "CanAttach") (Figure 5).

523

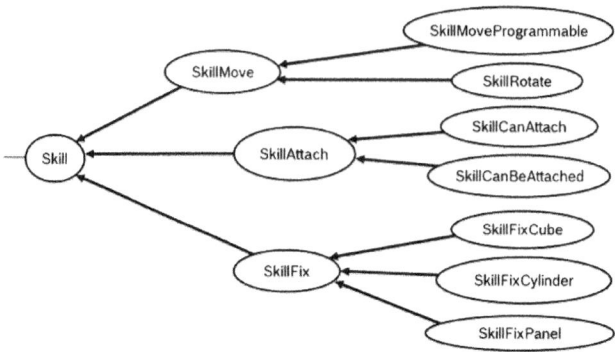

Fig. 5: Part of the ontology used at SMErobot[TM] to describe robot skills and to enable the generation device description for combined devices [4]

The devices register their specification to a central service ("Interconnector Modules"), which aggregates the executable processes and provides them to the user via an HMI. The user configures the robot task based on the provided processes using his wood-working expertise.

The application of the robotic cell is considerably simplified for the user with minimum programming skill in robotics, merely it must be programmed at the process level. Thus, the time to implement a program change due to new products or variants can be significantly shortened. However, classical construction work (mechanics, control technology) cannot be saved, since the robotic cell is a special machine.

2.3.3   Automatic Layout Detection: AutoPNP (fortiss, TU Munich)

The BMWi funded project "AutoPNP" (2011-2014) covered the versatility of machines within the factory layout. The customization of assembly-work plans to the current factory layout was carried out without manual intervention into the software. This included the consideration of material flow between individual workstations. [6]

The workstations were constructed as sliding units on a standardized carrier. Each station contained its own controller (PLC) with a unique station ID. Using an integrated optical sensor at each station carrier, the automatic recognition of neighboring stations was possible (Figure 6). This resulted in an automatic determination of the topology and layout. The setup of the communication between the stations and a central MES was carried out by the developed middleware CHROMOSOME. [6]

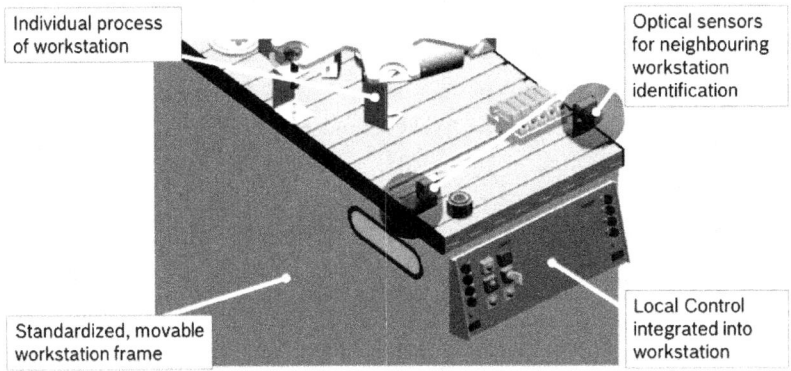

Individual process of workstation

Optical sensors for neighbouring workstation identification

Standardized, movable workstation frame

Local Control integrated into workstation

Fig. 6: The movable workstations can detect their neighboring stations using an optical sensor in order to exchange their unique station ID [7]

Executable functions (such as drilling, transportation, testing) were called "primitive operations". A combination resulted in combined functions, such as "sorting out" the combination of transport, testing and transport. The work plans were described using the terminology of these functions.

The automatic layout determination combined with the skill-based approach allowed for an automatic implementation of work plans with changes in the factory layout without additional construction or programming. To extend the concept, the following points were considered:

– Implementation of more comprehensive work plans.

– Consideration of machine states, such as error.

– Consideration of shared resources and multiple existing machines.

2.3.4 Conclusion Academic Examples

The academic examples above feature different approaches for the further development of today's state of the art in reconfigurable assembly systems. Essentially, these are reduction of design tools and needed expertise through plug&produce, automatic determination of factory layouts and description of assembly functions based on their skills.

The developed approaches so far cannot be found in current special machines. Possible reasons are:

- Approaches limited to simple assembly tasks.

- Complicated programming of devices (encapsulation) to demonstrate simple commissioning with plug&produce.

- Localization routines suitable for layout determination, but not accurate enough for handling tasks.

- Industrialization, for examples (accuracy, complexity of the tasks, work plan design) are inadequate/not satisfactory.

3 Development Fields and Potential for Versatile Assembly Concepts

Given one assembly task, the concept of special machinery is compared to the modular one. The costs produced over the phases design and implementation until start of production (SOP) followed by productive use will be examined. For simplification, a continuous change of the handling process is assumed, which is characterized by new variants and products.

The initial investment for engineering, sourcing, implementation and commissioning in modular machine concepts is higher compared to special machinery (Figure 7). The reason behind is planned flexibility: processes are implemented on a modular basis with duplicated components like controls, mechanics and supply on each module. The expected savings from standardization of interfaces and functionality will take effect during operation after SOP. New modules can be set up in parallel to operation and then be integrated with shorter downtime of the assembly line. The adaptation of the software to the machine is easier due to the standardized architecture and requires less expertise than a special machine.

Analysis of the ratio of costs related to hardware versus engineering personnel for design, programming, commissioning and administration shows, that it can be simplified to a ratio of 50:50.

By standardizing components, cost savings are expected in hardware sourcing due to volume effects. These savings will sum up over the duration of several projects. Conservative estimates expect savings of 20 %, which results in a total cost reduction of five percent (Figure 8).

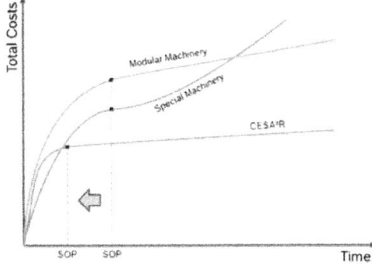
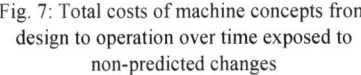
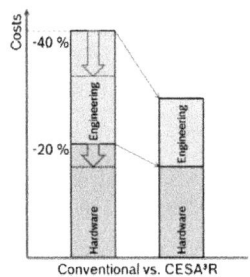

Fig. 7: Total costs of machine concepts from design to operation over time exposed to non-predicted changes

Fig. 8: Expected cost reductions of highly versatile systems compared to regular assembly system concepts

A significant impact on savings can be achieved within the planning and engineering process. Especially domains with special expertise like design, programming and commissioning require experts with expensive hourly rates. Optimizing the work done by experts by intelligent machine components could lead to 40 % savings. If changes during operation are considered too, the cost saving potential provided by highly versatile concepts is much higher due to much lower change engineering.

Development and utilization of independent, intelligent function modules capable of

– automatic layout and topology detection,

– capability-based programming,

– self-networking controls using plug&produce and

– automatic work plan reasoning triggered by the workpiece

will enable complex assembly systems to be configured quicker and easier. The resulting costs in planning, engineering and commissioning will be significantly reduced and a base for highly versatile assembly systems is build.

Following advantages emerge from the approach:

Reduction of engineering efforts (time and costs)

The assembly system is configured and compiled on the basis of standardized function modules. Engineering will be reduced to adjustments and configuration instead of design.

Reduction of commissioning time

The standardized function modules are delivered ready to run by the manufacturer. After their installation into the machine, an automated commissioning is executed in order to enable the integration into the control architecture.

Scaling of ramp-up lines with on demand automation

The machine's degree of automation can be adjusted by adjusting the number and types of individual functional modules. This enables scaling from manual to full automatic lines.

Reduction of investment risks through demand-based capacity adjustments

Adjusting the degree of automation allows an operation close to the optimum operating point.

Reduction of time between investment decision and start of production

Configuring an assembly system based on standardized modules reduces the engineering phase is significantly compared to conventional approaches. Hence, the period between investment decision and start of production is reduced as well. This allows for a more precisely definition of the machine in relation to product quantities and variants.

4 Description of the Pursued Approach

In order to prove the technical feasibility of highly versatile assembly systems, a conceptual demonstrator has been developed. It consists of autonomous functional modules which provide a set of plug&produce capabilities.

4.1 Requirements to fulfill Hypothesis

The initially defined categories lead to the following requirements:

The control system must be adaptable to the level of automation. Hard wiring has to be reduced in order to maximize versatility. Supplemented by connected industry techniques such as localization, index services and autonomous control systems, the needed expertise for configuration and operation will be reduced.

This also applies to the machine software: conventional, procedural programming has to be replaced by a skills-based programming model. This encapsulates the complexity and enables a simplified programming. Ideally, this reduces the programming-intense commissioning of a machine to a configuration.

Based on functional modules, any process can be implemented as long as it delivers the skill descriptions and connection to the central index service to supply administration and synchronizing functions.

In addition, the restrictive mechanical grid of modular machines for the integration of the functional modules has to be broken into a free one. This enables the free arrangement of each functional module. The size of the modules has to be freely selectable in order to allow the implementation of individual processes.

The number of functions implemented on a single module has to be implemented in a way that allows for reutilization in other assembly system configurations.

4.2 Versatile Assembly Station Operating Concept

The handling functions needed for automating a given assembly task are implemented as mechatronic objects (MO). These integrate all necessary functions for operation, like control, supply, communication and process. Using this approach, commissioning of an assembly task is as follows (Figure 9):

Phase 1: Configuration/Setup

The required MOs for the assembly task are arranged on a base plate, fixed and connected. An interactive screen supports the user during setup in finding appropriate positions.

Phase 2: Structural/Layout Recognition

Each MO performs a structural and layout recognition and reports the results to a central index service. This includes the physical position (origin) within the setup and technical data information about communication paths and addressing. The superposition of each origin with work ranges and work positions leads to the neighborhood relations of each MO.

Phase 3: Function Identification

Each MO transfers its virtual business card to the directory service. This comprises the functional descriptions (skills) and individual parameters, such as energy profiles, documentation, user advices and visual information. This information is optional but enhances the operation of the machine.

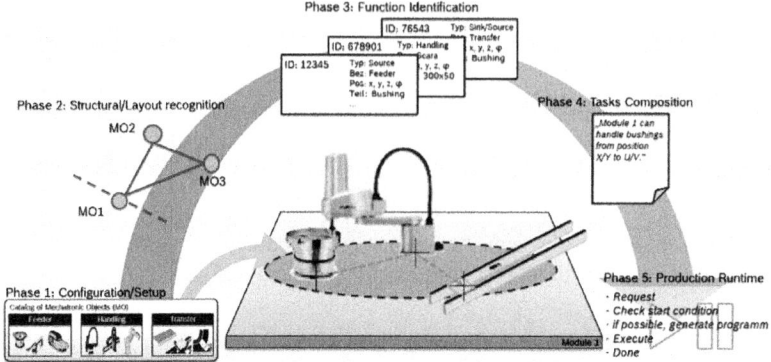

Fig. 9: Phases for commissioning an assembly task using the described approach

Phase 4: Task Composition

An IT-service analyzes the skills and layout of all available MOs and reasons them continuously to derive the executable process steps of the machine.

Phase 5: Operation/Runtime

The work plan is brought to the machine by the work piece, for example using an RFID media. Matching the requested process steps with the available ones generates the assembly sequence program and triggers its execution. An optimization of the sequence can be carried out on the basis of additional information such as energy data or throughput.

Additional value-added services such as visualization, optimization, data logging and diagnostics facilitate the user to influence both, the MO's and the executed assembly process behavior.

4.3 Technologies

To realize the autonomous behavior, each function oriented automation device must be equipped with the following technologies (Figure 10).

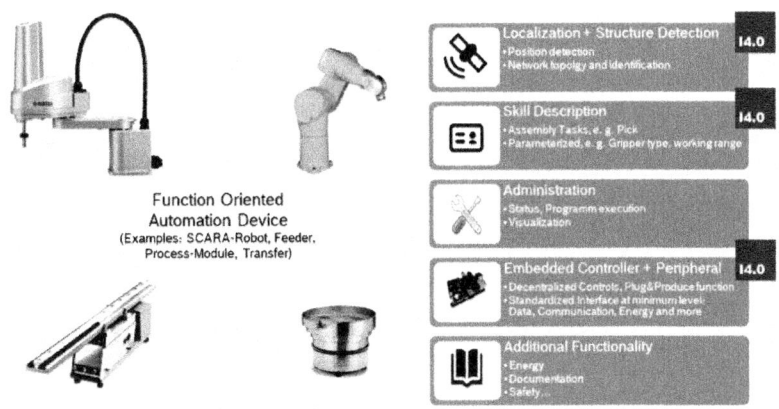

Function Oriented
Automation Device
(Examples: SCARA-Robot, Feeder,
Process-Module, Transfer)

Fig. 10: Technologies needed to enable autonomous behavior of function oriented standard automation device of versatile assembly systems

An integral part to enable automatic program generation and reduction of commissioning efforts is a **localization** technique. The spatial position detection must be precise enough for the parameterization and execution of assembly tasks. The network detection and identification must enable the automatic connection to the central directory service to allow automatic login and data exchange.

A **skill description** permits the Mechatronic Object (MO) to describe the provided processes using a virtual business card. Existing standards offer example classifications and terms to describe handling skills. These will be detailed using parameters.

The MO's **administrative** functions for operation are implemented by the manufacturer. This includes control of the process, sequence and state machines, operating states and visualization/user interaction.

The **embedded controller** ensures the autonomous execution of the MO program and algorithms. This includes the safe control of the process, as well as additional tasks such as communication, plug&produce and localization. A multi-level control architecture is used within the MOs deployed in the concept containing DSPs, multi-tasking OS and communication processors. The supply with energy, data and fluids is designed in terms of simplified reconfiguration.

Additional functions facilitate the machine operation by providing information such as energy data, operating instructions and system documentation.

5 Realization

The application of the demonstrator was carried out within the ForschFab-project on the research campus of ARENA2036. ForschFab's goal is to develop concepts for process and logistics in order to enable versatile production of new product types with a selectable degree of automation. Control and assembly concepts are analyzed on an experimental assembly line for car doors.

The realized process station consists of a pick-and-place application for handling screws used by the currently assembled door type. Three mechatronic objects have been implemented (Figure 11):

– MO Feeder supplies screws on a linear feeder

– MO SCARA robot handles screws by pick-and-place

– MO Transport provides a blister to store picked screws and transport them to the subsequent process station.

After installation and commissioning of the setup, each MO transfers its virtual business card to the index server using an Ethernet connection. A service running on the index server analyzes the available skills and spatial dependencies in order to generate the available assembly tasks.

An RFID tag at the transport system contains the work plan, which is then evaluated and executed by the service automatically.

The setup and commissioning of the assembly task can be done without special knowledge. In particular, no engineering tools are necessary. Due to simplicity reasons, the position of each MO could be aligned to a grid pattern in this first demonstrator. This enables the movement of the MOs in order to achieve an overlap with neighboring MOs.

The user has to acquire the position of each MO manually and save it to a control file on the MO. MO Transport serves as an interface to the assembly line and supplies the picked screw kits to the subsequent process station.

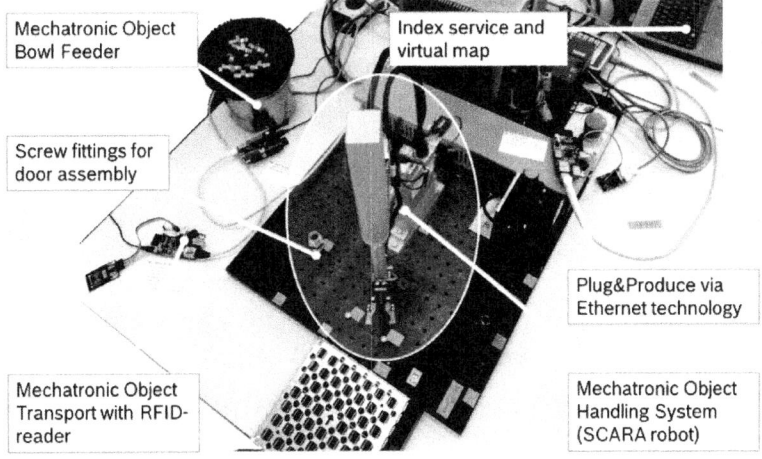

Mechatronic Object
Bowl Feeder

Index service and
virtual map

Screw fittings for
door assembly

Plug&Produce via
Ethernet technology

Mechatronic Object
Transport with RFID-
reader

Mechatronic Object
Handling System
(SCARA robot)

Fig. 11: Demonstrator setup for screw prepackaging using three mechatronic objects (MO)

6 Conclusion and Future Activities

The proposed approach for highly versatile assembly systems illustrates a vision for future assembly systems. It extends the conventional machine concepts special machine and modular machine with new technologies: localization, skill based auto-programming, decentralized control architecture and modularization of assembly functions. Thus, compared to conventional approaches, the aim of the reduction of planning, design and commissioning time can be achieved.

Future activities will detail the concept and develop technologies, which will include the development of an automated localizing solution, the detailing of a skill model, the implementation of an index service and of further mechatronic objects.

Work will be done in cooperation with partners from the ARENA2036 consortium at the ForschFab with the versatility enhancement of assembly lines in mind. Then, the concept will be applied to more different processes and scaling of both: size and complexity.

Acknowledgement

The content of this paper was supported by the BMBF-funded project 'The Research Factory for the production of the future' (ForschFab) within the research campus ARENA2036.

Bibliography

1. K. Feldmann, V. Schöppner, and G. Spur, *Handbuch Fügen, Handhaben, Montieren - Edition Handbuch der Fertigungstechnik.* München: Hanser, 2014, p. 439.

2. M. Litto et al., *Baukastenbasiertes Engineering mit Föderal: Ein Leitfaden für Maschinen- und Anlagenbauer.* Frankfurt am Main: VDMA Verl, 2004.

3. T. Arai et al., *Agile Assembly System by "Plug and Produce",* in CIRP Annals 2000: Manufacturing Technology: Annals of the International Institution for Production Engineering Research, Berne: Hallwag, 2000, pp. 1–4.

4. M. Naumann, K. Wegener, and R. D. Schraft, *Control Architecture for Robot Cells to Enable Plug'n'Produce*, in: Proceedings of the IEEE International Conference on Robotics and Automation: Roma, Italy: 10-14 April, 2007, Pitscataway, N.J: IEEE, 2007, pp. 287–292.

5. M. Naumann et al., *Robot Cell Integration by means of Application-P'n'P*, in Proceedings of the Joint Conference on Robotics: ISR 2006, 37th International Symposium on Robotics and Robotik 2006, 4th German Conference on Robotics; proceedings 2006, Düsseldorf: VDI-Verlag, 2006, pp. 93–94.

6. N. Keddis et al., *Towards adaptable manufacturing systems*, in 2013 IEEE International Conference on Industrial Technology (ICIT 2013), pp. 1410–1415.

7. N. Keddis, G. Kainz, and A. Zoitl, *Capability-based planning and scheduling for adaptable manufacturing systems*, in 2014 IEEE Emerging Technology and Factory Automation (ETFA), pp. 1–8.

Digital reality – A revolution for planning and scheduling of the smart factory

Bernd-Dietmar Becker, D. Wohlfeld

FARO

This manuscript is not available according to publishing restriction. Thank you for your understanding.

Linked logistics concepts for future automobile manufacturing using innovative equipment

K.-H. Wehking, J. Popp

IFT, Universität Stuttgart

Summary

In this article we present the necessary changes in production logistics of automobile manufacturing for the upcoming years. Based on the higher amount of car models and the continuing growth of variations in the premium car segment, it has become difficult to produce vehicles economically using Henry Ford's idea conveyor-based techniques. While this method has advantages regarding the high amount of output, changes will be necessary in order to cope with high variant productions tending to lot size one. Therefore the research project ARENA2036 (Active Research Environment for the Next Generation of Automobiles) and specifically the members of the "research factory" is focuses on developing new concepts for assembly and logistics including technologies such as human-robot interaction (HRI), virtual security fences and transparent information.

This paper describes the recent research results while illustrating upcoming changes in production logistics for the automotive industry.

Introduction

Automotive production faces the biggest change in the last 80 years. Production using conveyor line was a revolution at its time because, based on the alikeness of the model variants, the mass production was the most effective way to produce cars [14, p. 191]. But over the course of several years the car manufacturers determined that customers are interested in being able to select certain specifications themselves. Examples are not only the exterior color or the seat material but also certain specific technical systems. This resulted in a more complex situation for the production, especially for assembly and logistics. While the assembly planning has to make sure that the workers grab and mount the correct materials (e.g. displaying the difference between two black control boxes for diesel and gasoline motors) the logistics has to ensure on time delivery for more and more different materials. One path that has been taken in many assembly plants was the introduction of just-in-time-delivery and just-in-sequence (sequenced) material supply. While just-in-time-delivery helps to bring down the amount of stored material by delivering parts exactly in the moment they are needed, just-in-sequence focuses on handling high numbers of variant parts by sorting and providing them to the assembly worker in the identical sequence in which the cars pass through production.

This results in a more complex situation for production logistics [13, p. 107]. With low number of variants, logistics service providers and employees could still perform the additional requirements economically. But in the last decade, the further increase in variants became more complex. Not only more and more options were requested from marketing specialists but also the amount of different car models surged [1]. While fifteen years ago premium automotive brands such as Mercedes-Benz were offering 15 different vehicle models (defining model as vehicle series e.g. Volkswagen Golf) this number grew in the last years to 28 different vehicle models (without commercial vehicles) [15]. Together with other options this results in a (theoretically) very high number of vehicle variants which an automobile manufacturer offers its customer. For example the last VW Golf was possible in 10^{23} variants [3]. Although there is a large difference between the theoretical number and the actual number of ordered variants only two of 1.1 million produced Mercedes A-Class vehicles during a 5 year period were identical [4, p. 38-42]. This implies that for certain assembly stations logistics need to provide different parts for the assembly of almost every vehicle. With the use of picking supermarkets, serviced most of the time by logistics service providers with lower wage averages than the OEM workers, the rising number of variants and car models can be managed. However, a further rise in the number of variants or models will imply an exponential jump in the logistics cost. To cope with this future development, the Institute of Mechanical Handling and Logistics (IFT) has partnered with the Ministry of Economics Baden-Württemberg in a project with the aim of developing innovative and economic logistics processes for future automotive production. Since we presented three new concepts at the last FKFS-symposium, many aspects of the project have been further developed. [2, p. 185-196]

Linked logistics

In a production system defined processes for combination and conversion of production factors takes place [10, p. 2]. It is important to realize that the main requirements for manufacturing systems are compatible hard- and software and this hardware is not only located inside the enterprise but throughout the supply chain, therefore also encompassing logistics [11, p. 2]. While some researchers foresee the elimination of logistics processes, it is clear for others that even with the use of additive manufacturing, the production material somehow has to be transported from the place of the resources to the place of production. In the view of the authors we will still have a strong dependence towards logistics processes as long as we have physical production systems. Because of the expected changes in automotive production we focus on improving the logistics processes for this sector.

During the last year we developed and later on simulated three completely new logistics concepts for the automotive industry [12]. To be able to realize the new concepts, a different approach for material demand had to be installed. While the consumption-driven concept of KANBAN has its advantages for mass manufacturing, it is not ideal for variant productions [5 p. 175f; 8]. In the concept of our new information process we have leading information. Thus the car, which is in the process of buildup, knows its

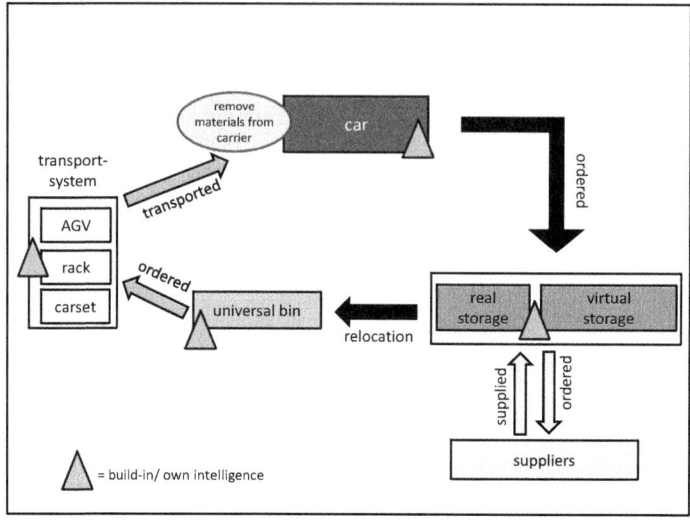

Figure 1: Information process

demand for all assembly stations it will run through. With a predetermined lead time the car orders its material directly from the responsible storage system. This storage can be real or virtual. This is due to the fact that some material is delivered just-in-time to the assembly stations and therefore gets stored in a warehouse managed by the automotive producer. After the storage receives the demand signal, the inventory check takes place. If the requested material is in the real storage, the universal bin is dispatched, while material from the virtual storage is dispatched directly from the supplier or logistics service provider. To keep the explanation simple, we assume that every universal bin carries only one part. After leaving the storage (for real storage materials) or arriving at the goods receipt zone (for virtual storage materials) the universal bin puts a transport request out to the "logistics market place" of this production building. The three logistics concepts AGV (automated guided vehicle), rack and carset apply with their individual costs to the tender. The most economic offer is selected from the universal bin and the transport to the assembly stations takes place. The paths for the AGVs and racks is planned using mixed navigation software [7, p. 1460]. After arrival at the assembly station, the universal bin does not leave the AGV, rack or carset frame because the assembly worker only removes the material from the transport box. Afterwards the universal bin is routed back to the supplier for further refilling. The described process is illustrated in Figure 1. The yellow triangles symbolize built-in or "own intelligence" and therefore enables the physical and computation processes, most of the time assisted by feedback loops with the goal of steering the physical process from the computation process and vice versa [6] [9, p. 1]. This logic is applied to our logistics equipment and while each part of the system directs its own orders, a central master control is dispensable. In this context the description "decentral decision making" describes best what is taking place. After defining a number of criteria's for the communication of the different intelligent systems, changes in the daily routine can be implemented much faster and with less expenditure. This is the case because the inner programming or changes of specifics for certain systems parts does not affect the other components in the logistics information process. This is the case as long as the communication protocols and work routine are continuously offered and fulfilled. One of the research questions that is worked on at the moment is the definition of this communication protocols as shown in Figure 2. The automated storage visualizes the virtual or real storage. While, in the view of the authors, there is a great deal of literature describing the difficulties and challenges in the automotive logistics, solutions and ideas for improvements have been of minimal effectiveness the last years. Therefore we focused not only on generating new ideas for logistics concepts but furthermore setting up the innovative equipment to put these new concepts into effect. The most promising logistics concept for variant productions, regarding an early economic point of view, is the rack concept which is described in detail below.

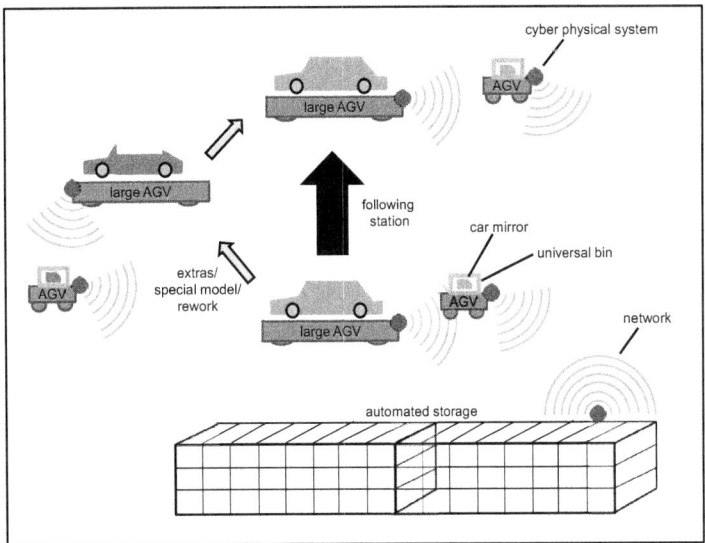

Figure 2: Communication among the logistics components

Innovative logistics equipment

During the last year of the project we developed and patented an innovative idea for a deployment design named the rack. The rack is a miniature automated storage system which accommodates at least 64 standardized universal bins (named KLT in the German automotive industry (e.g. size of 600 x 400 x 280 mm)). The rack is equipped with electrically steered, self-braking bottom rollers and features a lightweight and efficient design approach, as seen in Figure 3. By using an underdrive-AGV the rack can be moved anywhere inside the production building, requiring only level floor surface. Therefore it is very flexible regarding changes in the production layout or individual assembly station demands e.g. due to adaptions in the production program, sales fluctuations or production of special configurations.

Figure 3: Empty rack

At an assembly station, the miniature rack feeder is able to withdraw every universal bin from the rack (see Figure 4).

The general functionality follows a circle delivery schema. In the goods receiving area pallets stowed with universal bins are loaded from carriers' trucks directly into an automated feeding system (e.g. roll conveyors). Thereafter a handling system or a robot unloads the universal bins from the pallet and loads them directly into the rack. Empty pallets get stowed for the subsequent universal bin reverse logistics process. After the rack has been loaded completely, an underdrive-AGV transports the rack to the appropriate assembly station. From now on the miniature rack feeder is able to withdraw every universal bin from the rack (see Figure 4). It is important to note that two racks are always placed next to the assembly station. While both are accessible from the miniature rack feeder, one of the racks has been there first. In the following this is called the older rack. By contrast the rack which arrived later is referred to as younger rack. The principle for unloading is determined by asking the following questions: Is the required material available in the older rack? If the answer is yes, the universal bin is taken from this rack by the rack feeder. After offering the universal bin to the assembly worker and the extraction of the material, the empty bin is stored in the older rack. However when the required material is available only in the younger rack the miniature rack feeder withdraws the universal bin from there. After providing and reclaiming the material the empty universal bin is stored in the older rack. By using these procedures the older rack will, at a certain point of time, only be filled by empty universal bins. The older rack is transported back to the goods receiving zone by the underdrive-AGV. After arrival, the empty bins are unloaded from the rack and new filled bins are stowed. At this point the circle for delivery is complete. While the feasibility of the rack concept has been demonstrated through simulation by using real production data acquired from an OEM project partner, improvements will me made over the next couple of months. The first prototype of the rack is currently under construction, while completion is planned for May of 2016.

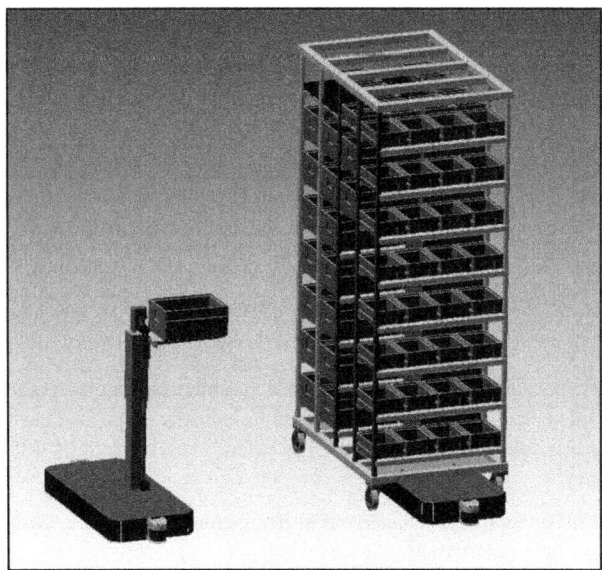

Figure 4: Rack concept: Miniature rack feeder and rack on AGVs

Conclusions

Changes in automotive assembly towards flexible and adaptable systems can only be successful if the right amount of thoughtfulness is brought to logistics processes. The path of the last years, to push expensive sequencing towards logistics service providers to be able to keep up with rising model and variant numbers has reached or will reach a final level soon. Companies that set up working alternatives to supply material to their assembly lines to cope with increasing amounts of different parts will achieve a strategic advance in their supply chain for the future.

Regarding logistics, this has implications on both the physical flow of goods as well as the flow of information. If vehicles order the required parts on their own, while they are being assembled, these orders must be fulfilled with the same accuracy as in today's centrally controlled supply chains. Therefore we presented concepts for future information and parts processes. The implementation can only be successful if new equipment is able to fulfill the altered requirements. Hence we continue to describe our material supply concept named rack in detail. This concept not only copes with further increase of variant parts but also reduces the amount of material handling steps. In preliminary studies we discovered that in today's logistics concepts for the automotive industry material is handled up to seven times between the supplier's factory and the assembly station in the factory of the OEM [12]. With the rack concept the number is reduced to four handling steps. This will reduce logistics cost and potentially enhance the flexibility in production logistics because the sequencing takes part only in the last step of the material handling. Further research should examine the innovative logistics concepts regarding their response time at the moment changes in the production program take place. Also the assumption that universal bins are loaded only with one part has to be extended towards loading uniform parts into one universal bin.

Bibliography

1. Schade, W., Zanker, C., Kühn, A., Kinkel, S., Hettesheimer, T., & Schmall, T. (September 2012). Zukunft der Automobilindustrie. Innovationsreport.

2. Krog, E.-H., & Statkevich, K. (2008). Kundenorientierung und Integrationsfunktion der Logistik in der Supply Chain der Automobilindustrie. In H. Baumgarten (Hrsg.), Das Beste der Logistik: Innovationen, Strategien, Umsetzungen (S. 185-196). Berlin: Springer.

3. Klug, F. (2010). Logistikmanagement in der Automobilindustrie: Grundlagen der Logistik im Automobilbau. Heidelberg: Springer.

4. Schlott, S. (2005). Wahnsinn mit Methode. Automobil-Produktion (1-2), S. 38-42.

5. Rahman, N. A. A., Sharif, S. M., & Esa, M. M. (2013). Lean manufacturing case study with Kanban system implementation. Procedia Economics and Finance, 7, 174-180.

6. Nof, S. Y., Morel, G., Monostori, L., Molina, A., & Filip, F. (2006). From plant and logistics control to multi-enterprise collaboration. Annual Reviews in Control, 30(1), 55-68.

7. Garibotto, G., Masciangelo, S., Bassino, P., Coelho, C., Pavan, A., & Marson, M. (1998, May). Industrial exploitation of computer vision in logistic automation: autonomous control of an intelligent forklift truck. In Robotics and Automation, 1998. Proceedings. 1998 IEEE International Conference on (Vol. 2, pp. 1459-1464). IEEE.

8. Surendra, M. G., Yousef, A. Y., & Ronal, F. P. (1999). Flexible kanban system. International Journal of Operations and Production Management, 19(10), 1065-1093.

9. Lee, E. A. (2008, May). Cyber physical systems: Design challenges. In Object Oriented Real-Time Distributed Computing (ISORC), 2008 11th IEEE International Symposium on (pp. 363-369). IEEE.

10. Zäpfel, G. (2000). *Strategisches Produktions-Management*. Walter de Gruyter GmbH & Co KG.

11. Chituc, C. M., & Restivo, F. J. (2009, June). Challenges and trends in distributed manufacturing systems: are wise engineering systems the ultimate answer. In *Second International Symposium on Engineering Systems MIT, Cambridge, Massachusetts*.

12. Wehking, K. H., & Popp, J. (2015). Changes in production logistics for automobile manufacturing. In 15. Internationales Stuttgarter Symposium (pp. 1193-1200). Springer Fachmedien Wiesbaden.

13. Boysen, N., Emde, S., Hoeck, M., & Kauderer, M. (2015). Part logistics in the automotive industry: Decision problems, literature review and research agenda. European Journal of Operational Research, 242(1), 107-120.

14. Masayuki, M. (2002). The end of the 'mass production system' and changes in work practices. Japanese Business Management: Restructuring for Low Growth and Globalisation, 181.

15. Mercedes-Benz (2015), Modellübersicht http://www.mercedes-benz.de/content/germany/mpc/mpc_germany_website/de/home_mpc/passengercars/home/new_cars/model_overview.html#_int_passengercars:home:model-navi:model_overview (last checked 29.01.2016)

Analytical challenge on Real Driving Emission

Hiroshi Nakamura, HORIBA Ltd.

This manuscript is not available according to publishing restriction.
Thank you for your understanding.

Introducing a method to evaluate RDE demands at the engine test bench

Dipl.- Ing. Jan Gerstenberg, Dipl. Ing. Helmut Hartlief, Dr. Dipl.-Ing. Stephan Tafel

Bosch Engineering GmbH

1 Introduction

Why RDE? The wording real driving emissions was created due to the growing discrepancies of vehicle emissions measured by the laboratory tests NEDC (New European driving cycle) and emission results from "real driving" on the road. In these NEDC tests on a roller test bench are no uphill driving, slopes, additional accessories like climatic control, seat or window heating included. This was neither regulated by the EU nor tested at a certification and let to discrepancies between test and real live emissions. Next to that, the average acceleration at the NEDC test procedure is calculated with 0.324 m/s² and a maximum speed of 120 kph is tested, which is not representative for normal driving in countries with higher speed limits.

Due to these circumstances the NO_X emissions and CO_2 measures at real live exceed currently the limits measured at the NEDC on the roller test bench.

2 RDE Test procedure

The RDE procedure today is not finally fixed. In January 2016, the European Parliament evaluated the proposal that emerged from the October meeting of the European Commission's Technical Committee on Motor Vehicles (TCMV) /1.

Starting with the Monitoring phase in 2016 technical services will test vehicles at real live conditions on the road using a portable measure system, PEMS. There are several boundary conditions for this test like ambient temperatures and the split of the test into urban, extra urban and motorway driving sections.

Even if the NEDC cycle consisting of the urban part UDC introduced by ECE R83 in 1970 and the extra urban part EUDC introduced by ECE R101 in 1990, is not representative for real driving any more, it gave a possibility to compare precisely the fuel consumption and emissions of different vehicles due to its high reproducibility on the roller test bench.

3 No reproducibility at RDE

There is no unique measure result of emissions and fuel consumption of real driving test. Both relays to driving behavior, ambient conditions and traffic situation /2.

Basically a specific RDE test processed by a technical service shows just a single chain of events based on ambient and engine temperatures, driving situations and road- and traffic circumstances. Due to this situation it will probably not be possible for an OEM (original equipment manufacturer) to test all possible driving events to validate every planned variant of engine and chassis combination. And even more severe, that there is no by law given driving- and track condition for the emission development. Economically it is a large challenge to cover all important driving events in a realistic time schedule.

4 Driving Events

The OEMs and Engineering service provider could tend to define its nearby road track to proof and develop its engines- and after treatment systems and so define their RDE-track. As a real driving test cannot be reproducible realized, it is not capable for development, but must be done for a final evaluation. Engineers need to see and qualify every single change in emissions due to software calibration changes and hardware adjustments and need a reproducibility of test situations (events).

This leads to the strategy to realize RDE tests on the roller- or engine test bench to create a reproducible engineering area. Or if there was no preparation to RDE challenges before, run a large number of test vehicles every single day to find out which variants are not performing well and engage a "firefighting" short before SOP (start of production) to improve the system on time.

5 Frontloading RDE in the development phase

Having a glance on an engine and vehicle emission development schedule, it will be general agreed, that a negative emission result from a variant of a vehicle model is not capable of meeting the RDE emission limits, cannot be accepted due to costs and time doing rework on soft- or hardware (Figure 1).

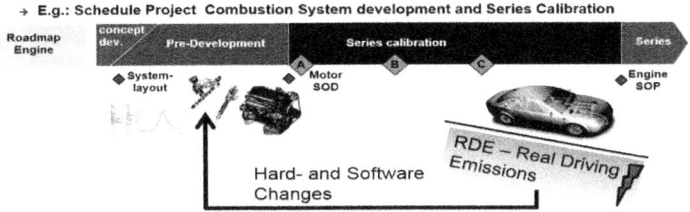

Figure 1: Iterations due to RDE results

Having this situation acknowledged, the strategy of frontloading must be realized to prove in an early stage of development the capability of the chosen hardware to fulfil the RDE challenges.

Of course in an early phase soft- and hardware are not finally fixed and at often no test vehicle is produced so far, but still first RDE tests on the engine test bench can be realized by using a predecessor software, hardware and information of the planned chassis. First checks parallel to the basic emission calibration could be done and severe problems can be fixed/discussed in a very early phase (front loading). The alternative way to wait for the first RDE test vehicle later on in the development phase cannot be recommended.

All problems which are realized just at the end of the development due to a real driving test are too late identified in the development process, since fundamental decisions have already been taken /4.

Late chances are always difficult to implement and rework can be predicted. In some cases it could lead to a delay in SOP, because rework in summer or winter tests can be necessary.

6 RDE Engineering at the engine test bench

To support our customers Bosch Engineering decided in 2014 to develop an engineering area which offers reproducible RDE tests at the engine test bench to handle variants regarding RDE development with a minimum of test vehicle usage.

This tool chain is based on a rapid engine exchange environment and a simulation of chassis variants and gearboxes, hybrids etc. in the loop with the real engine. Different simulated drivers will run the engines on gps-based road simulated tracks (Stuttgart RDE cycle, any free chosen track, ...) and the test bench will deliver the emission results (Figure 2).

Figure 2: Engine-in-the-loop-test-bench

The aim of this tool chain is not to reproduce a single RDE PEMS test from the real vehicle, but to visualize the emission range between different driving styles, traffic situations and road tracks. This range gives a clear answer regarding the robustness of the emissions of an engine combined with a simulated gearbox and chassis /5.

Additionally the fully equipped measurement systems of the test-bench including 24 hours testing with a rapid cooling system after the tests leads to a high efficiency in the preparing phase for RDE.

In any case, having a vehicle variant checked through such a tool chain, the final system for the first real PEMS evaluation in a real vehicle will be more successful and "emission failures" right before SOP can be reduced.

7 Preparing the RDE Engineering toolchain

7.1 Vehicle model

To start with the tool chain there are two possible scenarios. There is one vehicle variant for some days available to do a basic analysis of weight, cost-down, gearbox characteristics etc. or there are only data about the future vehicle available. In any case there is a good prediction about the robustness of the emissions possible, because it is not needed to meet some ppm like it is on a NEDC test. Target is to evaluate if the emission limit is exceeded heavily due to for example temperature drops at the aftertreatment system at low driving speeds or EGR (emission gas recirculation) mass missing at high loads. Nevertheless the high reproducibility gives a precise emission information about changes.

If there is a vehicle variant available, even if it is a predecessor the model can be improved by getting an idea of the real behavior of the chassis and gear-box. In that case an evaluation is done on the Bosch proving ground in Boxberg. Afterwards a comparing-test can be started at the test bench using the Boxberg gps-based track and the vehicle model. The model gets modified until the ECU (Electronic control unit) shows the same behavior like in the real vehicle. The simulated vehicle must accelerate, coast-down and show equal torques like the real vehicle on the proving ground.

The real vehicle can be compared to the model at engine test-bench with

a) ECU calculated torque at test track,

b) cumulated NO_X emissions,

c) exhaust gas temperature before catalyst (SCR).

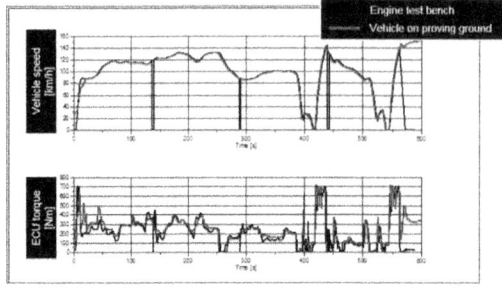 

Figure 3: Engine-in-the-loop-test-bench- Boxberg

As soon as the ECU behavior of the real test and the test-bench are comparable, the tool-chain is ready to meet all kinds of road-tracks and drivers. At Figure 3 there is a measurement of speed and ECU-torque of a real vehicle (red) and the test-bench system (black). The automatic transmission behavior has been realized in the AVL InMotion / IPG CarMaker environment.

In general any track can be copied into the system. In fact the quality of the altitude profile must be very precise in order to meet the same engine torque locally than the real vehicle.

Having the customers track measured once, it is saved into a host system and is available for automatic test-runs.

7.2 Comparison model- and vehicle behavior at the customers RDE track

To ensure that the track data and the ECU are successfully prepared with information from the driver (brake-, clutch, speed-, gear-, steering-signal, etc.), a special test is introduced. The calibrated model runs with the speed demand values from the real PEMS test at engine test-bench to evaluate the system behavior with the RDE track. At Figure 4 there are speed, ECU torque and cumulated fuel consumption shown in a driving distance-based diagram to demonstrate the accuracy of the combination of model and RDE track at test-bench.

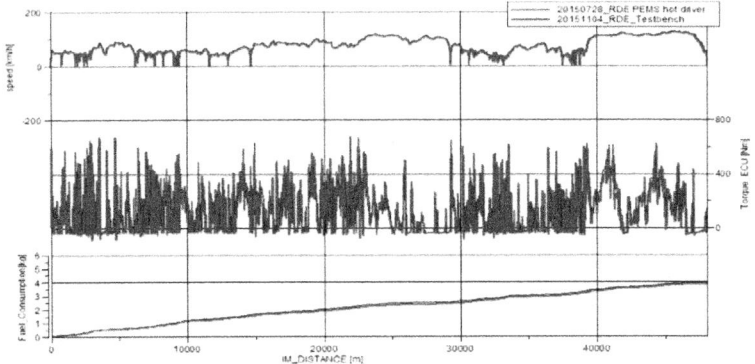

Figure 4: Fuel consumption: Real RDE test and test bench

The following Figure 5 shows vehicle speed, NO_X tailpipe and the cumulated NO_X emissions at a part of a RDE Test. The ECU measurement at the real test on road (red) and the engine controlled by the simulated driver (blue) seem to match very good regarding NO_X.

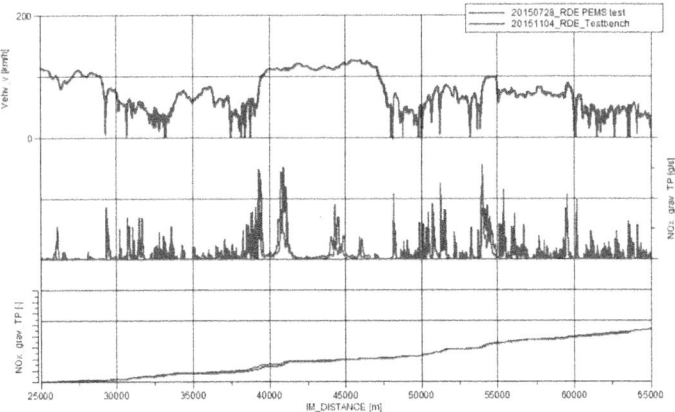

Figure 5: NO_X cumulated NO_X: Real RDE test and test bench

To be sure that the aftertreatment system at the engine test-bench are in a thermic condition like it has been on the real road, it might be necessary to realize a controlled temperature conditioning for the exhaust pipe.

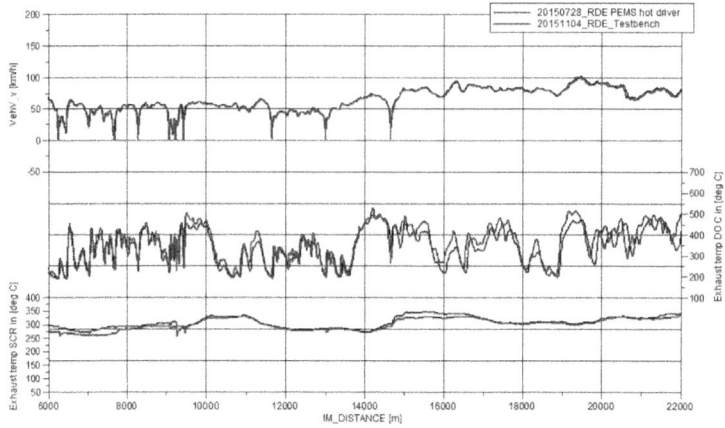

Figure 6: Exhaust gas temperatures: Real RDE test and test bench

Figure 6 shows speed, exhaust temperature before DOC (Diesel Oxidation Catalyst) and the temperature before SCR catalyst in a driving distance-based diagram. The vehicles temperatures (red) and the engine test-bench temperatures (blue) are suitable for further investigations. In the following the simulation is capable to produce measurements close to the real vehicle.

7.3 Choose variations for the vehicle variants testing

Preparing the 24 hours automatic test, the following combination can be done.

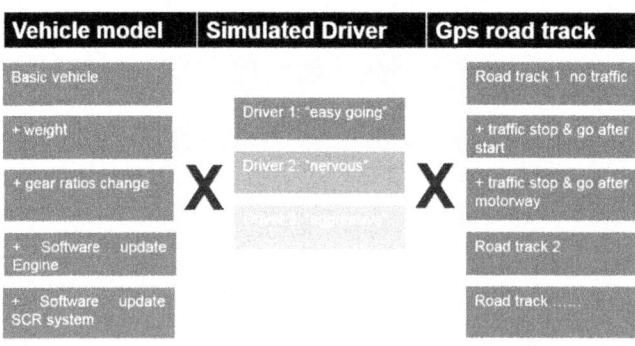

Figure 7: RDE Matrix for overnight tests

The vehicle model offers the possibility to vary weight, gear ratios and shifting strategies as well as gearbox types and chassis variants. It can be combined with a nervous, aggressive or normal driver. These combinations are tested with different tracks with or without traffic simulation (Figure 7).

Having this results of different drivers and the simulated vehicle on the chosen gps-track with the real engine and after treatment system a clear information about emission behavior in multiple driving events will be displayed.

Not only the special event which led to higher emissions is relevant, but also the thermodynamic history. For the analysis it is important to include the history before a special event regarding temperatures of the combustion chamber and catalyst system to draw the right conclusions /3.

8 Potential RDE test-bench evaluations

8.1 Reproducibility at test-bench

In figure 8 the reproducibility of the engine test-bench is demonstrated. The same simulated driver has been used to run two RDE tests after each other. The emissions

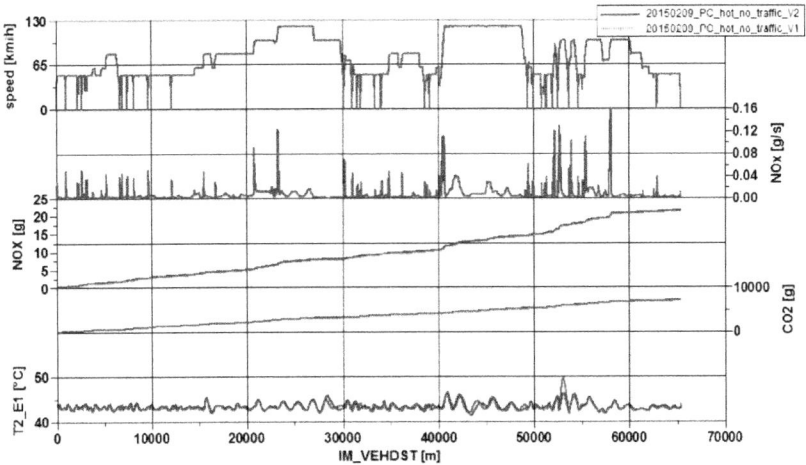

Figure 8: Reproducibility at engine test-bench

NO_X modal and cumulated as well as CO_2 and the air-intake temperature [T2_E1] provide a very good reproducibility. This offers an exact evaluation of possible hard- and software-changes, which cannot be realized on the real road nor with a real driver.

8.2 Influences of driving behavior

The behavior of the driver must be divided into two different factors. First factor is described by the driving dynamics; referring the acceleration of the vehicle. It is challenging for the emission relevant engine map.

The combination of an aggressive driver and a RDE Track with uphill accelerations lead to a relative high engine load like it is shown at Figure 9.

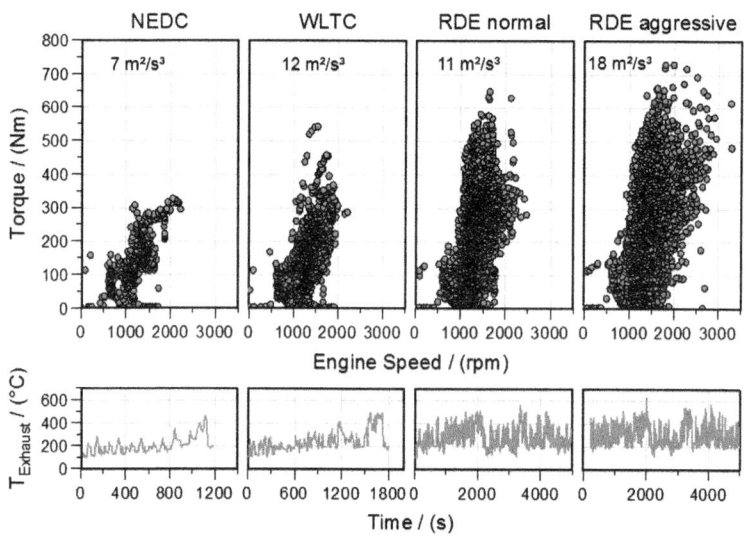

Figure 9: Emission Tests SUV

Second there is a parameter description of a nervous driver. A nervous driver can produce severe more engine-out NO_X emissions with only a slight influence on the CO_2 emissions (SUV example in Figure 10). Here the nervous driver (red) produces large NO_X peaks without a CO_2 penalty compared to the unsteady (blue) and normal driver (green). The evaluation shows, that a nervous accelerator-pedal behavior could be one of the worst case scenarios which has to be evaluated.

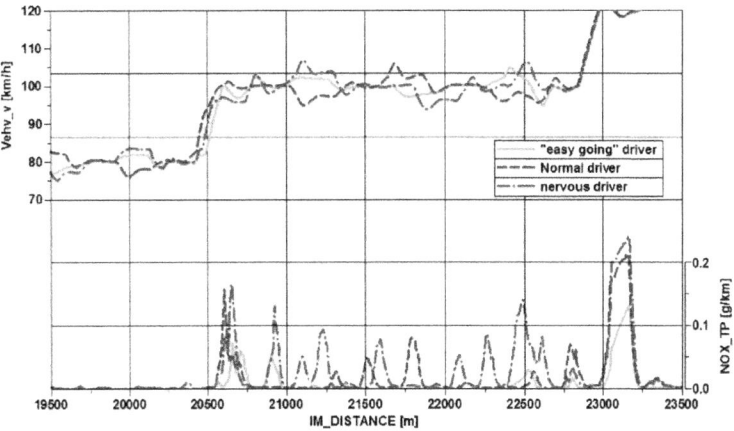

Figure 10: Nervous driving style

8.3 Emission robustness check at the engine test bench / driver behavior

To evaluate the emission robustness of an engine with SCR after treatment two RDE tests are prepared. An aggressive and normal driver are running the simulated vehicle with the engine- in- the- loop-system.

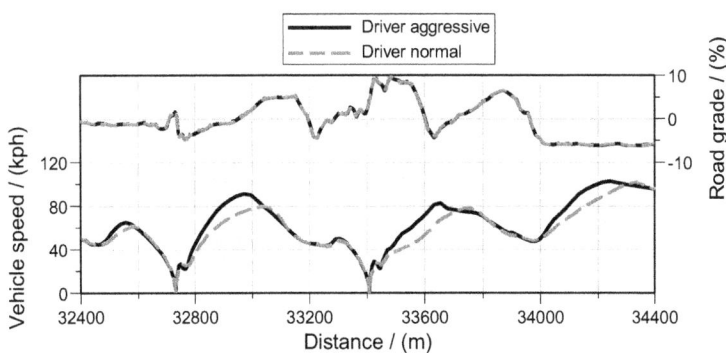

Figure 11: Driver behavior at RDE Test

Figure 11 shows the driving behavior at three accelerations.

The aggressive driver requires more engine torque for its acceleration. The first acceleration runs in the plane, second uphill and third downhill. This event evaluates the emission behavior between a normal and aggressive driver at uphill and downhill. Due to the fact that the driver is calibrated with a special acceleration-value, every vehicle can be accelerated the same way to give an opportunity to compare.

Figure 12 shows that the aggressive driver (black) leads to 2-3 times higher NO_X emissions at plane and downhill areas and three times higher at uphill than the normal (blue) driver at this engine.

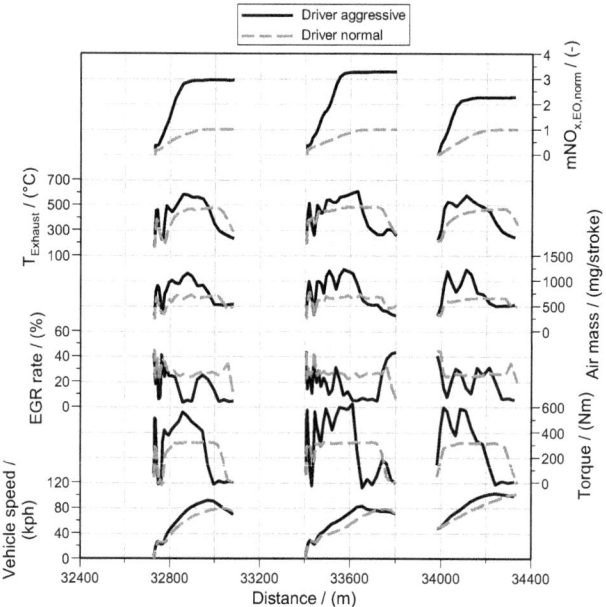

Figure 12: Comparison of accelerations

The air mass respectively the EGR-rate of the aggressive driver shows a necessary optimization of the air system. The EGR-rate is reduced to deliver the air-mass needed for the engine torque. The complete investigation shows that the SCR system is not capable of reducing the high engine-out peaks sufficiently.

8.4 Emission robustness check at test bench / traffic

Even if the RDE legislation askes for a speed based equal distribution of the RDE track into urban (v<60kph), extra urban (60kph<v<90kph) and motorway (90kph<v<145kph) sections, the engines must be capable of running through a long traffic without exceeding the emission limits. As we are evaluating worst case events, a large traffic through Stuttgart is tested even if it is not a representative RDE track.

For this evaluation the start-stop function has been deactivated to prove the system (engine + aftertreatment) in general for traffic before engaging the start-stop-strategies, which will be an extra evaluation.

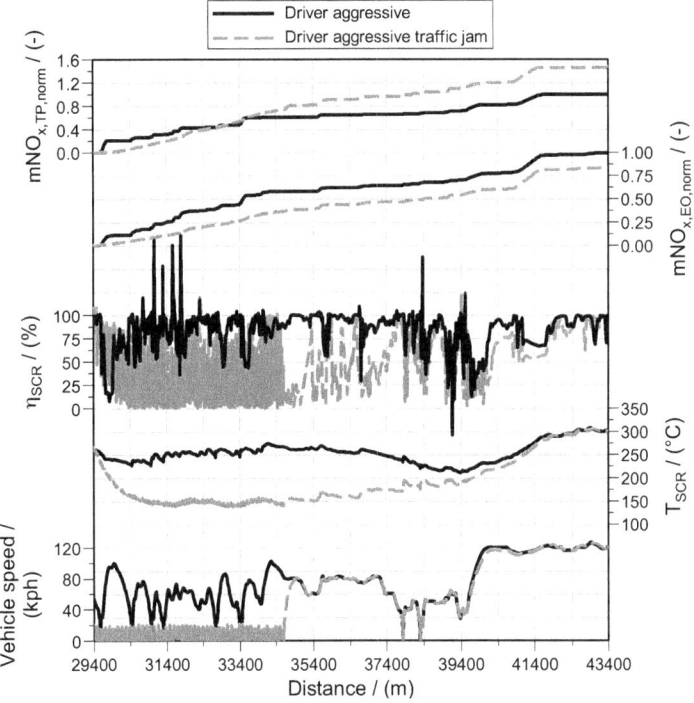

Figure 13: Traffic investigation

The tool-chain was setup for two RDE tests; one with traffic and one without.

In Figure 13 the RDE test with traffic (blue) is parametrized with a traffic driving behavior like it is defined in the first 400 seconds of the urban part of the NEDC. To analyze a systems performance, a worst case scenarios with 5 Kilometers traffic (UDC-style) is set up.

The value Tscr gives an idea about the temperature of the SCR catalyst cooling down to 150°C after two kilometers. The performance of the SCR drops and mNO_X-TP (normalized tailpipe NO_X) of the traffic test has a higher positive gradient than the test without traffic. After the traffic the test with traffic emitted 1.5 times higher NO_X emissions.

For these situations the system is not robust and need a SCR heating strategy.

Summary

A diesel engine with SCR has been set-up on the engine test-bench and connected to a vehicle model, gps-based RDE track and parametrized drivers. This engine-in-the-loop system was used to measure the robustness of the emissions of an engine at a simulated RDE environment.

In this case an automatic test with an aggressive and normal driver as well as simulated traffic situation was realized.

The analysis has shown that a modification of the air system and a heating strategy for the SCR system might be necessary for the first step.

This tool chain enables to run many RDE tests automatically at the engine test-bench. Next to that the high reproducibility gives precise answers to the effects of soft- or hardware changes on emissions.

Offering a possibility to run a "front-loading" at the calibration development process and reducing the amount of testing vehicles, not only the costs can be reduced, but fire-fighting direct before SOP can be avoided.

Bibliography

1. Vicente Franco, Peter Mock, ICCT, "The European Real-Driving Emissions Regulation", 28.12.2015 press release

2. VDA Press Release, „Wissmann fordert klare Messbedingungen für RDE-Straßentests", 2015

3. Dr. F. Wirbeleit; H. Hartlief; J. Gerstenberg, Bosch Engineering GmbH „Realisierung von Hochdynamik auf einem Motorprüfstand, VPC+ Conference Hanau", 2014

4. H. Maschmeyer, D. Schmidt, Prof. C. Beidl, TU Darmstadt, "Simulation and Test Methodology fort the development of Power Train Systems under RDE Boundary conditions, International Symposium on Development Methodology", Wiesbaden 2015

5. Dr. S. Tafel, H. Hartlief, J. Gerstenberg, „RDE Entwicklungsumgebung am hochdynamischen Motorprüfstand", ATZ 09.2015

Model supported calibration process for future RDE requirements

M. Steinbach, D. Neumann, T. Kutzner, A. Lehmann, V. Kassem, M. Dreiser

IAV GmbH Berlin

1 Introduction and motivation

The new test procedure [7], which is planned to be introduced for the new legislation for passenger cars in September 2017, has significant impact on future calibration approaches. Especially the measurement of exhaust emissions on the road leads to a major increase of boundary, ambient and testing conditions to be considered.

The diversity of driving situations, driver's behavior, traffic and driving routes have a big influence on the emission results. Figure 01 shows the influence of an exemplary driver on NO_x tailpipe emissions in the same vehicle on the same route for 6 different RDE tests with PEMS measurement equipment. All tests are valid due to the proposals for the legislation [7] and were carried out with a robust, series intended calibration. The experimental reproduction of emission results or the quantification of calibration steps on emission results is hardly possible.

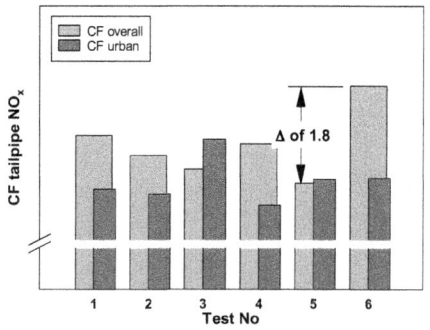
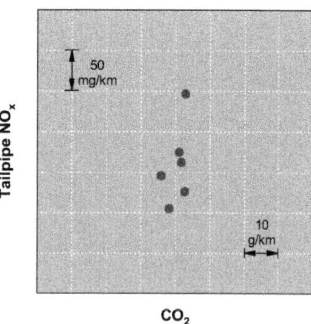

Constant
- Vehicle
- Calibration
- Route
- Driver
- Driving style (subjective)
- Ambient conditions

Unavoidable spread due to
- Traffic
- Traffic routing (traffic lights)
- Driver performance

Figure 01: Scatter band of RDE-testing

A second stimulus for innovations in the development methodology is surely to be seen in the further increasing number of vehicle variants in the future. Considering the limited time and resources available for the determination of all relevant data as well as the restrictions in the availability of prototype test vehicles, a pure experimental-based calibration methodology is insufficient.

Very high requirements regarding the calibration maturity and robustness as well as a high number of vehicle variants have to be met demanding an integrated model-supported approach in powertrain calibration.

This article will show the model components and their validation, followed by the combination of these models to a toolchain. With this toolchain the combination of virtual and experimental calibration methods is presented. An example of the application of this toolchain in a model-supported engine and exhaust aftertreatment calibration for RDE concludes this article.

2 Model components and their validation

2.1 Engine control unit model and dosing control unit model

The models of the engine control unit (ECU) and the dosing control unit (DCU) of the emission aftertreatment system (EAT) are basically Simulink® models of the control unit functions. They are calibrated by inputs (datasets) of calibration tools like INCA®.

2.2 Engine model

The emission core of the engine model is created using a dynamic DoE-based data survey, thus enabling mapping of the entire engine operating range. Also, different operating modes of the engine can be measured as part of the survey. The DoE tests used are designed of dynamic excitation sequences: sinusoidal or APRBS (amplitude modulated pseudo random binary signal)[6]. Figure 02 shows the excitation sequences and the model in- and outputs.

The benefits of dynamic compared to steady-state measurements are the significantly reduction of the measuring time and the representation of the dynamic behavior closer to reality. A further advantage is that measuring points and measuring results are not averaged over a time interval. In this way, each measurement point has full influence on the modeled result. The inputs of the engine model must be defined according to the variables to be configured in the calibration process (see figure 02).

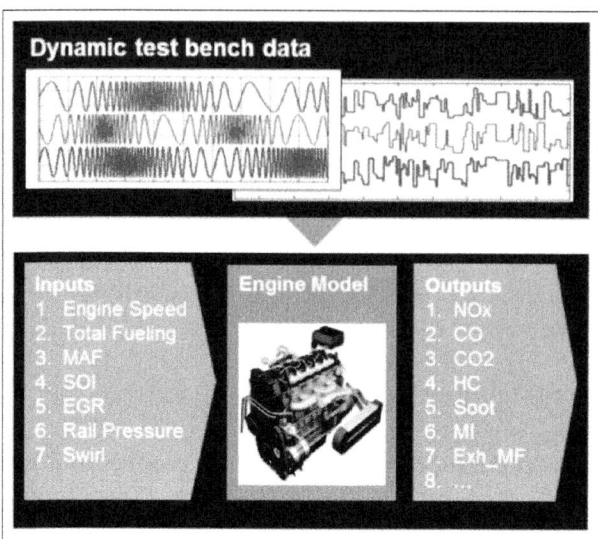

Figure 02: Excitation sequences and engine model in- and outputs

The models are trained with the help of data recorded during the measurement. Thus, the engine model is following a purely data-driven approach (Parametric Volterra Series, NARX model, ...). Resulting models are able to capture internal engine dynamics (e.g. air path, temperatures, ...) and can predict transient engine behavior [6]. Additionally it is possible to link an air path model which is created using physical approaches. Nevertheless, the dynamic model can still be used for steady state prediction. The emission models are mainly trained to predict NO_x- and CO_2-emissions due to the fact that the main effort in calibration has to be invested to limit these exhaust components. For NO_x- and CO_2-emissions the models are highly accurate. For HC, CO and soot emissions, the accuracy still needs to be evaluated. In any case, qualitative statements concerning soot emissions, as well as a good correlation to test bench measurement equipment are within reach.

Figure 03 shows the very good correspondence of a measurement and a simulation result of the NO_x-raw emissions (engine-out NO_x) for an RDE route. The upper diagram plots the vehicle speed for this RDE test.

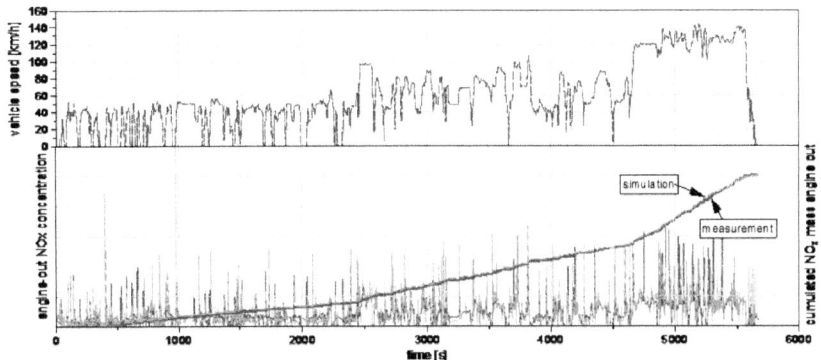

Figure 03: Validation of the engine model in a RDE route

2.3 Exhaust aftertreatment modelling

The exhaust aftertreatment (EAT) model is created by using the commercial software Axisuite® and is based on physical and chemical modelling of the EAT system [2,3,4]. Therefore, the aftertreatment components are specifically tested on synthetic gas test bench (SCAT) to collect experimental data for the calibration of the reaction scheme and its corresponding parameters. Finally, the calibrated models are validated with transient data from engine (ETB) or roller test bench (RTB). By this approach accurate, robust catalyst and EAT system models are created for all important emission components, also taking into account secondary emissions such as NH_3 which are valid in the complete RDE range. Figure 04 shows the comparison between measurement and simulation of tailpipe NO_x in a WLTC for an SCR only EAT system.

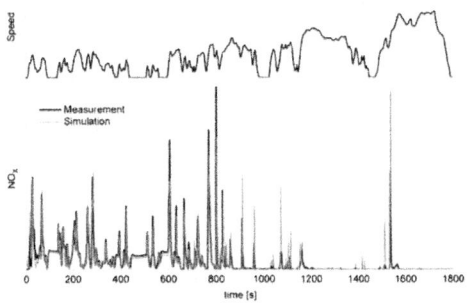

Figure 04: Validation of the exhaust aftertreatment model in WLTC

569

2.4 Longitudinal dynamics model

The longitudinal dynamics of the vehicle are described by a simulation model that can be set up from vehicle data. For example, coast down parameters from the roller-dynamometer, the vehicle weight, gear ratios and final drive (see figure 05).

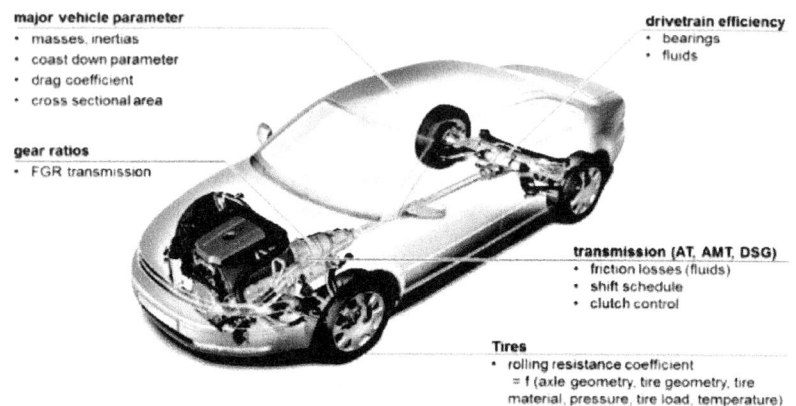

major vehicle parameter
* masses, inertias
* coast down parameter
* drag coefficient
* cross sectional area

gear ratios
* FGR transmission

drivetrain efficiency
* bearings
* fluids

transmission (AT, AMT, DSG)
* friction losses (fluids)
* shift schedule
* clutch control

Tires
* rolling resistance coefficient
  = f (axle geometry, tire geometry, tire material, pressure, tire load, temperature)

Figure 05: Parameters of the longitudinal dynamics model

This model includes a driver model which is basically a PID-controller based on the simulated vehicle speed and the target speed. Different driving styles (e.g. soft, normal, aggressive) can be realized by parameter setting (see also chapter 4).

3 Model based calibration toolchain

The adequate toolchain presented here consists of several model components that can be used as modules. The model combinations are used in different constellations in different stages of the calibration development. Figure 06 shows the complete toolchain with the in- and outputs for the combination of the models.

Figure 06: Simulation toolchain

The platform VeLoDyn [1] is the necessary tool to integrate the individual parts of a model. With this platform the networking of the models is considerably simplified by means of a customized bus system and the parameterization of the overall model.

3.1 Virtual engine test bench

The combination of the models for the engine and exhaust aftertreatment can be called a virtual test bench as illustrated in figure 07. This simulation environment can also be used as such with the help of automation tools; so engine maps, trade-offs and other dynamic tests can be simulated like in the experimental test bench operation.

The resulting opportunities are versatile - especially considering this simulation tool calculates faster than real time - and thus can bring advantages for the calibration process and significant savings of time.

Figure 07: Virtual engine test bench

3.2 Virtual vehicle

The overall model, consisting of the engine, exhaust aftertreatment and longitudinal dynamics model, can be called a virtual vehicle (see figure 08). This model makes it possible to simulate various variants of a vehicle by different parameterizations of the longitudinal dynamics model.

Figure 08: Virtual vehicle model integrated in VeLoDyn interface

Inputs of the model are environmental conditions such as temperature and the speed-time profile of the simulated cycle. So it becomes possible to simulate critical cycles for all variants of a vehicle, including vehicles that are physically not available. In addition, a wide range of real-road measurements can be used to validate the results. These measurements (speed-time profiles) can be repeated and optimized in a reproducible way. By this model-based approach less prototype vehicles are required.

3.3 MiL-Desk (Model in the Loop on your Desktop)

MiL-Desk is IAV's tool to compile models into an executable file. The compiled model is both real time, as well as acceleration capable and therefore able to shorten the processing times.

The standard calibration tool INCA® from ETAS can be connected directly to the MiL-Desk executable. Consequently, the environment of the model ensures its operability and easy implementation with the help of well-known calibration tools. The calibration work can be carried out effectively applying the usual software with existing tools and evaluation scripts.

4 Combination of virtual and experimental calibration methods

By combining the individual models, support can be generated for every step of the calibration process. The detailed configurations of simulation as well as the experimental effort necessary are variable for every step depending on several influences. The first factor to explore the individual share of virtual and experimental work might be determined by the availability of physical components (engine, vehicle, …) in comparison to such of adequate models. Additionally the decision might depend on available ambient conditions, number of planned variants of vehicles or required robustness of the calibration. Figure 09 shows the calibration process with variable combinations of virtual and experimental tasks.

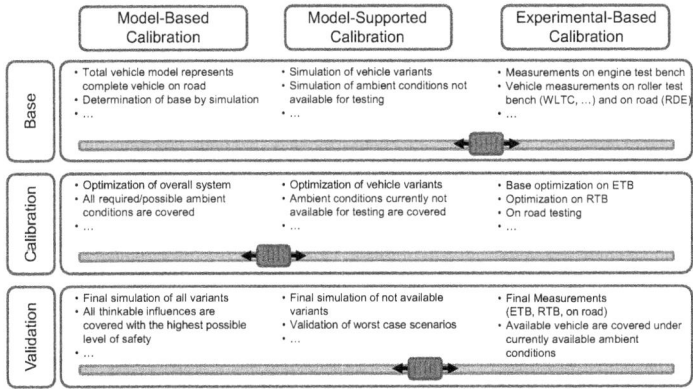

Figure 09: General calibration process with variable combination of virtual and experimental activities

Whereas all these decisions to determine the perfect process for every calibration project had already to be made in the past in a comparable way, the additional requirements of RDE legislation intensify the needs of virtual calibration methods.

As one of the first steps the determination of most critical variants of a vehicle is carried out [5]. This task is possible by simulating the load point shifting by various vehicle-generating variants. Thus, the effort of experimental measurements, which would be necessary to determine the critical variants of a vehicle throughout an experimental-based calibration process only, can be reduced significantly. The overall model of the vehicle also offers the opportunity to consider planned variants of a vehicle even before the first prototype samples have been built.

To determine the calibration of a variant of a vehicle, both experimental and simulative steps are necessary. However, due to the different models, parts of the calibration process can be well supported. Results for trade-offs and engine maps can be generated on the virtual test bench in shorter time duration than using real test bench measurements. Furthermore, by using the vehicle model the datasets can be tested in many different cycles. Here, also the great advantage appears that measurements can be reproducibly analyzed and optimized with different calibrations, which is not possible by on-road measurements only. In addition, datasets that were created for the respective vehicle variant can be transferred to other variants and be tested by using the full model of the vehicle.

Finally, in order to increase the robustness of the calibration to an adequate quality the usage of a virtual vehicle model becomes essential. A wide range of ambient conditions combined with almost unlimited compositions of driving patterns leads to a nearly countless amount of test situations to be covered. The combination of a full vehicle model with automated cycle variation and generation allows to define a worst case cycle for a single vehicle variant with an actual state of calibration. Each change in calibration might lead to a change in this cycle which can be easily covered by recreation of the worst case cycle.

The configuration of a driver controller is a very important task since it has a huge impact on the entire model chain. This variance allows simulating different driving behaviors on the same RDE route by using different driver controller settings. Thus, the expected range of emissions of a vehicle variant on defined RDE routes under various ambient conditions can be determined. Figure 10 shows the variance of three different driver controller parameterizations in a WLTC.

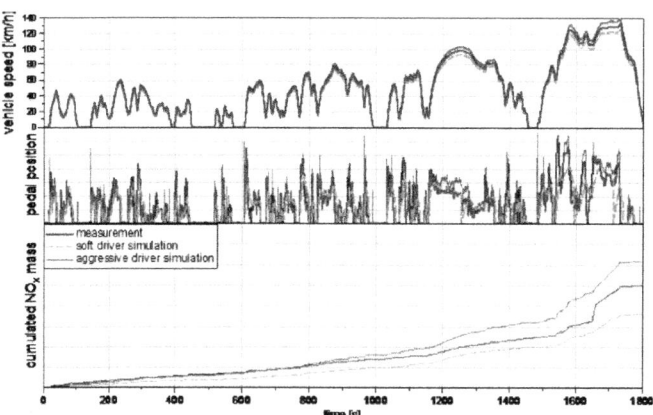

Figure 10: Different driver parameterization for WLTC

Overall, the measuring effort can be significantly reduced by the use of highly accurate simulation models. Using a model based approach also offers the possibility to implement targeted optimization by using partial algorithms for maps and map areas.

5 Model-supported engine and exhaust aftertreatment calibration for RDE

To test the model-supported calibration, a study on a sample vehicle equipped with an SCR aftertreatment system was conducted. The interactions of experimental calibration methods and new options by the simulation models have been tested. The procedure for this study was selected as shown in figure 11.

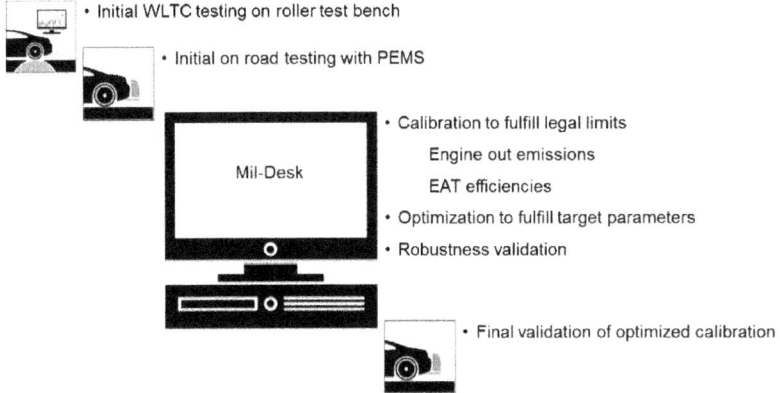

Figure 11: Procedure for RDE case study

Starting from a base survey in WLTC on the dynamometer and several RDE-cycles on the road, the calibration of the ECU and the dosing control unit DCU was adapted and optimized for engine and exhaust aftertreatment.

The loops of optimization for the calibration of ECU and DCU were performed by simulation only and initially processed separately for each model (engine and exhaust aftertreatment). To gain information about the possible range of results the calibration work in this study was done in two separate groups. The goal of every group was to fulfill the expected legal limits for valid RDE-cycles under consideration of defined limits for DEF (Diesel Emission Fluid, e.g. AdBlue®) consumption and CO_2-emission. This led to one calibration of the ECU for minimal engine-out NO_x and to a second calibration of the DCU for maximum EAT system performance as shown in figure 12. This figure also represents the conflict of targets between DEF-consumption and CO_2-emission. The strategy for the final calibration needs to be defined depending on market situations, customer requirements and vehicle characteristics.

After completing these steps, the results were compared. Both calibration groups joined together to generate a calibration output with a compromise of DEF consumption and CO_2-emission under consideration of fulfilling the NO_x legal limit. Figure 12 shows the different calibration steps with equal tailpipe NO_x emissions and a final result which could be named a balanced compromise. Every change of the engine calibration also has side effects on other parameters that have to be taken into account during the complete process, such as noise, soot mass and drivability. For the shown case study these side effects were not considered.

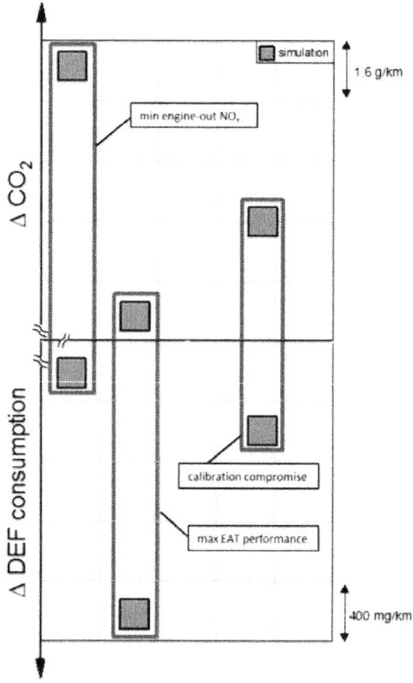

Figure 12: Simulation results for different weighting of limiting values

The calibration compromise was used to perform validation measurements on the road. Therefore, the influence of different drivers, routes and traffic conditions had to be considered. As visible in figure 13 the measurement results could not match the calibration compromise simulation exactly due to the above mentioned influences, but are within all given targets. The key aspect is that the results proved that the output of the simulation process correlates to the final results of final evaluation on the road.

To verify the quality of the simulation model the experimental measured speed-time profiles were simulated with very good correlations to these measurement results. An example of the agreement between measurement and simulation is shown in figure 13.

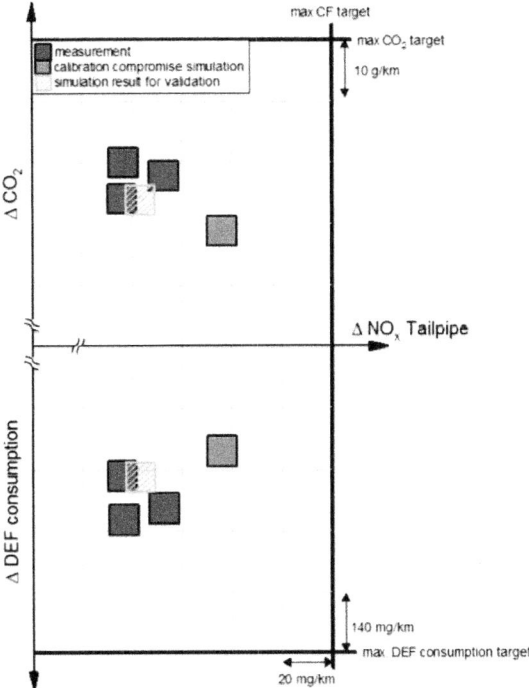

Figure 13: Validation of simulation results of final calibration

The model-supported calibration of diesel engines and aftertreatment systems with its potential of time and resources saving abilities is a sensible method and can be used for offline calibration without major deviations from the real measurement. During calibration work the reproducibility of RDE testing was one of the main advantages of model supported calibration that allowed determining influences of calibration changes without interference due to changed boundary conditions. The possibility to simulate RDE tests with speeds faster than real time increased the speed of calibration progress significantly. Finally the usage of different driving styles allowed checking the calibration robustness frequently and therefore being sure to meet a level of robustness needed for real on road testing.

6 Conclusion

For the future RDE legislation the approach to apply simulation models to support the calibration process is an important step towards more robust and high-performance system calibrations.

The simulation toolchain presented in this article and the associated methodology provides a way to generate a reproducible statement about the impact of calibration changes in various cycles. The components of the model chain were validated and show high accuracy. Exemplary a virtual calibration step is presented, which matches the tailpipe emission results on the road very well.

A simulation model as shown could be applied in many settings of tasks. The optimization of shifting points, the representation of different driving styles on a route with various driver controller settings, the function and algorithm development of control functions and the possibility to screen calibrations for many variants of vehicles and variants that are not physically available.

Due to the faster calculation time of simulation models, algorithms for optimization are conceivable as an approach for initial engine calibration or as a further auxiliary means for the calibration process.

The opportunity to calculate tailpipe emissions for hundreds of road cycles in various variants of a vehicle for different virtual driver types under all required ambient conditions generates a high maturity and robustness for a calibration dataset.

Bibliography

1. Lindemann, M.; et. al.: "VeLoDyn – Ein Werkzeug zur Triebstrangsimulation von Kraftfahrzeugen", HDT (2003)

2. Koltsakis, G.; et. al.: "Modelling of diesel de-NOx aftertreatment components", 1st MinNOx Conference (2007).

3. Schrade, F.; et. al.: "Physico-Chemical Modeling of an Integrated SCR on DPF (SCR/DPF) System", SAE Int. J. Engines 5(3):(2012). doi:10.4271/2012-01-1083

4. Adelberg, S.; et. al.: "Model based diesel exhaust aftertreatment development for EU6-RDE", JSAE (2014)

5. Fukuhara, K.;et. al.: "Model Based RDE Development", 2015 IAV Powertrain Calibration Conference, Tokyo (2015)

6. Richard D Burke, Wolf Baumann, Sam Akehurst and Chris J Brace: Dynamic modelling of diesel engine emissions using the parametric Volterra series. Journal of Automobil Engineering. 2014, Vol 228(2) 164–179

7. Technical Committee – Motor Vehicles: Commission regulation (EU) amending Commission Regulation (EC) No 692/2008 as regards emissions from light passenger and commercial vehicles (Euro 6), Brussels (2015)

Methodological development from vehicle concept to modular body structure for the DLR NGC-Urban Modular Vehicle

Dipl.-Ing. Marco Münster, Dipl.-Ing. Michael Schäffer, Dr.-Ing. Ralf Sturm, Prof. Dr.-Ing. Horst E. Friedrich

German Aerospace Center (DLR), Institute of Vehicle Concepts, Stuttgart
Deutsches Zentrum für Luft- und Raumfahrt e.V. (DLR), Institut für Fahrzeugkonzepte, Stuttgart

Abstract

Three new vehicle concepts in the field of transport are being developed at the German Aerospace Center (DLR) as part of the Next Generation Car META-project: Urban Modular Vehicle, Safe Light Regional Vehicle and Interurban Vehicle. In the submitted contribution, the focus lies on the development and application of a method for determining the vehicle concept and body in white construction for the Urban Modular Vehicle.

With the electrification of cars, there is an opportunity to redesign the vehicle concepts and architectures of future vehicles. In the development of electric vehicles, the integration of new components (e.g. the volume and mass-intensive batteries), provide increasing demands on the overall vehicle design and vehicle packaging.

By matching the arrangement and the integration of components for the Urban Modular Vehicle, a functionally integrated and modular body structure design is being developed. In the structural development phase, the methodological approach, which describes the complete development, employs various optimisations for the load path aligned design and modular structure design. The body structure concept from the Institute of Vehicle Concepts in Stuttgart is based on the purposeful use of different materials for the purposes of multi-material construction and can show advantages over conversion designs.

1 Next Generation Car

The German Aerospace Center is conducting research in the field of transport on the two META-themes Next Generation Car (NGC) and Next Generation Train (NGT). The objective here is the interconnection of different technologies, methods and tools for the holistic development of the vehicles of the future. The NGC project is divided into six working groups: Vehicle Concepts, Vehicle Structure, Energy Management, Drive Train, Chassis and Vehicle Intelligence (Figure 1). Various technologies are implemented in demonstrators and tested in the necessary research infrastructure. The three vehicles that were defined at the beginning are: Urban Modular Vehicle (UMV), Safe Light Regional Vehicle (SLRV) and Interurban Vehicle (IUV). The development method for vehicle concepts and structures is carried out using the example of the UMV and summarised in excerpts in the following paper.

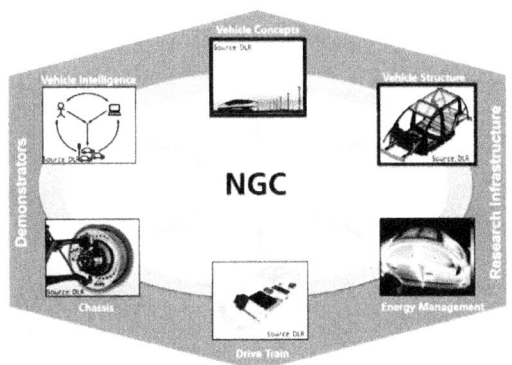

Technologies, methods and tools for integrated development of road vehicles of tomorrow

Figure 1: Organisational structure of the Next Generation Car (NGC) project with the six working groups: Vehicle Concepts, Vehicle Structure, Energy Management, Drive Train, Chassis, Vehicle Intelligence and the three vehicles: Urban Modular Vehicle (UMV), Safe Light Regional Vehicle (SLRV), Interurban Vehicle (IUV)

2 Challenge and motivation

The major topics of the future: resource shortage, climate change, carbon dioxide limits, industry 4.0, automated driving and digitalisation of the vehicle exert a great deal of influence on the future development of vehicle concepts and structures.

Today's vehicles are greatly influenced by the drive topology of the combustion engine. In the traditional configuration (Figure 2 a)) an internal combustion engine with gearbox is located in the front, which is usually in a transmission tunnel in the middle of the vehicle. The components, such as the chassis and body, are specifically designed for the drive topology of the combustion engine. The vehicle building blocks have thus evolved with a focus on the internal combustion engine. Through the electrification of the vehicle, new drive topologies are available that change the vehicle package entirely (Figure 2 b)). The drive (electric motor) may, for example, be arranged in three different configurations as a mid-engine, close to the wheels, or in-wheel motor. One of the key components is the battery. This is very volume- and mass-intensive. It may not be positioned in crash-relevant areas of the vehicle. Hence, only the T-shape, double-floor, or an under the rows of seats arrangement are possible applications [1]. For vehicles specifically designed as electric cars, the double floor solution has proven to be the best solution. All of the points listed show that, through the development of new electric vehicle concepts, the components and concepts change. To this end, a holistic development methodology for vehicle concepts and body structures has been developed and is described in more detail later.

Figure 2 a): Traditional combustion engine vehicle concept with disassembled components [2]

Figure 2 b): Traditional bodywork style with components for an electric vehicle [on the basis of 2]

3 Holistic development methodology for vehicle concepts and body structures

The holistic development methodology for vehicle concepts and body structures is divided into two main phases – phase 1 'Concept phase' and phase 2 'Body structure development phase' (Figure 3). The concept phase starts with the definition of the parameters describing the vehicle concept, such as the number of seats or the use/purpose of the vehicle, as well as the desired/required range. The parameter selection is based on various studies [3,4] and DLR research policy guidelines. The geometrical and simulative vehicle design begins with a preliminary definition of a rough specification. Here the volume- and mass-intensive components are dimensioned in the simulative part with the aid of the driving resistance equation. In the geometric part, the dimensions of the passenger compartment and the front/rear of the vehicle are estimated. A vehicle package is determined from the arrangement of the components, as well as the construction area proposal. In the next step, a design is developed and a parametric CAD model constructed with which it is possible to control different areas of the construction space parametrically. A more detailed package study is conducted with this [6].

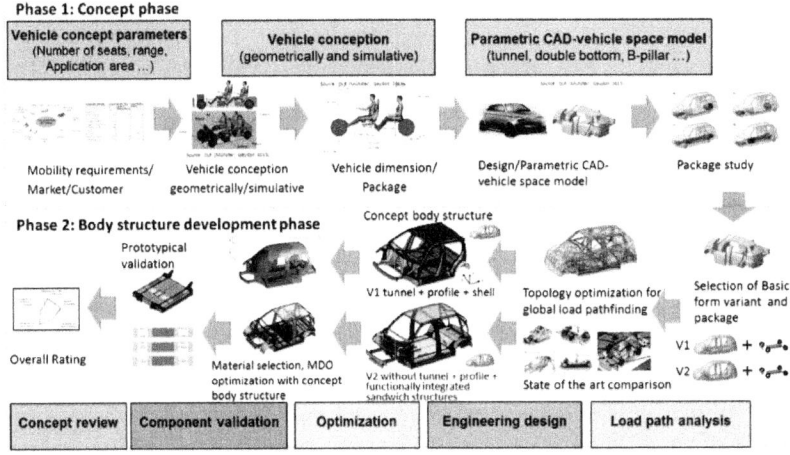

Figure 3: Holistic development methodology for vehicle concepts and body structures with Phase 1 Concept Phase and the Phase 2 Body Structure Development Phase

Phase 2 of the methodology starts with the selection of a basic shape variant and a drive package from phase 1. With the selected construction space, e.g. Variant 2 Double floor, a study of the global load paths is carried out [5]. Using this, the topology optimisation and the assessment with the state-of-the-art comparison database, a vehicle body structure is developed. More detailed materials are assigned to the body in white concept in the next phase and a first calculation and evaluation of the body in white concept is performed on static and dynamic loads, from which two body variants are subsequently derived. (In Chapter 4, the two body types developed are presented in more detail, variant 2 is then described in detail). To validate the overall vehicle concept, a component prototype is built and tested in the next step. An overall assessment takes place at the end.

4 Development of the vehicle body structure for the electric Urban Modular Vehicle

As described in Chapter 3, two body styles – variant 1 and variant 2 – are now discussed in more detail. Then, variant 2 is presented in detail, on different levels, up to prototype validation of a vehicle floor.

4.1 Body in white variant 1

Figure 4 a) shows the basic shape of the construction space for body variant 1. The construction space model offers a construction space design with a tunnel, seat cross member and a double floor for topology optimisation. Nine crash load cases and three static load cases were defined for the analysis of new load paths in the topology optimisation: 100 %, 40 %, 25 % front crash, pole crash side, IIHS side crash, 100 % rear crash, 100 % front crash rigid wall, bumper test front, bumper test rear, Torsion, Bending and Roof impression test [5,7]. A result of the topology optimisation is shown in Figure 4 b). Various optimisation levels are used for the load path analysis: a) unlimited space model (without components), b) limited space model (no-design-space for major package components), c) limited space model (no-design-space for major package components), partially erased space model of dynamic loaded body areas. The topology results are interpreted with load path derivation rules, for which the following rules are defined: unbranched structures to profiles; flat area with medium-sized element densities to shell/sandwich; branched structures to node elements. In the derivation of the body structures (Figure 4 c)), the proposals are always compared with a state-of-the-art comparison database. Body in white variant V1 (Figure 4 c)) shows the classic components of the body in the Greenhouse and a new structure in the roof. Novel structures can be found in the floor area of the body where the new electrification components are housed. Here, a tunnel shape and a new cross structure have been implemented.

With the calculation of the first eigenfrequency trimmed body of 34.4 Hz for variant 1 there is a first good value compared to the prior art of bodies in the A/B segment (average first eigenfrequency trimmed body 31 Hz) [8].

  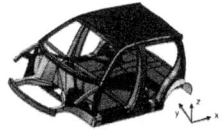

| Figure 4 a): Construction space model V1 tunnel, seat crossbar and double floor | Figure 4 b): Topology optimisation for the derivation of new load paths V1 | Figure 4 c): Derived body structure V1 |

4.2 Body in white variant 2

For the development of the second body variant, methodically, as in the example above, only the input of the available construction space was selected once again. Figure 5 a) shows the construction space model for the load path study. In the floor area, the construction space model offers a large double floor. The topology optimisation result is shown in Figure 5 b); compared to Figure 4 b), in the floor area there are now no branched areas, but rather a large base plate for the double floor. The results were compared with state-of-the-art bodies, and the body variant (Figure 5 c)) was developed. The same design guidelines were applied for the development of the body in white, as described in 4.1. Since variant 2 enables more innovative structural approaches, it was selected for a more detailed consideration.

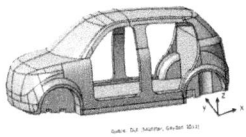 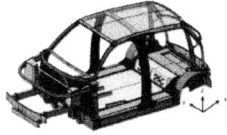

| Figure 5 a): Construction space model V2 double floor | Figure 5 b): Topology optimisation for the derivation of new load paths V2 | Figure 5 c): Derived body structure V2 |

4.2.1 Design philosophy

The main features of the design philosophy are an aluminium-intensive frame structure consisting of profiles and nodes (similar to the ASF [7]) and function-integrated thrust areas as sandwich structures and surface components in FRP (see Figure 6). The sills and longitudinal beams are designed as aluminium profiles, which provide both length variability and adaptability in the cross-section and function integration. The transition from the floor module to the front/rear of the vehicle is achieved using large node structures. The node structures offer a very good opportunity to integrate as many functions as possible at this construction group interface, as well as to create module interfaces for different versions. Sandwich structures are used, above all, in the area of the double floor and the battery box. This combines mechanical loads and functions (e.g. isolation) in one. A novel sandwich crash concept to protect the occupants and the battery can be found in the side crash area, between the sills and inner longitudinal beams. In the Greenhouse variant of the body shown, both profiles and classic shear fields can be found.

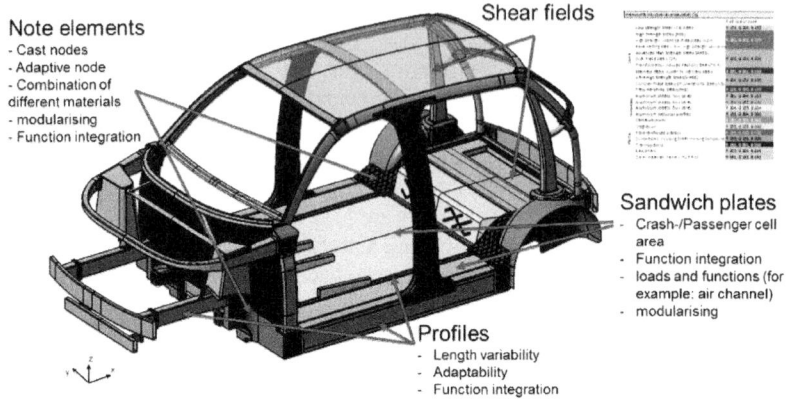

Note elements
- Cast nodes
- Adaptive node
- Combination of different materials
- modularising
- Function integration

Shear fields

Sandwich plates
- Crash-/Passenger cell area
- Function integration
- loads and functions (for example: air channel)
- modularising

Profiles
- Length variability
- Adaptability
- Function integration

Figure 6: Design philosophy of the Urban Modular Vehicle body in white: Aluminium-intensive frame structure with sections and nodes with function-integrated thrust surfaces as a sandwich and FRP surface components

4.2.2 Modularity of the body in white

The node-intensive aluminium frame structure provides an excellent opportunity for body modularity:

- Construction variability
- Length variability
- Drive variability
- Chassis variability
- Interior variability
- Etc.

Figure 7 shows how such length variability is possible through the simple adaptation of profiles and sandwich panels in the floor module. That means that it is possible to design a body for the urban reference body structure with a length of 3600 mm (Figure 7 a)), but also an extended 4000 mm (Figure 7 b)) or shortened 3000 mm variant (Figure 7 c)).

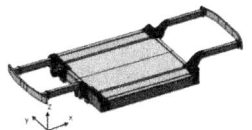
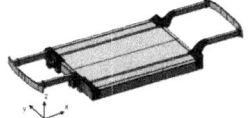
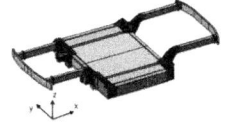

Figure 7 a): Modular floor module of the UMV base reference structure, length = 3600 mm

Figure 7 b): Modular floor module of the UMV, length = 4000 mm

Figure 7 c): Modular floor module of the UMV, length = 3000 mm

The construction using node structures provides further possibilities for modularisation by attaching different front and rear modules (Figure 8). For example, the front-end module with steerable axle can be combined with the driven rear chassis module or the steered driven front end with the non-driven rear end.

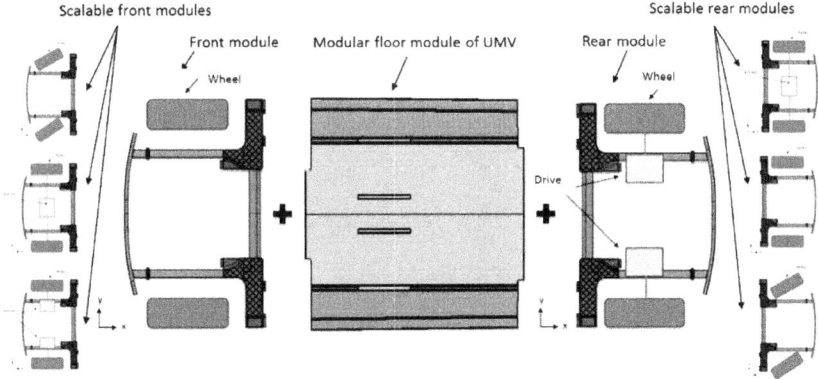

Figure 8: Modularisation possibility of different front and rear modules in the design of the UMV, combination example: front-end module with a steered front end with a driven rear chassis module (selected configuration for the basic variant of the UMV)

The variant for the UMV is the front-end module with a steered front end with the rear end module with the drive close to the wheels. The following sections focus on the urban basic body structure with a length of 3600 mm (Figure 6).

589

4.2.3 Complete vehicle crash simulation

The Urban Modular Vehicle basic version was first simulated on static load cases for torsion, torsional frequency, bending and dynamic loads (pole, 25%, and 40% front crash). Only the crash results are described, not the static results.

The complete vehicle crash model of the UMV was calculated with the explicit LS-DYNA Solver 7.11. A variety of techniques are used for the joint and joining technology. Various welds, screw and rivet connections, as well as structural bonding are used. The bonds are calculated using cohesive elements and the welded joints with spot welds.

Figure 9 a) shows the simulation of the pole crash before (a)) and after (b)) the crash. It can be easily seen that the crash concept absorbs the energy in the side area extremely well and prevents excessive intrusions. There is no damage to the battery module. The maximum intrusion of the pole experienced is approx. 220 mm.

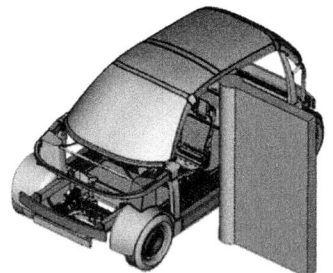 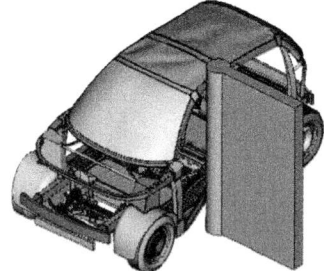

Figure 9 a): Simulation model of the UMV before the pole crash

Figure 9 b): Simulation model of the UMV after the pole crash with an intrusion of 220 mm

For the frontal crash with 40% coverage it was demonstrated that the first concept approach is functional. The Small Overlap Crash is also conceptually absorbed well by the laterally bevelled cast joints of the body structure and the chassis. The body and chassis absorb some of the impact energy and then sliding takes place.

In a first step, as part of the NGC project, the virtually designed body in white was not fully assembled and tested as a complete vehicle prototype. It is initially only possible to select an area of the body and validate it as a prototype. For this purpose, the floor module was selected with the pole crash validation load case.

4.2.4 Prototypical validation of the floor crash concept

The structure of the floor module of the UMV is shown in more detail in Figure 10 a). The aluminium extrusion profile and the front and rear large nodes form the frame structure with the transverse beams. The battery box (Figure 10 b)) is inserted into the frame structure from underneath. The battery box consists of a large sandwich base plate in which functions such as cooling, wiring and mechanical power transmission, and the connection of the battery pack are integrated. The battery box is enclosed at the top and on the sides by a lightweight housing. The battery module is fully integrated through the use of continuous longitudinal beams in the floor module frame structure and a top sandwich plate. Two sandwich crash elements are located in the crash area between the longitudinal beam and the sills. The crash elements are manufactured from an aluminium trapezoidal sheet and two aluminium covering layers, attached with structural adhesive and screws. The crash concept provides for a buckling in the door sill profile and then buckling of the two trapezoidal sandwich absorbers. The deformation zone ends just before the zero intrusion zone on the longitudinal beam. The trapezoidal absorber extends over the complete length of the floor module, which provides optimum protection for the occupants and the battery module. In the event of a pole crash, safety can be guaranteed over the entire width of the floor module.

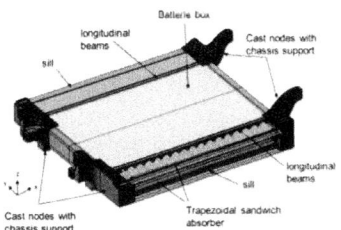 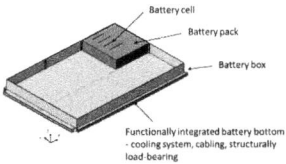

Figure 10 a): Aluminium extrusion profile floor module of the UMV, large nodes at the front and rear, battery box, and crash concept with aluminium trapezoidal absorbers

Figure 10 b): Battery box made of a large sandwich base plate with functions such as cooling, wiring and mechanical power transmission, and connection of the battery packs

Three levels of components have been defined for the validation chain of the prototype floor crash test for the complete side crash concept.

- Trapezoidal sandwich (I)

- Trapezoidal sandwich with edge profiles (II)

- Complete crash concept of the floor module (sills, trapezoidal absorber and replacement structure for longitudinal beam/node bearing) (III)

Trapezoidal sandwich (I)

Figure 11 shows the component level I consisting of a trapezoidal sheet with Al 6082, sheet thickness t = 1 mm and the outer layers of Al 6082, sheet thickness t = 1 mm. The trapezoidal absorber is attached using Dow Automotive BETAMATE 2098 structural adhesive with a bond line thickness of approx. t = 0.4 mm. Figure 11 a) shows a CAD model of the simulation and b) a real sample. In the simulation, different angles, sheet thicknesses and sandwich heights were varied with a multidisciplinary design optimisation and the most promising solutions were tested using prototypes.

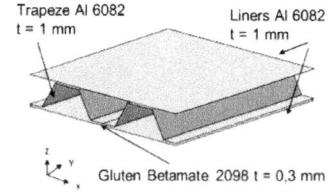

Figure 11 a): Trapezoidal sandwich absorber CAD model, sheet thickness t = 1 mm, outer layers Al 6082 sheet thickness = 1 mm. Dow Automotive BETAMATE 2098 structural adhesive bonding, bond line thickness of t = 0.3 to 1 mm

Figure 11 b): Trapezoidal sandwich absorber as a real prototype

A comparison of the simulation and test results in the force-displacement diagram (Figure 12) shows an initial good match between simulation and test. However, it also shows a relatively high force peak that could be somewhat mitigated with different trigger actions. For this, the trapezoidal sheet was equipped with, for example, folding aids.

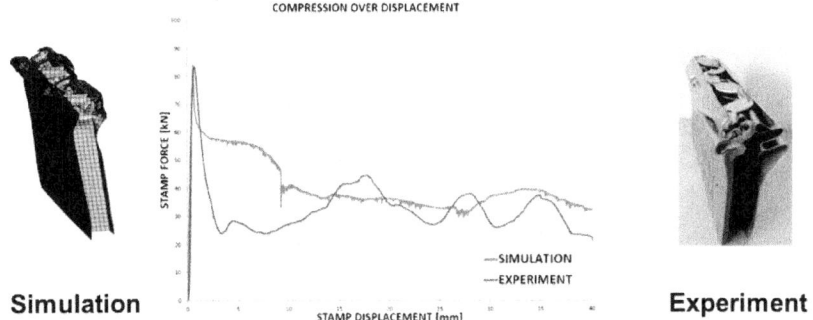

Simulation   Experiment

Figure 12: Force-displacement diagram of the simulation and test of a trapezoidal absorber

Trapezoidal sandwich with edge profiles (II)

The sandwich absorbers with edge profiles were bonded and screwed for validation level II. Figure 13 shows an undeformed sample in the test machine (a)) and a deformed sample (b)). It can be seen that the trapezoidal sheet folds robustly and the outer layers remain attached to the sandwich absorber by the screws and the edge profiles.

Figure 13 a): Undeformed trapezoidal sandwich crash absorber, bonded and screwed with edge profiles

Figure 13 b): Trapezoidal sandwich crash absorber, bonded and screwed with edge profiles, deformed by 40 mm

Complete floor module crash concept (III)

Figure 14 shows the test set-up as a virtual CAD model (a)) and the crash simulation (b)) of the test rig. The test objects are the sill profile and the two trapezoidal crash absorbers. The test framework consists of two lateral replacement devices for the node structure and a rear attachment plate, which represents the inner longitudinal beam (see Figure 10a)). During the crash simulation of the test rig, it is clear that the first to deform was the sill, and then the trapezoidal absorber.

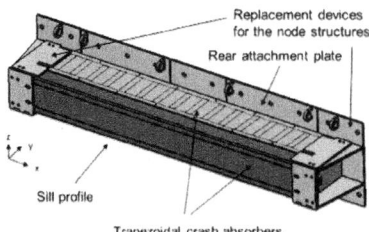

Figure 14 a): CAD model of the UMV floor module crash concept test set-up; test objects: sill profile and two trapezoidal crash absorbers; test set-up: two replacement devices for the node structures and rear attachment plate

Figure 14 b): FEM model of the UMV floor module crash concept test set-up; deformed with maximum intrusion

5   Conclusion and outlook

The holistic development methodology for vehicle concepts and body structures was discussed in the previous chapters. Two bodies were subsequently presented, whereby the second body variant was detailed with an aluminium-intensive body frame with sections and nodes with functional integrated sandwich structures, as the most promising solution. The UMV body in white provides excellent modularity, in particular in the realisation of different structures or the length variability. The UMV body in white variant with a length of 3600 mm was virtually computed and partially validated using a prototype. The selected floor crash concept is being implemented from component to floor section. The crash test will subsequently be conducted in the in-house crash test facility.

In future, it is planned to detail the UMV body further. For this purpose, different construction concepts are being developed and other components implemented and tested using prototypes.

The new components and tools created in the overall NGC project – for example in the area of vehicle intelligence, chassis, power train and man-machine interface – will be progressively included in the UMV development process.

References

1. A. Kampker, D. Vallee und A. Schnettler, Elektromobilität – Grundlagen einer Zukunftstechnologie, Berlin; Heidelberg: Springer-Verlag, 2013.

2. Manfred Klangwald, Joachim Staat, 2015, VW Up: 100.000-Kilometer-Dauertest, AutoBild.de, http://www.autobild.de/bilder/100.000-kilometer-im-vw-up-5574029 html#bild39

3. Robert Follmer, Dana Gruschwitz, Birgit Jesske, Sylvia Quandt, 2010, Infas Institut für angewandte Sozialwissenschaft GmbH [Infas Institute for Applied Social Sciences], DLR Institut für Verkehrsforschung [Institute of Transport Research]: Mobilität in Deutschland (MiD) [Mobility in Germany] 2008, Results report, Bonn and Berlin, February 2010, PN 3849, http://www.mobilitaet-in-deutschland.de/pdf/MiD2008_Abschlussbericht_I.pdf

4. Till Gnann, Patrick Plötz, Florian Zischler, Martin Wietschel, 2012, ISI Fraunhofer, Elektromobilität im Personenwirtschaftsverkehr – eine Potenzialanalyse [Electromobility in commercial passenger transport – a potential analysis], http://www.isi.fraunhofer.de/isi-wAssets/docs/e-x/de/working-papers-sustainability-and-innovation/WP07-2012_Wirtschaftsverkehr.pdf

5. Münster, Marco und Schäffer, Michael und Friedrich, Horst E. (2015) Development of body structure concepts for electric vehicles using the topology optimization for global load pathfinding. 2015 European Altair Technology Conference, 29.09. – 01.10.2015, Paris, France.

6. Münster, Marco und Friedrich, Horst E. (2015) Vorgehen zur Grundkonzeption eines Elektrofahrzeugs und Entwicklung leichter und neuartiger Karosserie-Strukturen. In: 4. Symposium Elektromobilität Tagungs-CD. Ostfildern: Technische Akademie Esslingen. 4. Symposium Elektromobilität, 23. Juni 2015, Ostfildern. ISBN 978-3-943563-19-1.

7. Hans-Hermann Braess, Ulrich Seiffert, Publisher, 2013, Vieweg Handbuch Kraftfahrzeugtechnik [Vieweg Manual for Automotive Technology], Edition Number 7, ISBN 978-3-658-01690-6, Springer Germany Wiesbaden.

8. Dipl.-Ing. Marco Münster , Dipl.-Ing. Michael Schäffer, Dipl.-Ing. Gerhard Kopp, Dipl.-Ing. Gundolf Kopp, Prof. Dr.-Ing. Horst E. Friedrich New approach for a comprehensive method for urban vehicle concepts with electric powertrain and their necessary vehicle structures. 6th Transport Research Arena, April 18-21, 2016, Warsaw, Poland.

Quantification concept for vehicle packaging

Arthur Frick, Roland Müller, Thomas Blauß
Daimler AG

Prof. Dr.-Ing. Dieter Schramm, Universität Duisburg-Essen

Abstract

The limited installation space in the vehicle presents a challenge within the framework of the total vehicle design already in the phase of predevelopment. The complexity increases because of additional new components and systems, and the question of their integration into the vehicle as a whole increases this challenge. As a result, the objective appears to be a quantification of possible package variations in order to determine an optimal solution from both the technical and economical as well as the customer's point of view.

Factors influencing the packaging in the design of the complete vehicle are the design features like the vehicle concept including the dimensional concept and ergonomic requirements. The functionality of the system to be integrated has also be a given as well as the function of the entire vehicle. Requirements in regards to service, production capability of possibly already existing facilities and costs are other factors influencing the package design.

The assessment model introduced in this paper includes five model groups: design, function, system, costs and market. These are further divided into main- and subgroups each, which eventually define the valuation parameters that are used to determine the overall grading of the vehicle package. This allows an overall package quantification.

1 Motivation and Objectives

The total vehicle design has the task to allocate a position to all the components in the vehicle. The limited availability of installation space in the vehicle presents a package design challenge already in the predevelopment phase. The addition of numerous new systems increases the complexity of the integration even more. Despite the extensive use of mechatronic product layout in new systems, which usually comes along with an advantage in terms of the required installation space compared to purely mechanical systems, there is no redundant installation space in most vehicles (see Figure 1). A glimpse under the hood of a penultimate generation vehicle and of an automobile that is under development today shows how the degree of utilization of installation space has changed.

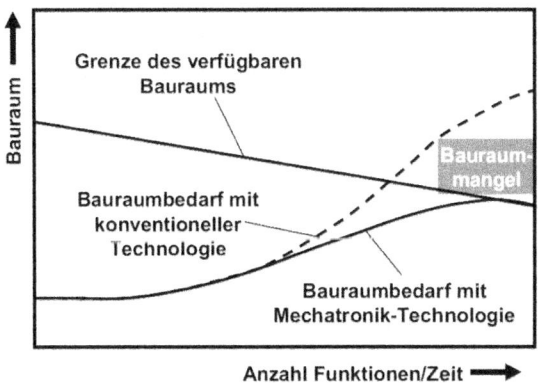

Figure 1: Lack of installation space despite the increase in mechatronic system layout [4]

In addition to the purely geometrical integration, functional interdependencies resulting from positioning also have to be considered. That means that the position of the component influences its own functionality along with the functional properties of the complete vehicle.

Vehicle dimensions cannot grow arbitrarily, and the lack of installation space is a fundamentally existing problem. Hence an intelligent package concept that considers a wide range of requirements and constraints is the goal in the course of the vehicle construction process. At this point, the question arises, "What is a good vehicle package?" As a result, the objective becomes a quantification of possible package variations in order to determine a solution that would be optimal from both the technical and economical as well as the customer's point of view.

The complex interrelations become increasingly difficult to treat with each innovation [2]. The objective of the present concept is therefore a vehicle based dependency analysis and a system integration evaluation.

2 Factors Influencing the Package Design

As mentioned in the beginning, a number of different parameters are relevant to the package design in order assess the value of the package. These are separated into five groups, described in more detail below.

2.1 Design Features

The group of design features consists of the vehicle concept, dimensional concept and the dimensions of the components. The geometrical position of the component and the size of the system to be integrated can influence the dimension concept parameters that are decisive for the complete vehicle concept. For example, if a six-cylinder in-line engine should be installed in a vehicle with a transverse engine front-wheel drive instead of a four-cylinder while retaining the same dimensioning chain, it will have direct influence on the vehicle width. It would also become apparent that a geometrical adjustment of the longitudinal elements is required because the space between the longitudinal elements is not sufficient. The constructive adaption of the vehicle frame would be required. This influences the functionality of the complete vehicle, like e.g. the crash behavior.

2.2 Functional Features

The functional features are related to the complete vehicle. Depending on the position of the component to be integrated, its functional properties can change. One of the features of this group is ergonomics. A negative impact on the ergonomics should be avoided whenever possible because it instantly has a negative impact on the customer value. The passive security is another component of the complete vehicle functionality that significantly depends on the package concept. The vehicle construction determines such crash-relevant factors as blocking, deformation behavior, energy absorption, and passenger acceleration. If crash simulations show that the component assembly is not expedient, and the requirements set cannot be achieved, the package has to be adjusted in such a way that the desired behavior takes place, as e.g. with the A-Class, where the integral sub frame is an essential feature of the deformation concept (Figure 2).

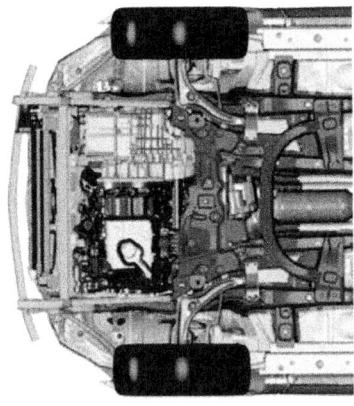 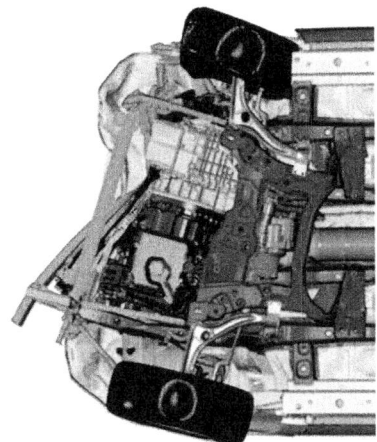

Figure 2: Crash behavior of the MB A-Class [1]

The requirements for the drivability of the vehicle are an interplay of the vehicle components and the features resulting from the vehicle package such as center of gravity position and axle load distribution. For example, a change of the axle load distribution occurs in plug-in hybrid vehicles with a heavy HV-Battery in the rear compared to a conventional non-hybrid vehicle.

Further features are comfort and thermal protection requirements. Customer expectations on the NVH behavior can lead to positioning restrictions of components, which can on the other hand cause a negative impression on the passenger through sounds and/or vibrations. Compared to that, thermal protection is irrelevant for the customer if the impeccable operation of the vehicle is provided. The important part here it is to determine the temperatures on the components by using simulation and to ensure that there is no thermal stress. If the specified limit values are exceeded during simulation, measures leading to an acceptable solution have to be taken. For example, a shielding of the relevant component (under consideration of the necessary installation space and additional costs) or a different positioning in a non-temperature-sensitive area could be a solution.

2.3 System-specific Features

The new system to be integrated into the vehicle must be evaluated for both geometrical and functional integration. The geometrical aspects are the positioning, the necessary minimum clearance and the location of the connections. As a result, the cable routes to a component and its accessibility are also an important aspect of the package design. Mechanical, electrical and thermal system requirements are also critical for the functional integration.

2.4 External Requirements

A package concept does not achieve its objective, if its feasibility is not guaranteed, or the assembly sequence does not allow it. Therefore, a dialogue with the production plant colleagues responsible for feasibility is essential in the course of designing the package, in order to take into account their technical constraints in time.

An additional challenge presents itself when a succession vehicle is to be built in an existing plant on an existing equipment.

Furthermore, in order to enable customer-friendly and cost-effective service during the lifecycle of the vehicle, service requirements present yet another factor to be taken into account in the early phases. For this, accessibility to wear parts and units requiring regular maintenance is required.

Legal rules and regulations as well as marginal conditions from ratings are yet another external requirement that has conceptual relevance.

2.5 Costs

Cost aspects have a still increasing impact on the package design. Development- and cost-intensive components that can be used in a wide range of designs make derivate-specific execution ineffective. As a result, an installation location might be necessary that must meet certain system requirements and geometric features, like for example the position of power or cooling connections. The environment and the extent of modifications of the components concerned must be considered here. This leads to changes in costing for the relevant components caused by the new package.

System costs, vehicle costs and external costs are considered as main groups. External costs can arise e.g. in the production because of the requirement of an additional handling unit.

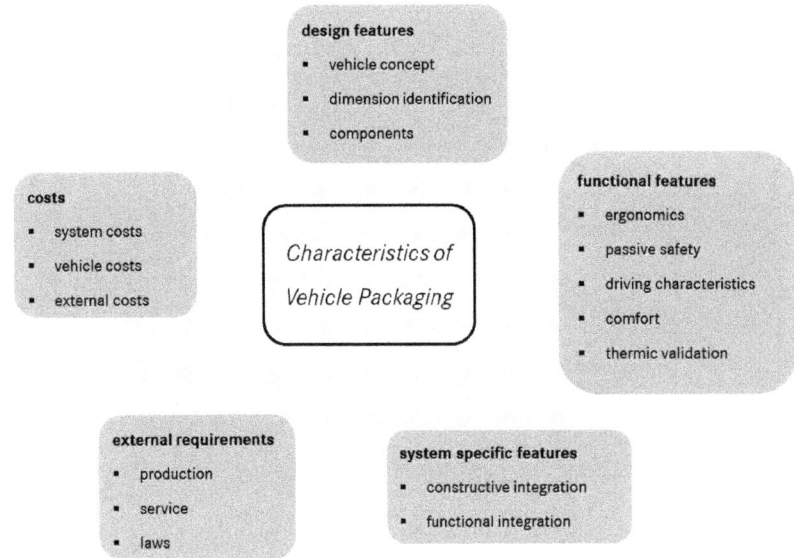

Figure 3: Factors influencing the vehicle package

3 Structure of the Assessment Model

The assessment model consists of five model groups: design, function, system, cost and market. These are further divided into main- and subgroups each, which eventually define the valuation parameters that are used to determine the overall value of the vehicle package. This allows an overall package quantification.

Each model group is shown as a main group containing features as seen in Figure 3. These features build the subgroups containing the respective parameters. Example: Figure 4 is showing the structure of the Function model with the two main groups "Ergonomics" and "Passive Safety". The aspects of anthropometric vehicle design are seating, view, control- and display components, sense of space, entry and exit, loading, and service [3]. In the subgroup "Seating" there are such parameters to be evaluated as "pedal position" and "steering wheel position". If, for example, a component is positioned in the tunnel hence resulting in a shift of the pedals, it becomes relevant for the functional assessment of the package. Therefore, the structure of the model allows for a good overview of where the integration of a system changes will be necessary.

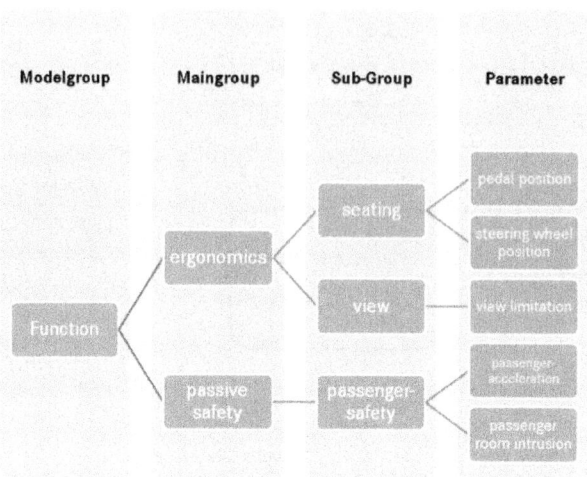

Figure 4: Structure of the assessment model

4 Valuation System

In the first step, the component is positioned in the vehicle. Thus, the geometrical position data is known (Figure 4). Depending on the position, dimensional concept parameters can become relevant, but they may not necessarily have to be changed, for example, the L114 measure (the space between the middle of the front wheel and SRP [5]). Furthermore, an environment analysis determines components present in the surroundings of the new component. The components involved are then transferred to a design structure matrix (DSM). Because not all parameters in the assessment model are needed everywhere, a valuation parameter set is generated based on the position data of the new component as well as the affected components.

Finally, value ranges are assigned to the individual parameters, based on which a valuation can be made. The parameters are normed and multiplied by the weighting factor of the subgroup. In the next step, the numerical values of the sub-group are calculated with the weighting factor of the main group. In the end, the main groups are provided with the respective weighting and summarized in the model group. Thus, the package can be quantified and positioning variations can be evaluated and compared.

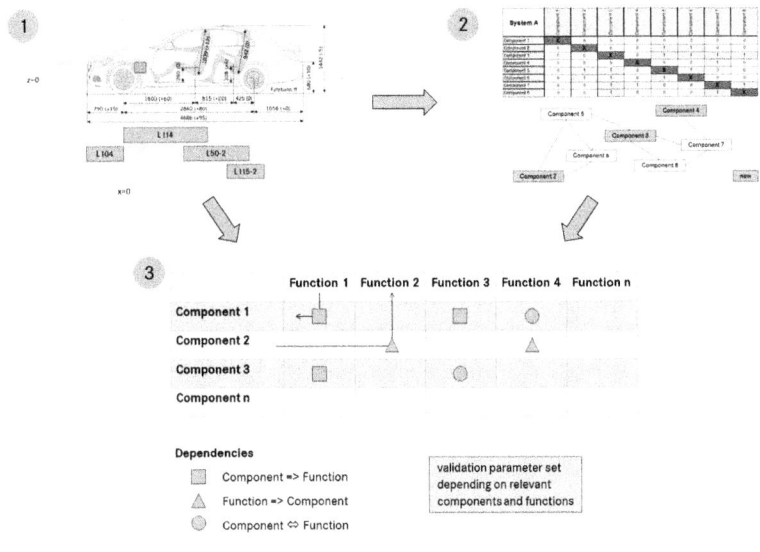

Figure 5: Procedure for component integration

Model group	$\dfrac{HG1 * x + HG2 * y + HG3 * z}{100}$; $\quad \sum x + y + z = 100$
Main group (HG)	Evaluation points$_{UG}$ * weighting factor$_{HG}$
Sub-group (UG)	$\sum$ normed parameter evaluation points * weighting factor$_{UG}$
Parameters	Identification of the numerical values / points in defined domains 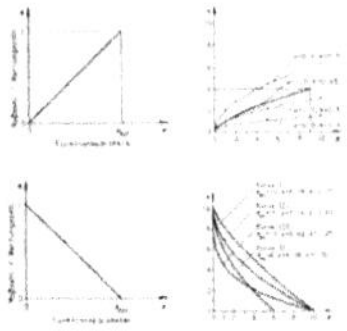

Figure 6: Validation composition

5 Summary and Outlook

This paper presents a quantification concept for the vehicle package design. The increasingly intensified demands on installation space call for efficient and intelligent packages that meet the diverse requirements in the overall vehicle development.

The package influencing factors are design, functional and system-specific features as well as external requirements that are requested on the vehicle, and the costs. The assessment model consists of the parameter in whose value range the assessment is made, the sub- and main group, and finally the model group with the valuation measure for the respective group. The assessment process is based on the geometric integration of the component in the vehicle, in order to extrapolate the relevant functions and components in the environment for the selection of the valuation parameters. The parameters are then provided with the respective valuation measure and lead to the model group by means of the relative weighting of the sub- and main group. As a result, the impact of the new position variations can be quantified.

Next steps in the development of this methodology includes an implementation in a software tool, further detailing of the cross-linking and dependencies in the vehicle, and an automation of the design procedure.

Literature

1. ATZextra (2012), Die neue A-Klasse von Mercedes-Benz, Springer Automotive Media, Wiesbaden

2. Breiing, A.; Knosala, R. (1997), Bewerten technischer Systeme, Springer Verlag Berlin Heidelberg

3. Bubb, H.; Bengler, K.; Grünen, R.; Vollrath, M. (2015), Automobilergonomie, Springer Fachmedien Wiesbaden

4. Gevatter, H.-J.; Grünhaupt, U. (2006), Handbuch der Mess- und Automatisierungstechnik, Springer Verlag Berlin Heidelberg

5. Norm DIN 70020-1 (1993): Straßenfahrzeuge – Kraftfahrzeugbau. Teil 1: Personenkraftwagen, Begriffe, Grundlagen, Bestimmungen, Maßkurzzeichen, Beuth Verlag, Berlin

Analysis of hydraulic brake systems with regard to the requirements for future vehicle concepts

Dipl.-Ing. Christian Riese, Prof. Dr. Frank Gauterin

Robert Bosch GmbH, Institut für Fahrzeugsystemtechnik (FAST) KIT

1 Introduction & Motivation

In recent years vehicles were going through massive changes with regard to power-train structure and driver assistance systems. The percentage of hybrid electric vehicles and electric vehicles is slowly but steadily increasing. In contrast, the architecture of the brake system and its basic layout has only been subject to minor changes over the last 60 years. A centralized architecture, which was driven by the combustion engine as solely energy supply source on the one hand and the driver directly connected through the brake pedal on the other hand, has been established and optimized over the last decades.

The key driving factors during the development of brake systems have been safety and pedal feel. The requirements for the brake system have changed rapidly during the last years through new impulses given by alternative vehicle concepts and highly automated driving, figure 1. Due to that fact new aspects, like for example brake system efficiency or active pressure build up, become more and more important. This work analyses and studies the state of the art hydraulic brake system with an electro-mechanical brake booster, with regard to its efficiency. A state of the art brake system architecture of a small electric vehicle with an electromechanical brake booster (iBooster) is studied and defined key figures for its efficiency are derived in simulation and experiment.

Figure 1a): Emergency braking city [1]

Figure 1b): Hybrid electric vehicle with regenerative brake system [1]

2 Requirements for brake systems

The requirements for brake systems, shown in figure 2, can be divided into three main groups. The first group represents the unmodified requirements, these are the ones which stayed almost the same over the last decades and undergo only minimal changes. The second group represents the requirements that already exist but undergo changes and adaptions for example due to development progress regarding these existing requirements. The third group represents the new requirements, which come up since only a short time and have been subject for the requirements only recently. These new requirements are due to bigger technological changes or impulses by new technologies.

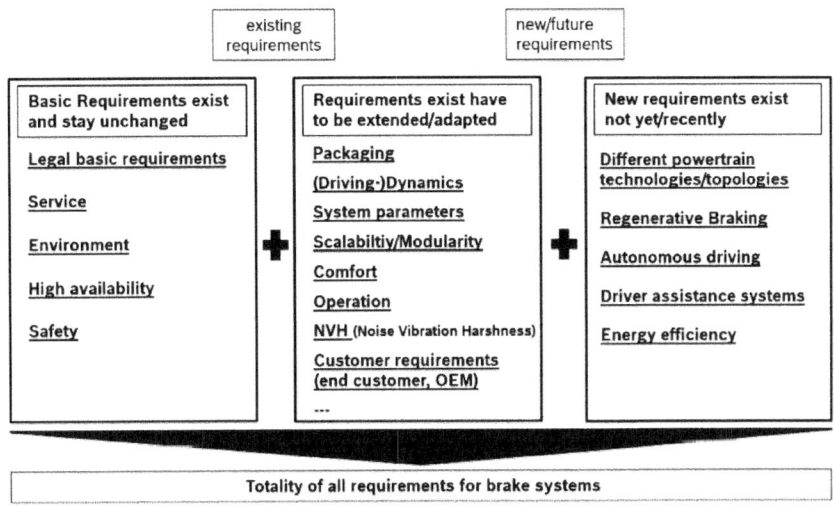

Figure 2: Overview of requirements for brake systems

The basic requirements for brake systems are for example the requirements by law, like documented in the ECE-R 13H [2] or other governmental or industrial standards. As the brake system is a safety critical component the most important requirement was and still is due to safety aspects. This includes the ability to decelerate the vehicle properly in case of a failure, like booster failure or a power outage of the onboard power system. As brake systems have been developed for decades, the largest group of the requirements is the group of existing requirements that have to be changed and adapted to the continuous progress of technology and the demands of the OEMs and customers. An example of this group is the continuously growing number of segments demanded by the customers and offered by the OEMs. This requires the brake system to offer a certain modularity, so that components can be used over a wider range and size of vehicles. Another example is the packaging, the existing design space, which can be covered by the brake system. Due to the increasing number of auxiliary devices and packaging aspects, this design space is shrinking more and more, so that a more and more compact design of the brake system, brake booster master cylinder combination is desired. Completely new requirements are introduced by new technologies entering the automobile, like e.g. autonomous driving. As these technological impulses change and affect driving a vehicle strongly, they generate new demands for the systems. In the case of autonomous driving for example an autonomous pressure built-up is required. Further on the way to autonomous driving, a vast variety of driver assistance functions is implemented into the vehicle. To assure a good rating in future testing and evaluation, like NCAP, a high pressure built-up dynamic is asked of the brake system even autonomously to optimize functions like automated emergency braking. The other big new driver for brake systems is the impulse given by new and alternative powertrain technologies. For hybrid and electrical vehicles the maximization of recuperation functions should not be limited by the brake system. As the battery is still the limiting factor, with regard to the operating range of electric vehicles, developers attempt to minimize the energy consumption by auxiliary devices. So far this approach is mainly focused on non-safety relevant components and comfort functions, like for example the air conditioning and heating. Due to this reason the efficiency of brake systems is an interesting aspect worth studying, as it influences not only the energy consumption but also the dynamic behavior and other characteristics.

3 State of the art brake system: iBooster

Due to the demand for vacuum-free on demand brake booster systems, electromechanical brake booster start replacing the vacuum booster systems established over decades.

Electromechanical brake boosters are provided by a variety of brake equipment manufacturers, the iBooster shown in figure 3a) is the BOSCH version of such a booster concept. The inner design is kept pretty similar to the state of the art vacuum boosters, except for the air valves for the vacuum booster control. The driver pushes the pedal, which is transferred using the lever for amplification onto the input rod, which is connected to a plunger inside the brake booster. The boost force generated by an EC-motor is transferred to the brake system by using a boost body which is surrounding the valve body coaxially. The valve body and the plunger, both push on the reaction disc, a small rubber disc, which is connected to the output rod of the booster. One reason for keeping this design is the good pedal feel characteristic of such a layout. Another reason is that it offers the possibility to generate a combined force, consisting of driver pedal force and booster force to act on the brake fluid column and help pushing the liquid to the wheel cylinders.

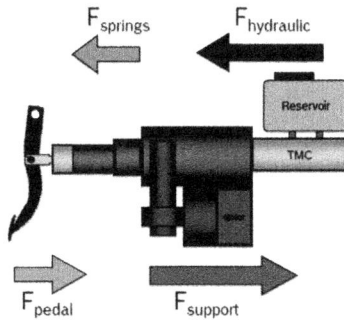

Figure 3 a) :iBooster-electromechanical brake booster with master cylinder [1]

Figure 3 b): Schematic acting forces [1]

The needed support or booster force, compare figure 3b), depends on:

- the spring forces $F_{springs}$ which assure that the system is reset, even in case of a failure,

- the hydraulic force $F_{hydraulic}$ which depends on the actual and target pressure, which is demanded by the driver and the dynamic of the application,

- supported is the booster as typical for an assisted brake system by the driver foot force F_{pedal}.

Therefore the forces, that the booster has to overcome, neglecting friction, are:

$$F_{support} = F_{springs} + F_{hydraulic} - F_{pedal}.$$

4 Tools for the analysis of the brake system

To study the characteristics of state of the art hydraulic brake systems a test bench, including a temperature test chamber and hydraulic simulations, using AMESim, are used. Vehicle test drives are also performed, but mainly used to generate pedal travel and pedal force profiles.

4.1 Test bench

The basic advantage of using a test bench instead of vehicle measurements is the reproducibility of the measurement conditions. Relevant temperatures can be set and the system can be studied repeatedly at the same condition. Due to the properties of brake fluid (DOT4) the temperature influence plays a major role for hydraulic brake systems. Usually tests at very low, room and high temperature are conducted as the kinematic viscosity and the compressibility of the brake fluid vary strongly. These conditions can then be recreated in simulation and used to identify the characteristics of the system depending on temperature and help creating a proper model of the studied brake system. Furthermore it is possible to equip the test system with much more sensitive and accurate sensors compared to vehicle measurements, like for example the pressure sensors. By that it is also possible to conduct tests like pressure build-up dynamic measurements, which need a time resolution high enough to deliver proper results. The test bench, shown in figure 4, consists of two main parts, the servomotor linear actuator combination and the hydraulic brake system itself. The servomotor linear-actuator combination is used as a braking robot and is attached to the outside of the temperature chamber. The brake system itself consists of the booster and the hydraulic part of the system mounted on a compact module, as can be seen in figure 4. The actuator is used to give the corresponding pedal commands.

Figure 4: Test bench inside temperature chamber with iBooster and hydraulic module

The linear-actuator pushes directly the input rod of the booster. In addition it is also possible to skip the brake booster, due to the high force potential of the servomotor and directly push the master-cylinder plunger. This offers the possibility to study different speed profiles like for example ramps, or even use it for the development of future booster systems by purely modelled boosters implemented in the control software of the actuator. To measure the force acting on the booster or master-cylinder a force sensor is mounted inside an adapter at the tip of the actuator. Another force measurement adapter is located at the pushrod of the booster, so it is possible to measure the amplification by the booster as well. The hydraulic brake system is mounted on the module to fit inside the temperature chamber. The hydraulic unit (ESP) is used in passive mode. Therefore no Electronic Brake Distribution (EBD) is active to limit the pressure at the wheel brakes. The hydraulic brake system is equipped with up to five pressure sensors. The complete system is operated by using LABView RealTime and FPGA. The analog signals are recorded by the FPGA, transferred to the real time system and finally to the host pc. The servomotor is controlled using CAN-open (industrial CAN) protocol. The mounted encoder delivers a resolution of minimum 0.01mm accuracy. The booster data is collected by using an XCP-protocol interface of a special iBooster version for testing. Further the highly accurate position signal makes it possible to use the test bench to measure pressure over volume characteristics for single elements as well as the complete hydraulic brake system. For this the actuator travels at minimum speed to generate quasistatic behavior inside the system and eliminate all dynamic side effects.

4.2   (Co-)Simulation

To gain better understanding of the relevant effects inside the brake system simulation models are set up and validated with measurement data. These validated models can then be used for further studies. Using simulation makes it also possible to conduct parameter studies or basic changes in the layout, which actually would be quite time consuming when done on the test bench system itself. Another big advantage of the simulation is that it can be extended to a complete vehicle simulation. This is done by coupling the brake system model to a vehicle dynamics model including a simple tire model. For the coupling MATLAB/Simulink is used as master and is coupled to the hydraulic brake system model in AMESim, as can be seen in figure 5. For that the co-simulation interface provided by AMESim is used. In that manner both simulation tools can work with their own solver efficient for the specified use case. To study the hydraulic components solely AMESIM is used standalone. Furthermore it is possible to overcome limitations of the test bench, as it is only a brake system test bench with no rotating parts and a passive hydraulic unit. Effects like EBD and ABS cannot be studied properly on this type of test bench. But by using the co-simulation and the validated hydraulic model in combination with the vehicle model this shortcoming can be overcome partly. Using MATLAB driving cycles are imported into the model and can be used to benchmark and analyze the efficiency of the system over various driving cycles.

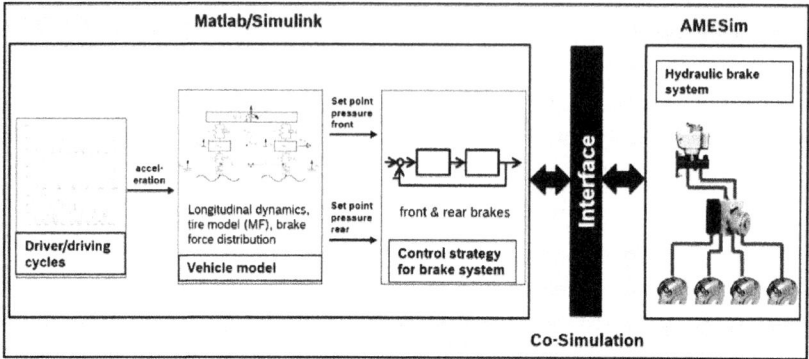

Figure 5: Co-Simulation environment

Further simulation offers the possibility to study the influence of single elements of the brake system on the overall brake system performance and efficiency. Models for different system degrees can be set up and analyzed. Starting with an ideal system, taking into account only the fluid compressibility and the pipe roughness. Extending this model step by step until the completely detailed model is reached.

5 Key figures to characterize brake systems

To determine and compare the performance of different brake systems key figures are used. As the brake system is a safety critical component, meant to generate maximum deceleration in emergency situations, mainly the dynamic for building up pressure is taken into account. Other key figures regarding for example efficiency are barely mentioned.

5.1 Reaction time for pressure build-up

The most relevant number is and will be the dynamic response of the system in a panic/emergency braking maneuver. For this criteria the time the system needs from panic braking initiation until a certain pressure is generated is taken into account. Typically the target pressure is the pressure which is needed to generate a brake torque, which is big enough to lock the brake disc, or in case of drum brakes the drum [3]. This key figure is referred to as time to lock (TTL).

5.2 Efficiency

So far the efficiency of brake systems plays only a minor role in their development. The brake system itself is in this case defined as the actuator (the booster) and the modulator (hydraulic unit). Influences by the hydraulic network, due to its centralized architecture, are not really studied so far. Emergency braking events represent only a very small amount of the total braking events over life time. Most of the braking maneuvers are in the range of comfort braking, which means low target pressure and low dynamics. Especially with alternative powertrain technologies like electric vehicles, a huge amount of the brake events can be covered by the electric motor running in generator mode and recuperating the kinematic energy. Due to that the overall distribution in the load spectrum is shifted. As many of the braking maneuvers of the low and medium pressure range can be covered purely by recuperation, only the standstill events of this group are shifted to low pressure, as the generator operation has, due to its characteristic, a lower limit of approximately 5-10km/h. Therefore the amount of low pressure blending events to standstill will increase. Nevertheless as the goal for the development of battery electric vehicles is to optimize the energy consumption of auxiliary devices the energy efficiency of the brake system has to be studied, especially at relevant operating points.

As the brake system can be regarded as one large pump, an approach similar to that of determining the efficiency of hydraulic pumps is used to study its efficiency. The degree of efficiency can be divided into a mechanical efficiency η_{mech} and a hydraulic efficiency η_{hydr} [4]. The mechanical efficiency η_{mech} describes how efficient for example the booster generates a force from the used energy of the onboard power system of the vehicle. The hydraulic efficiency η_{hydr} describes how efficient the hydraulic network transfers the generated hydraulic pressure and flow from the tandem master cylinder to the wheel brakes. The hydraulic losses can be divided into two main parts, one by pressure losses due to friction and resistances in the hydraulic network and the other one by the elasticities due to the volume consumption of the single elements of the brake system, like for example brake hoses. Together they can give an idea of the total efficiency η_{total} of the studied brake system.

6 Results for studied brake system

In figure 6 a fast actuation, typical for an emergency braking event is shown. The locking pressure for the studied vehicle under normal conditions, no heat induced fading effects, is around 90bar. In the shown test case the system has a time to lock t_{TTL} of around 160ms. It can be seen that there is quite a difference between master cylinder pressure and wheel brake pressure. As the total time to lock used to be something up to 500ms this difference was not that relevant. With an increase in dynamics and a drop of the characteristic time to lock to below 300ms, and for most new systems even below 200ms, here the difference of 40ms between locking pressure at the master cylinder and pressure at the wheel brake makes a difference of about 25%.

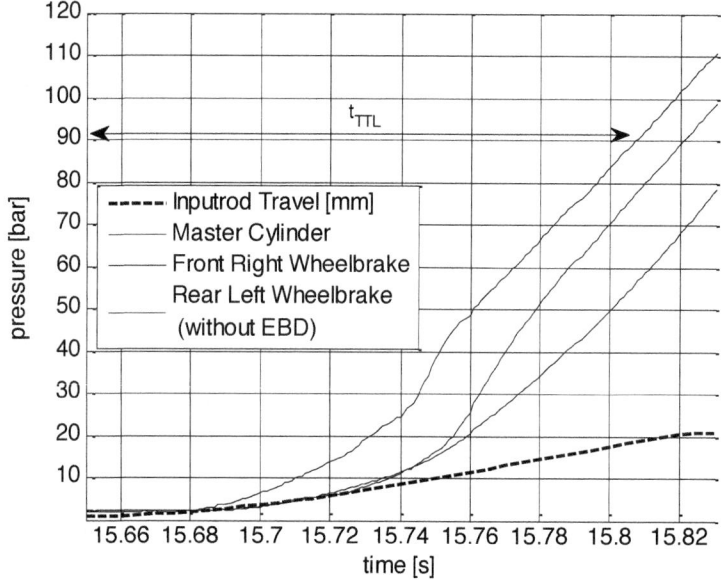

Figure 6: Time to lock for high dynamic braking maneuver at room temperature v=150mm/s, electronic brake force distribution (EBD) not active

In the case of low dynamic braking maneuvers the measured pressure is almost the same at every measurement point all over the system. As pictured in figure 7 master cylinder and wheel brake pressure plots are congruent. The efficiency of the system starts at zero, as the usable energy at the wheel also depends on the pressure and as it takes a certain amount of time before an increase can be detected there. Due to the determination method, which uses also interpolation, the very first part, up to 0.1s, delivers no valid results. During that phase the system is only consuming energy without generating any usable energy in form of wheel brake pressure. The mean efficiency from the generated motor torque of the booster to the measured wheel brake pressure for the shown test case is around 40%. It should be mentioned, that the pedal force is included in this analysis.

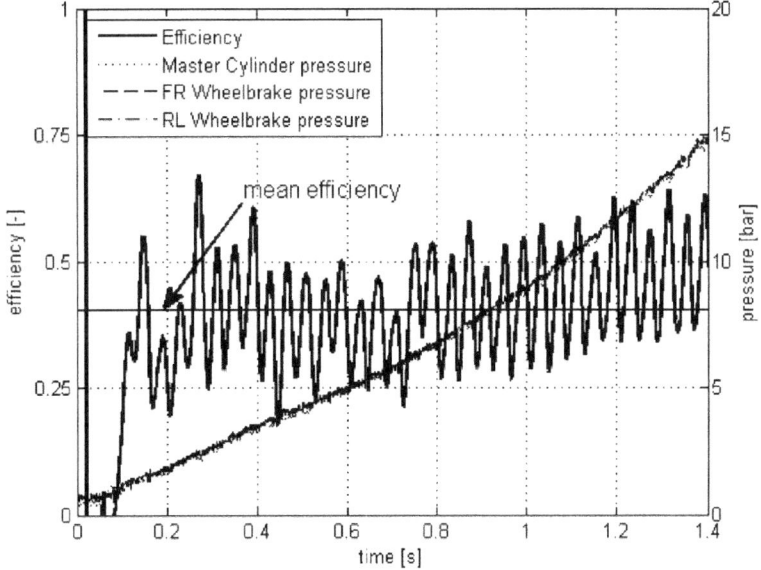

Figure 7: Efficiency booster to wheel brake for one circuit FR&RL for a low dynamic, low pressure pressure built-up, target pressure 15bar

7 Conclusions & Outlook

Autonomous driving and alternative powertrain technologies have a strong influence on the requirements for brake systems and their future development. They create a new group of fast growing criteria into the existing requirement portfolio. These criteria have to be fulfilled already to a certain degree by state of the art brake systems like for example systems with electromechanical brake boosters. Reasonable tools for the investigation of the characteristics, especially in an early development stage, are simulation and test bench studies. By combining these two tools, limitations of one tool can be overcome by the other tool. These tools help determining and evaluating key figures to characterize brake systems and make them comparable to other competitor systems, but also other concepts and architectures. For dynamic braking maneuvers the time to lock (TTL) is still the relevant key figures, but as the dynamic of the systems has increased so much over the last years a clear definition of the place where the pressure is measured has to be made. This is due to the latency of the pressure propagation in the hydraulic network, which starts to become more relevant with more and more decreasing TTL values. A new aspect which has not been studied in depth so far is the efficiency of the hydraulic network and the brake system in total. With a shift of the braking maneuvers load spectrum, due to the electrification of the powertrain, certain low pressure braking maneuvers will become more relevant for energy optimization of the system. For the presented low dynamic, low pressure test case shown here a mean efficiency for the studied brake circuit of 40% was found.

In future work these key values will be studied further for varying temperatures and test cases.

Bibliography

1. Bosch Mediaspace; http://www.bosch-mediaspace.com

2. ECE-R 13H/00 Suppl. 16: Braking (2015).

3. Breuer, B.; Bill, K. H. (2012): Bremsenhandbuch. Hg. v. K. H. Bill: Springer Vieweg.

4. Matthies, Hans Jürgen; Renius, Karl Theodor (2012): Einführung in die Ölhydraulik: Vieweg+Teubner.

Formula Student – A successful part of engineering education

Helena Ortwein, Simon Heußner

Rennteam Uni Stuttgart e.V.

1 Introduction

„The Formula Student competition is for students to conceive, design, fabricate, and compete with small formula-style racing cars."[1]

Formula Student is the European answer to the American Formula SAE™ that was launched in the United States in 1981 by the 'Society of Automotive Engineers' as college competition series. Based on the American model the Formula Student was launched in 1999 in Great Britain. The rules are very similar to those of the Formula SAE so that the teams can participate in events of both competition series. The European competition series also added the areas of project management and project presentation. Since 2005 also Italy is running an event and in 2006 the Formula Student Germany took place for the first time on the Hockenheimring. In the moment there are ten official events worldwide. The significant increase of participants around the world and especially in Germany proof the success of Formula Student.

2 Rennteam Uni Stuttgart e.V.

Rennteam Uni Stuttgart e.V. was founded in 2005 by Michael Kissling, Conrad Paule and Sebastian Seewaldt, who were students at the University of Stuttgart at that time.

Figure 1: Logo Rennteam Uni Stuttgart e.V.

They accepted the team management in the first two years and first tried to enlist support at the university. After that support was granted, 30 students could be won for the project – the "F0711-1" was born. Since then a team of around 35 students are working every year on a new race car for the Formula Student series. In 2015 the tenth race car – the "F0711-10"– was built.

After ten years of successful participation more and more team members finishing their study courses at the University of Stuttgart and started working in the industry. All Alumni received deeper technical knowledge and special skills during their participation at Formula Student which are required upon young professionals.

---

[1] Quote from international Formula SAE Rules

2.1 Race season schedule

Rennteam Uni Stuttgart separates the race season into different phases (Figure 2). Firstly the team creates in collaboration with the previous team a new race car concept. Rule changes or new technological ideas are feasible for the new concept. Secondly the design and development phase starts. The whole car is designed within three months – afterwards a complete 3D-CAD-Model of the new race car is ready. The manufacturing of the car starts normally in January. If all parts are available the assembly starts. The first shake down is a highlight of every season. After the shake down the race car is ready for the testing phase. In around 2 months the performance and of course the driving experience will be improved. The last phase of every year is the most nerve-racking: The competitions!

CONCEPT	CONSTRUCION & DESIGN			FABRICATION & ASSEMBLY				TESTING		COMPETITION		
September	October	November	December	January	February	March	April	May	June	July	August	September

Figure 2: Race season schedule

3   Engineering education

The amount of first year students increases year by year – alike the number of female students. [2]

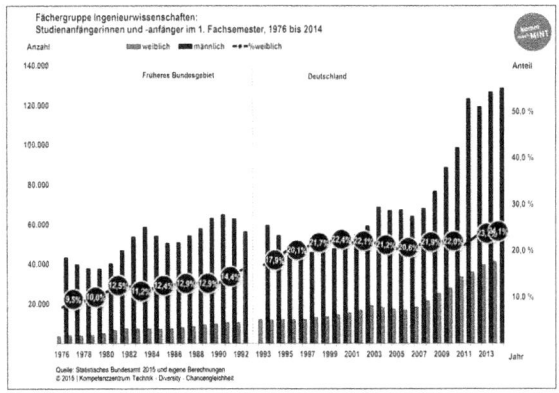

Figure 3: First year students in engineering courses

---

[2] http://www.komm-mach-mint.de/Service/Daten-Fakten/2014/Studienanf-FG-Ing-Studienjahr-2014

3.1 Education at the University of Stuttgart

The classic engineering course at the University of Stuttgart is subdivided into a multi-annual Bachelor and Master study. A study plan is given to every student, which serves as a guide throughout his studies. Figure 4[3] shows for example the study plan of the Bachelor course for automotive engineering at the University of Stuttgart.

Makrostruktur B.Sc. Fahrzeug- und Motorentechnik - PO 2015

Grundstudium				Fachstudium	
1. Semester (WS)	2. Semester (SS)	3. Semester (WS)	4. Semester (SS)	5. Semester (WS)	6. Semester (SS)
Höhere Mathematik I + II 9 LP	9 LP	Höhere Mathematik III 6 LP	Numerische Grundlagen 3 LP	Pflichtmodul "Messtechnik - Fahrzeugmesstechnik" 3 LP	3 LP
Experimentalphysik mit Physikpraktikum 2 LP	1 LP	Schlüsselqualifikationen (fachaffin) Technische Akustik / Technische Schwingungslehre 3 LP	3 LP	Pflichtmodul mit Wahlmöglichkeit Fahrzeug- und Motorentechnik I Wahl aus Kompetenzfeld I 6 LP	
Werkstoffkunde I + II mit Werkstoffpraktikum 3 LP	3 LP	Technische Thermodynamik I + II 6 LP	6 LP	Pflichtmodul mit Wahlmöglichkeit Fahrzeug- und Motorentechnik II Wahl aus Kompetenzfeld I 6 LP	
Technische Mechanik I 6 LP	Technische Mechanik II + III 6 LP	6 LP	Technische Mechanik IV 6 LP	Wahlpflichtbereich Wahl aus Kompetenzfeld II oder nicht gewähltes aus Kompetenzfeld I 3 LP	3 LP
Konstruktionslehre I + II mit Einführung in die Festigkeitslehre 6 LP	Pflichtmodul mit Wahlmöglichkeit (Konstruktionslehre III+IV / Konstruktionslehre III/IV-Feinwerktechnik) 6 LP		6 LP	Wahlpflicht - SQ (fachübergr.; Uni-Kat. Ber. 1-3) 3 LP	Schlüsselqualifikationen (fachübergreifend) 3 LP
Fertigungslehre mit Einfg. i. d. Fabrikorganisation 3 LP	Einführung in die Elektrotechnik 3 LP		Pflichtmodul Regelungs- und Steuerungstechnik 3 LP		Pflichtmodul Technische Strömungslehre 6 LP
Schlüsselqual. (fachaffin) Grundz. d. Angew. Chemie 3 LP	Schlüsselqualifikationen (fachaffin) Grundlagen der Informatik I + II 3 LP	3 LP	Schlüsselqualifikationen (fachübergr., Projektarbeit) 6 LP		Bachelorarbeit 12 LP
Summe: 29 LP	Summe: 31 LP	Summe: 33 LP	Summe: 30 LP	Summe: 30 LP	Summe: 27 LP

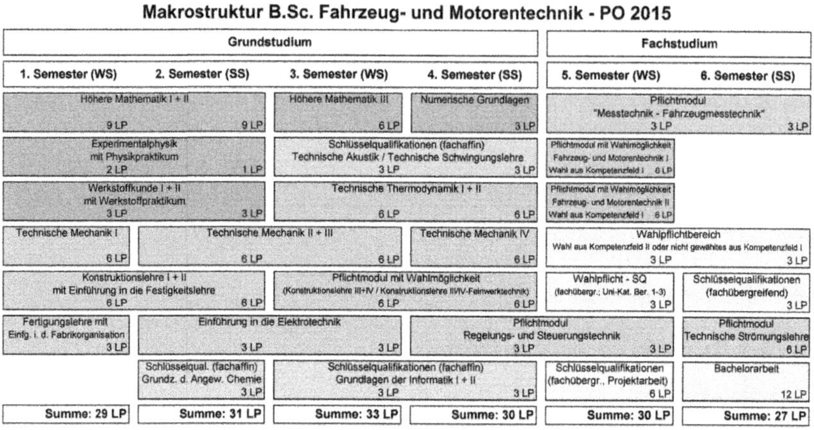

Figure 4: Showcase: study plan

Besides the general engineering fundamentals – Mathematics, Mechanics, Thermodynamics, Material science, Electrical engineering, etc. – students have the opportunity to choose between modules of their self-interest. The studies at the university are overall theoretically arranged leaving little to no chance for practical activity. Therefore, many students choose to attend internships, academic assistant's activities in the pursuit of gathering practical experience. Furthermore, student groups in all universities exist, which provide others the opportunity to collect the wanted experience. For this reason, the interest of engineering students and especially those who are studying automotive engineering increases in attending such a group, which participate in the Formula Student. At the University of Stuttgart, students have the opportunity to do so in the combustion team: Rennteam Uni Stuttgart e.V. and in the electric team: Greenteam Uni Stuttgart e.V.

---

[3] http://www.uni-stuttgart.de/bologna/modulhandbuecher/studienverlaufsplan/SVP_BSc-FMT_V15-PO2015.pdf

3.2 Education within Formula Student participation

Approximately 35 students participate at the Rennteam Stuttgart each year. Most of them are engineering students. A small amount studies management or marketing.

3.2.1 Deeper technical knowledge

All team members use, similar to the industry, many methods, processes and tools for the cars development. The best performance of the race car is given by a lap time simulation. With this lap time simulation an analysis of key parameters is possible as well as a comparison of parameters.

Firstly the whole car with all details will be constructed within a CAD Tool like Catia or Siemens NX. Therefore a lot of time is needed.

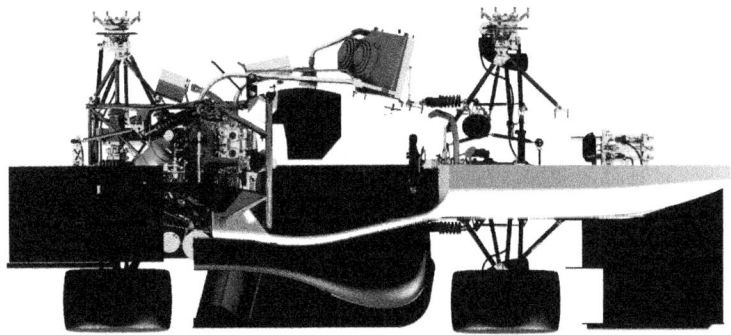

Figure 5: CAD model of "F0711-10"

At the end for every single part a drawing exists. All drawings are derived by oursel-ves. Many parts will be produced in-house. In our workshop are many machines available – for example a milling machine. With loving detail the parts and finally the complete vehicle will be manufactured and assembled.

Besides the construction and manufacturing of the car simulation is also an important point.

Every year team members are working with several simulation programs. These are used for applications such as simple component optimization but mainly for complex 3D aerodynamic research or combustion engine simulations. For every new member the topic of research is at most cases uncharted territory which means they have to firstly come to grips with the topic. In this way, basic knowledge learned thought out the studies becomes detailed expert knowledge in the length of the season. Such topic-experts are the ones pursued right now by the industry.

Figure 6: Simulation model of "F0711-10"

With the support of the FKFS Stuttgart, the Rennteam Stuttgart had the opportunity on the "Tag der Wissenschaft" (20.06.15) at the University to validate their simulation results in the wind tunnel.

Figure 7: "F0711-10" in the FKFS wind tunnel

The results of the simulations as well as the validation data of the wind tunnel subsequently become documented in a report and function as discussion topics at the competition. In the static discipline "Design event" team members try to advocate the concept of the race car in a discussion with industrial and motorsport engineers.

Additionally, during the production and assembly of the race car even more knowledge and experience gets gathered.

3.2.2 Working within a team

The Formula Student Team of every University changes every year. A small amount participates a second or third year. Normally these team members will get the chance to grasp a leader position. The Rennteam Stuttgart is organized as following:

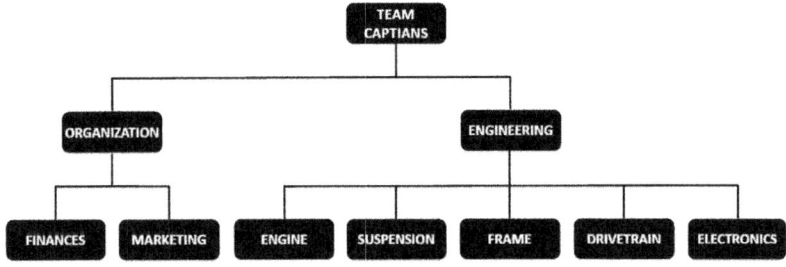

Figure 8: Structure of Rennteam Uni Stuttgart

Experienced team members become team and sub team leaders. All other further work packages in the particular sub teams get written out and every student interested in the particular work package has the chance to apply for this precise position. After an intensive selection process, the new team is formed. After that, a team building event takes place which its structure is based on the actual structure of the team's season, divided in four main phases. Assisted by miscellaneous team building processes, the mixed group of people becomes a high performing team. "All for one and one for all" is alive throughout the whole season.

Rennteam Uni Stuttgart follows the classical teambuilding process[4]:

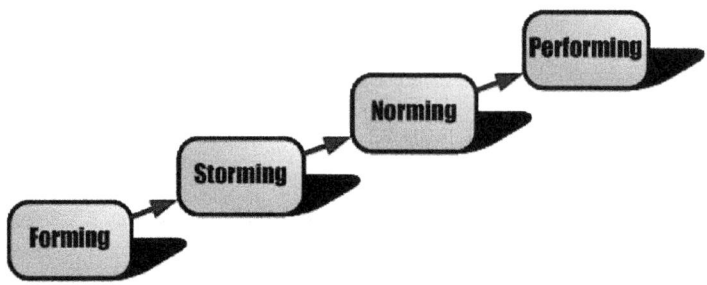

Figure 9: Team building process

3.2.3 (Sub) Team Leader Position

The (sub) team leader positions are a great challenge. Team leaders are responsible for the overall goals and the management of the whole team. The Rennteam Uni Stuttgart follows the goal „Complete – Finish – Win". Derived of these overall goals the sub team leader set the goals for the sub team. All sub teams working closely together to keep the overall goals alive.

Especially the motivation wobbles during the year. It is the task of the team leaders to motivate each team member – if it's necessary. All participants are volunteers – lastly there is no greater reward possible as the victory of a Formula Student competition.

4 Requirements upon young professionals

Every new students target is for him to successfully end his studies, generally followed by the search of the first professional employment. The demands of today's industry are diverse. On top of the technical knowledge, personal properties and soft skills round up the job profile. The following properties/experiences are most commonly requested: Experience in project management, the ability to work in a team, resilience, flexibility, willingness to travel and the knowledge of foreign languages.

For this, the students get exposed to additional commitment that has to be made. On many occasions this becomes the opportunity to establish these appropriate abilities. Participating in the Formula Student is optimal for doing this, during this activity the

---

[4] http://leadingtechnicalteams.com/you-can-survive-and-thrive-in-the-4-team-stages/

student are able to experience besides the technical enlightenment, soft skills which could afterwards be of importance for his professional career.

Many companies recognized the potential of Formula Student participation. Reasoning is an investment into marketing and early binding of participants.

5 Feedback of Formula Student Alumni

Numerous Rennteam Stuttgart Alumni are looking gladly back to their participation at Formula Student. Some reached very special challenging jobs at companies or in motorsport. They will never forget the experience of Formula Student. The F0711-10 organization team wrote a jubilee book for the tenth Rollout in 2015. Inside this book are interviews and quotes from different Rennteam Uni Stuttgart participants.

Several Alumni answered the following question „In my opinion Formula Student is …":

- a striking example for what is possible with new knowledge, highest motivation and strongest team spirit
- a great thing and probably the best project I have attended. during my study course
- the best example to improve motivation, ambition and passion
- lots of fun with friends – and by the way the start of my career as motorsport engineer
- passionately preparation for my personal career
- the plattform of my life
- to cope challenges within a teamn, assume responsibility and outgrow itself
- in total a successful story, branded with love of detail, will for victory, sleepless nights, team spirit and fun!
- an interesting and helpful time during my studies and meanwhile the best pool for best educated and motivated young professionals

6 Summary

Since the establishment of Rennteam Uni Stuttgart every year a new race car is designed, constructed and manufactured pursuant to the objective "Complete, Finish, Win". The Rennteam Uni Stuttgart counts to one of the most successful teams worldwide. A high-performing race car with a high performing team is just one side of the story. Without all the support of our numerous sponsors the project would not be possible. Our sponsors give us every year the opportunity to experience invaluable adventure, which enrich the whole course of our studies. For this reason, we would like to thank all of our sponsors and supporters.

Formula Student is definitely a successful part of engineering education!

Figure 10: "F0711-10"
Official FSG 2015 team photo

Electromobility in daily life – Are you still exploring or riding already?

Matthias Vogt, bridgingIT GmbH

Motivation

What drives an IT-consulting company as bridgingIT to convert its fleet to the greatest possible share of electric cars? What do you need therefore and what has to be considered? Which challenges arise? What is the economics of this plan look like? May this even be more economically than expected, since the operation of electric cars actually is cheaper?

Hand over of keys for the electric bridgingIT fleet

Those questions came up at the beginning of 2014 when the managing director Klaus Baumgärtner with a team of employees starts with the project. The unobstructed view of the reality of electromobility still shows, that actually every participant involved is waiting for each other – either on special tax offers or on attractive products of OEMs.

However, we did not want to wait, but just get started.

Confirming the motto: "Practice what you preach" we didn't want to longer talk about electric mobility at that time, but get experience with it at first hand. One reason for that was that our consulting company offers special services under the keyword "Connected mobility". Digital information and the connection of traffic participants, infrastructure and services, will be the drivers for the mobility and the development of entire urban space in future. For many years already we advise other companies in the context of Smart Energy and Smart Mobility and so it is important for us to show that electric mobility today is not only working on short distances, but also on the long distances. We wanted to prove that it is possible to implement electromobility in fleets – not only as pool vehicles for city driving or working traffic. We wanted to be a positive example and encourage other companies to follow us with electrification of their fleets. At bridgingIT we always have had a great interest in sustainable interaction, which is also reflected in our value system. As a consulting company with relevant travel activities moreover we can reduce our own CO_2 footprint in mobility at the same time.

Three premises should be observed to in this project

- There will always be a "1: 1 exchange" of combustion car against an electric vehicle place. A back-up is not provided. Even half-pregnant solutions such as plug-in hybrids, were not possible.

- This should not be a Board project with the typical publicity effects of changing one or two board vehicles to green. All employees can be part of the project, especially also Sales colleagues with high mileages.

- The electric-vehicles will drive for three years leasing without any subsidy pot existing.

Which benefits persist for bridgingIT?

If the only thing to be considered are the total cost of ownership, today electromobility in fleets will be economically only in very rare cases. Why it is worth to do this effort anyway, even if electric cars appear uneconomical and far too expensive at first glance?

In addition to pure economic figures, the investment in electromobility will pay back for bridgingIT in short and long term by positive effects such as credibility, sustainability, image, marketing effects and motivation of employees.

Credibility already was mentioned as motivation reason. Through the use of electric vehicles, we can provide a real proof of our innovation attendance and present ourselves as a company, that shines through modernity and sustainability and act as a pioneer and multiplier that encourages other companies to take this step as well.

Another aspect for using electric vehicles are marketing effects. Therefore, the popularity of our medium-sized company could be increased by press releases to a dimension, that would have been impossible to reach or been very expensive with usual marketing methods such as advertisements and advertising. Also the "viral marketing" effects should not be underestimated.

Additionally based on the electric cars it is much easier to start a conversation with potentially customers. It was possible to get a six figures assignment with a new customer and two further contracts are in preparation for example with its origin in the electric fleet.

Recruiting

An underestimated advantage is to have a flexible and attractive mobility concept with electric cars. This is a very effective recruiting instrument for hiring new or obtaining existing employees. As bridgingIT is a young company with an average age of 38 years, the employees are part of a generation, for which the topic sustainability is very important.

In the IT industry, good junior staff is rare and there is a "war for talents". Because of the publicity of our electric fleet, many applicants put their attention on us and in the meantime several new employment contracts already could signed, which have their origin in the electric vehicles visibility.

In addition, the longtime employees observe the electric car as a relevant differentiation feature that increases satisfaction of the employees and can be a reason for long-term loyalty to the company. Maybe this is also a reason for a turnover rate below average.

Consulting offer

As already mentioned, we want be a good example and encourage other companies or public fleet operators to do this step as well.

We want to support interested companies in introducing of electric vehicles into their car fleet. Our big advantage is the practical first-hand experience with implementation into fleets, as we know what we are talking about.

Operation scenario of the car fleet

For better understanding of the project context, a few facts will be helpful. In the IT-consulting company bridgingIT more than 400 employees currently working at the locations Mannheim, Karlsruhe, Frankfurt, Stuttgart, Cologne, Munich, Magdeburg and Zug (Switzerland). The target group includes companies in the upper mid-sized and large customers (corporations). BridgingIT supported its customers in the implementation of corporate strategies and the use of modern IT technologies.

In order to meet customer requirements and the resulting high mobility demand there are several different mobility solutions such as the classic "motivation fleet", TrainCard

50 and 100 and the car-sharing available. Momentary our company car feet has about 140 predominantly personalized company cars.

In addition to the existing 2 short-range electric cars (BMW i3 and Smart ED) in first step 10 long-range electric vehicles of Tesla Motor company were purchased and delivered in December 2014. After that the electric fleet has increased in 2015 for further 8 Tesla up to the current number of 21 full-electric vehicles, which corresponds to a share of approximately 15%. In 2015 there were completed over 500,000 km and saved 75t CO_2 with it. In 2016 we expect approx. 60,000 km can be electrically driven per month and thereby saved over 10 tons of CO_2 every month.

The user scenario of the vehicles is a typical application for a consulting company with several locations in Germany. The employees use the vehicles as company cars and drive them privately – that's why there is a mobility demand, which is comparable the use of a classic combustion car.

All vehicles are used in the classical field service. In business use there are primarily customer appointments and projects. Because all drivers also use their company car privately, one way or another holiday and excursion will be added. However, it is important that the train system is used for transport and we want to clearly emphasize that the electric fleet is no signal "pro individual traffic". But if traffic cannot be When the water-but can not be avoided, so it should cause so low CO_2 as possible.

The typical distances of a business trip are between 150 and 300 kilometers a day. This daily distance is possible without charging the car. For longer distances of 400 or more kilometers, you can use the Tesla Supercharging network free of charge, where the car is charged in 20 to 30 minutes up to 80%.

In the meantime, our employees have traveled very different kind of journeys – skiing in Switzerland or Stubai valley, trips to South Tyrol, the Côte d'Azur, northern Germany or to Amsterdam.

Challenges and fields of action

During the electrification project we quite quickly recognized, that this is more than replace combustion cars with electric cars and several fields of actions had to be mastered and issues to be addressed.

Reactions of customers and colleagues

Initially there were some concerns within the company in relation to the representation of our company and the image of our drivers. We did not want that our customers misinterpret the large vehicles as a luxury project. Nor should any envy discussion within the company start. The selected model type is certainly in the category of a high-priced vehicle and therefore it was feared that the driver can be reduced only to the size and the performance of the vehicles.

The actual motivation, sustainable mobility and innovation, can get very quickly in the background. But the fact is, there is no alternative car to Tesla Model S available. It is currently the only long distance capable electric vehicle on the market. Because of the transparency throughout the project and communications, the concerns of our employees could also be extinguished and responses now are very positive. At least when employees themselves have had the opportunity to feel electromobility themselves, the view has changed significantly.

Everybody understood, that reasoned by the contemporary and sustainable mobility options we are also in future an attractive and interesting employer. On the customer side, the picture is a little more differentiating. The majority appreciated our commitment and takes it quite well and creditable, but we must continue to explain a lot for not being reduced to a "expensive" car fleet.

Integration in processes and functional units

By fleet electrification different functional units are affected and should be involved. Therefore you have to talk with fleet manager, as well as the recruiting, the HR department and the Marketing department.

But the process owner is and remains the fleet manager. All processes, such as switching from fuel to a charge card or the installation of personal charging infrastructure at the employees home have to be clarified, before any company car user orders an electric vehicle. Firstly, the issue appears complex. But if you start with replacing the uncertainty by your own experience, you learn that this complexity can be managed. If you don't dare to do this by yourself or not want to operate this effort yourself, it is advisable to take professional support.

With this knowledge not only singular parts of the fleet can be electrified, but also change the entire company car policy on green processes and achieve an unexpectedly large penetration of electric cars.

A mammoth task one might think, but it is all a matter of planning and the will to really try it consistently. The central slogan is: The fleet manager is no longer a TCO manager, but becomes a Mobility Manager. This ensures that the switch from combustion cars to battery cars does not end in frustration for those who are the addressees of this green philosophy: the company car users.

Expectations of drivers

Often, the actual user is not involved enough in intense changes. However, the transition to electric mobility must be well prepared. Otherwise, the whole thing ends for company car drivers with great frustration. The driver has to be informed also about the disadvantages. In addition, electric cars require changes in processes and should be integrated accordingly. This begins when switching from fuel to a charging card and extends to the establishment of personal charging infrastructure.

Is electro mobility for all employees suitable? No, not at all. It is clear that the driver has to change his travel behavior. This is not possible for everyone and not everyone car driver is also ready to change his behavior. An electric car is more than just another company car, the switch should be monitored and the driver will be informed and educated before deciding on the changes in his personal mobility. Therefore, the company should consider carefully in advance, how the decision fails. There must be people who are willing to get involved with all the consequences on the topic. Otherwise, the first initial fascination can be fast over.

How it was decided which employees can get an electric vehicle? Everybody who is entitled to a company car with a particular remaining term for the company car. A total of 40 employees were eligible. All of those were asked, how they think about an electric car and 18 of them were interest in it. Ten people finally have decided for an electric car, for the other eight employees it just did not fit the use aspect.

In many companies, the decision for electric mobility was taken without sufficient basics that often leads to anger and resentment. Anyhow, the employees decide for an average of three years for their new company cars and have to pay tax for a approx. 85,000 EUR expensive car. This tears a large hole in the cash, so details are relevant. Such as whether the company accepts a portion of the extra costs. Finally, the car is supposed to have an added value for the employees and motivate him.

Most company car drivers often are initially impressed by the technics of electric vehicles, but this fascination may fade away after two or three months again. Then the capability for everyday life is counting. As the car in terms of sacrificed compensation for company car users represent a high value, an employee cannot be forced to conversion to electric vehicles, including their disadvantages. Otherwise, the opposite of the intention will be achieved.

How can you prevent someone wants an electric car, which is either personally or because of the ambient conditions not suitable for it? First, the emotional aspects should be clear. Is the driver ready for e-mobility at all? For this, the fleet manager must not only tell the benefits, but also precise information about possible restrictions. For example, that you have to put a dirty charge cable in the trunk sometimes wearing a suit or you simply will not find free and functioning charging station. To find the right candidates the responsible must be look very closely. We have developed a special questionnaire, which can be used to classify each driver individually.

Technics and Types of vehicles

Which vehicles are possible as personal company car? We have considered various models and tested the BMW i3 in continuous operation – for more than twelve months. The experience showed that the i3 is working fine as a pool car or for company traffic, but not as a personal company car. The i3 has the same weakness as most other offered electric vehicles when driving in bad weather or with higher speeds – the range is just too small. For schedule-driven company car users, this is not suitable for everyday use scenario.

Currently only the Tesla Model S meets all the requirements for a long-distance vehicle in this category. Finally, the employees take an average of 35,000 kilometers per year. Only the Tesla achieved with more than 300 km reasonable range for use in the field. So far, there is unfortunately no alternative all-electric models from other manufacturer.

Further vehicles with range extender or the plug-in hybrid engine are not in our intention. These models either enable because of the additional combustion engine a greater range, but also combine the disadvantages of two drive systems such as higher weight, higher costs for acquisition and operation and double expenses in fleet management. In addition, the pragmatism of the driver should not underestimated, so that the supposed eco-advantage in practice can quickly turn into fizzles and even in a disadvantage.

Charging at home and at the company

One of the key figures relating to electric cars is where and how they can get charged.

Following that, it has proved to be absolutely right and important that we have looked in advance intensively with the charging infrastructure and how to purchase power. So we could adapt processes quickly and created charging options. Thus, the transition for our drivers designed easily.

As most of our car fleet are personal company cars the responsibility for the loading management is up to the driver.

So one of the selection criteria for the appropriate drivers was, that they have the possibility to install a home charging capability for the electric car.

The experience has proven that about half of the charges done at home. Another large part of charges made by the Superchargers supplied by Tesla.

Charging at Tesla Supercharger

At our locations in Munich, Stuttgart, Karlsruhe and Frankfurt charging stations have been installed, which can be used from everybody with an electric car. Of course electromobility is only really sustainable, if it is powered by renewable electricity. As the cars are often charged at home, the drivers have committed themselves to choose a green electricity provider for their household.

Economic feasibility

The costs were one of the biggest challenges in this project. In euros and cents, a fleet of electric cars compared to conventional vehicles today is usually still more expensive, it doesn't matter whether purchased or leased. Also with encouragement by participating in a research project and the eco-bonus of Tesla is not preventing, that electric cars are still more expensive than conventional cars. So unusual and new business models to cover the costs must be found. All efforts around the electrification of the fleet in the company has to get a value. The internal currency must be allocated to all cost centers in the company. E.g. if due to the positive response to the electric vehicles an Advertisement for recruiting can be saved, these saves has to get transferred from Human Resources in the fleet budget.

In the complex calculation, each single screw can be easily varied, so that electric mobility will quickly get unprofitable. We counted all the expenses to obtain all the relevant cost positions. There was a funding gap of 250 Euros for each car. From a driver's perspective, there is also still a taxation issue due to the higher list price, and the one-time investment for personal charging infrastructure. Overall, these are monthly 368 Euro additional costs for each driver. This issue was implemented in the personal objectives of the employee.

Basically we expect an improved business success, which can offset to the higher costs of electrified fleet. This ambitious step requires not only courage, but also financial concessions of the drivers. Most of colleagues are willing to do this – more prestige is the reward.

Experiences

Everyday capability

After more than one year and 500.000 km there are a number of individual and singular experiences at each driver.

Overall we can say: Electromobility is working also on the long distances perfectly – as long as the car and the charging infrastructure concept fits and the drivers are willing to question their personal mobility behavior and - where necessary – adapt it. The rewarded for electric driving on the long distances is great experiences, that nobody wants to miss any more.

Capability of electric car in winter conditions

By our own surprise we haven't been able to find any real disturbances. There is still no meeting cancelled, all planned trips took place and still nobody has had to spend an extraordinary night in a hotel or a car due to lack of charge.

Of course, the driver ahead has to plan where and how he can charge the car. Most of this planning is covered by the navigation system of the vehicle, the remaining plans can be done easily by the electric car drivers itself with a little experience, good charging infrastructure information Apps and a positive attitude. So far nobody of the employees yet regretted the exchange or would like to switch back to a combustion car. Moreover, there are even more interested colleagues for electric vehicles.

On the vehicle side, we are surprised by the high level of technical maturity of the relatively young vehicle technology and it seems to be true: as electric cars have much less parts than combustion cars, less parts can get a defect. So in 500.000 km there was just one car broke down on trip and one vehicle would not start in the morning at the home. This is absolutely competitive compared to a conventional car fleet.

The biggest obstacles always showed up, if the driver had to rely on the public charging infrastructure. Because of the fast growing Tesla Supercharger network this wasn´t necessary so often, but still was occasionally. To be able to charge at a public charging station at the scheduled time indeed is and will remain a gamble and depends on several factors. Is the charging station occupied? Is the charging station parked with a combustion car? Does the charging station work either? Do I have the right charging card or is it able to do roaming? Because of these many uncertainties it happened in several situations, that vehicle could not be as charged up as foreseen.

Changing attitudes and travel behavior

Apart from the question that electromobility works in everyday life, some colleagues changed their traffic behavior and also their personal attitude on various sustainability topics.

The fact that you consciously think about the effects of the own mobility on the environment, behavior and attitude changes have the shown that these colleagues consume conscious and move around and are more consider the impact of their actions on the environment than before. Therefore, you have to plan for example, a travel and trip ahead completely, to see whether and where to charge up and how much time you have to provide therefore.

It also appears that an electric car in the stress of everyday life and transport can help to slow down and drive more relaxed and less aggressive, as you look more ahead. Thus, a deceleration of the driver could be determined in the charging breaks, which must be carried out while driving an electric vehicle in a certain regularity.

An often underestimated comfort feature is the silent driving without humming noise and vibrations. For example, some drivers who were previously fascinated of the V8 engine sound now feel this as ugly noise and appreciate the sound of silence in the Tesla.

Only when driving in parking lots or in pedestrian areas the drivers wish to have a noisy horn similar to a bike bell to make dreamy pedestrian attention.

Digitalization and connectivity

Another fascinating development is the digitalization of the vehicles and their connection with the environment. E.g., the Tesla is always online via integrated SIM card, the driver is able at any time to carry out different actions like opening and closing the car, starting, watching the charging status, settings, operate climate control, and locate the car remotely via cell phone app.

In case of any problem, the technical support has access from anywhere to the diagnostic data of the car, carry out an initial diagnosis and improve support. For this reason in many cases trips to a workshop will not be necessary and saves the customer time. The workshop process, which was the same for decades, will change dramatically with the new possibilities. In addition, the combination of remote support and diagnosis will reduce the necessary service network and save many costs.

Particularly drivers with small children appreciate in the cold winter and hot summer the pre-climate function (heating and cooling). It is possible to adjust the comfortable temperature for the small passengers and in just few minutes, the car is warm or cold.

A further disruptive highlight are the SW-Updates over the air. So in less than one year the cars turned from vehicles with standard cruise control as the only one assistance system to a self-drive car with autopilot functions with many assistance systems just 10 months later. This development in such a short time was very impressive and game-changing and shows, what customers are expecting in the future of vehicles of all brands.

This leads to, that the cars get new functions and are getting better, as older they become with each software update. Probably we will see this also in higher resale value as known for similar cars.

As IT company we know the rapid development of IT and telecommunication industry, so we are sure this will be just the beginning of digitalization. For example on the big touch display there is still a lot of space for new apps and functions. The playground is huge and is just waiting for new, appropriate business models.

Conclusion

After more than a one year, we can say that it was absolutely the right step at the right time for us, with which we do justice to our own claims in terms of sustainability. The use of electric cars for us is no longer a project, but a regular operation. We also get a lot of positive response from customers and other companies and it turns out that many are interested also to electrify part of their fleets. Further it became apparent that our employees are proud that we not just talk about innovation, but live them actively. This has of course a positive impact on motivation and our image.

Preview

Based on our experience, we want to give some recommendations and new thoughts. These recommendations are directed on the one hand to the car manufacturers, who also wants to sell cars in future and the politics, how to push the goal of sustainable mobility solutions.

Car Manufacturer

Every driver who get experience with an electric car for longer time won´t go back to combustion cars and is mostly lost for this market.

So the offers of road capability electric cars has to grow fast. Once here a competition arises, a lot of improvements and economies effects can be expected, which will activate the market in disruptive way. This will result in electric cars will be better and cheaper in a few years than comparable combustion cars and the market changes very fast. Maybe you remember: The first home flat TV was presented in 1999 for 13.000 €, in 2006 more flat TV were sold than CRT TV. The point when nobody wants to buy a combustion car will come, just be prepared.

Politics

If the market for electric cars will be boosted seriously, then electric cars should have no significant disadvantages in the TCO compared to combustion cars. This shows the way how to push the market.

Electric cars need a tax framework, which makes them equal, but not better in comparison with combustion cars. For example, the tax on company cars could still more reduced or the tax advantages for diesel fuel should get eliminated.

Also it will help a lot, if a nationwide public charging infrastructure with easy access and moderate cots will be available to ensure the mobility of electric cars.

System-integrated data acquisition in validation and operation of electric vehicles

Dipl.-Ing. Katharina Bause[1], Dipl.-Ing. Florian Munker[1], Dipl.-Wirtsch.-Ing. Nicolas Reiß[1], Dipl.-Ing. Armin Rupalla[2], Dr.-Ing. Matthias Behrendt[1], Prof. Dr.-Ing. Dr. h. c. Albert Albers[1]

[1] IPEK – Institute of Product Engineering at Karlsruhe Institute of Technology (KIT), Kaiserstr. 10, 76131 Karlsruhe, Fon: +49 721 608 46992, Fax: +49 721 608 46966

[2] RA Consulting GmbH, Zeiloch 6a, 76646 Bruchsal, Fon: +49 7251 3862-0, Fax: +49 7251 3862-11

1 Introduction and Motivation

The success of electro mobility depends on its sustainability for daily use. The integration of electro mobility will be supported by the benefits compared to existing mobility solutions and infrastructure. Currently, integrated concepts, models and methods for systematically evaluation of electro mobility's adoption under different criteria are missing. Thereby, regarding at electro mobility as well as neighbour domains is important.

Regarding at fleets, the aim is avoiding bottlenecks or oversupply. For an efficient distribution and charging of electro mobility fleets, typical fleet processes have to be modelled and simulated. Fleet operators, like car-sharing provider, make heterogeneous car fleets in urban environment available. In context of electro mobility, new issues arise, e.g. limited usability caused by limited range and required charging times between potentially bookings.

Providers aim to regain the previous car availability. Caused by current usage rate conservative estimates of range are sufficient. For increasing usage rates, the accuracy and reliability of range forecast have to be improved. Rapid charging concepts displace the problem of low energy density but do not solve it. Thus, realistic planning of individual electro mobility therefore requires more precise and generalizable prediction models with respect to the energy requirements for a ride, but also to the expected usage time, type and duration of vehicles. The focus of further research is:

- improve forecasting quality through the use of field data and inclusion of different usage scenarios (commercial, private, shared),

- evaluation of energy consumption and different usage and booking scenarios,

- incorporating standing and charging times in forecast of the vehicle availability,

- generalization of energy demand forecasting as well as optimization strategies for different electric vehicles over their lifetime.

Starting point for the aspects mentioned above is to improve the underlying data, which is to be created with a customizable field data library.

Three cluster projects (Cluster electric mobility south-west) „IeMM – **I**ntermodales **eM**obilitäts**m**anagement", ELISE – Autonom**e** **L**adeeinheit und **s**ystemintegrierter Daten-Gateway für **E**lektrofahrzeuge" and BiE – **B**ewertung **i**ntegrierter **E**lektromobilität" deal with the partial aspects of the issues mentioned above. This paper discusses the implementation of an integrated system for data acquisition and its potential for the vehicle validation and usage. Additionally to data acquisition, this means the use of data for optimal design, operation and monitoring of mobility services.

2 State of the art

2.1 Conventional data acquisition

Conventional data acquisition in vehicles acquires signals from internal sensors or from additional applied sensors. This data acquisition systems can be system-integrated or add-on devices.

There are two significant systems in the automotive domain, which are relevant for this case: add-on data loggers and system-integrated telematic solutions of OEMs.

Data loggers are sophisticated add-on devices, which can be connected to the vehicle's bus systems and acquire data in an array of additional applicable sensors. Their use case is acquiring data for engineering in the automotive development process. However, the cost of 2.500 – 5.000 € per device is suitable for this case but too expensive for scaled use in fleets. These devices are designed for acquiring and logging data, they are not capable of streaming live data to a server backend.

System-integrated telematics solutions of OEMs are integrated in the head unit and are capable of streaming live data. They can be connected to the bus systems of the vehicle but are not designed to be extended with additional sensors. Configuration and datasets are accessible to the OEM only.

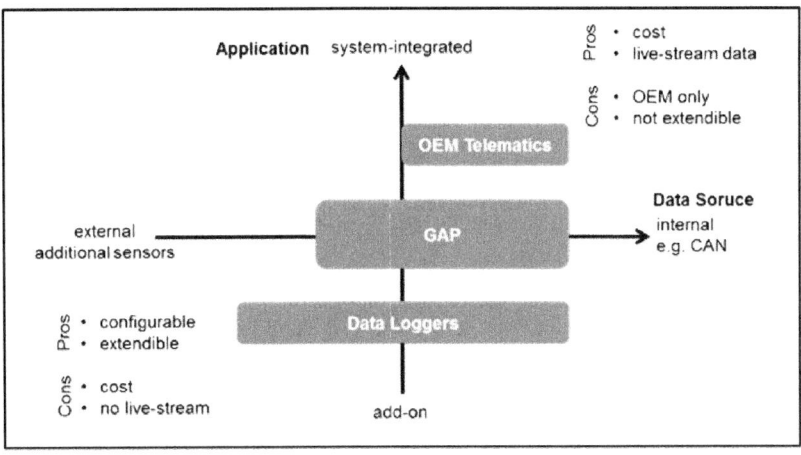

GAP in automotive data acquisition

645

2.2 System-integrated data acquisition in BEV

There is a gap in conventional data acquisition solutions concerning affordable integrated applications for fleet use. These need to be capable of live-streaming vehicle data and configurable and extendible to meet requirements of future use cases. The data acquisition has to work in fleets with vehicles of different OEMs.

In the project ELISE, these requirements have been considered and a system-integrated data acquisition system has been designed which closes the gap[1]. It consists of a control unit box in the vehicle and a server backend. The control unit box is an embedded system based on the existing product "Flea Box" of CarMediaLab GmbH which can be configured remotely and extended with additional sensors. It connects to the vehicle's bus systems and to the web application of a server backend through an encrypted connection via cellular network. A mobile application has been developed demonstrating the live stream from the vehicle on a mobile device.

"Flea Box"[2]

2.3 Consumption models and range forecast

In practice, the fleet operators previously use conservative estimates of the range of electric vehicles. These are data, which are given by manufacturers or base on standardized cycles. For further integration of electro mobility by increasing user acceptance and thus achieving higher use rates, more reliable range forecasts are required.

The previous used, conservative estimations do not consider a variety of factors. Experience shows, that range reduces fast if heating or cooling are working.[3][4] Besides the

---

[1] Munker et al. 2014

[2] CarMediaLab

[3] Conradi 2014

[4] ExxonMobil Central Europe Holding GmbH 2011

obvious influencing factors like auxiliaries, even the interactions and their interdependencies with environment parameters like temperature or track profiles are well known from various validation activities (chapter 6).

The development of consumption models and range forecast for one vehicle in private usage is state of the art. Based on the influencing factors and the interactions, predictive consumption models are developed. Regression models describe the influencing factors and environment parameters. Past trips are evaluated and thus speed-based driver models are obtained.[5] This kind of modelling provides good results for the case of private usage: one driver uses one car under similar usage conditions. However, it is not very suitable for fleet operators, who have to deal with various vehicles, drivers and usage scenarios. That is why a generalization of the prognosis models is desirable, so that the models can deal with different vehicle types and usage scenarios in multiple environmental contexts. The environmental contexts are specified for example by weather conditions influencing operation of air condition and vehicle operating points.

In further work, generalization of models will be done by identification of relevant parameters using self-learning algorithms respectively real-time information originating from the electric vehicle. The major challenge is to maintain the goodness of the range forecasts.

2.4 IPEK – X-in-the-Loop

Different characteristics of validation activities resulting from the principle according to ALBERS, „Validation [...] has to be systematically and continuously carried out from the start to the end of development [..].“[6]. Depending on the status of the development process, the system exists in different stages of maturity or even in subsystems. Validation of the entire system is not sufficient, thus, the subsystems in the context of the overall system must be validated. Because functions are often fulfilled in system network, validation is always done in the context of interactions with the overall system.

Following figure shows the IPEK X-in-the-Loop approach and describes the understanding of system to be developed ("X") in context of the interacting systems: restsystem, environment and driver. All systems including environment and driver can exist physical, virtual and mixed physical-virtual.

In the present case, first knowledge of the system "electric vehicle" in the field respectively in field-like behaviour is gained. These analyses are the basis for the development

---

[5] Grubwinkler et al. 2014
[6] Albers et al. 2016

of consumption models for range forecast in future works. Hence, the validation activities are mainly used for gaining knowledge, so in this context the "X" is called system under investigation (SuI).

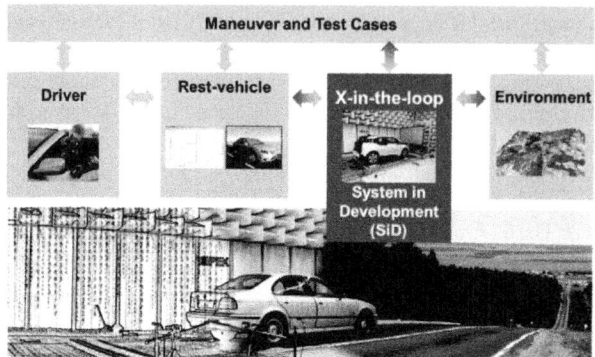

IPEK X-in-the-Loop Approach[6]

The system to be examined can exist on different system levels: according to the following figure of IPEK XiL framework starting at functional contacts (WSP – working surface pair) through the subsystem up to the complete vehicle. This framework, which is generalized for vehicle development, will be adapted for the present validation case.

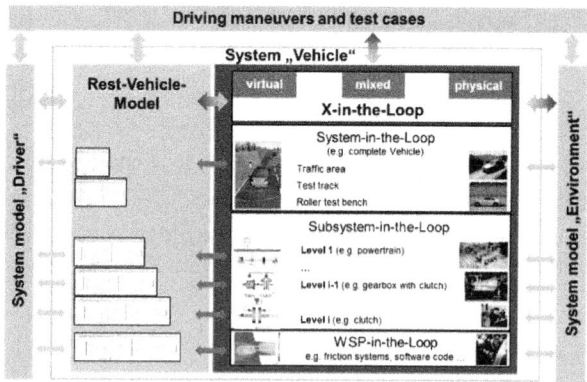

IPEK X-in-the-Loop Framework[6]

In this paper, in a first step the plausibility of field data is proven, described in chapter 5. Second, initial knowledge about the auxiliaries' influence on the range of electric vehicles should be gained, described in chapter 6. In both cases, investigations will be done at first level of XiL framework: Thus, a real electric vehicle (system under investigation) is driven on a roller test bench, which simulates the rest vehicle (wheel-street-interaction) as well as the environmental conditions (e.g. track profile, temperature, …). The driver behaviour can be modelled by a driving robot, which is integrated in the real vehicle and depresses the driving pedal.[6 7]

3 Methodical approach

Our methodical approach for evaluating electro mobility systems benefits from field data, model and simulation and vehicle test. A fundamental aspect is system-integrated data acquisition which is used for acquiring field data from fleets of BEV. As a benefit, it can also be used for efficient acquisition during vehicle test.

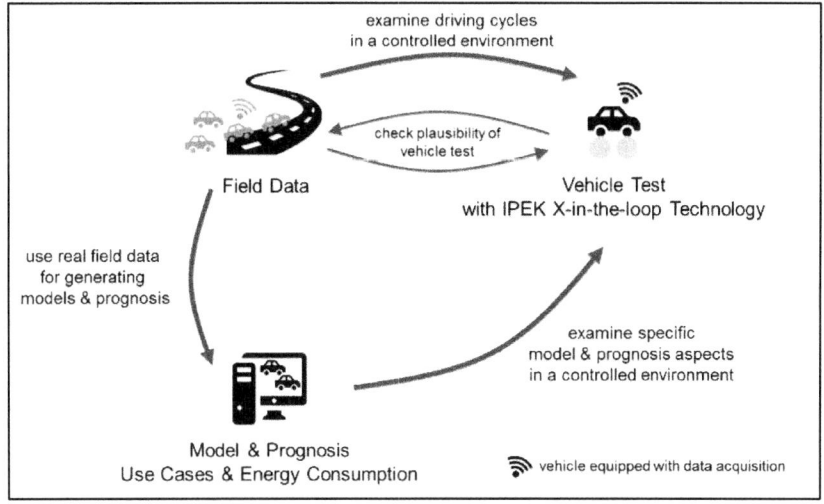

Approach for validation of electro mobility systems

Validation has to consider the needs of users in application. Therefore, the initial focus of this approach is field data from fleets because they represent user behaviour. This

---

[7] Matros et al. 2015

data is used for modelling & simulation of mobility systems. The vehicle test can examine either specific factors of the simulations or driving cycles captured by field data by the use of a controlled environment. Especially in this case, the plausibility of the vehicle test has to be checked to ensure a valid knowledge transfer from observed phenomena on vehicle test to real field behaviour.

Along this methodical approach, three exemplary cases are conducted. These are acquisition of field data, a plausibility check of the vehicle test on a roller test bench and a sensitivity analysis of factors for improving the model & simulation. They are described in further detail in the following sections.

4 Acquisition of field data

ELISE was implemented in a BEV car and connected to the CAN bus and configured remotely. A route had been selected which includes sections of urban traffic, interurban road and the autobahn. The GPS signal was used to measure height and therefore slope of terrain. GPS signals were plausible in interurban and highway sections but shadowed by surrounding buildings in city traffic. Since the city of Karlsruhe, Germany is build on flat terrain, the implausible height measurements were revoked. The route is 30 km long and was named „Karlsruher Runde" (Karlsruhe tour).

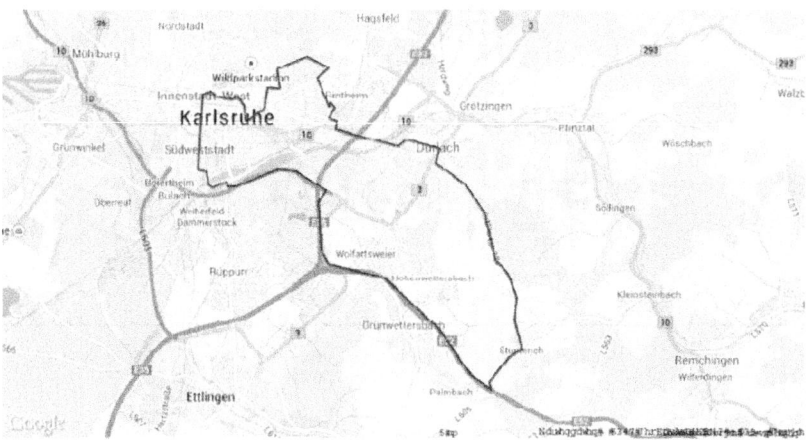

Karlsruher Runde (Karlsruhe tour), 30 km

A driving cycle was recorded including datasets from the BEV drivetrain including speed, GPS, SOC, depression of the driving pedal, motor torque, battery voltage and current and more. The acquired data will be used for further analysis and prognosis models in current and future research projects.

5   Plausibility check of vehicle test

Further works – using field data for generating generalized consumption models and range forecasts and its evaluation – require checking the data quality and the validity of roller test bench. The vehicle (chapter 6.1), which also was used in field data acquisition, runs on roller test bench. The cycle profile is calculated from the GPS signal, which is also logged by the flea box. The environmental conditions and driving resistances are simulated by the test bench; the driver is represented by a driving robot, which has to follow the vehicle speed. As well as in the field, the flea box records all data.

The speed progression obtained on roller test bench match with the progression obtained in field. The following figure shows three curve progressions: field data (dashed), data on roller test bench with high amount of recuperation called "EcoPro+" (full line) and data on roller test bench with minor amount of recuperation called "EcoPro" (dotted), which was also used for field data acquisition. (The recuperation modes' names are originating from BMW.) The progression clearly shows the speed regulation in the "EcoPro+"- mode. In general, the energy consumption in this mode is less than the energy consumption in the "EcoPro".

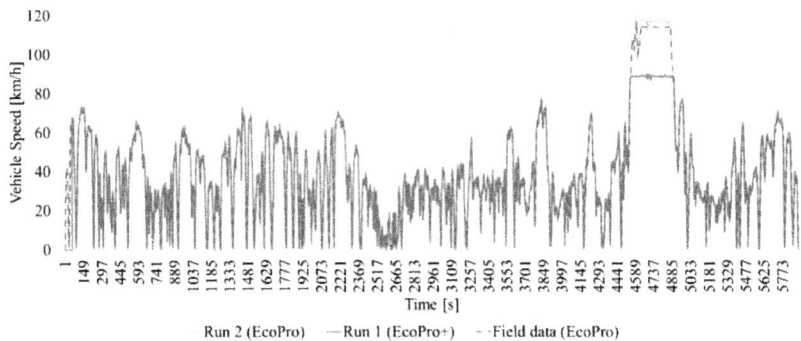

Comparison: Speed profile gained from field data and tests on roller test bench using different recuperation modes

In order to evaluate the test bench results, the comparison of speed progressions is not sufficient. One uncertainty is, for example, that the modelling of track profile is not realistic. Thus, other parameters like pedal depression or the progression of energy consumption have to be analysed. The following figure shows an excerpt of pedal depression over time. One curve shows the pedal positions' progression date back to field data (dashed). The other shows the progression resulting from driving robot's behaviour on roller test bench, which represents driver behaviour (dotted). The overlap is sufficient accurate, so that field behaviour of vehicles can be reproducible modelled by the roller test bench. Thus, later consumption models can be examined on roller test bench resulting in valid results.

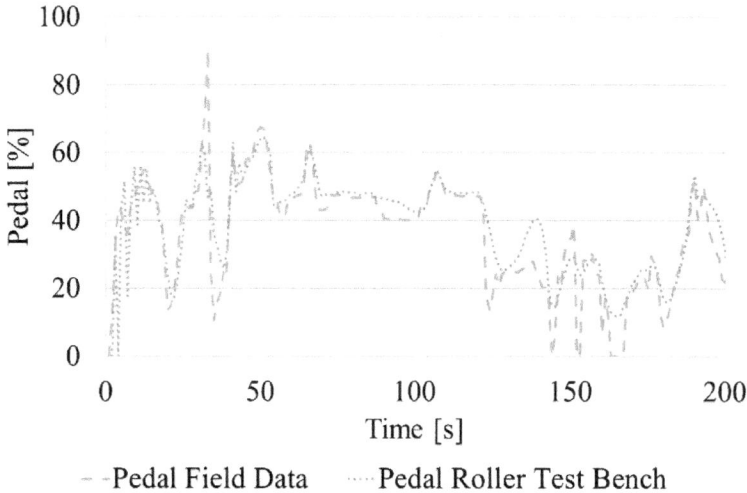

Comparison: Pedal position gained from field data and from tests on roller test bench

Engine torque and battery current naturally correspond with the pedal position. The obtained data (engine torque and speed, battery current, voltage, temperature) allows conclusions about the state of power electronics. This information gives an insight about the load progression and environmental conditions and contribute to the estimation of battery ageing, which is important for the range forecast.

6 Sensitivity analysis for range forecast

6.1 Method

The following chapter describes the procedure for the assessment of the range forecast of electric vehicles. The aim was an exemplary sensitivity analysis to determine the percent coverage losses due to additional consumers using the internal vehicle data acquisition. Thus, knowledge about the energy needs of consumers in the vehicle can be obtained. At the beginning, the test environment with relevant parameter was defined. In this case the various attributes of the different parameters were varied or kept constant. The parameters, which will be examined, are: outside temperature, vehicle lighting, inside temperature, multi-media, velocity, mode, driving profile.

Experimental setup:

Reproducible road tests are often associated with high costs. Therefore, it is often useful to move the investigations of the consumption on roller test benches. As part of the project IeMM the sensitivity analysis for range forecast was exemplarily applied with the BMW i3 with 125 kW + Range Extender. Therefore, the validation experiments were conducted on a four-wheel roller test bench as shown in the figure below.

Electric vehicle at roller test bench at IPEK – Institute of Product Engineering

Test procedure:

The goal was the well-founded description of interactions and influences of environmental and vehicle parameters to the required range of an electric vehicle. For this purpose, reproducible cycles on the four-wheel roller test bench were driven. The vehicle was driven in the experiment in a steady operating state. Next, the test sequence is described:

1. battery completely loaded until SOC = 100%

2. acceleration to defined speed

3. commissioning of the relevant consumers

4. start the measuring method in the stationary operating point at SOC = 80%

5. documentation of the interior temperature (frequency 0.1 seconds)

6. documentation of the number of kilometres (frequency 0.1 seconds)

7. end of the measurement method at SOC = 50%

A total of 12 operational cases were analysed, which are examples of: summer driving, winter driving, city driving, urban driving, highway driving:

Test cases, vehicle and environment parameters:

Testcase	T1	T2	T3	T4	T5	T6
outside temperature	20°C	20°C	20°C	20°C	20°C	20°C
vehicle lighting	off	off	on	off	off	off
inside temperature	20°C	20°C	20°C	10°C	30°C	20°C
multi-media	off	off	radio	off	off	navigation system, radio
velocity	100 km/h	100 km/h	100 km/h	100 km/h	100 km/h	100 km/h
mode	comfort	comfort	comfort	comfort	comfort	comfort

Testcase	T7	T8	T9	T10	T11	T12
outside temperature	10°C	30°C	20°C	20°C	40°C	20°C
vehicle lighting	off	off	off	off	off	off
inside temperature	20°C	18°C	20°C	25°C	15°C	20°C
multi-media	navigation system, radio	navigation system, radio	radio	radio	navigation system, radio	off
velocity	100 km/h	100 km/h	138 km/h	138 km/h	138 km/h	50 km/h
mode	comfort	comfort	comfort	comfort	comfort	Eco Pro

6.2 Results

The listed chart visualizes the test results of the selected test cases.

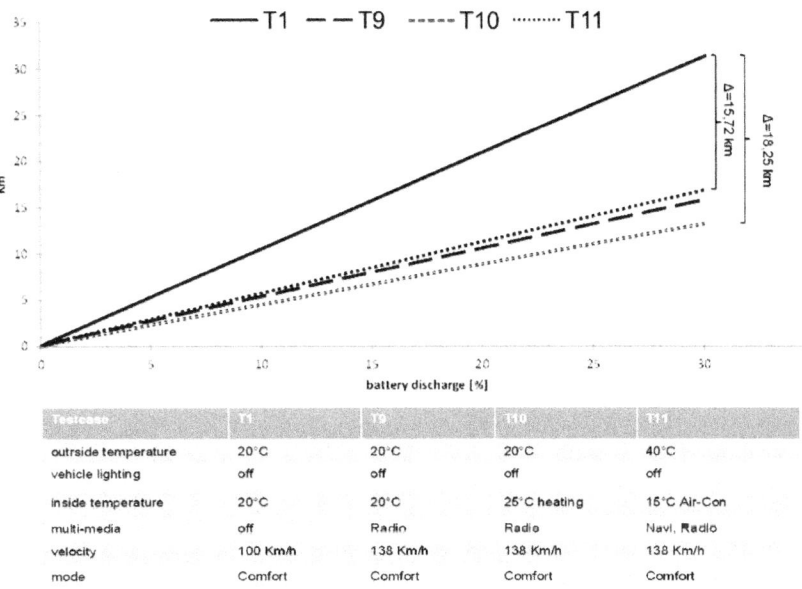

Testcase	T1	T9	T10	T11
outside temperature	20°C	20°C	20°C	40°C
vehicle lighting	off	off	off	off
inside temperature	20°C	20°C	25°C heating	15°C Air-Con
multi-media	off	Radio	Radio	Navi, Radio
velocity	100 Km/h	138 Km/h	138 Km/h	138 Km/h
mode	Comfort	Comfort	Comfort	Comfort

Range over remaining SOC depending for parameter variation

In the diagram, the distance traveled in km (ordinate) is plotted versus the battery discharge in % (abscissa). The battery discharge is calculated in % by the following formula since the start of the experiment:

$$([i]-SOC[1])*(-1), \text{ with } SOC[i] = \text{SOC at the measuring time i} \tag{1}$$

In all three test cases with high speed (T9-T11) the discharge is much faster than at 100 km/h. In the heating mode (T10) the discharge is much faster than in cool air mode (T11). The chart shows that the cool air mode has less influence on the discharge even so T11 uses navigation system as well as the speed and the heating mode.

Some aspects are consistent with the literature[8]. As overall conclusion of the range forecast study can be said:

– The measurement method is suitable for the collection of test cases.

– The influence of the cool air mode is less than the operation of the heating.

– If the operation of the navigation system no or only a minimal effect on the range.

– The outside temperature has a strong influence on the battery discharge, which no longer is linear.

– Greatest impact on the reach, the driving style and the speed.

7 Summary and Outlook

Actually, concepts, methods and models for evaluating integrated mobility solutions in fleet use are missing. The introduced approach (chapter 3ff.) ought to help filling the gap. Main aspects are: gathering field data, building generalized consumption models and verify these models on roller test bench. This paper focusses enabling gather field data and proving its suitability to generate generalized consumption models as well as the validation of test bench.

Therefore, the system-integrated data acquisition of the project ELISE was used and integrated in electric vehicles, which are intended to collect data in field. In a first step the plausibility of field data was proven. In a second step, few records of real world data are used to repeat the cycles on roller test bench. The results show that field behaviour can be modelled mixed physical-virtual on the roller test bench. Thus, in a third step, a sensitivity analysis increases knowledge about influences on driving range. These investigations of driving ranges considering different test cases increase the understanding of vehicle behaviour in fleets.

Based on that, consumption models for range forecast can be developed and can be validated on the roller test bench. Within this controlled environment, the functionality of models' generalization and the correlated accuracy of range forecast can be checked. Specific model and prognosis aspects can be examined. In the future, fleet providers are able to optimally design, operate and monitor BEV operating in fleets.

---

[8] Liebl et al. 2014

8 Acknowledgements

This paper presents excerpts of work in the projects IeMM (Intermodales eMobilitäts-management, funding code: 13N12274), ELISE (Autonome Ladeeinheit und Systemintegrierter Datengateway für Elektrofahrzeuge, funding code: 13N12390) and BiE (Bewertung integrierter Elektromobilität, funding code: 16EMO0043). The authors are grateful to the German Federal Ministry of Education and Research for funding these projects related to Cluster electric mobility south-west.

9 Bibliography

1. F. Munker, T. Kotschenreuther, M. Behrendt, A. Rupalla, „ELISE – Ladeeinheit und Datengateway für Elektrofahrzeuge". ATZ extra, Ausgabe „Elektromobilität", Band 90, Sept. 2014.

2. CarMediaLab GmbH Bruchsal, Germany, www.carmedialab.com.

3. P. Conradi, „Reichweitenprognose für Elektromobile", in *Vernetztes Automobil*, W. Siebenpfeiffer, Hrsg. Wiesbaden: Springer Fachmedien Wiesbaden, 2014, S. 179–184.

4. ExxonMobil Central Europe Holding GmbH, „Energieprognose 2011-2030 – Schwerpunkt: Wie viel Zukunft steckt im Auto von heute?", Hamburg, 2011.

5. S. Grubwinkler, M. Lienkamp, „Energy Prediction for EVs using support vector regression methods", Intelligent Systems'2014: Proceedings of the 7th IEEE International Conference Intelligent System IS'2014, September 24-26, 2014, Warsaw, Volume 2: Tools, Architectures, Systems, Applications.

6. A. Albers, M. Behrendt, S. Klingler, K. Matros, „Verifikation und Validierung im Produktentstehungsprozess". In: Lindemann (Hrsg.), Handbuch Produktentwicklung, Carl Hanser Verlag, München, 2016; ISBN 978-3-446-44518-5; „to be published 04/2016".

7. K. Matros, F. Schille, M. Behrendt, und H. Holzer, „Manoeuvre-based validation of hybrid powertrains", *ATZ*, Nr. 02/2015, 2015.

8. J. Liebl, M. Lederer, K. Rohde-Brandenburger, J.-W. Biermann, M. Roth, H. Schäfer, „Energiemanagement im Kraftfahrzeug - Optimierung von CO_2-Emmissionen und Verbrauch konventioneller und elektrifizierter Automobile", Springer Verlag, 2014.

Optimized operating strategies for electrified taxis by means of condition-based load collectives

Raphael Pfeil, Prof. Hans-Christian Reuss, Dr. Michael Grimm

Research Institute of Automotive Engineering and Vehicle Engines Stuttgart (FKFS)

1 Motivation

Alternatively-powered vehicles are coming into the focus of public awareness, not least because of the VW emissions scandal. The emissions standards for conventionally-powered vehicles (ICE) are becoming ever stricter, in order to achieve stringent environmental and climate protection targets. In order to adhere to future emissions standards, the demands on exhaust gas aftertreatment processes continue to increase. This is at the expense of the "Total Cost of Ownership" (TCO) as well as the specific fuel consumption. Initial projections predict a competitive disadvantage for compact-class diesel vehicles of up to 5000 euros (over an assumed life cycle of 10 years) when compared with petrol, hybrid and electric vehicles [1].

In comparison to using conventionally-powered vehicles, running a taxi business with alternatively-powered vehicles presents new challenges for taxi drivers and business owners in the planning and implementation of their operation. The operating strategy with conventionally-powered vehicles is primarily influenced by the demand for transport and the transport capacity. For alternatively-powered vehicles, the range, charging period and charging infrastructure are also relevant for the operating strategy [2]. Table 1 [3, 4] lists the energy storage capacities, the charging periods, and the energy consumption determined using the NEDC and, based on this, the calculated ranges for the Mercedes Benz (MB) E-Class 200 Blue TEC and the MB B-Class ED. The range for the MB E-Class 200 Blue TEC in the NEDC is a factor of six larger than that of the MB B-Class ED. To make a journey of 100 km, the MB E-Class 200 Blue TEC must be filled with fuel for approximately half a minute. The MB B-Class ED must be charged for more than 100 minutes. The range of the purely battery-driven (BE) B-Class ED and the reserve range of the MB E-Class 200 Blue TEC are almost the same.

Table 1: Range comparison of the MB E-Class 200 Blue TEC and the MB B-Class ED *NEDC (combined)

	E-Class 200 Blue TEC	B-Class ED
Energy storage capacity	59 l	28 kWh (-)
Of which reserves	8 l	(no data)
Energy consumption*	4.8 l/ 100 km	16.6 kWh/ 100 km
Range	1229 km	200 km
Reserve range	167 km	(no data)
Charging period	approx. 5 min	approx. 3 h

For taxi operation with a BE vehicle such as the MB B-Class ED, the taxi driver must make do with the reserve range of an ICE vehicle, such as the MB E-Class 200 Blue TEC. In certain weather conditions (high/low temperatures), or with an ineffective driving style (e.g. low recuperation level), the energy consumption for alternatively-powered vehicles increases more rapidly than for conventionally-powered vehicles, due to the characteristics of the system. This means the range conditions shift further to the disadvantage of alternatively-powered vehicles. Remedies could be, for example, a larger energy storage capacity, a more effective vehicle system with lower driving resistances and lower weight, as well as interim charging. However, a larger energy storage capacity and optimization of the vehicle (e.g. more expensive materials for lightweight construction) would increase the TCO. Interim charging not adapted to the taxi operation sequence reduces the time the vehicle is available for transporting customers. In these cases, the economic viability and profitability will suffer. From an economic point of view, it makes more sense to analyze and optimize the taxi operating strategy regarding job planning and scheduling. This approach enables the requirements for alternatively-powered taxis to be derived, and increases the effectiveness (and therefore the profitability) of the entire taxi fleet. By optimizing job planning and scheduling, it is not only possible to adapt interim charging to the waiting times, but also reduce it by using the waiting times and empty runs.

2   Methodology

The following methods (see Fig. 1) are discussed and applied for the analysis and optimization of technical operating strategies for taxis. Here the focus is on the job-oriented operation sequence and appropriate vehicle use. The analysis phase assumes that the relevant operating parameters for running the EV taxis have already been determined. These are recorded using specific measurement equipment. Subsequently, relevant measurement data from real-life taxi operation are collected from a defined fleet with different drivetrain configurations (ICE, EV, hybrid). These measurement data are converted, compressed and stored in a database. Using data consistency and plausibility tests, the data quality is determined and the necessary data preparation activities defined and implemented. The database is the basis for the determination and evaluation of the taxi operation sequence with the associated vehicle load profiles. The model is the result of physical laws (driving resistances, etc.), the limiting conditions present (charging infrastructure, taxi rank location, etc.) and expert knowledge of the operating procedure (driving behavior, driver habits, etc.). The models are parameterized using the collected data. The simulation model is validated using known techniques, e.g. cross-validation or sensitivity analysis. These are used to check the simulation model for potential aberrations and modeling errors.

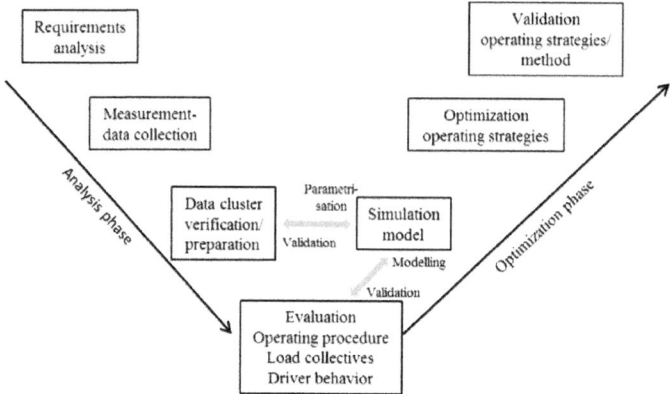

Figure 1: Methods for analyzing and optimizing operation strategies

In the optimization phase, taxi operation is simulated for a defined fleet size, using the identified load profiles such as routes driven, demand for transport etc. Here the taxi distribution and distances for occupied and unoccupied taxis are changed iteratively until e.g. a global maximum is reached for the capacity under defined limiting conditions. The limiting conditions include the energy storage capacity, charging period, range, shift hours and driving times, as well as the taxi transport capacity. Conversely, the simulation can be used to find the optimum distribution of drivetrain types (ICE, hybrid, EV) for a taxi fleet, for an existing or optimized operating strategy. One goal here is to service the demand for transport with the largest proportion of EV taxis. Further goals include the optimization of a taxi's individual operating strategy and the task-oriented design of the drivetrain for different applications, such as, for example, taxi or police work.

3 Analysis

The relevant outer operating parameters for determining the job and load profiles are split into three clusters: Use, Infrastructure and Environment. The data cluster Use includes e.g. customers, driving profile with vehicle trajectory and loading. Infrastructure includes data such as coordinates and capacities/loading of depots, taxi ranks and charging points. Data regarding the ambient temperature, light intensity and traffic density is included in the Environment cluster. The outer operating parameters interact with the vehicle and its subsystems such as the drivetrain, comfort systems and use-specific systems (see Figure 2). So, for example, energy is removed from the energy store to move the taxi forwards, or the energy store is replenished with energy from charging and recuperation. The drivetrain subsystem includes components such as the traction battery and the electrical drive unit. The comfort system includes systems such as air-conditioning and lighting. The specific systems comprise e.g. the taximeter and taxi radio. [2]

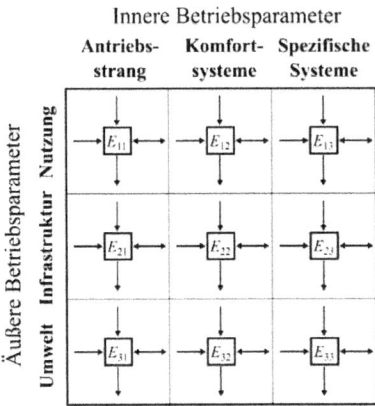

Figure 2: 3x3 Operating parameter matrix [TAE]

3.1 Data Acquisition

To determine the job and load profiles of taxis, data is acquired for the reconstruction of movement profiles (GPS) and vehicle states, e.g. free, occupied, etc. In this case, the Fleet Management System (FMS) from Austrosoft® Weiss Datenverarbeitung GmbH is used for this. The system was originally developed for dispatching taxis via the central taxi dispatch office, using taxi radio. In addition, chronological parameters regarding the vehicle state and vehicle position are recorded and transmitted to the central taxi dispatch office, where they are stored temporarily. For efficiency reasons and due to the limited bandwidth of mobile data transmission, the measurement parameters are recorded approx. every 30 seconds and transmitted to the intermediary server. In addition, to dispatch EV taxis without interruption, the residual range of vehicles is recorded and transmitted to the central taxi dispatch office. This can be used to prevent taxis picking up passengers without having sufficient range to complete the trip. Data concerning e.g. energy consumption, recuperated energy and the battery's state of charge (SOC) are also recorded for EV taxis, based on the minimum datasets[1]

---

[1] Defined scope of measurement parameters for uniform data acquisition and evaluation of the vehicles, to help showcase electro-mobility.

of the Phase I model region [5]. The measurement parameters of the minimum data set are output as an aggregate for each trip (ignition change as trigger).

3.2 Data Analysis with Data-Processing

The first raw database considered in this report contains measurement data from four EV taxis (vehicle type Mercedes Benz B-Class ED) between November 2014 and October 2015, with approx. 1.2 million data values. These data have not been anonymized. Therefore the individual taxi routes (from the start of the shift until its end) can be assigned to a particular EV taxi over the entire period. The second raw database of the mixed taxi fleet (ICE, hybrid) includes approx. 700 vehicles over a period of one week. All taxis in the mixed fleet have a combustion engine in common – therefore this database will be referred to as the ICE database from now on. The ICE database is anonymized and contains around 3 million data values. For this reason, analysis of the movement profile of a certain taxi over several days is not directly possible. However, the data points are ordered chronologically per taxi and day, according to the associated routes (shift start and shift end). All the data recorded comes from taxis working for a taxi company in Stuttgart.

3.2.1 Data Analysis

The two databases differ in the number of vehicles and the time period. The EV database contains four vehicles. The mixed fleet contains around 700 vehicles. The precise number of taxis in operation varies, for example due to taxi concessions being discontinued or sold. The time period for the EV fleet data ranges over 365 days. The time period for the ICE fleet is over seven consecutive days. Due to the ratio of the number of vehicles to the investigation period, the order of magnitude of the data points is comparable (EV approx. $1.2 \cdot 10^6$ / ICE fleet approx. $3 \cdot 10^6$). Individual taxi routes are only contained in one database respectively. However, the investigation periods of both databases overlap. The operation of the EV taxi fleet influences the ICE fleet and vice versa, by undertaking customer journeys, standing at the taxi rank etc. Due to the proportional size of the EV fleet (4 vehicles) compared with the ICE fleet (700 vehicles), this dependence can be ignored. Before processing the data, the data is analyzed with a plausibility check and the quality of the data is evaluated. Implausible data points are filtered out. Additionally, it is checked whether, for example, the GPS points are within Europe, or the timestamp t of a taxi route increases sequentially:

$$t_n < t_{n+1} \quad n \in \mathbb{N} \tag{1}$$

Using a quality factor of 0 to 1, the taxi data for each vehicle and shift are correspondingly weighted for the further evaluation results. The following information and evaluations are incorporated in this factor [6]:

- Sequence and frequency of the taxi state compared with the local data

- Length of the taxi trajectory as a ratio of the shift time

- Vehicle speed (average, min, max) compared with the speed limits and other taxis currently in the area

By removing implausible data points, the amount of work for the subsequent data-processing is reduced. The quality factor can be used to appropriately weight and interpret the evaluation and optimization results.

3.2.2 Re-routing

Due to the lengthy sampling rate of the FMS (approx. 30 seconds), the data density of the GPS points is increased using a re-routing procedure. In Figure 3, A shows the initial state of the GPS data record made by the FMS before re-routing. Using map-based algorithms, additional GPS points are created synthetically between two data points (see Fig. 3 B). The "Open Source Routing Machine" (OSRM) procedure is used for this. The artificially generated GPS points are distributed throughout the route. The density of the GPS data is higher on curves than for straight stretches of road. To calculate the shortest route between two points, the procedure uses the "Contraction Hierarchies" algorithm. Here the relevant nodes (crossroads, etc.) are reduced by hierarchical consolidation. This means that this method calculates much faster and more efficiently than previous routing algorithms such as "Dijkstra's algorithm". Furthermore, the OSRM procedure takes into account traffic routing rules such as one-way streets [7].

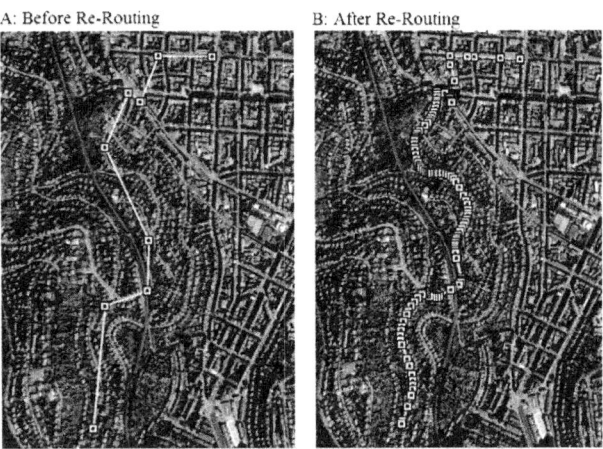

Figure 3: GPS data processing using re-routing [Map: Google Earth]

If the density of GPS data is low and there are large spatial distances from one GPS point to the next, the actual taxi trajectory can no longer be determined with absolute certainty. Different trajectories are possible, especially in urban areas. In contrast, the problem is significantly reduced when traveling on cross-country roads and motorways. Here road junctions are further apart than in urban areas, resulting in fewer alternative routes. To evaluate the quality of the re-routing, the re-routing trajectories are combined with the available information (taxi state, etc.) and the quality factor is recalculated using a similar method to the one used in Section 3.2.1.

The distances driven can be calculated more accurately using the replicated trajectories. The distance between two GPS points (squares) is calculated using the "Haversine formula" as follows, taking the curvature of the Earth into account [8]:

$$d = 2r \, \sin^{-1}\left(\sqrt{\sin\left(\frac{\phi_2-\phi_1}{2}\right)^2 + \cos(\phi_1)\cos(\phi_2)\sin\left(\frac{\lambda_2-\lambda_1}{2}\right)^2}\right) \qquad (2)$$

The surface of the Earth is assumed to be an idealized ball with a radius of 6378.1 km. Φ represents the latitude and λ the longitude of both points. The total distance d_{ges} of a route results from the summation of the individual distances d as follows:

$$d_{ges} = \sum_{n=1}^{n} d_n \, n \in \mathrm{N} \qquad (3)$$

The total length of the route shown in the example in Figure 3 is 3.259 km before re-routing and 3.705 km after re-routing. Using the additional geo-information from re-routing, it is possible to calculate the driven taxi routes more precisely. Furthermore, the average speeds between two data points ($\Delta t \approx 30$ s) in the original data can be calculated more accurately with the re-routing data. The accuracy of the calculations depends on the driving area (city, cross-country or motorway), the precision of the GPS position, rounding errors and the locality-dependent deviation from the assumption that the Earth's surface is an idealized ball. For the calculation of high resolution speed and acceleration profiles which are as realistic as possible, re-routing alone is not sufficient. Although the degree of detail regarding the taxi locations is improved by re-routing, the location-time relationship remains at the same level of detail as in the original data set, as shown in Figure 4.

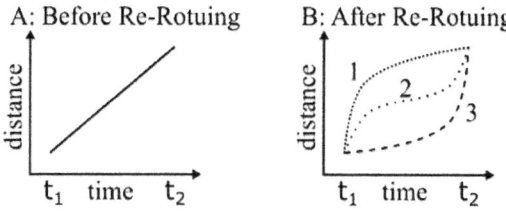

Figure 4: Location-time problem of re-routing.

In re-routing, depending on the route, additional data points are inserted between two coordinate data points from the original data ($\Delta(t_2 - t_1) \approx 30$ s). The time at which the taxi is on the trajectory between two original data points is unknown. Figure 4 B shows three of the infinite possibilities for the location-time relationship. The real location-time relationship cannot be determined after the journey has taken place.

3.3 Job and Load Profiles

The evaluation of the job and load profiles is carried out using the previously-analyzed and prepared databases for the EV and mixed fleet. Figure 5 shows the GPS data records of the ICE fleet in CW 29, 2015. It can be seen that the taxis travel throughout the entire Stuttgart region and the surrounding areas. Furthermore, passengers were also carried to more distant destinations in Baden-Württemberg, such as Mannheim, Karlsruhe and Ulm, as well as beyond the state boundaries, such as to Basel and Munich.

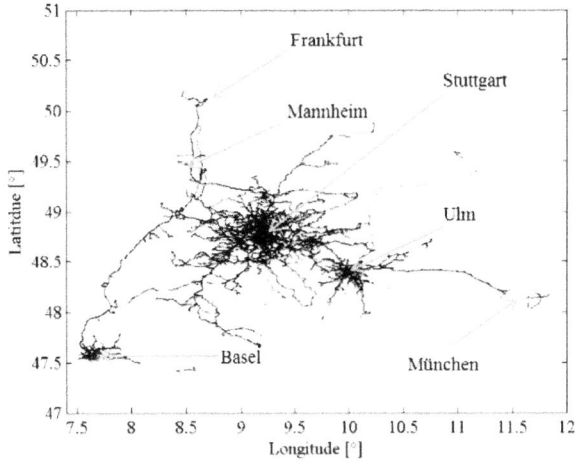

Figure 5: Movement profile of the ICE fleet in CW 29 2015.

Most trips made by the EV fleet were within the Stuttgart city area. Occasionally, trips to destinations in neighboring areas such as Ludwigsburg, Reutlingen, Göppingen and Nagold were also made [2]. Trips outside the state border were not served by the EV fleet.

3.3.1 Distance from Stuttgart Town Hall

The numerical evaluation of the taxi locations confirms the visual impression. Figure 6 shows that approximately 77 % of ICE-fleet taxi movement was within the Stuttgart city region, in a 10 km radius around the town hall. More than 90 % of the passengers carried by the ICE fleet were making trips in the Stuttgart, Esslingen, Filderstadt and Leinfelden-Echterdingen areas (20 km radius). Only 1 % of the taxi trips have destinations outside Baden-Württemberg, such as Munich and Basel. The maximum distance from Stuttgart town hall (as the crow flies) is approximately 350 km. At more than 98 %, the EV taxis carry passengers in the Stuttgart city area significantly more frequently compared to the ICE fleet. The frequency of trips to more distant destinations such as Ludwigsburg, Reutlingen, Göppingen or Nagold is less than 1%. The maximum distance from Stuttgart town hall is almost 63 km. It can be assumed that, due to the limited range, the charging infrastructure and charging period, the EV taxis only rarely leave the city limits.

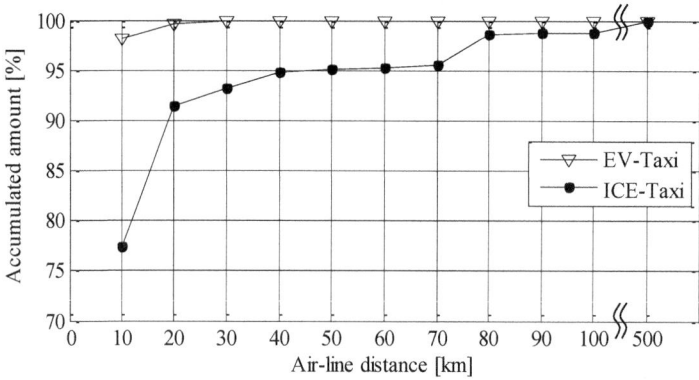

Figure 6: Distances of the taxi locations from Stuttgart town hall.

3.3.2 Daily Distances Traveled

The daily distances traveled by the ICE fleet have an average of 154.5 km. The result is compatible with, and therefore validated by, the "Report in compliance with § 13 Sec. 4 PBefG regarding the functional capability of the taxi trade in the state capital of Stuttgart and in the towns of Filderstadt and Leinfelden – Echterdingen" by Linne and Kraus [9]. This report specifies an average daily distance of approximately 150 km. The maximum daily distance of the ICE fleet is around 425 km. The average daily distance of the EV fleet of approx. 80 km is roughly half that of the ICE fleet. The maximum journey distance of the EV fleet is 203 km. This is less than half the longest trip made by the ICE fleet. Figure 7 shows the accumulated frequencies of the daily distances for both fleets. The daily distances of the ICE fleets are distributed relatively evenly (an almost constant increase) ranging from 75 km to 250 km. This is reflected in a high standard deviation of 71.6 km. This means that the daily journeys of the ICE fleet are very inhomogeneous. The standard deviation of the EV fleet is, at 38.27 km, significantly less than for the ICE fleet. This is due to the limited range, number of vehicles and therefore also the taxi drivers. Here the daily distances are more affected by seasonal variations for the maximum ranges (due to the energy requirement for air conditioning) and the operating strategy – such as interim charging, etc.

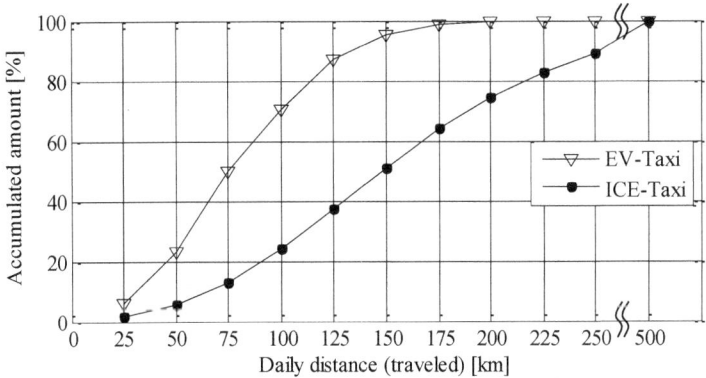

Figure 7: Daily distances traveled by the Stuttgart taxis.

3.3.3 Overall Distance vs. Distance with Customer

The EV fleet travels an average of 39.81 km per day with customers. Based on the average overall kilometers traveled, the capacity utilization of the EV fleet is approx. 50 %. In comparison, the taxis in the ICE fleet travel an average of 62.39 km per day with a customer, resulting in a capacity utilization of approx. 40 %. The difference in capacity utilization of approximately 10 % is reflected in the cumulated frequency of this ratio (see Fig. 8). Therefore, regarding the distance traveled, the EV fleet is used more effectively than the ICE fleet. It is highly probable that the reasons for this lie with the limited range of the electric drive – promoting a more conscious use of the vehicle. Another reason is that the ICE fleet makes trips outside the Stuttgart tariff zone more frequently. Outside the tariff zone, the Stuttgart taxis are not permitted to take on new customers, meaning these taxis make the return trip without passengers.

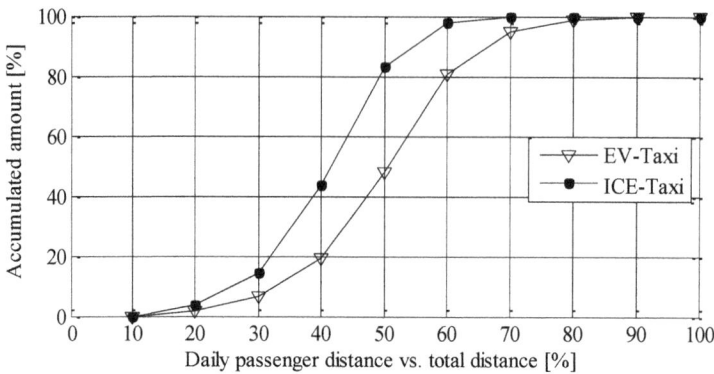

Figure 8: Ratio of kilometers with a passenger to total kilometers traveled per day by the Stuttgart taxis

4 Optimization of the Operating Strategy

The optimization of the operating strategy can take place on multiple levels. One level is to optimize the operating strategy of a fleet with a defined fleet size and distribution of drivetrain configurations (ICE, hybrid, EV). An input parameter for the simulation is the demand for transport with the static job and load profiles shown. Here the total size of the fleet and the distribution of the drivetrain configurations are varied iteratively. To take into account the energetic processes with the limited energy storage capacities and the recharging periods, the simulated taxi movements are coupled with realistic vehicle models. The simulation results include the degree of feasibility of the defined simulation scenario, the capacity utilization of the taxi fleet and the fleet driving strategy (time, short/medium/long journeys etc.).

Another level is to optimize the individual operating strategy from the point of view of a single taxi during its shift. Here, based on the demand for transport and the transport capacity, the ideal sequence is determined for the start of operation, driving area with taxi rank selection, breaks and recharging periods (see Figure 9). Furthermore, during running taxi operation, focused recommendations to go to the next taxi rank or transport area are specified – to minimize empty trips between journeys with customers.

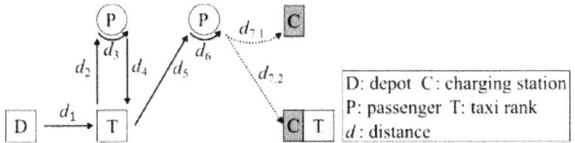

Figure 9: Example of a taxi operating sequence [TAE]

The limiting conditions include driving time, energy storage capacity, taxi rank capacity and vehicle capacity (one, two or three shift operation). Furthermore, taxi driver-specific habits are incorporated in the optimization process, such as preferred driving area and taxi stands. For the simulation, the taxi operating sequence (see Figure 9) is abstracted and coupled with the appropriate vehicle models, in a similar way to the simulation for the entire fleet. The parameterization of the vehicle model is adapted continuously based on the history. This enables the energy consumption to be predicted more reliably. The result is an operating strategy recommendation for an individual vehicle – either coupled with information on the entire fleet, e.g. via the central dispatch office, or independent of the fleet from the point of view of the vehicle.

One level below is the specific optimization of the drivetrain design for taxi operation, such as energy storage capacity, drive power etc. Here the component types, arrangement and dimensions are varied, depending on the job and load profiles. Furthermore, the effectiveness with regard to TCO, CO_2 emissions etc. is optimized by the iterative simulation process and the feasibility demonstrated in a simulation.

5 Summary and Outlook

In this article a method was presented for optimizing the taxi operating procedure and the drivetrain design, using state-dependent load profiles. This method permits the composition of the fleet (ICE, hybrid, EV) to be analyzed and optimized – taking into account capacity utilization and strategies for the fleet and individual vehicles. The approach is integrative and, alongside the technical aspects such as data acquisition and evaluation, it also takes sociological factors such as the behavior and habits of the taxi drivers into account. After all, the success of the optimization measures is dependent on their implementation by the taxi driver. Furthermore, this method can be applied to other vehicle fleets, e.g. for the police or distribution companies.

More research is required regarding data analysis with determination of job and load profiles. The location-time problem still needs to be solved for data preparation topics such as re-routing. This is to enable the job and load profiles to be presented with sufficient quality for subsequent optimizations. Currently the degree of detection of the location-time relationship is at the same level as for the original data. One approach is to increase the detection level using statistical correction factors for crossroads and curves with different radii of curvature.

Initial evaluations of the job and load profiles show a very inhomogeneous operating behavior for the taxis. The daily routes, distance from the city center and amount of capacity used varies significantly between the individual vehicles. Based on a daily average range of 80 km for the EV taxis, from a statistical point of view an estimated 20 % of the ICE fleet can be replaced by EV taxis, without optimizing the operating strategy. The initial fleet simulations are very promising and also show considerable potential for optimization of the operating procedure. This would allow the EV fleet component to be increased further.

References

1. Wirtschaftswoche Online: Dieselmotor rechnet sich in der Golfklasse nicht mehr. URL: www.wiwo.de/unternehmen/auto/folgen-des-vw-skandals-dieselmotor-rechnet-sich-in-der-golfklasse-nicht-mehr/12547478.html, accessed 08.11.2015.

2. Pfeil, R.; Reuss, H.-C.; Grimm, M.; Krützfeldt, M. S.: Identifikation und Analyse einflussrelevanter Parameter des E-Taxibetriebs – Erste technische Projektergebnisse aus GuEST. 4. Symposium Elektromobilität, Ostfildern, 23. June 2015.

3. Deutsche Automobil Treuhand GmbH: Leitfaden über den Kraftstoffverbrauch, die CO2-Emissionen und den Stromverbrauch. Ostfildern, 4. quarter 2015.

4. ADAC e.V.: ADAC Autotest MB B-Klasse Electric Drive Electric Art. München, April 2015.

5. Schallaböck, K. O. et al.: Modellregionen Elektromobilität – Umweltbegleitforschung Elektromobilität. Wuppertal Report, Wuppertal Institut, 2012.

6. Gawlik, W; Litzlbauer, M.; Schuster, A. et al.: ZENEM – Zukünftige Energienetze mit Elektromobilität. Neue Energien 2020 – final report, TU Wien, 2013.

7. Mapbox: Smart Directions Powered by OSRM's Enhanced Graph Model. URL: https://www.mapbox.com/blog/smart-directions-with-osrm-graph-model, accessed 20.11.2015.

8. Robusto C. C.: The Cosine-Haversine Formula. The American Mathematical Monthly Vol. 64, No. 1, Washington, January 1957.

9. Linne + Krause: Gutachten gemäß § 13 Abs. 4 PBefG über die Funktionsfähigkeit des Taxigewerbes in der Landeshauptstadt Stuttgart sowie in den Städten Filderstadt und Leinfelden – Echterdingen. URL: www.landkreis-esslingen.de, accessed 19.05.2015.

Vehicle simulation of an electric street sweeper for substitution analysis

M. Sc. Rene Budich, Prof. Dr.-Ing. Manfred Hübner

Hochschule für Technik und Wirtschaft Dresden (HTW), Germany

673

1 Abstract

Small street sweeping machines are important for a clean city. A permanent use guarantees clean sidewalks and public spaces and shapes thus considerably the overall picture of a modern metropolis. However, these uses have significant impact on the environment. There are not only CO_2, NOX and PM_{10} ejected, it will also increase noise emissions. The switch to electric mobility can locally reduce these emissions and allows additionally low-noise operations.

But the changeover to electric mobility carries high risks and costs. To estimate costs and possibilities a full vehicle simulation can be helpful (e.g. for designing battery size, planning tours).

In the project EBALD, a small electric street sweeper will be realized by the Laboratory of Electric Mobility from the HTW Dresden.

Based on collected field-data, collected during the project a simulation tool for substitution analyzes was realized. Using this tool, it's possible to say, which conditions must be fulfilled (e.g. position of charging stations) to realize environmentally friendly and future-oriented cleaning concepts.

This paper includes a structural approach of such a full vehicle analysis.

2 Introduction

Due to the steady trend towards urbanization in recent years, air pollution has become a pressing issue for large cities and urban centers. For the health of the inhabitants and for the long-term goal of CO_2 reduction, it is necessary to develop and implement adequate measures at an early stage.

The limited amounts of fossil fuel resources like oil and the economic and political uncertainties that influence its availability, make it necessary to ensure a mobility concept that is independent from other countries and their resources. A promising solution to solve these problems is the widespread introduction of electric mobility. Germany has started a major shift from fossil to renewable energy in 2011 (first spoken about in 1980) and since then has become more and more independent of resources like oil. By 2050, at least 80 percent of the electric power supply and 60 percent of the total energy supply will originate from renewable energy sources. In addition to the installation of more renewable energy plants, the introduction of electric mobility is an important part of this strategy [1].

When speaking of electric mobility, we primarily think of passenger-carrying electric vehicles used to meet individual mobility needs. But, electric mobility also contains vehicles that are used in municipal or public service like street cleaning or garbage disposal, which do not focus on passenger transport.

Looking at street cleaning, street sweepers are promising candidates for electrification. These vehicles generally follow fixed routes and work schedules so that their use can be precisely planned which makes them ideal for electrification. In this paper, a small street sweeper, also called compact street sweeper (CSS) was investigated.

Small sweeping machines are regularly used in busy places that are heavily frequented by many people, e.g. sidewalks, public places and parks. Therefore, it is desired to minimize the discharge of emissions (exhaust gases, noise, dust). Furthermore, it can be expected that authorities will install more laws and regulations to restrict these polluting contaminants.

One way to reduce pollution is to divert traffic from city centers. However, this is not applicable to street sweepers, because they have to work exactly at these places. Therefore, only substituting solutions, i.e. replacing a combustion engine with an electric one, are possible.

One problem of introducing such an environmentally friendly street cleaning concept is, that it is difficult to illustrate and project it's monetary gains. Therefore, many municipalities avoid the introduction of environmentally friendly concepts. In order to estimate the long-term benefits of different options, useful tools are needed.

In the project, "Elektromobilität in Bereichen der Abfallwirtschaft der Landeshauptstadt Dresden (EBALD)" the Laboratory Electric Mobility HTW Dresden tackles these problems, e.g. by examined which possibilities arise towards improvements of noise, exhaust and particulate pollution by electrification of a street sweeper.

The substitution of an internal combustion engine (ICE) by an electric one is difficult because in advance it is impossible to clarify whether this is useful or not. The economic risk at this point is very high as the use of new technologies is always expensive and often does not yet justify an environmentally-friendly implementation.

One way of estimating the usefulness of electrification, is the use of a full vehicle simulation when correspondingly accurate parameters and conditions are known. With it, costs and uses can be planned, and savings compared to ICE can be determined. The corresponding measurement data could be collected in the project EBALD, by long-term tests at the sanitation department Dresden (SRD).

Through the use of electric vehicles, however, other problems, such as not comprehensively-developed infrastructure, a limited amount of energy storage or energy-intensive air-conditioning of the interior emerge.

This is accompanied by a limited operational capability (load restrictions and coverage limitations). The more efficient the stored energy can be used, the more independent the vehicle is from public infrastructure.

Many of these problems are associated with the limited energy capacity available on the vehicle, so a solution must be found that uses available energy as efficient as possible.

To this end, the concept of an intelligent sweeper with distributed individual drives and specific management system was developed (this is patent pending [2].)

In addition, a strategy to recharge energy as quickly as possible must be found. For street sweepers, two principle solutions are apparent: The first one is to exchange an empty battery with a charged one, the second one is to design the battery capacity large enough to sustain a full working-shift. However, the second approach of energy storage has to be adapted to the particular cleaning environment in order to not oversize batteries, which would make them unnecessarily expensive.

The approach of a cascading battery, which is flexible, interchangeable, and can be adapted to the respective task, represents the best approach for this scenario. All these different approaches can also be checked with a simulation environment.

3 Small street sweepers

Small electric street sweepers are mainly used for the cleaning of sidewalks and pedestrian areas. In this section, the basic structure and functions of a small electric street sweeper are described.

3.1 General structure of a small street sweeper

Figure 1 shows the general structure of a small street sweeper. It consists of a brush unit with 2–3 disc brushes (1) that loosen the dirt, a suction port (2) which carries dirt and debris through the suction tube (3) to the waste container, and a suction fan (4) which generates the necessary vacuum.

These assemblies are summarized in 3 units: a soil release device, a dirt collector, and a dirt holding device.

These structural modules can be assigned to specific functions, which are introduced in the following. The units also serve as the basis for the structuring of the overall vehicle simulation (see. Section no 4.1).

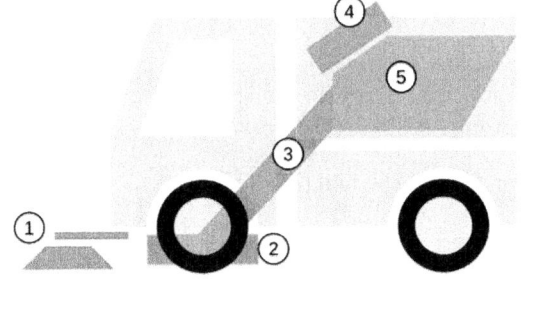

1. Brush unit (disc brushes)   4. Suction fan   ▬▬▬ Soil release device (1)
2. Suction mouth.   5. Debris hopper   ▬▬▬ Dirt collector (2,3,4)
3. Suction tube   ▬▬▬ Dirt holding device (5)

Figure 1: General structure of a small street sweeper [3]

Suction (suction mouth / suction tube/ suction fan): Belongs to the dirt collector. Vacuums the sweepings and transports them into the hopper.

Sweeping (circular disc brushes, soil release device): Loosens the dirt and transports the waste to the suction device.

Air-condition: Interior air-conditioning of the driver's cab (heating and cooling).

12V electrical system: For lighting and additional consumers like fans, radio, etc.

Steering (drive): Power steering for steering assistance of the driver.

Lifting and lowering (dirt collection device): Emptying of the container. Lifting, lowering and left-right movement of the brushes.

Drive: Comply bring serves firstly, the implement to the destination (transport mode) and, secondly, the optimal cleaning speed (working mode).

A street sweeper is a complex system of various functional groups, which are all interlinked and influence each other. Therefore, the system must be considered in its entirety.

3.2   Example of a typical small street sweeper

To illustrate the examination and retrofitting of small street sweeper, we chose a basic model of a German manufacturer. Due to its configuration, the machine can be used as a general representative for compact street sweepers.

3.2.1   Construction and energy function diagram

The main drive of the vehicle is powered by a typical diesel engine that also feeds a complex hydraulic system consisting of several hydraulic pumps, valve manifolds, cylinders and hydraulic motors. Via this hydraulic system, the required energy for the main processes (creating the vacuum, sweeping, driving, etc.) is provided.

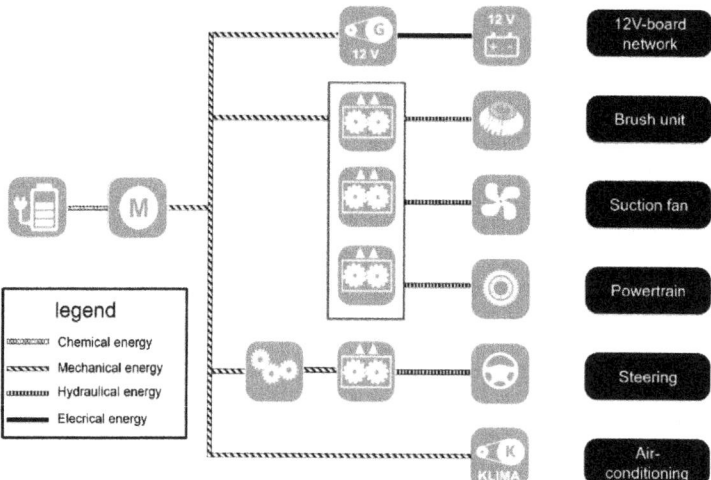

Figure 2: Energy flow scheme of a conventional small street sweeper [4]

Figure 2 shows an overview of the necessary energy conversions. Through the internal combustion engine (ICE) chemical energy is converted into mechanical energy (1) that supplies the various available auxiliary units via a shaft. In addition, the hydraulic block converts mechanical energy into hydraulic energy to drive the disc brushes (brooms), the suction unit and the traction motor. Another hydraulic circuit is responsible for the steering, as well as the lifting and lowering of the brushes and the dirt collector. An air compressor, which is mounted directly on the mechanical shaft, ensures the air conditioning.

3.2.2 Energy flow representation

To illustrate the power flow, Figure 3 shows a Sankey diagram of an example cleaning drive at 4km/h.

It is easy to see, that efficiency is strikingly poor. The internal combustion engine losses 72% of the original power through heat, while the subsequent hydraulic unit adds another 17% to these losses. Thus, we see a considerable optimization potential in the substitution of the internal combustion engine and the electrification of the whole system and its individual units. This measure also leads to a reduction of the hydraulic unit to a smaller and lighter aggregate.

The data was collected as a part of the project EBALD at Stadtreinigung Dresden (sanitation department). For this purpose, it was necessary to develop special measurement technology as there were no suitable solutions in the market for such applications. This measurement technology has been integrated into several test vehicles and allowed the collection of long-term operational data including specific situations like starting on an incline or brushing on the curb.

The obtained data was analyzed with MATLAB/Simulink and form the basis for the simulation, both for modeling and for validation.

For the validation of the components, special test equipment was developed and built: A driving engine test bench that supports different drive modes, a fan test for different suction capacities, and a brush bench with both different surfaces, as well as downforce and broom types. The basic construction of the brush test rig is shown in Figure 8 and 9 [5].

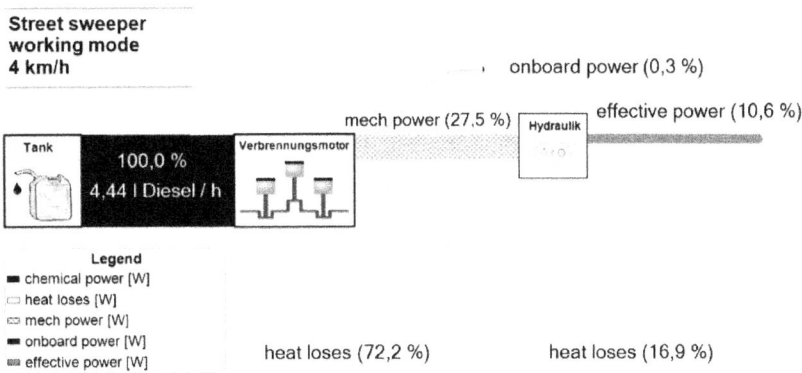

Figure 3: Sankey diagram of a conventional small street sweeper [5]

3.3 Small electric street sweeper

Based on the collected data and additional survey results [3], the conversion concept in Figure 4 was developed. Here, all units are electrified, which is expected to significantly improve the efficiency of the overall system.

Currently, a conventional street sweeper is converted into a fully electrical setup by the HTW Dresden (university of applied science). During this conversion, the internal combustion engine and a large part of the hydraulic system was removed.

For the hydraulic motors, permanent magnet synchronous machines are used. All translational movements, which rarely occur in the workflow in comparison to the rotational movements, continue to use hydraulic components (steering, lifting and lowering).

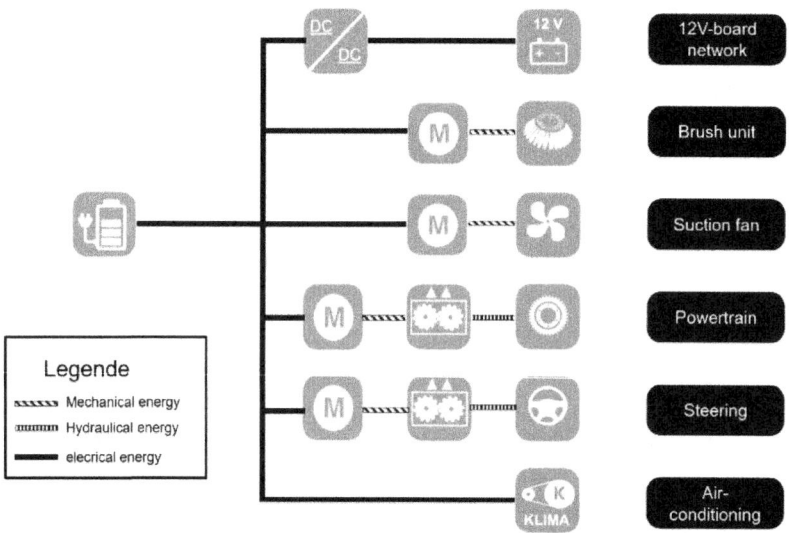

Figure 4: Energy flow scheme of an electrical street sweeper [4]

4 Simulation of a small electric street sweeper

The goal of simulation is to recreate real behavior as close as possible. For electrical simulations, three general approaches can be identified: conservative systems, non-conservative systems, and digital systems. Each of these principles is suitable for a specific problem area.

Non-conservative systems use signal flow diagrams with individual input and output terminals and are thus, particularly suitable to solve control technical problems (e.g. with software like MATLAB / Simulink). Where conservative systems are used specifically for physical problems (e.g. VHDL-AMS, Spice derivate) [4].

For the simulation of our examined street sweeper, which is a mechatronic system, physical and technical control (signal flow) approaches are necessary. Therefore, a non-conservative (signal flow) and a conservative (physical) simulation is needed. There is the option of a co-simulation, which is difficult to handle because one solver has to wait for the others. Furthermore, the realization of interfaces between two simulations is challenging.

The simulation is done using the non-conservative tool MATLAB / Simulink. In addition, the toolbox Simscape, which includes its own simulation language for physical problems (similar to VHDL-AMS) is used. So we don't need a second simulation tool and also the interface problem is solved. In Figure 7, a conversion from physical level to control technical - level is shown (circled blocks).

Simulations allow the evaluation of extreme situations which cannot be realized in reality, or would require a very high effort (for example worst-case scenarios such as high ambient temperatures, over- or under loading the battery or high tilt angle of the travel path) to do so.

Our main goal with the full vehicle simulation of street sweepers is to give estimates of energy consumption in different scenarios, e.g. different terrain topologies or driver types in advance. These results will be used to give recommendations regarding the substitution of ICE- machines. The decision to switch to environmentally-friendly concepts will be easier for municipal companies.

4.1 Simulation overview

Figure 5 shows the structure of the simulation on the highest level of abstraction. The diagram corresponds to the power scheme from Figure 4. The entire system has a modular design and can be separately tried and tested. Each simulation block consists of up to 4 sub-blocks, where the last block is always the physical layer (Simscape).

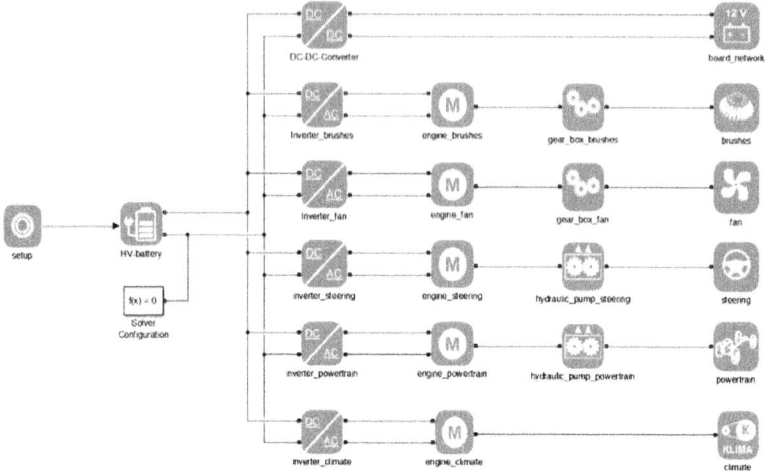

Figure 5: Simulation overview (MATLAB/Simulink)

For each branch (function group), an individual test environment (test bench) was developed with which the simulation model has been tested and validated. The validation of simulation results is done with the data from the long-term studies and the measured data from the test benches.

Furthermore, there is the possibility of hardware in the loop (HIL) testing, where hardware components can be integrated into the simulation. This allows to test the behavior of certain hardware in specific states, e.g. during the start-up process or while responding to an error.

In the next section, the brush unit is exemplarily investigated. The methodology corresponds to the structured approach for the other simulation models.

4.2 Brush unit

The real brush unit consists of an inverter, which is fed from a high voltage battery, a permanently excited synchronous machine, an outboard bearing, and the brush head. The setup of the brush unit is shown in Fig.9.

The brush should be initially considered isolated from the overall system. For this purpose, a separate simulation test bench was designed. Subsequently, the data was compared and validated with measurements from a real test bench.

4.2.1 Simulation of the brush unit

Figure 6 show the setup of the simulation test bench. The test environment consists of a DC voltage source (1) an inverter (2), the engine control and power cable (3+4), the permanent magnet synchronous machine (5) and the brushes (6). Data having a directional flow (non-conservative simulation) is symbolized by arrows, e.g. speed data transmitted by a controller. Data without arrows (bubbles) symbolize physical ports.

Via the setup ('*setup*') block, simulation parameters like battery size, attributes of the brush, machine parameters and so on, can be adjusted. The HV-battery provides a DC voltage of 650V. The inverter transforms, the voltage for the motor. In the simulation we use a DC motor to represent the real engine. The motor converts electrical energy into mechanical energy. Subsequently, the rotational speed and the torque is adjusted via the transmission. The brushes create a counter-torque depending on contact pressure and brush type. All simulation data are collected in the block '*brush_data*', where they are evaluated. To illustrate the use of Simscape and Simulink, the red encircled subsystem will be explained in more detail in the following.

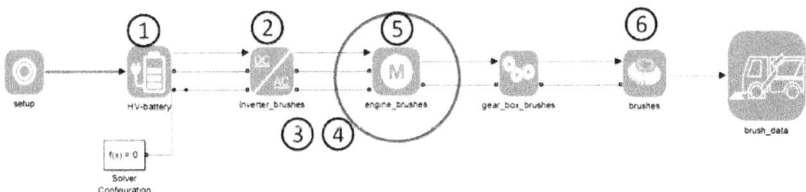

Figure 6: Brush branch of the vehicle simulation

Figure 7 shows the physical part of the simulation (lowest level). It illustrates the simplified structure of an electric machine with its armature resistance, machine constant and inductance. Next you can see 2 transducers that transfer the data from the Simscape environment (physical simulation) into the Simulink environment (signal flow simulation) [5]. At the input terminals 1 and 2, the output of the inverter is connected (see also Figure 6).

At the output 3 (mechanical shaft) mechanical energy is forwarded to the transmission. The block '*rotational electromechanical converter*' converts electrical energy into mechanical energy.

The data obtained is bundled by a MUX-block and transferred via a common data output ('*output_data*') to the outside. This has the advantage, that different data can be sent via a single output port. Thus, a consistent interface can be realized throughout the system. One signal flow port for input, one for outputs and 2 ports for each physical signal.

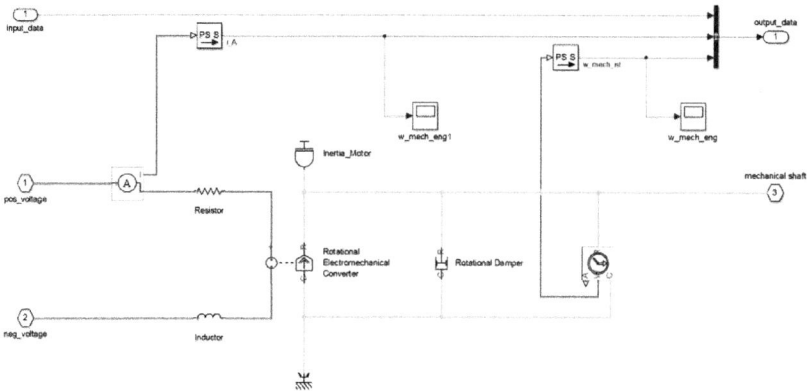

Figure 7: Simscape model of an electric brush engine [6]

4.2.2   Test bench of the brush unit

In order to validate simulation results, real measurement data is required. Since the entire system is modular, it is also validated in a modular fashion, i.e. by branch. For this purpose, a test was developed which allows to test different settings, e.g. different speeds, contact pressure, kind of brushes, etc.

Figure 8 shows the draft of the test rig. The construction of the brush test bench is shown in Figure 9 (all components was described in 4.2.1.). By adjusting screw connections, the angle of attack of the brush can be changed and contact pressure can be varied. To change the rotation speed, a set point speed is given to the inverter (2) via a bus system. It is also possible to test various surfaces like concrete tiles, grass pavers or asphalt on this test bench. To determine the input power, measuring devices were installed as shown in Figure 8. The output power is calculated by the inverter. The data is made available via a CAN-bus system.

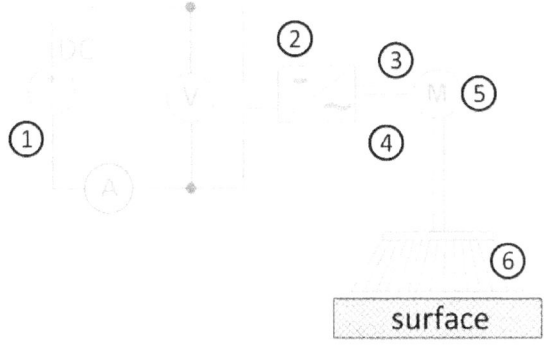

Figure 8: Simplified structure of the brush test bench

Figure 9 shows the brush test bench. You can see the DC-Source (1), the inverter (2) for the brush motor (5), the resolver cable (3), the power cable (4) and the brush head (6).

Figure 9: Brush test bench (HTW Dresden) [7]

4.3 Results of full vehicle simulation

To compare vehicles with each other, driving cycles are used. In the domain of passenger cars, the New European Driving Cycle (NEDC) is often used for this purpose. However, small street sweepers cannot be tested with this reference (they do not reach the necessary speed and working for example). Therefore, a separate cycle was developed and is shown in Figure 10. This cycle was designed, based on long-term studies in the SRD and contains three driving phases (I, III, V), intermitted by two cleaning phases (II, IV). The overall duration of one cycle amounts to 1200s.

At an average speed of 12.5 km/ h, a distance of 4.21 kilometers will be covered during the driving cycle. The maximum speed amounts to 40 km/h, this speed is accelerated with a maximum of 0.8 m/s^2 and delayed with a maximum of 1.71 m/s^2 [8].

First results of the study are shown in Figure 11 and 12. The simulation has been validated with real data.

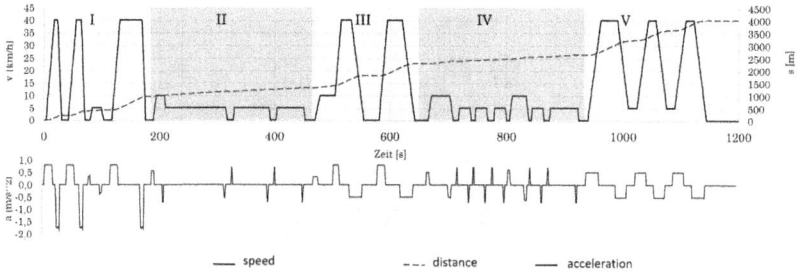

Figure 10: Driving cycle for small street sweeper [7]

Figure 11 illustrates the simulated performance history during the driving cycle over time. It is clearly evident that the greatest amount of power is required during acceleration (max. 60 kW). The average power consumption during the cleaning operations amounts to approximately 10 kW.

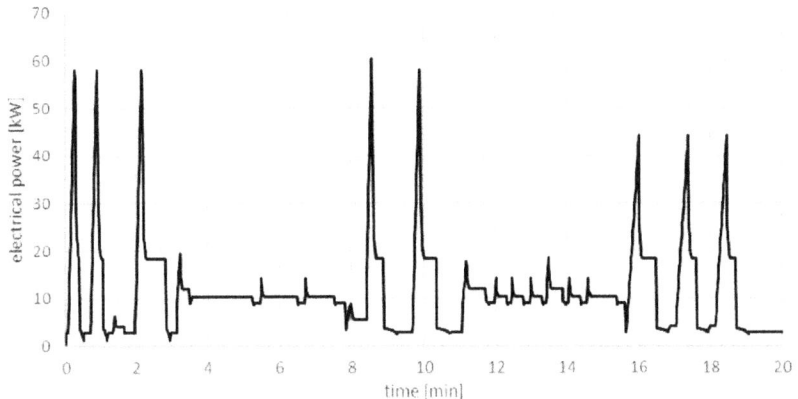

Figure 11: Performance history during simulation the driving cycle [7]

Looking at the energy distribution in the Sankey diagram in Figure 12, we find that the greatest losses occur in the drive system. There is an energy conversion from electrical to mechanical to hydraulic energy and then back into mechanical energy. This is very inefficient and should be replaced by in-wheel hub motors in a future project, as the current projects budget was not sufficient for this task. Assuming an efficiency of about 90% for electric wheel hub drives the entire system could be improved by ca. 30%.

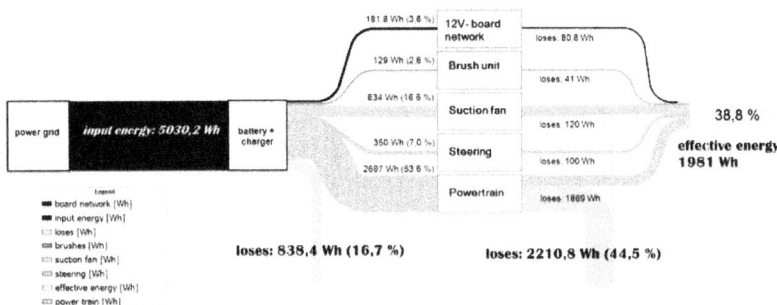

Figure 12: Sankeydiagramm of the simulation result (street sweeper driving cycle)

5 Conclusion and Outlook

The use of electric sweepers can lead to the improvement of air quality in metropolitan areas. Not only the local CO_2-neutral transportation is crucial, but also measures to minimize noise. As the EU calls for the reduction of particulate matter emissions, appropriately equipped small sweeping machines can contribute to this effort.

With a simulation, it is possible to make preliminary calculations and thus substitution analysis. This helps to answer questions on the scope and application options in advance. Versatile investigations were made both at ICE-sweepers as well as an electric sweeper to collect data [2].

Based on the data base, collected by long-term studies in the SRD, a simulation environment has been developed and tested.

The resulting selected concept is currently being implemented by the Laboratory Electric Mobility (HTW Dresden). In further steps, the system will be optimized and further operation strategies will be tested.

By equipping a fully electrified sweeper with communication interfaces and environment sensors, autonomous reactions to the environment can be enabled. In the following, we describe future development approaches for our concept of a fully electrified sweeper with decentralized drives.

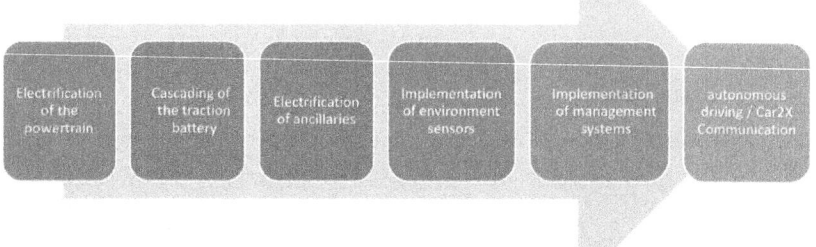

Figure 13: Development stages of an intelligent street sweeper

By electrifying the drive train of the sweeper it provides all the advantages of an electric car like CO_2 neutrality and quiet driving (Fig. 13 step one).

By cascading the battery an optimal adjustment in terms of energy storage and weight to the work effort can be achieved (Fig.13 step two), as each battery module can be loaded separately. In addition, simultaneous loading of all individual modules is possible. This can be a great advantage for fleet management, i.e. exchanging battery modules between different electric vehicles. The potential of cascading batteries has to be investigated in a follow-up project. Currently, only theoretical investigation using simulations have been made.

Using intelligent ancillaries can further reduce energy requirements (Fig. 13 step three). This leads to a reduction of the battery or greater ranges. In addition, individual cleaning concepts are also possible in which the aggregates are only switched on when they are needed. This electrification is the basis for the implementation of intelligent management systems [2], [9]. This part has already been realized.

By utilizing environmental sensors, the texture of the road surface can be detected. Based on these readings, the degree of contamination can be determined and the accessories (brushes and suction unit) can be optimally adjusted (Fig. 13 part four). In addition, these sensors can be used in the implementation of a warning system for pedestrians and drivers.

Based on the information derived from the sensor data, the following intelligent operating strategies can be implemented, for saving energy (Fig. 13 part five) [2]:

- Intelligent Energy Management (IEM)
- Intelligent Cleaning Management (ICM)
- Intelligent Security Management (ISM)

As a final innovation step, a Car2x communication system can be implemented (Fig.13 part six). Thus, an optimal cleaning can be realized even in problematic areas, by utilizing Car2Infrastructure. This allows the realization of autonomously adjusted cleaning.

6 References

1. https://www.deutschland.de/de/topic/politik/deutschland-europa/was-sie-ueber-die-energiewende-wissen-sollten

2. Budich R. (2015) Patentanmeldung, Elektrisch betriebene Arbeitsmaschine zur mobilen Reinigung

3. Jordan M. (2014): Diplomarbeit, Konzept und Durchführung messtechnischer Untersuchungen von Straßenreinigungsmaschinen mit Verbrennungskraftmaschine (VKM) und Elektroantrieb, HTW Dresden

4. Budich R. (2015) Symposium TAE-Esslingen: CO_2-neutralen Reinigung von Großstädten

5. Budich R. (2014), 2. Jahrestagung des Schaufensters Bayern-Sachsen, Elektromobilität im Bereichen der Abfallwirtschaft der Landeshauptstadt Dresden

6. Glöckler, Michael: Lehrbuch, Simulation mechatronischer Systeme, Springer Fachmedien Wiesbaden GmbH

7. Hüther J. (2015): Diplomarbeit, Gesamtfahrzeugsimulation einer elektrischen Straßenreinigungsmaschine (Kleinkehrmaschine) mit MATLAB/Simulink, HTW Dresden

8. Peterhänsel, K. (2015): Diplomarbeit, Untersuchung von Emissionen einer Kleinkehrmaschine mit Dieselantrieb als Grundlage einer Elektrifizierung, HTW Dresden

9. Nikoleizig M.: Diplomarbeit, Entwicklung von Managementsystemen für eine intelligente Elektrokehrmaschine, HTW Dresden

Reliability-oriented simulation in the engine development process

M.Sc. Stefan Jetter[1], Dipl.-Ing. Frank-Oliver Müller[1], Dr.-Ing. Ralph Weller[1], Dipl.-Ing. Michael Zöllner[1], Prof. Dr.-Ing. Bernd Bertsche[2]

[1]Daimler AG, Stuttgart

[2]Institute of Machine Components, University of Stuttgart

1 Motivation and Objective

Individuality and lightweight are essential strategies in the automotive industry, which have to be considered in each part of the development process [1]. For reaching the request for individuality and being efficient, the use of modular systems and product families is inevitable. Focused on the engine development this means, that the application range of an engine and the requirements for the construction increase significantly. An engine is nowadays used for several vehicle series whereby a wide performance range has to be covered as well as the materials have to be used up to their limits for reaching the lightweight targets, but not risking any damages or failures. Furthermore the engines get more complex and more powerful, which can be seen in figure 1 and figure 2. Figure 1 shows a 4-cylinder engine from the 1980s with a maximum performance of 72 kW and 230 Nm. Figure 2 shows an actual 4-cylinder engine with about 150 kW and 500 Nm. In this comparison the increasing specific power and the lightweight engineering in newer engines is well illustrated.

Figure 1: 4-cylinder engine in the 1980s [2]   Figure 2: 4-cylinder engine today [2]

For ensuring the statistical reliability of these complex engines, extensive testing has to be done. In future, this procedure shall be supported by a reliability-oriented simulation, by which high quality statements, regarding damage or failure can be gained in early stages of the development process. Up to now this is not an established standard and so only testing with the original component and under realistic conditions can deliver statistical reliable statements. For this issue, the following work gives a possibility to do statistical testing within the simulation process, focused on the crankcase; especially the lower part of the crankcase with the bearing cap and the main bearing bolts. In figure 3 the crankcase of the actual 4-cylinder engine (figure 2) is shown. Clearly visible is the complex structure of the crankcase and also the quantity of connectors and bores to attach components. For simulating the reliability or the possibility of failure, a large amount of variations has to be calculated before a statement can be given [3]. Especially in the calculation of the crankcase, the current simulation method, the finite element method, reaches its limit due to temporal reasons. Compromises have to be done concerning the resolution of the analysis or the complexity of the model. Up to now these compromises were not very serious, because the critical load cases were well-known and interactions between attached components did not have this influence; so smaller models with less resolution could be used. With the increasing efforts in the field of lightweight engineering and the building of product families, the critical loads cannot be defined as exactly as before and the influences of attached parts have to be analyzed by complex models.

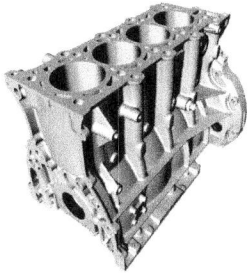

Figure 3: Crankcase of an actual engine [2]

For these reasons there is the question how the future design and calculation of crankcases has to be set up with regard to the increasing requirements mentioned above. A promising method in this case is the elastic multibody simulation. It combines the possibility to regard the interaction between different attached components and the requirements for a fast computing time. Out of the elastic multibody simulation, a reliability-oriented simulation can be set up, which consists of four different steps (figure 4). The first step is an analysis of the usage of the engine. The second step is a statistical analysis of the emerging strain (out of step one) and the tolerant strain (material characteristics). Also the preparation of the data for the third step, which is the simulation process, is done in step two. The last step is an interpretation of the results regarding the reliability.

In this work, some parts of this process will be presented and therefore the theory of the elastic multibody simulation and its application at the crankcase (step three) is shown in chapter 2. A central element of this chapter is the comparison with the finite element method, especially with regard to the stress analysis. In chapter 3 the introduction of the reliability and the necessary steps are described. The main focus in this chapter is set on the influence of the loads from driving operations (step one) and the analysis of the material characteristics (step two). In chapter 4 the topic is critically evaluated and an outlook is given for open issues.

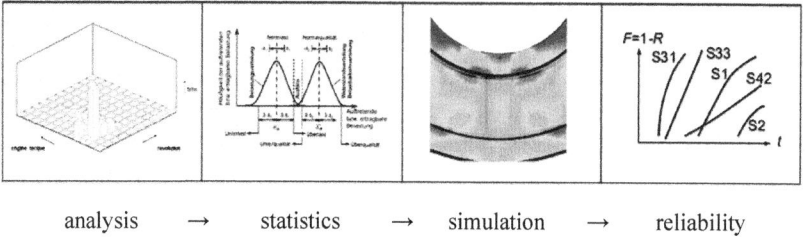

analysis → statistics → simulation → reliability

Figure 4: Process of the reliability-oriented simulation

2 The elastic multibody simulation applied to the design for crankcases

An established standard in the numerical design of a crankcase is the finite element method. But the challenges mentioned in chapter 1 take this method to its limits and it needs to be extended. More detailed calculations concerning the resolution of the calculation steps and the interaction between other parts of the engine have to be carried out. Regarding the computing time, the elastic multibody simulation seems to offer noticeable advantages due to the reduction process [4, 5], which will be presented shortly in the following paragraphs. Furthermore the application of the elastic multibody simulation to the crankcase will be described and a comparison with the finite element method will be given in this chapter.

2.1 Theory of the elastic multibody simulation

The elastic multibody simulation is an extension of the classical multibody simulation which mainly consists of rigid bodies and massless coupling elements [6, 7]. In contrast to the classical multibody simulation the elastic multibody simulation can also consist of flexible bodies and consider deformation und stress. The basis of the flexible body is a model which is the same like used in the finite element method. The difference is the generation of a substructure, which is made out of the basic model and then integrated in the multibody system (sec. 2.1.1).

2.1.1 Reduction and substructure generation

Before a flexible body can be integrated in the multibody simulation process it is necessary to create a substructure out of it. In this step all dispensable or not explicitly retained degrees of freedom of the model are eliminated [8].

Basis of the substructure generation is a modal analysis of the model for including dynamic effects [4, 8]. In the elastic multibody simulation the calculated eigenmodes in the substructure generation are used to perform the deformation or the stress of the body (sec. 2.1.2) [4]. So it is very important to know which eigenmodes have to be considered. The choice of them is a compromise between time and accuracy. If more eigenmodes are selected, the results are getting more accurate, but the computing time will increase.

2.1.2 Calculation of stress

For gaining the stress and deformation it is necessary to calculate the eigenfrequencies and eigenmodes of the model. Therefore equation 1 is used, whereby K is the stiffness matrix, M the mass matrix, ω the eigenfrequency and ϕ the eigenmode [9].

$$(K - \omega^2 M) \cdot \Phi = 0 \tag{1}$$

With a superposition (eq. 2) of the calculated eigenmodes (ϕ_i) any time-dependent deformation u(t) can be calculated [9]. Therefore a scaling factor $s_i(t)$ is determined.

$$u(t) = \sum_{i=1}^{n} \Phi_i \cdot s_i(t) \tag{2}$$

Figure 5 and 6 show an example of two different curves of $s_i(t)$.

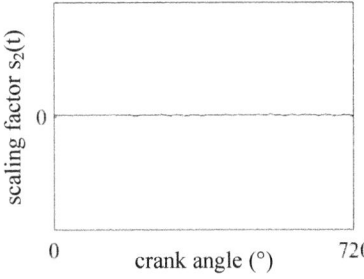
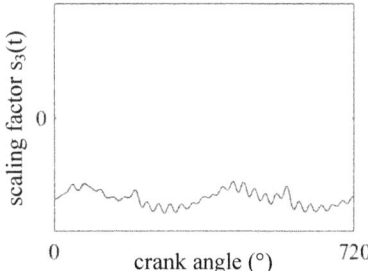

Figure 5: Calculated scaling factor of the second eigenmode in the bearing cap

Figure 6: Calculated scaling factor of the third eigenmode in the bearing cap

Figure 5 shows the scaling factor for the second eigenmode from a model of a main bearing wall (figure 9). As it can be seen, the factor is noticeably smaller than the scaling factor for the third eigenmode (figure 6). This means that the third eigenmode has more influence to the deformation of the body and so the dynamic behavior of the system is mainly characterized through it. In this way, significant or dispensable eigenmodes can be determined and the performance of the calculation, with regard to accuracy and time, can be optimized.

Beside the deformation also the stress can be determined with equation 2. For each selected eigenmode (ϕ_i), the corresponding stress state has to be calculated and inserted into the equation. So any stress state over the time can be simulated. In figure 7 and 8 the stress states of the second and third eigenfrequency are shown. Figure 7 shows the stress state for the second eigenfrequency, figure 8 for the third eigenfrequency. The combination of the stress states and the scaling factors yields to the final result, which will be presented in section 2.2.2, figure 11.

max. tension

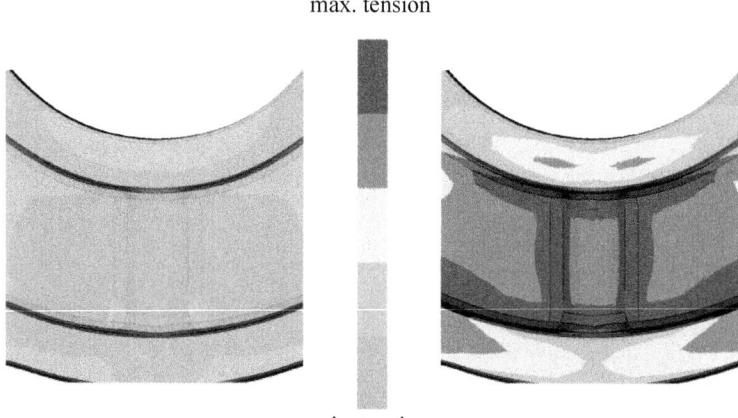

min. tension

Figure 7: Calculated stress in the bearing cap by the second eigenmode
Figure 8: Calculated stress in the bearing cap by the third eigenmode

Additionally to the modal stress states also stress states from unit loads (p) on explicit nodes can be used for the calculation (eq. 3) [9].

$$(K - \omega^2 M) \cdot \Phi = p \tag{3}$$

Especially for areas where loads are initiated or in contact areas, these modal states are important for calculating exact results.

2.2   Comparison between the elastic multibody simulation and the finite element method

For evaluating the elastic multibody simulation, it has to be compared with the finite element method, which is well established in the calculation of the crankcase. The main criteria in this comparison will be the computing time and the result quality of the stress analysis. Within the computing time also the time for creating the model will be considered.

2.2.1 Modeling and calculation

For comparing the methods it is necessary to have the same basics. For this, a model of a bearing wall of a crankcase is built. It consists out of the bearing wall, the bearing cap and the main bearing bolts, like illustrated simplified in figure 9.

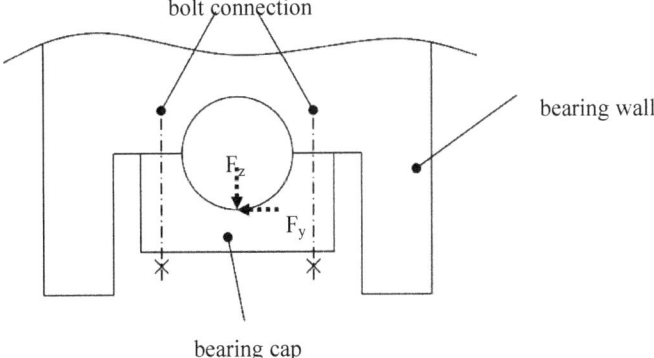

Figure 9: Schematic illustration of the model of the main bearing wall

The load of the combustion is initiated via the bearing cap with the forces F_y and F_z, which is also a simplification in this comparison.

Different to the modeling, the calculation is not identical within both methods. Due to reasons of time, the analysis with the finite element method cannot be as high resolute as the elastic multibody simulation, regarding the calculation steps. With the elastic multibody simulation a resolution about twelve to 15 times higher is possible, so more states can be calculated within an adequate time (sec. 2.2.2).

2.2.2 Comparison

As mentioned in the beginning of this chapter the main comparison between the finite element method and the elastic multibody simulation is the computing time and the quality of the results, regarding stress.

The first criterion in the comparison is the computing time. As described in the previous sections, the elastic multibody simulation must be faster than the finite element method due to the generation of a substructure with less degrees of freedom. In the comparison, the computing time for calculating the main bearing wall with the elastic multibody simulation is about 100 times faster than calculating it with the finite element method. But it has to be mentioned that additionally unique time is needed for the substructure generation (sec. 2.1.1) and the calculation of the stress (sec. 2.1.2), which is about the same time range as one finite element calculation. So the whole computing time for one variant is identical for both methods. If more variants have to be calculated, the elastic multibody simulation shows its temporal advantages.

With regard to the fact, that in future more variants have to be calculated, like said in chapter 1, the elastic multibody simulation could be a good alternative to the finite element method. Furthermore the modeling of the bodies is about the same effort.

The next comparison is the stress analysis. For the strength of the construction this is one of the most important criteria. Therefore different regions of the model are regarded. In this comparison the finite element method is set as the benchmark, because this method is established for a long time and well validated by testing. It is supposed that the results from the finite element method match with the real stress states in the bearing wall under the assumed conditions.

The first region to compare is a part of the bearing cap, which is in alignment to the initiation of the load (see figure 9). The corresponding results are shown in figure 10 and figure 11. Figure 10 shows the calculated stress of the bearing cap with the finite element method whereas figure 11 shows the calculated stress with the elastic multibody simulation. Altogether it is noticeable, that the calculated von-Mises-stress in the elastic multibody simulation is smaller than in the finite element method, but the difference is less than 10%. Regions of high or low tension can be identified exactly with both methods. For example the maximum tension in both is near the load initiation. Also the compressive stress in the left and the right side of the bearing cap matches.

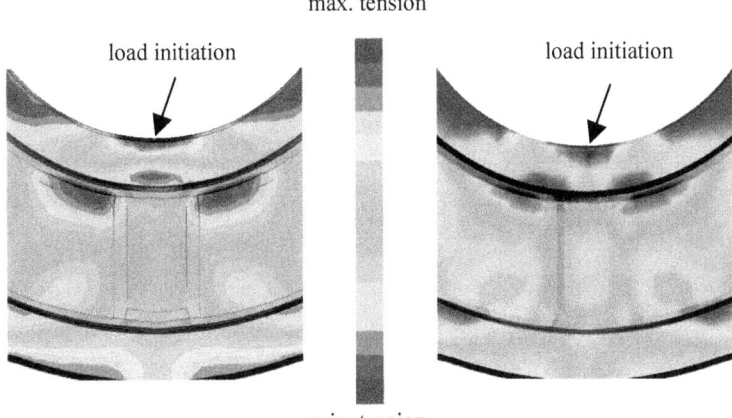

max. tension

load initiation   load initiation

min. tension

Figure 10: Calculated stress by finite element   Figure 11: Calculated stress by elastic
method in main bearing cap   multibody simulation in main bearing cap

The second region to compare is the contact surface of the bearing wall to the bearing cap. In figure 12 the calculation of this contact area with the finite element method is show, figure 13 shows the same area calculated with the elastic multibody simulation. It has to be said, that the contact areas are special to compare, because the methods use different techniques to simulate the contact. In contrast to the elastic multibody simulation, the finite element method can calculate a "hard contact" which means that no penetration between the bodies is allowed [8]. Due to numerical reasons of the program, the elastic multibody simulation cannot calculate a "hard contact" [9]. The comparison demonstrates that the different techniques consequently lead to very different results. Mainly in the right side of the contact area, two times higher stress states are calculated with the elastic multibody simulation than with the finite element method.

The comparison between the finite element method and the elastic multibody simulation shows that the finite element method has still some advantages in the calculation of stress in contact areas, but in the remaining structure the results are in a comparable range. Furthermore the elastic multibody simulation has the possibility to analyze in a higher resolution and to consider attached components. Also is has noticeable advantages by calculating different variants due to the temporal reasons. This can be used for regarding the reliability of the crankcase in different applications.

max. tension

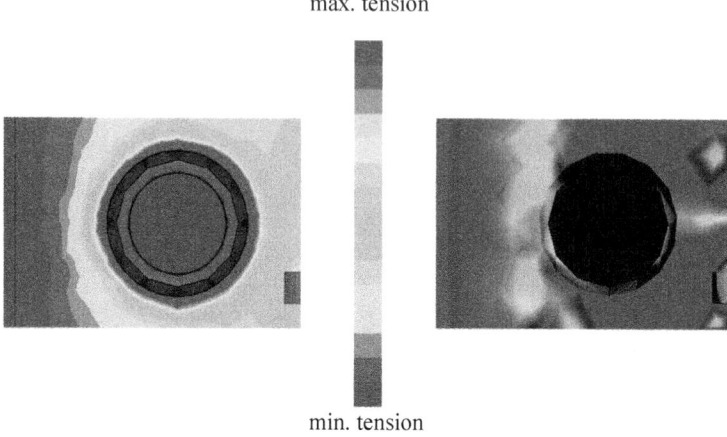

min. tension

Figure 12: Calculated stress by finite element method in the contact area

Figure 13: Calculated stress by elastic multibody simulation in the contact area

3 Integration of the reliability into the design process

Due to the temporal advantages of the elastic multibody simulation it is possible to calculate a crankcase with a variation of different loads (emerging strain). This can be necessary if the engine gets installed in several vehicle series which have different kind of costumer groups and consequently get strained in different ways. So with the method of the elastic multibody simulation it gets possible to specify the probability of default more accurate for each costumer group or each vehicle variant. Critical or oversized areas can be found more efficient. Furthermore the opportunity to calculate different testing cycles and analyze them in detail is given. So the cooperation between simulation and testing can be supported.

Beside the knowledge about the emerging strain, also the knowledge about the tolerable strain is necessary for gaining information about the strength of a construction. While the emerging strain can be calculated out of the data given by analyzing test cycles or costumer groups, the tolerable strain has to be determined through material testing.

3.1 Tolerable strain

The tolerable strain is the upper limit of load which a construction can resist without getting broken. So for the design of a component it is essentially to know this limit and also its distribution as precisely as possible [3]. For this reason the material characteristics are determined via oscillating tests on the concrete structure of the construction. This has advantages concerning the accuracy of the results, because no correction factors are needed. The disadvantage is that the tolerable strain can only be determined exactly for this special investigated spot. Other interesting points have to be tested on their own. This procedure ensures that all effects, like notches or surface roughness, are taken into account [3]. So it is ensured that uncertainties can be avoided and the construction is tested like in an actual operation. With the oscillating testing, stress-cycle diagrams and S/N curves can be generated.

Because of the complex testing of the critical areas in the concrete structure it is very important to know exactly the interesting and critical points. Therefore it is necessary to investigate the failures which occurred so far and gain knowledge out of them. Also a deep analysis of each structure is indispensable for the construction. Therefore a Failure Mode and Effects Analysis (FMEA) can assist [3].

In this work also the statistical spreading of the material characteristics is of interest. Therefore crankcases out of the serial production are taken and tested. This ensures that a realistic statistical distribution can be described like it would be in the actual use of the costumer.

3.2 Emerging strain

The emerging strain is the strain out of the forces and torques acting in the engine. In this case only the loads reacting out of the combustion are considered, like the forces on the bearing, the piston side forces or oscillating effects. External forces, like shocks from the road, are neglected in this model.

For getting information about the reliability of the crankcase it is necessary to know the actual or the most harmful forces and torques in the engine. Like mentioned in chapter 1, the most harmful states in newer engines are not as clearly as before. So it becomes necessary to calculate different driving states in a high resolution for ensuring that the most critical can be found. Such driving states are gained out of the combination of revolution and engine torque, which can be well recorded in testing cycles. Out of these parameters, the load and pressure in the cylinder can be calculated and so all forces and torques in the engine can be determined [10, 11]. For each driving state one whole cycle (720° crank angle) gets simulated and thereby the damage of the crankcase in this state can be determined.

In figure 14 and figure 15 the damage in the bearing cap of the calculated bearing wall (sec. 2.2.1, figure 9) is shown. Figure 14 shows the damage at a low level of revolution and engine torque, whereas figure 15 shows the damage at a high level of revolution and engine torque. The results demonstrate that a driving state at a high revolution and engine torque level is more harmful than a driving state at a low revolution and torque level. First, this is not very surprising and also well-known from former calculations, because the forces and torques out of the combustion at a high level of revolution and engine torque are higher than at a low level. But obviously the strain in former calculations distinguished to a higher extend because mass forces or the strain out of oscillating effects had been smaller. In newer and lighter constructions these strains increase significantly.

max. damage

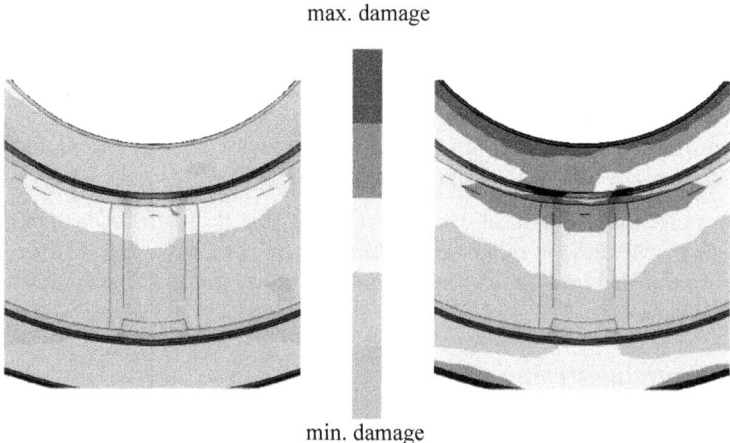

min. damage

Figure 14: Damage at a low strain level   Figure 15: Damage at a high strain level

Also with regard to the residence time of the different driving states, the stress states have to be evaluated in a new way. As it can be seen in figure 16, the time slice of a driving state by a high revolution and engine torque is minimal, whereas the state at a low engine torque and revolution level has a high time slice.

With the combination of the time distribution and the damage to the according driving state, a new evaluation of the harmfulness of each driving state gets necessary. High level driving states are hardly driven and so the percentage of the damage for the regarded construction is minimal, whereas low level driving states are driven to a high extent and so need to be especially validated.

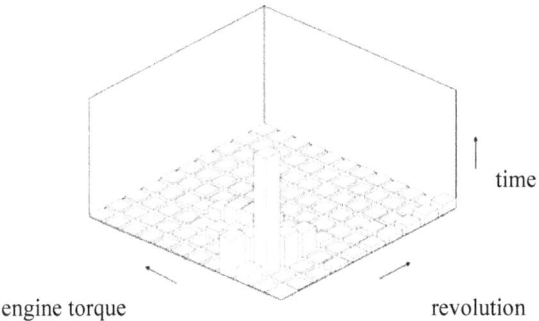

engine torque   revolution

time

Figure 16: time slice of a test cycle

With the combination of simulation and analysis, the percentage of damage and the time slice of the driving states can be evaluated more in detail and so oversized or critical areas can be determined exactly. Out of this and with the knowledge about the material characteristics (sec. 3.1) a good basis for lightweight constructions is given.

4 Summary and Outlook

With the increasing efforts in the field of lightweight engineering also the simulation methods have to be adapted. Faster methods with a higher resolution are needed to analyze the engine in detail, especially with regard to the increasing number of applications of the engines in different vehicle series. In the presented work, the elastic multibody simulation showed a lot of advantages regarding the mentioned purpose. In the direct comparison to the finite element method, the elastic multibody simulation was about 100 times faster in the calculation of a bearing wall and had the same result quality with the same effort in the modelling process. Only in the contact area the results differ and the elastic multibody simulation has to be optimized. But especially the temporal advantages of the elastic multibody simulation predominate. Therefore the method can be used to calculate variants and so set up a reliability-oriented simulation. With the possibility to calculate different driving states and to generate knowledge about the real strains in the engine, critical or oversized areas can be detected efficiently.

The investigation in this work showed, that driving states at a high level of revolution and torque strain the engine noticeably more than driving states at a low revolution and torque level, but are hardly driven. So the percentage of damage for the engine is hardly existent. To evaluate the strength and reliability of the crankcase as a whole, also the statistical spreading of the used materials is necessary. Therefore oscillating tests at the critical spots of the crankcase are made.

The whole procedure of the reliability-oriented simulation, presented in this work, makes it possible to evaluate the strain of the crankcase in detail and optimize its structure for being efficient and fulfill future lightweight targets. Therefore the method will be further developed and so the simulation and testing will grow more together for achieving highest quality in the reliability of the crankcase and in the engine development process.

5 References

1. Friedrich, A.: Gemeinsam in Stuttgart für nachhaltigen Automobilbau und Produktionswelt der Zukunft. Festliche Grundsteinlegung für Forschungsfabrik ARENA 2036. Stuttgart 2015.

2. Daimler AG, Global Media Site. Date 13.01.2016

3. Bertsche, B.: Reliability in automotive and mechanical engineering. Determination of component and system reliability. Berlin 2008.

4. Gasch, R.; Liebich, R.; Knothe, K.: Strukturdynamik. [s.l.] 2012.

5. Wilhelm, K.: Strukturdynamische Analyse von Kurbelwelle und Motorblock mit elastohydrodynamischen Wechselwirkungen, Dissertation. Aachen 1999.

6. Rill, G.; Schaeffer, T.: Grundlagen und Methodik der Mehrkörpersimulation. Wiesbaden 2010.

7. Schiehlen, W. O.; Eberhard, P.: Applied Dynamics. Switzerland 2014.

8. Dassault Systems: Abaqus 6.14 Online Documentation. URL: http://abaqus.software.polimi.it/v6.14/. Date 15.12.2015.

9. Simpack AG: SIMPACK Documentation 2014.

10. Braess, H.-H.; Seiffert, U. (Hrsg.): Vieweg-Handbuch Kraftfahrzeugtechnik. Mit 50 Tabellen. Wiesbaden 2013.

11. Merker, G. P.; Schwarz, C.: Combustion engines development. Carburation, mixture formation, combustion, emission and simulation. Berlin, Heidelberg 2012.

Variable valve timing of intake valves of a heavy-duty Diesel engine as a way to improve fuel consumption

Dr.-Ing. Simon Schneider, Dipl.-Ing. Sascha Naujoks

MAHLE International GmbH

1 Introduction

Valve train variabilities at Diesel engines are a current topic both in light vehicle applications as well as for commercial vehicles. In the last years, both applications were investigated by Forschungsvereinigung Verbrennungsmotoren [1], [2]. There are series applications for light vehicle engines and for medium duty truck engines [3]. In heavy-duty truck engines, variable valve trains are currently only used with engine brakes.

The Atkinson-cycle or combustion process, realized with a conventional crank train by Miller valve timings (early or late intake valve closing, EIVC / LIVC), is a variant of the four-stroke combustion process that enables the expansion ratio to be slightly higher than (and not identical to) the compression ratio. This is in theory beneficial for cycle efficiency, but the fuel-saving potential depends on turbocharger efficiency, exhaust gas temperature and the charge air pressure available [5]. If the Atkinson-cycle shall be realized with identical air mass flow as the baseline, then a higher charge air pressure is required, as the compression work is partially shifted from in-cylinder compression to the turbocharger compressor stage. This enables – at the same charge air cooler temperature level – a more effective charge air cooling, which leads to a reduced temperature of the cylinder charge at start of combustion and as such can be beneficial for reducing NO_X emissions [4].

In the medium-speed marine Diesel engine sector, fixed Miller valve timings are successfully applied in series production. With high-efficiency two-stage turbochargers, no EGR, no exhaust-gas aftertreatment and a narrow band of engine operating speeds, both significant fuel consumption savings and NO_X emission reductions can be realized simultaneously [6].

Current Atkinson-cycle investigations for truck engines like [4] or [7] focus strongly on thermodynamics and the testing is thus performed on single-cylinder engines with external supercharging. This leaves an uncertainty about the interaction with the real truck turbocharger in terms of scavenging gradient for EGR and turbocharger efficiency. Some of the theoretically required charge air pressure levels for the Atkinson-cycle might be challenging to achieve with current single-stage heavy-duty Diesel turbochargers.

2 Approach of the MAHLE investigations

MAHLE investigated the real-world fuel consumption reduction potential of different valve train variabilities for a modern full size Euro 6 truck engine. This investigation realizes variability via the valve train, which is adjustable or switchable back to baseline valve timings.

This requires considering the limitations of the real turbocharging in full engine tests as well as the cross-influence with high pressure EGR, which is already applied at the base engine to control NO_X emissions. For the first step of engine investigations, the turbocharger specification was kept identical to the baseline, although modifications might offer even further potential for future applications of the concepts.

MAHLE performed simulation work to investigate production feasible valve train variabilities with constant intake valve opening, constant maximum valve lift and variable valve closing for both intake valves. That is a valve timing matching the Atkinson-cycle with late intake valve closing (LIVC). Bearing a potentially adjustable system in mind, LIVC was considered to be easier for production engines compared to EIVC, with comparable thermodynamic benefits to be expected for both variants.

Based on this investigation, the most promising valve closing timing was selected, produced as a fixed timing prototype camshaft and back-to-back tested against the series configuration in complete engine tests. Combustion calibrations were optimized to increase the fuel consumption benefit while keeping NO_X emissions on baseline levels.

The testing was done on a high precision thermodynamics engine test bench with conditioning units for all engine media and the MAHLE flexible ECU as controller. The engine in this investigation was a 6-cylinder heavy-duty engine with approximately 13 l displacement, EURO 6 emission level calibration and cooled high pressure EGR.

3 Simulation work and selection of most promising intake valve timing

The potential of intake-side valve train variability is investigated using 1D engine simulation with a well-correlated engine model. A variable system shall be used that is able to generate a variable intake valve closing time.

Previous studies on a passenger car Diesel engine showed that gas exchange through only one opened intake valve leads to increased pumping work. Therefore, in the investigated heavy-duty Diesel engine setup, a CamInCam® technology based valve actuation system was used. This enables that both intake valves have a LIVC timing and identical valve lifts with constant baseline intake valve opening time (IVO). This leads to a highest possible air charge when LIVC is set.

Ideally, the ramp of the valve lift profile should remain identical to the series lobe, with acceleration of the valve lift staying unchanged. The second half of the valve lift profile can be phased towards late intake valve closing via the system, forming a plateau at maximum lift. In Figure 1 the investigated baseline and maximum retarded (60 °CA LIVC) lift profiles are shown. With the CamInCam® system, the valve closing time can be freely varied between these two extremes.

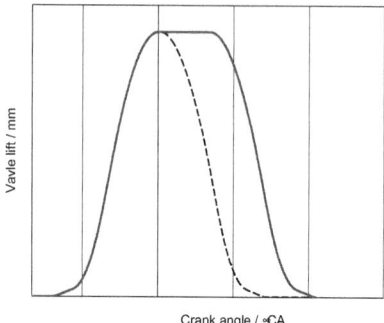

Figure 1: Proposed valve lift profiles for baseline and LIVC operation

The intake valve closing time was swept between 0 and 60 °CA in simulation whilst limiting the cylinder peak pressure to the baseline value as an indicator for similar engine-out NO_X emissions. To increase efficiency, the 50% mass fraction burned (MFB50) is adjusted by shifting start of injection (SOI) until the baseline maximum cylinder pressure is matched. The wastegate is closed wherever possible, to keep the drop in air mass flow moderate.

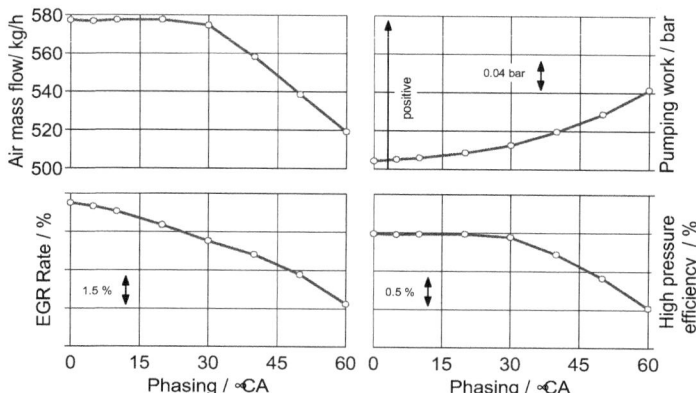

Figure 2: Gas exchange and combustion parameters for a sweep of the intake valve closing time (simulation result)

Figure 2 shows the results of the sweep for the ESC (European Stationary Cycle) A50 operating point, which is one of the most frequent operating points in long-haul truck applications. From left to right, the intake valve closing time is gradually retarded. Starting from baseline, the air mass flow can be held constant by closing the wastegate until 30 °CA phasing. Further on, the wastegate is fully closed. The EGR rate decreases although the EGR valve is at its maximum opening position. Pumping work is continuously improving with LIVC. The high pressure cycle efficiency follows the trend of the air mass flow and is also constant up to 30 °CA phasing. The highest net efficiency is found at approximately 30 °CA phasing for this operating point.

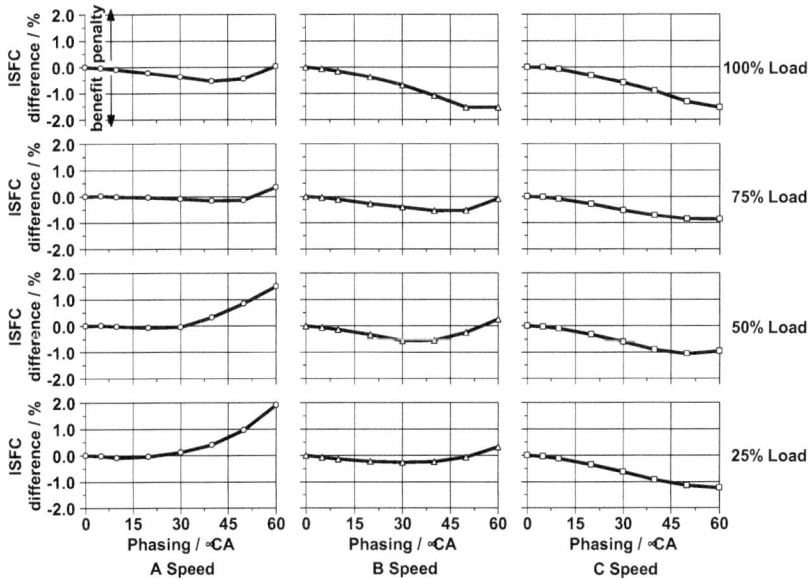

Figure 3: Fuel consumption results for ESC operating points (simulation result)

Figure 3 shows the simulation results with the 0 to 60 °CA LIVC variation for all ESC operating points. Depending on the operating point, the maximum fuel consumption benefit is found between 20 and 60 °CA phasing. The maximum improvement can be achieved at 100% load with B- and C engine speed. Operating points A25 and A50 show penalties in fuel consumption, both at maximum valve phasing. At C speed, the improvements follow a similar trend for all engine loads. The ISFC advantage generally rises with increasing LIVC phasing at all C speed loads.

Based on the operating profile of a long haul truck, operating points between A25, A50 and A100 are most relevant for overall fuel consumption. With this target in mind, a 40 °CA phasing was chosen for the LIVC camshaft prototype for the full-engine tests.

4 Engine testing

4.1 Test bench, engine setup and testing procedure

The MAHLE heavy-duty engine test bench features engine oil and coolant conditioning units (with constant temperature settings for all operating points) and high speed cylinder pressure indication is implemented at all cylinders for these tests. The charge air cooler outlet temperature is held constant at a value depending on engine speed. Following the simulations, engine tests were performed to directly compare baseline against the fixed timing LIVC camshaft. The procedure of testing the LIVC camshaft is described in Figure 4.

The boundary conditions for the tests are similar engine-out NO_X emissions and cylinder pressure gradient compared to the baseline camshaft. The engine is run with the MAHLE flexible ECU [8], which opens up the possibility to calibrate the engine parameters to fit these limits and to reach the lowest fuel consumption. During engine testing, mainly the following parameter settings were optimized: wastegate position, start of injection (SOI) and EGR valve position.

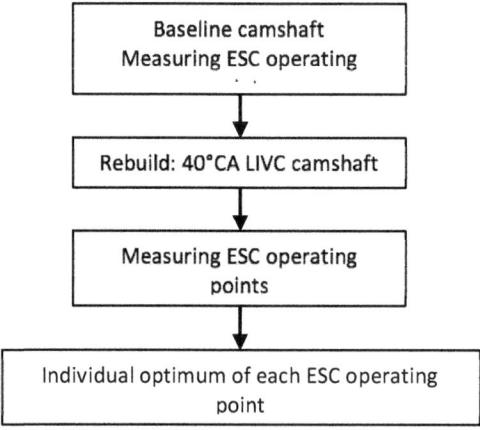

Figure 4: Procedure of the comparison of baseline and LIVC setup in engine testing

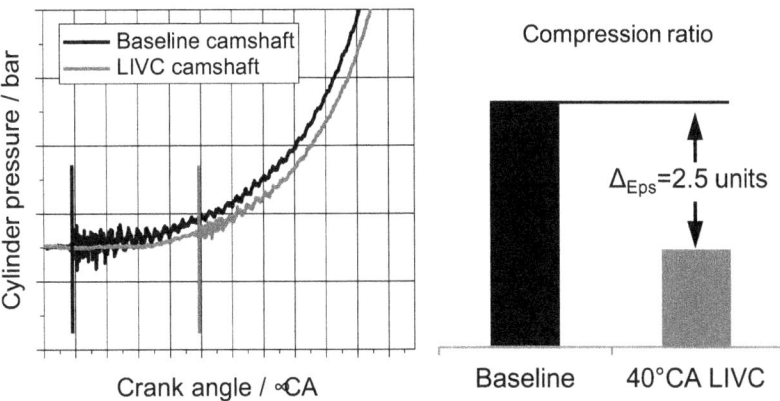

Figure 5: Start of compression and difference in effective compression ratio by the Atkinson valve timing

To confirm the proper function of the new camshaft, the pressure traces were careful-ly checked at start of the tests and the begin of the compression phase was confirmed to begin at a 40 °CA delayed crank angle. The shift in valve closing time and the later start of the compression is directly visible in the high speed pressure signals (Figure 5). Atkinson valve timings are known to change the effective compression ratio. This difference was calculated via the cylinder volume at intake valve closing and also checked via motored engine run. With the 40 °CA LIVC camshaft, the compression ratio decreases roughly 2.5 units compared to the baseline.

4.2   Reduction of fuel consumption in engine tests at ESC points

For each of the ESC operating points, an optimization was done at the test bench to find the most appropriate combination of parameters as described before. The result of this testing is shown in the Figure 6. It is the comparison of each optimized operating point with LIVC camshaft against the baseline. The percentage of improvement in the indicated specific fuel consumption (ISFC) differs, depending on engine speed and load.

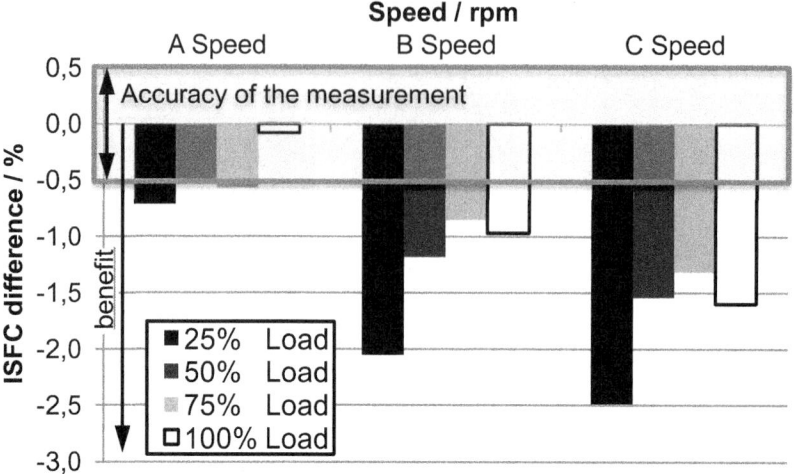

Figure 6: ESC operating points individual optimization: fuel consumption result (0 % = baseline) (engine testing)

At "A speed", the measured improvement is within the accuracy of the measurement. It is nearly independent from engine load. At "B speed", the highest reduction in fuel consumption is at 25% load and it decreases towards higher loads. At the C25 operating point, the largest overall reduction in fuel consumption is observed. All optimized operating points have the same NO_X level as their corresponding baseline points. Depending on the operating point, the smoke emissions of some LIVC points are slightly increased over the baseline given a lower air-fuel ratio.

4.3 Optimization steps at operating point A50

As an example, a detailed analysis of the A50 operating point is presented hereafter. Figure 7 shows the steps to achieve baseline NO_X emissions with the best specific fuel consumption. The steps are named by an index (see labels 1-6 at the X axis in Fig. 7).

No.1 is the result of the baseline camshaft. With the LIVC camshaft and baseline ECU settings (No.2), the air mass flow is reduced. With the new camshaft, the pressure difference between exhaust and intake is lower, which reduces the EGR rate despite maximum opened EGR valve. The EGR rate is calculated based on CO_2 emissions in the exhaust gas flow and in the intake manifold. The closing of the wastegate (No.3) leads to a marginal increase of air mass flow and lambda. Due to an earlier

SOI in combination with the closed wastegate (No.4), the exhaust enthalpy is lower, which can be recognized in a slightly lowered turbo speed.

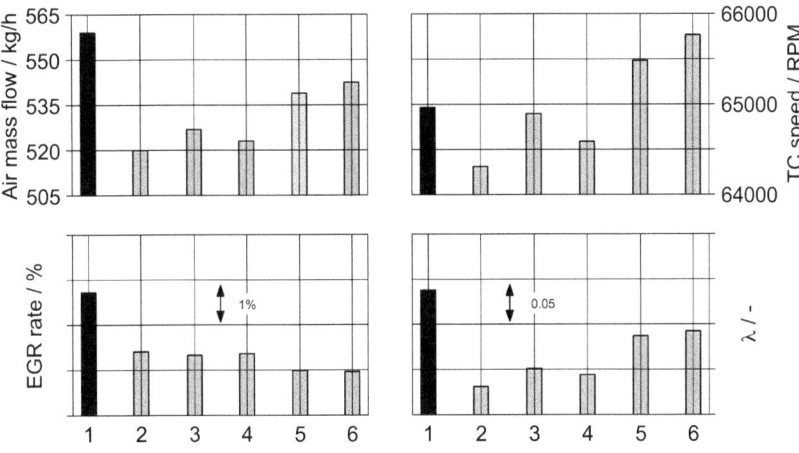

Figure 7: Steps from baseline cam lobe (black, No.1) over LIVC lobe with baseline settings (bright grey, No.2) towards optimized operating point (No. 6) (engine testing)

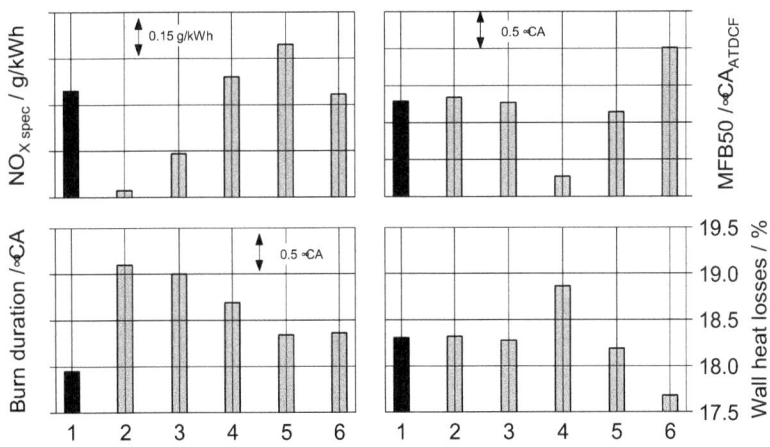

Figure 8: More parameters for the optimization steps at A50 operating point (engine testing)

Alternatively, reduced fuel consumption can be achieved by closing the EGR valve (No. 5) within a small scope and with baseline SOI timing. This results in a higher air mass flow and as a consequence lower EGR rate. In addition, the exhaust enthalpy is higher, which speeds up the turbocharger a little. To decrease the NO_X emissions, the SOI is then shifted towards 1 °CA late (No.6), shown in Figure 8 top left. This is the setting with best specific fuel consumption that matches all boundary conditions. The fuel saving for the A50 point is thus 0.5 %, as shown in Figure 6.

4.4 Detailed loss analysis via pressure trace analysis at A50

To investigate in more detail which effects bring a fuel consumption benefit or a penalty, the losses were analyzed by pressure trace analysis for the A50 operating point as shown before. In Figure 9, this analysis is put as the difference between the baseline camshaft (No.1 from Figure 7) and the LIVC camshaft with baseline settings (No.2 from Figure 7).

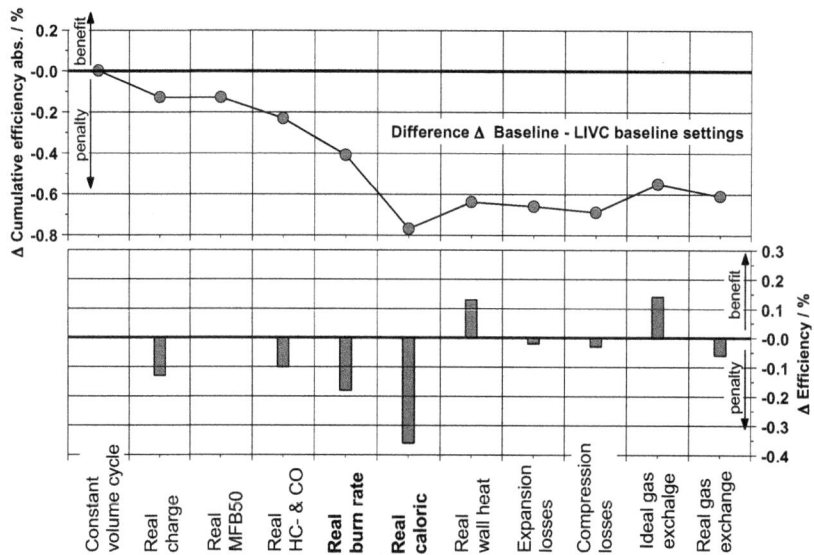

Figure 9: Split of losses difference baseline vs. LIVC baseline settings at A50 operating point (pressure trace analysis)

This comparison directly shows the influence of the late intake valve closing. The upper diagram shows the cumulative difference of the absolute efficiencies, whereas the lower diagram shows the individual influence of each step. From left to right, the contributions ranging from constant volume cycle, real charge, real MFB50, real HC- & CO-emissions, real burn rate, real gas behaviour of the cylinder charge (real caloric), wall heat losses, expansion losses, compression losses, ideal pumping work ("ideal gas exchange") and real pumping losses ("real gas exchange") are listed (see axis labels).

The first influence factor is the real charge with a disadvantage for the LIVC camshaft. No difference is visible in MFB50 because of the same SOI setting. Then there are the primary influences of the LIVC cam timing: the real burn rate und the real caloric, both of which are detrimental to the cycle efficiency. An improvement for the LIVC camshaft comes from lower wall heat losses, but this is only a small contribution. The reduced wall heat losses come from the lower combustion pressure level, the reduced mass flow and the slight increase in cooling power of the charge air cooler that come from the Atkinson-cycle. As a conclusion, there remain some disadvantages for the LIVC using baseline ECU settings.

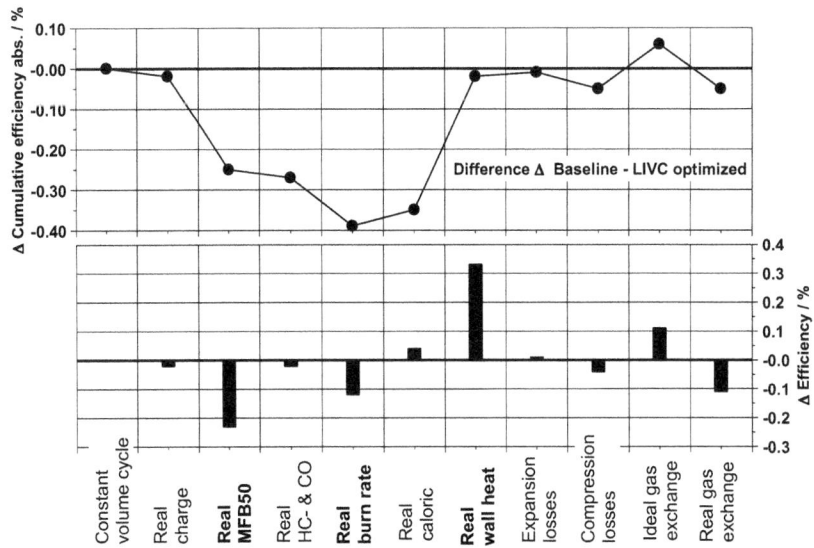

Figure 10: Split of losses difference baseline vs. LIVC optimized point at A50 operating point (pressure trace analysis)

Figure 10 shows the comparison of the baseline camshaft (No.1 from Figure 7) and the optimized operating point with LIVC camshaft (No.6 from Figure 7), again for the A50 operating point. Due to the late SOI, the LIVC camshaft has a disadvantage in MFB50 and in real burn rate due to longer burn duration. These contributions are compensated with a positive influence of lower wall heat losses. This reduction of wall heat losses is the biggest single effect in this split of losses and comes from Atkinson cycle (compare Figure 9) and from late MFB50. At the end, both results of the two tested camshafts are nearly identical in overall efficiency as seen by the pressure trace analysis (whereas the fuel measurement, section 4.2, pointed out a small benefit of the LIVC setup at A50).

4.5 Map-dependency of the fuel consumption saving potential

As can be seen in Figure 6, the highest reduction in fuel consumption can be reached at the C25 operating point. The optimized variant of the C25 operating point has an ECU setting with 0.5°CA earlier SOI and a completely closed wastegate, compared to the baseline C25 settings. The following section looks in more detail at why there is the possibility for reduction of fuel consumption at C25 and not at A50. The key parameters for that are shown in Figure 11.

In the direct comparison of these two operating points, both LIVC camshaft results show a reduction of the wall heat losses. At A50, this advantage is partly caused by the late MFB50 as a result of later SOI, which is a thermodynamic drawback. At C25 however, with LIVC camshaft and baseline ECU settings, there is a NO_X emission advantage. To fully exploit the baseline NO_X value, the MFB50 can be even shifted towards higher thermodynamic efficiency. Despite of that shift, there remains a positive contribution of wall heat losses at C25.

The burn duration of the results with the LIVC camshaft seems to show an contradictional behavior, it increases at A50 while it decreases at C25. One reason for this large difference is the level of air-fuel ratio at both operating points. At A50, it is already at a very low range with baseline camshaft, with high impact on combustion efficiency, whereas at C25, both baseline and optimized lambda are so high that the lambda reduction has no relevant influence on combustion.

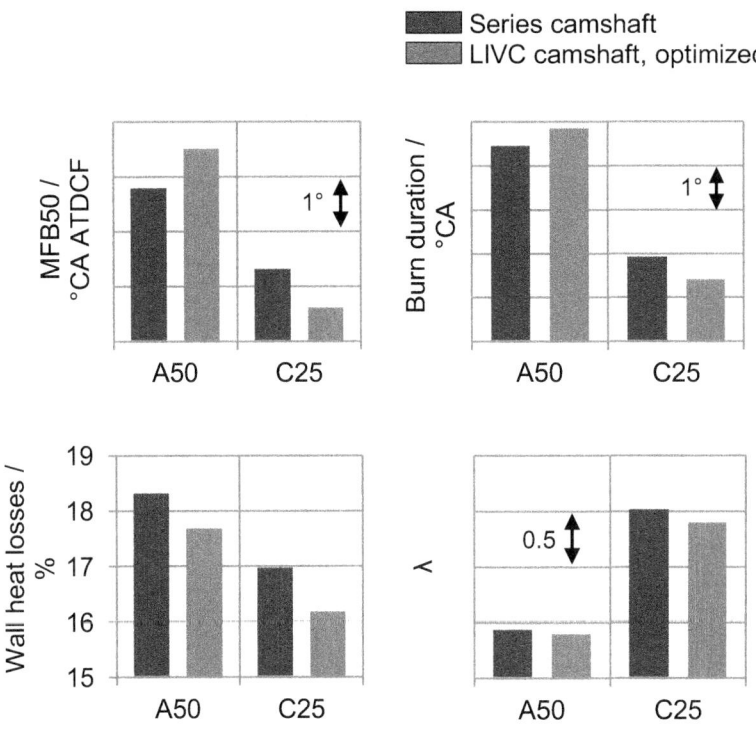

Figure 11: Difference between A50 and C25 operating point, baseline and optimized LIVC operating point each (engine testing)

In Figure 12, the same schematic of the split of efficiencies out of the pressure analysis like in Figure 9 is shown. Now, the difference between LIVC camshaft with optimized parameters and baseline valve timing is shown for the operating points A50 and C25 in comparison. This analysis confirms the previous explanation of the thermodynamic effect. The biggest impact is the opposite behavior of the burn duration (column "real burn rate"), with further contributions by MFB50.

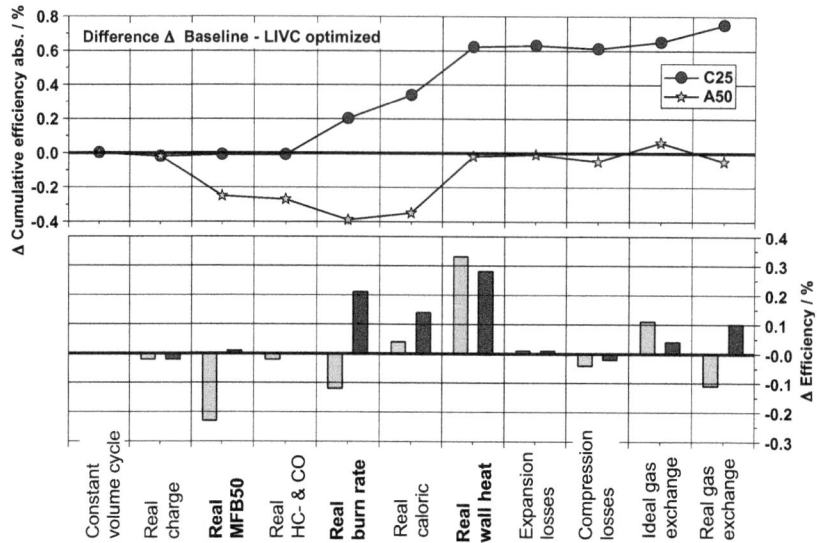

Figure 12: Split of losses difference baseline vs. LIVC optimized point at A50 and C25 operating point (pressure trace analysis)

4.6 Thermal management

Apart from reduction of fuel consumption, fulfilling the emissions regulations is another big challenge for heavy-duty engines. To meet the emission limits, the exhaust aftertreatment system temperature has to be held at operating temperature. Especially low-load operating points can be critical with exhaust temperatures post turbocharger falling below 250 °C.

With the LIVC camshaft, it is possible to increase the turbocharger outlet temperature at operating points with lower loads without a negative impact on the fuel consumption. Exhaust gas enthalpy stays constant, compared to the baseline, but an exhaust gas temperature rise is possible by approximately the same degree as the exhaust mass flow is lowered due to Atkinson cam timings. This also implies that this effect helps maintaining the temperature of the aftertreatment unit, but doesn't help for heating it up, as no additional enthalpy is brought to the system.

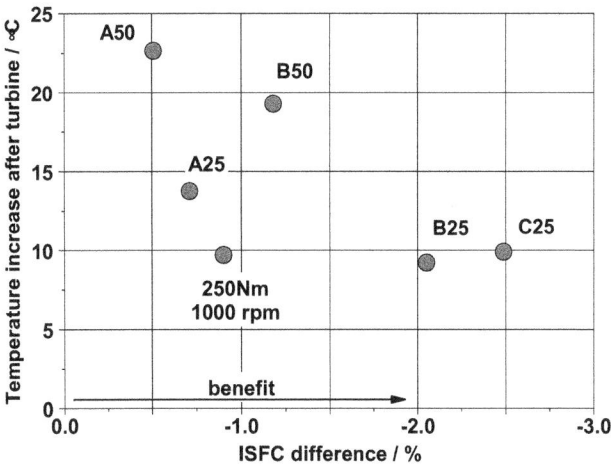

Figure 13: Increase of exhaust temperature with LIVC

In Figure 13, several medium and lower load operating points are shown. The Y axis shows the increase of exhaust gas temperature post turbocharger. The benefit in specific fuel consumption is displayed on the X axis. It is worth mentioning that the gain in exhaust gas temperature comes in all tested points as an additional effect to the fuel consumption savings. The highest increase appears at A50 within an increase of just above 20 °C. At 250 Nm and 1000 RPM, an operating point occurring at HD diesel engines in slow moving traffic or congestions, the temperature can still be increased by about 10 °C.

5 Summary and Outlook

MAHLE investigated an Atkinson-cycle with late intake valve closing by simulation and full-engine testing on a EURO 6 heavy-duty Diesel engine. A simulation study pointed out a promising valve timing for the sample testing on the engine.

The engine test results confirm the potential for fuel consumption savings in the range of 0.5 to 2 percent, varying across the engine map, for realistic boundary conditions in a full engine test. The potential is low in the main operating region of the engine, where the turbocharger air massflow is perfectly exploited by applying the lowest air/fuel ratio with good fuel consumption in series application. However, in regions with some reserves of the turbocharger compressor, a part of the NO_X emission reduction of the EGR can be substituted by the Atkinson-cycle effect, leading to fuel consumption benefits while still meeting baseline emissions.

As the Atkinson-cycle uses up the reserves of the turbocharger, a switchable system might be more appropriate than a simple fixed LIVC valve timing. This helps operating the engine in elevated altitudes and gives some additional freedom in the choice of the LIVC timing. There are additional effects like a slight exhaust gas temperature increase that can be useful in some applications.

For application of LIVC valve timings at other engines, modifications of the turbocharger are of interest, which have not yet been implemented within the scope of this investigation.

References

1. Kovács, D.; Gehrke, S.; Eilts, P.: Downsizing durch Luftpfadvariabilitäten am Nfz-Motor. FVV-Forschungsvorhaben Nr. 1065, Abschlussbericht, Heft 1060, Frankfurt am Main, 2015

2. Honardar, S.; Deppenkemper, K.; Nijs, M.; Pischinger, S.: Potenziale von Ladungswechselvariabilitäten beim Pkw-Dieselmotor. MTZ 75 (2014), Nr. 9, S. 64–69

3. Herrmann, H.-O.; Nielsen, B.; Gropp, C.; Lehmann, J.: Mittelschwerer Nfz-Motor von Mercedes-Benz — Teil 1: Motor- und Abgasreinigungskonzept. MTZ 73 (2012), Nr. 10, S. 730–738

4. Theißl, H.; Kraxner, T.; Seitz, H. F.: Miller-Steuerzeiten für zukünftige Nutzfahrzeug-Dieselmotoren. MTZ 76 (2015), Nr. 11, S. 18–25

5. Eilts, P: Das Miller- und das Atkinsonverfahren an Verbrennungsmotoren, ATZ live Konferenz Ladungswechsel im Verbrennungsmotor, Stuttgart, 2014

6. Tinschmann, G.; Holand, P.; Benetschik, H.; Eilts, P.: Potenziale der zweistufigen Aufladung am Großdieselmotor 6L 32/44 CR von MAN. MTZ 69 (2008), Nr. 10, S. 818 – 828

7. Schutting, E.; Neureiter, A.; Fuchs, C.; Schatzberger, T.; Klell, M.; Eichlseder, H.: Miller- und Atkinson-Zyklus am aufgeladenen Dieselmotor. MTZ 68 (2007) 6, S. 480–485

8. Mahle Flexible ECU information on http://www.mahle-powertrain.com/

New product aimed to optimize air intake system for low end torque enhancement

V. Raimbault (1), J. Migaud (1), D. Chalet (2), Q. Montaigne (2), H. Buhl (3), H. Fuchs (3), A. Bouedec (2)

MANN+HUMMEL France SAS (1)
Ecole Centrale de Nantes, LHEEA Lab. (ECN/CNRS), Nantes, France (2)
MANN+HUMMEL GmbH (3)

Abstract

Low end torque enhancement for turbocharged engines is now more important than ever due to upcoming new regulations after Euro 6 with real drive emission application. With further trend to reduce the weight of next cars generation, there is the need to adapt and reduce continuously the size of engine and its components to find good trade-off between engine size, vehicle performances, fuel consumption and costs. This paper is describing how air intake systems can be optimized taking into account the particularity of small turbo charged Diesel engines of passenger cars. The novelty of the paper is to address the benefit value thanks to the optimization of interaction between all stand-alone components like turbo chargers, charge air coolers, ducts and air intake manifolds. It is described and demonstrated thanks to both simulation and testing, how switchable volume and length can help the performance improvement while reducing the engine displacement. As resonance charging is the physics used here to help low end torque improvement up to more than 10 percent for 4-cylinder engines, and even more for 3-cylinder engines, a special new design of a switchable air intake system is shown, giving good performances in terms of pressure waves propagation and pressure loss. Tests on prototypes made on engine bench are showing the benefits of such a concept compared to state of the art. The complete air intake system composed of plastic parts routing the air from the charge air cooler to different charges air ducts geometry will offer at the end a new possibility for increasing volumetric efficiency thanks to resonance charging also for small turbocharged engines, at a reasonable cost. Also for new turbo charged gasoline engines, as real drive emission could impact scavenging and air/fuel ratio control in a wider engine operating range. In this case again, resonance charging thanks to air intake system geometry could be a way to compensate lack of performance especially at very low engine speed.

Introduction

Following the changes in emissions regulations, with introduction of real driving emissions (RDE) tests for passenger cars, the trend for Diesel cars should be more to reduce dramatically the NO_x emissions thanks to usage of EGR even at full load during transient operations. It could be used in combination with SCR for further exhaust NO_x emissions reduction, this combination can help to reduce urea consumption over time. At high load operating points, and especially at low engine speed when turbocharger action is limited, it will be then needed to introduce more gas inside cylinders to achieve an acceptable trade-off between NO_x emissions, and performance. It will be shown first by engine tests in this paper how resonance charging - using active device and special geometries in air intake - is a possible way to increase volumetric efficiency at a reasonable cost for both 4-cylinders and 3-cylinders Diesel engines. In transient conditions for the vehicle, we can even consider enhancing the acceleration thanks to instantaneous increase of low end torque; it will be shown thanks to system simulation made with engine data. A detailed description will be then made on a so called "Active Charge Air Duct" using a "Rotary Shifter" to commute between a torque mode and a power mode.

This device can also be used with a special control strategy at part load to optimize BSFC, by reducing pumping losses. The Rotary Shifter can bring also good efficiency to bypass the intercooler and regulate the inlet temperature, and decrease risk of low pressure EGR condensation in very cold conditions. Based on experience with Diesel engines, some preliminary work has been also initiated with modern turbo charged gasoline engines to prove benefits of resonance charging. In this case challenges are more to reduce fuel enrichment at full load, and reduce need of scavenging due to new RDE emissions. Again, here, and even more especially for new modern small highly downsized gasoline turbocharged engines, resonance charging can be used to compensate lack of performance.

1 Resonance charging using switchable ducts

1.1 Product description

During the last decades many solutions have been found to enhance the air mass trap in the cylinder obviously one of the most efficient is the turbocharger as it is using part of waste energy from the exhaust to compress the air at the inlet and thus increases the air mass trap in the cylinder [1-2]. The main drawback of such technology is the lack of performance at low engine speed when the exhaust enthalpy is not sufficient. The engineers developed also for naturally aspirated engines some solutions to improve the power by increasing the air mass trap in cylinder by tuning the air intake manifold in order to control the pressure waves, leading in an added air mass in cylinder at the intake valve closing [3-6].

The proposed products developed by MANN+HUMMEL consist in mixing those solutions to get a good performance at low engine speed as well as efficient behavior at higher speed. Therefore, it is possible to tune the intake system for resonance charging not only in the air intake manifold. Because of the long pipe are needed to reach the right tuning at low engine speed the complete air intake has to be considered from the turbo outlet to the intake valve.

In order to be able to well characterize the intake system, it is mounted on a dedicated test bench so called dynamic flow bench. The first measurement allows getting the impedance of the complete system, which will give the resonant frequencies. Then it is possible to tune it by adding duct length or by changing the duct diameter [7].

Therefore, different geometries have been pondered, the simplest one named ACAD (Active Charge Air Duct) [8-9] consists in using two ducts with different length and diameter. The first is a long duct with a small diameter which allows getting resonance around 1500 RPM and thus helps to compensate the lack of enthalpy at low engine speed. At higher engine speed the duct is designed with a larger cross section and a shorter length in order to get both a lower pressure drop and a higher frequency resonance. To switch from one configuration to the other a dedicated product so called "Rotary Shifter" has been designed and will be presented in a later section. Furthermore, in order to get strong control of the position of the reflection in the air intake line and ensure a good flow stream, a volume is placed just downstream the charge air cooler as described on the picture Figure 1.

Figure 1: ACAD (Active Charge Air Duct) concept description

Another solution consists in using additional volumes and ducts. The so called DRS (Double Resonance System) has been tested on 3-cylinder Diesel engines.

In order to understand interaction of resonance charging with complete engine and turbocharger, a primary evaluation of the potential has been handled with GT Power simulation model. Thanks to that study it has been noticed the further benefit of resonance charging on a turbo charged engine compare to naturally aspirated one. Indeed there is a loop effect when the right wave action gives an increase in air mass trap in cylinder the enthalpy at the exhaust raises. This will lead to more available energy at the turbine being converted whether on reduction of the backpressure or more advantageously in an enhancement of the boost pressure giving again an added mass air trap in the cylinder. It's finally a kind of virtuous cycle. With this loop effect we measured that it can be possible to get twice the improvement due to the lonely pressure wave action.

1.2 Engine tests description

The ACAD geometry has first been sized thanks to GT Power simulations, and 2 loops of prototypes have been used. The fine tuning has been carried out with impedance measurements on dynamic flow bench [5]. The engine used for the test is a 1.5 liter 4-cylinder, turbocharged Diesel engine.

The engine has been installed in Centrale Nantes's engine dyno and equipped with different sensors in intake and exhaust lines, including high performance pressure transducers (cylinder, inlet and exhaust manifolds) for high sampling rate data acquisition. For each test, the air-fuel ratio has been kept constant and the efficiency of the system has been evaluated by comparing the output brake torque for three main configurations:

- ACAD system with a long duct (torque mode)
- ACAD system with a short duct (power mode)
- Water Charge Air Cooler (WCAC) directly connected close to the inlet manifold entrance as "a stand-alone WCAC" solution.

Tests have been realized in steady state at full and part load, in order to evaluate the impact of the three different architectures on engine performances.

Then, load transient tests have been performed to evaluate the impact of resonance charging on torque response with step demand.

1.3 Steady state engine tests

As showed in Figure 2, engine tests have been performed in full load configuration. The output torque difference between the long duct and the short duct configuration is higher than 10 percent at low engine speed due to beneficial resonant charging effect. As depicted in Figure 3a, the pressure waves amplitudes are quite different for each configuration and depend on the engine rotational speed. At 1800 RPM, the pressure waves amplitudes are very important for the long duct configuration in comparison to the short duct configuration. For this reason, it is possible to improve the air mass flow. Furthermore, the tests were realized with the same air-fuel ratio. As a consequence, it is possible to increase the engine torque. It can be noticed that the pressure waves amplitude are smaller for the stand alone WCAC close to the air intake manifold. This is why the engine output torque is also smaller than the two other configurations.

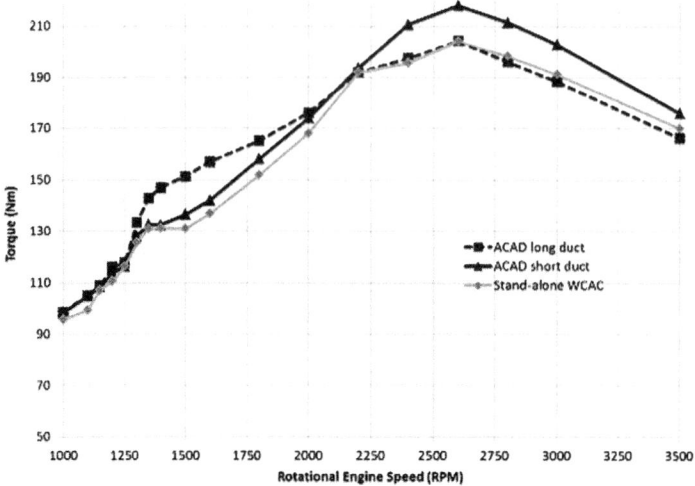

Figure 2: Engine torque evolution

As depicted in Figure 3b, a wave supercharging effect can be obtained with the short duct as well for a higher engine speed for instance at 2600 RPM. In combination with a reduction of pressure drop due to a larger cross section, the engine torque raises around 10 percent. The standalone WCAC with a short travel to the air intake manifold configuration provides lower pressure but the benefit compare to the long duct is not that much compared to the opportunity of tuning provided by the short duct.

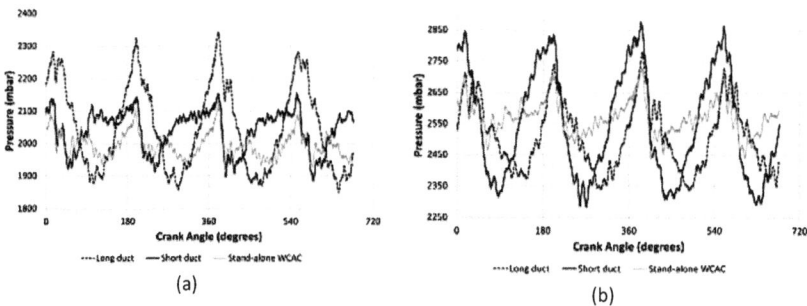

Figure 3: Instantaneous pressure signals at 1800 RPM (a) and 2600 RPM (b)

1.4 Non-steady state engine tests

The ACAD gives the possibility to increase the engine volumetric efficiency thanks to the resonant charging effect. It is quite different from transient case using only a turbocharger, with the well-known lag effect.

Transient tests were realized at constant rotational engine speed. The initial torque is chosen in order to obtain a Brake Mean Effective Pressure (BMEP) close to idle conditions. An acceleration step demand is applied. It is then possible to measure the torque evolution for each configuration (long duct, short duct and stand-alone WCAC close to the inlet manifold). Two experimental results are presented in this paper (as depicted in Figure 4): 1750 RPM and 2500 RPM.

For this kind of experiment, it appears that the long duct configuration gives the possibility to reduce the time necessary to increase the torque for low engine rotational speed. The same kind of conclusion is observed for the short duct for higher rotational engine speed. These results are in agreement with the steady state results.

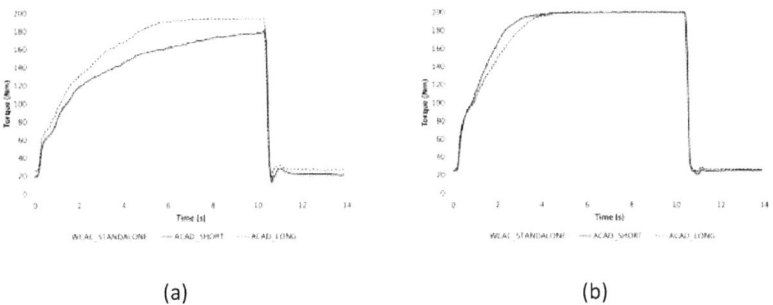

(a)   (b)

Figure 4: Torque evolution during load transient tests at 1750 RPM (a) and 2500 RPM (b)

2 Resonance charging effect using switchable volumes

2.1 Simulation

The resonance charging effect can be obtained by using different volumes placed in the air intake. One of this can be advantageously the charge air cooler and then different valve can connect or disconnect another volume to the main pipe. The following figure shows the different steps that have been considered to optimize the volumetric efficiency.

Definition of System Test Stages

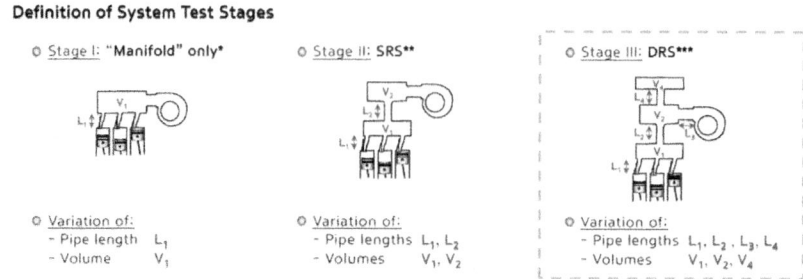

Figure 5: Double resonance system set up

Actually there are two main architectures for the current air intake: either the charge air cooler is integrated in the air intake manifold (Stage 1) or the cooler is a stand alone with direct or indirect cooling system (Stage 2). For the first version as described in Figure **5** there is few opportunities to tune the system for resonance charging as there is only a short duct between the valve and the volume of the charge air cooler. The second version provides further opportunities to play with geometry as already explained with the ACAD. The version 3 is a step ahead as there is an added volume especially to get a new resonance which will offer new engine speeds where the system can be tuned for.

Indeed with both volume and duct length there is plenty of opportunities to optimize the volumetric efficiency. Therefore, there are four lengths and four volumes we can play with in order to achieve the expected performances in a given space available around an engine.

In order to evaluate the impact of each configuration, a GT Power has been set up and all parameters have been evaluated. For instance at 2000 RPM the following simulation results (Figure 6) showed a kind of optimum when the length between the volume V2 and V4 is about 750mm and at the same the length between the volume V2 and the air intake manifold plenum is also 750 mm while keeping all other parameters constant.

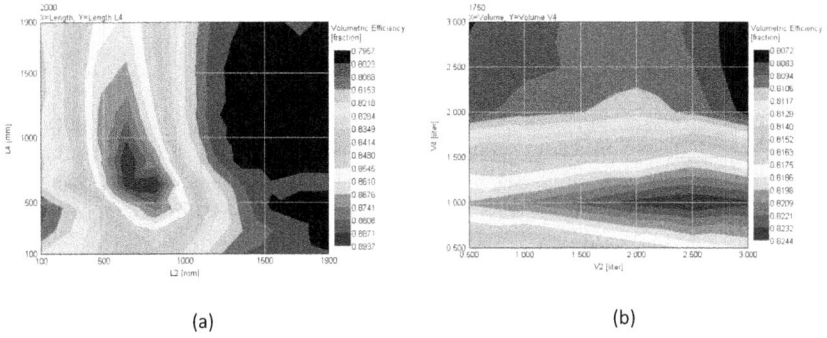

(a)   (b)

Figure 6: Volumetric efficiency variation with ducts length (a) and with volume (b)

In the same way it is possible to change the volume V2 and V4 to evaluate the conse-quence on the volumetric efficiency. For instance Figure 6 shows the impact at 1750 RPM.

After having defined the right layout for all volumes and the different duct lengths, some engine tests have been carried out with prototypes. The considered engine is a 3-cylinder Diesel engine. Mainly the transient behavior has been evaluated starting from low load and with an abrupt request in torque until full load in 0 s. Test results for 1500 RPM are showing in the following Figure 7.

Volume V2 can be further reduced to improve packaging volume situation without losing too much performance.

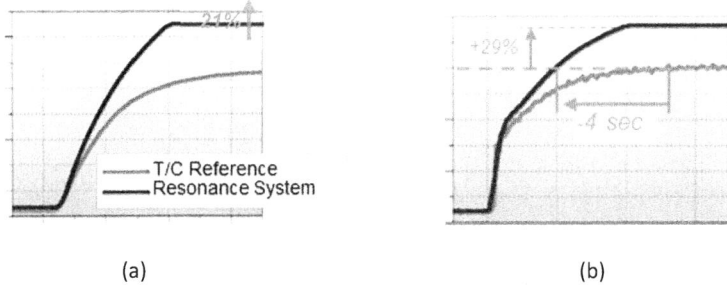

(a)   (b)

Figure 7: Effect of DRS on transient behavior (a) boost pressure (b) output torque

Improvement is pretty impressive as it is possible to reach a defined high torque really faster and in the same time to increase the maximum torque by 30 percent.

3 Presentation of new switchable intake system

3.1 Product description

In order to be able to shift from one configuration to the other a dedicated product has been developed. It has to be resistant against high pressure pulsation level, to be tight, and offer a good permeability. The air path section is kept constant as it is known that expansion volume can affect wave propagation introducing reflection or dampening.

The product developed is composed of two thermoplastic injected shells, respectively a housing and a cover that are assembled by a classical spin welding process (Figure 8). The housing provides two open ports, one for the power mode duct and the other used for the torque mode whereas the outlet is located in the cover. Inside there is a rotary ball with an inner air path to shift from one configuration to the other.

Figure 8: Rotary Shifter design

Compared to the state of the art using flap and shaft crossing the air path, the Rotary Shifter concept has lower pressure loss and is less sensitive to the pressure pulsation. In the case of resonance charging, there is no need to work in intermediate position, and a simple vacuum actuator can be used to drive the rotary ball.

Because it has to be produced in plastic an innovative way to compensate tolerances has been developed and applied making the product able to ensure a high level of air tightness.

3.2 Specifications set up

It has been shown that the BSFC can be improved at part load with the short duct [9]. Indeed the increase of the intake pressure help to reduce the pumping losses as described in Figure 9.

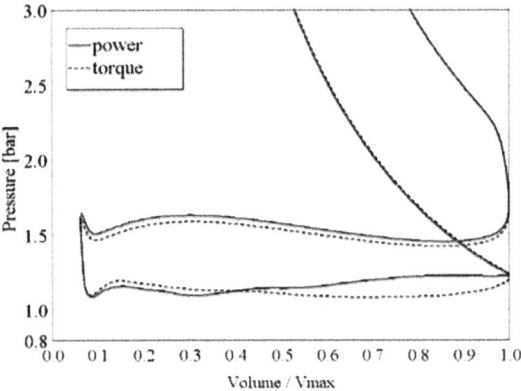

Figure 9: Low pressure PV diagram at 1600RPM and 30 Nm.

This will define the strategy to shift from one duct to the other meaning that the long duct would be used only close to full load condition and for low engine speed. Then, considering the WLTC as a more realistic reference for real driving conditions it is possible to assess the number of activation (Figure 10). Actually on the WLTC, the car runs around 23 km and the number of activation should be around 11. According to this kind of usage the part would be activated approximately every two kilometers.

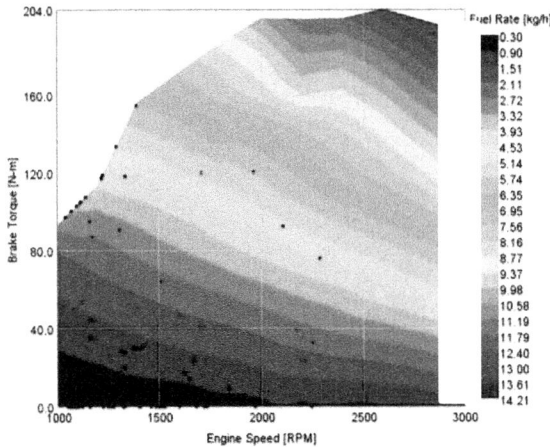

Figure 10: Fuel consumption map with main WLTC points.

4 Fun to drive simulation

The next step in this study is to define another possibility to use the short and long duct system. In this case, a vehicle simulation was made by full load acceleration from 90 km/h to 120 km/h in 5th gear. The aim is to reach the final vehicle speed as fast as possible. The experimental maps for the BFSC and FMEP feed a GT Suite model.

GT POWER simulations using DOE have been conducted with an ACAD (it is possible to switch between the long and the short duct) with a variation of the vehicle final drive ratio. The results are compared to the ones obtained with a classical passive intake line, named reference. The results are presented in Figure 11. A modification of this last parameter gives the possibility to change the engine rotational speed and, as a consequence, the engine torque. For example, with a final drive ratio equal to 2.0, the engine torque is more important with the ACAD system (left circle). The vehicle reaches the final speed faster with the long duct. For a final drive ratio equal to 2.4, the average torque is the same with the two configurations and the same time is obtained in order to reach the final vehicle speed (centered circle). Finally, if the final drive ratio is equal to 3.4, it is necessary to select the short duct configuration and the results are better with the ACAD system (right circle).

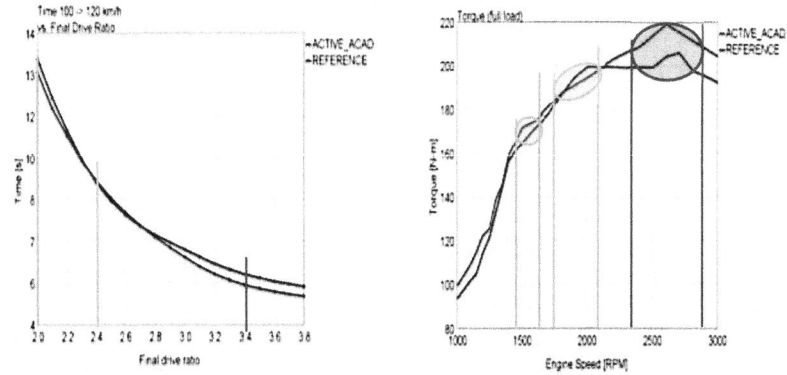

Figure 11: Fun to drive simulation results

In conclusion, it is very important to take into account the complete vehicle definition to get the maximum benefit of the ACAD. It actually gives the opportunity to combine fun to drive with a reduction of the fuel consumption and an improvement of the engine torque.

5 Application to Gasoline engine

For applying such resonance charging effects to modern gasoline engines, it is needed to consider interaction with the complete system thus impact on knocking, exhaust temperature, turbocharger matching and valve control for cylinders scavenging.

First, due to new regulation coming with at least introduction of RDE, emissions control will have to be performed in higher range of engine speed, load and more in transient case. This will limit as a consequence usage of scavenging, and fuel enrichment.

The risk is to see degradation of performance due to impact on compression ratio to stay under the knocking limit, or to see lack of energy in turbine at low engine speed leading to poor boosting effect in transient. The need is even more important for strongly downsized engine such as 3-cylinder engines. System like double resonance system as described previously can be adapted to gasoline engines need. Even if the potential can be limited by knocking, and temperature increase due to heat transfer effect at inlet ports in cylinder head, more enthalpy can be reused at turbine, and it can be a virtuous additional gain for increasing in average the boost pressure out of the only advantage given by resonance charging alone.

Conclusion

In the content of this article, it has been demonstrated some possible improvement in terms of performances or fuel economy for Diesel engines. Basic phenomena to be used is not more to get energy from the pressure waves due to opening and closing of inlet valves, propagating and then amplifying pressure and density inside engine cylinders. The benefit seen on dyno with a small 3-cylinder Diesel engine has even reach high amount of torque improvement close up to 30%. To really get benefit on the whole engine map, it is needed to activate either volumes, either duct lengths for charge air ducts ; this is now possible with good performance (leakage and pressure loss) thanks to a new concept so called "Rotary Shifter". Made of standard plastic raw material, compact, it can be easily integrated with parts around, as intake manifolds, or charge air ducts.

Added value for end customer can be basically either performance improvement as time to torque reduction for improving the fun to drive, torque and power improvement, either fuel consumption reduction in the range of a few percent, thanks to downspeeding and activation of active air intake system also at part load.

Out of Diesel application, it is now engaged in new activities dealing with resonance charging to be used for modern gasoline engine using turbochargers and direct injection. Due to new regulations (RDE and Euro 6.d), resonance charging combined with an optimal turbocharger matching can be a low cost solution to compensate lack of performance due to new strategies needed for emissions and knocking control.

Acknowledgements

The work in this article is done in a joined International Teaching and Research Chair entitled "Innovative Intake and Thermo-management Systems" between MANN+HUMMEL and Ecole Centrale de Nantes.

Bibliography

1. WINTERBONE D.E., PEARSON R.J. – Theory of Engine Manifold Design: Wave Action Methods for IC Engines – Professional Engineering Publishing, 2000.

2. BOREL M., – Les phénomènes d'ondes dans les moteurs (Wave phenomena in engines). Paris: Publications de l'IFP, Editions TECHNIP, 2000.

3. OHATA A. and ISHIDA Y. – Dynamic inlet pressure and volumetric efficiency of four cycle four cylinder engine, SAE paper 820407, 1982.

4. BROOME, D. – Induction ram – Part 3: wave phenomena and the design of ram intake systems, Automobile Engineer, pp. 262-267, 1969.

5. CHALET D., MAHE A., MIGAUD J., HETET J.-F. – A frequency modelling of the pressure waves in the inlet manifold of internal combustion – Applied Energy, Volume 88, n°9, pp. 2988-299, 2011.

6. KROMER G., POLTZ H.W., THUDE M, LEITNER P. The new Audi V6 engine. SAE paper 910678, 1991

7. MIGAUD J., RAIMBAULT V., MEZHER H., CHALET D., GRANDIN T. – A New Fast Method Combining Testing and Gas Exchange Simulations Applied to an Innovative Product Aimed to Increase Low-End Torque on Highly Downsized Engines – ATZ –VPC.plus Simulation un Test für die Antriebsentwicklung, 16 MTZ-Fachtagung, Hanau, Germany, September 30 – October 1, 2014

8. MIGAUD J., RAIMBAULT V, CHALET D., MEZHER H. – Variable Charge Air Duct for Low-End Torque Enhancement and high Speed Performance _ MTZ Worldwide, Volume 76, n°1, pp. 20-25, DOI 10.1007/s38313-014-1006-y, 2015

9. MIGAUD J., BÜHL H, RAIMBAULT V., KORN A, CHALET D, MEZHER H, MONTAIGNE Q, PRÉTOT P-E – A model based system approach to innovative smart intake products: CO_2 savings and specific performance – 15th Stuttgart International Symposium, Automotive and Engine Technology, Forschungsinstitut für

Kraftfahrwesen und Fahrzeugmotoren Stuttgart – FKFS, Volume 2, pp. 403-420, Stuttgart, Germany, March 17-18, 2015

Real-time simulation of the effects of catalyst on automotive engines performance

Agostino Gambarotta; Marco Crialesi Esposito Industrial
Engineering Department – University of Parma PARMA,
Viale delle Scienze 181/A – I-43121
Tel.+39 0521 90-5064/5172 – Fax.+39 0521 905705
Email: agostino.gambarotta@unipr.it

Panayotis Dimopoulos Eggenschwiler, Francesco Lucci
EMPA-Swiss Federal Laboratories for Materials Science and Technology
Überlandstrasse 129, 8600 Dübendorf, Switzerland

Abstract

Today restrictions on pollutant emissions are forcing more and more the use of catalyst-based after-treatment systems both in SI and in Diesel engines. The application of monolith cores with a honeycomb structure is an established practice: however, to overcome drawbacks such as poor flow homogenization, the use of ceramic foams has been recently investigated [1,2,3] as an alternative showing better conversion efficiencies (even if with higher pressure losses).

The scope of this paper is to analyse the effects of foam substrates on engine performance. For this purpose a 0D "crank-angle" real-time mathematical model developed by the Authors [4,5] has been enhanced improving the heat exchange model of the exhaust manifold to take account of thermal transients and adding an original 0D model of the catalytic converter to describe mass flows and thermal processes.

The model has been used to simulate a 1.6l turbocharged Diesel engine during a driving cycle (EUDC). Effects of honeycomb and foam substrates on fuel consumption and on variations of catalyst temperatures and pressures are compared in the paper.

1 Introduction

1.1 The use of fast-running mathematical models in the design and management of Engine Systems

In the last decades the constant need to reduce pollutant emissions from automotive Internal Combustion Engines (ICEs) led OEMs both to enhance existing subsystems (e.g., fuel injection, valve actuation systems, etc.) and to introduce innovative solutions (with particular reference to after-treatment devices). As a matter of fact in order to allow these technologies to be really effective a proper and concurrent design of plant layout, control systems and management strategies is needed.

The complexity of systems and the large number of control variables require a deep understanding of processes that determine the behaviour of the controlled powertrain as a system on the whole. The design of system architecture and of its control devices definitely need a solid theoretical support from physical models to outline system overall behaviour, which especially in automotive applications is mostly non-linear and therefore difficult to predict. Mathematical models are powerful tools to estimate the influence of system layout and control strategies on the final result thus shortening the way from design specifications to on-road testing [6].

The application of fast mathematical models in the design of powertrains and related management systems is well known from more than a decade and several examples can be found in the literature [5]. A comprehensive scenario is outlined in [6]. Typically Filling-and-Emptying (F&E) and Quasi-Steady Flow (QSF) approaches are used to build up 0D, lumped parameter models that are used both for the intake and exhaust systems and for in-cylinder processes, which still allow for "real-time" simulations [4,7]. Even if chemical and physical processes which take place in the cylinder are very complex, "fast" models require simplified 0D single-zone approaches where combustion is considered through the definition of a proper fuel burning function [8] and pollutant formation reactions through very simplified mechanisms or –more often- with black-box models [6]. Most commercial tools are based on these methodologies (as reviewed in [5,7]).

This scenario highlights the significant role of fast mathematical models in the simulation of complex systems, whose overall behaviour arises from the interactions of different components and processes in a complex and unpredictable way. Following this consideration, and in order to investigate the effects of different catalyst substrates on powertrain performance, a model of the after-treatment system has been developed and coupled with a "crank-angle" engine model [7]. Interesting results were obtained, as shown in the paper.

1.2 Open cell foams as catalytic reactors

Open cell foams are attractive materials characterized by high porosity, low density, high mechanical strength and large surface area. In recent years they have been considered for various industrial applications like filters, heat exchangers and catalytic reactors. As catalyst substrates they present several advantages over honeycomb monoliths and packed beds. The open cell structure allows higher flow uniformity which is a critical factor for the pollutant conversion efficiency and for the catalyst durability [9,10,11].The tortuous flow path is expected to enhance the mixing and the heat/mass transfer, and the high specific surface will yield more compact catalysts [12]. In automotive applications a critical parameter is the pressure drop, which affects engine efficiency. Foams have higher pressure drop compared to a monolith with the same dimensions [2,13]. This can be compensated by an increased mass transfer that allows to downsize the catalyst [14] or by different geometrical reactor configurations [15].

It is not straightforward to quantify the influence of catalyst substrate structure on engine performance due to the different dynamic behaviour of honeycombs and foams during transients and to the high non linearity of the overall engine system. To compare the influence of honeycomb and foam substrates an original 0D mathematical tool has been used to model an actual 1.6l turbocharged Diesel. Simulation results during an EUDC driving cycle are reported in the paper showing the effects of these supports on catalyst thermal transients and on fuel consumption.

2 Real-time modelling of the Engine and exhaust Aftertreatment system

2.1 "Crank-angle" engine model

For the purpose of this work the model described in [4,7] has been used. Considered engine layout is sketched in fig.1. In-cylinder and gas exchange processes were described using a QSF approach for intake and exhaust valves and a F&E method for manifolds and cylinders. Combustion is considered defining a proper Heat Release Rate (HRR) and pollutant formation is estimated through black-box sub-models. An original algorithm has been developed for the integration of conservation equations in the cylinder with a suitable time step (tuned to keep an angular step of approx. 1deg CA for any engine speed n), while keeping a larger overall time step for intake and exhaust systems. Fuel system model takes account of the fuel rail dynamics (through its bulk modulus), of injectors flow characteristics and of leakages and allows to calculate injected fuel flow rate from rail pressure p_{rail} and Energizing Time ET. Black-box map-based models have been used for compressor and turbine.

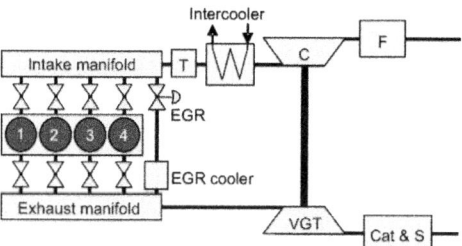

Fig.1 – Layout of the considered engine.

The model and its causality scheme are described in [4,7]. After calibration and validation, the proposed model has been used in an original PC-based Hardware-in-the-Loop (HiL) system developed by the authors [16] showing good capabilities.

2.2 Exhaust system and Catalyst models

Heat transfer models in the exhaust system are an important aspect in the simulation of ICEs performance due to the significant influence of exhaust gas temperature on aftertreatment systems efficiency. Therefore a careful description of heat exchange processes is of fundamental importance to cope with critical transients (e.g., catalyst "light-off", particulate trap regeneration). For this reason, albeit with the limitations imposed by a 0D approach, the paper is mainly concerned with the simulation of thermal behaviour of the exhaust system.

Model of the exhaust manifold

The mathematical model of the exhaust manifold has been developed following a F&E approach: temperature and pressure are obtained from the equations of conservation of mass and energy applied to the manifold considered as a 0D volume. Working fluid is considered as a mixture of perfect gases (defined through a vector of mass concentrations X_{mi} referred to 7 chemical species, i.e., N_2, O_2, CO_2, H_2O, CO, H_2 and NO). Estimating heat flow through manifold walls as suggested in [6], energy conservation equation for exhaust gases inside the manifold can be written as follows:

$$\frac{dU}{dt} = \dot{m}_{exh}h_{exh} - \dot{m}_{tur}h_{tur} - \dot{m}_{EGR}h_{EGR} - \dot{Q}_{in} \tag{1}$$

where $\dot{Q}_{in}$ is the heat flux from the gas mixture to the manifold.

Changes in manifold walls temperature can be estimated from the following differential equation:

$$\frac{dT_w}{dt} = \frac{1}{m_w \cdot c_w}\left(\dot{Q}_{in} - \dot{Q}_{out}\right) \tag{2}$$

knowing manifold thermal inertia $m_w \cdot c_w$ and convection heat flux between gas stream and walls and convection and radiation heat flux between walls and ambient air respectively. Radiation inside the manifold and conduction through the walls are neglected since corresponding thermal resistance is significantly low.

A specific correlation suggested in the literature for intake and exhaust systems of ICEs has been used in the following form [18]:

$$Nu = a \cdot Re^b \cdot Pr^c \tag{3}$$

The term Pr^c often assumes a value close to 1 and values of a and b are defined from measurements. The value of Nu was estimated from the Gnielinski correlation reported in [19,20] introducing as suggested a suitable Convective Augmentation Factor for the internal Nu to take account of unsteadyness and turbulence. From the value of Nu, convection coefficients and heat flow rates can be calculated through well-known correlations where thermodynamic properties are estimated with reference to the exhaust gas temperature in the manifold (assumed to be uniform).

The estimation of convective heat flux from manifold walls to ambient air is more difficult due to the component geometry and of external flow pattern. For the sake of simplicity, manifold geometry has been assumed as cylindrical and external flow field uniform and related to the vehicle speed. The model is based on the correlation proposed in [19,20], thus estimating Nu as follows:

$$Nu_{out} = 0.3 + \sqrt{Nu^2_{out\_lam} + Nu^2_{out\_tur}}, \quad 10 < Re < 10^7 \qquad (4)$$

where Nu_{out_lam} and Nu_{out_tur} are functions of Re and Pr numbers. Convection heat flux can be then calculated from h_{out} and external manifold area A_{out}. Thermodynamic properties are estimated with reference to the film temperature (i.e., at the average temperature between manifold walls and surrounding external air).

Radiation heat flux outside the manifold is estimated considering the external surface as a grey surface in a cavity of infinite extension. Heat flux can then be calculated through well-known relationships [17] based on the Stefan-Boltzmann constant and on the total hemispherical emissivity of the grey body. The total heat flux from the collector was then calculated from the convection and radiation values.

Catalyst Model

A catalytic converter is a complex component from the point of view of both gas flow pattern and of chemical reactions. Fluid dynamics, heat and mass transfer processes have a significant role in its behaviour and should be carefully considered. Due to the scope of the presented work, neither a 3D (eg., [2,21,22]) nor a 1D modelling technique (e.g., [23,24]) could be used. A 0D approach has been followed assuming a uniform spatial distribution of thermodynamic parameters and applying conservation equations with empirical correlations where required. The developed model proved to be able to simulate catalyst behaviour and its influence on powertrain performance over significant transients (e.g., driving cycles) with very short calculation time and taking account of layouts, sizes and control strategies during transients.

The model has been developed according to the schematic reported in fig.2. Two volumes were considered (in green upstream and downstream of the catalytic core) following a F&E approach. The model of the core (in yellow) was based on a QSF procedure (i.e., assuming no accumulations of mass and energy). Proper correlations were used to model honeycomb or foam substrates taking account to their flow characteristics. At each time step mass flow and temperature changes through the core were estimated with two set of algebraic equations, describing gas flow and thermal dynamics of the substrate respectively. These two processes are linked through heat exchanges between exhaust gas and substrate walls according to the diagram in fig.3.

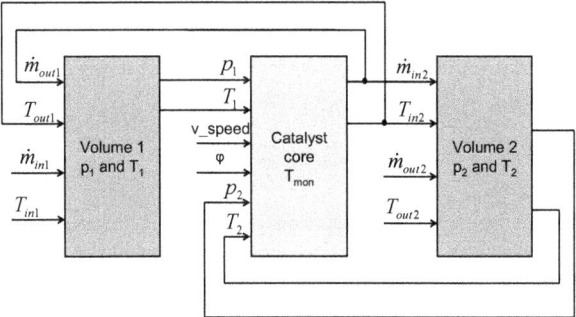

Fig.2 – Schematic and causality of the catalyst model.

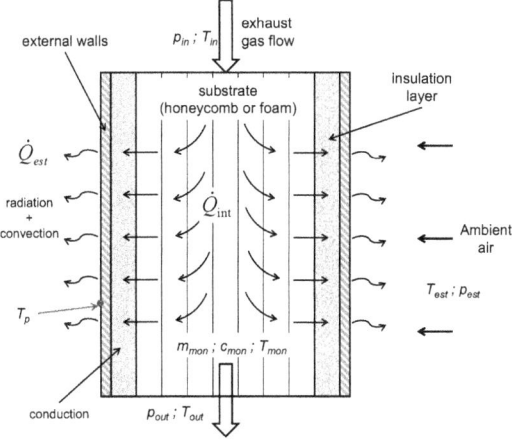

Fig.3 – Schematic of flow and heat exchange processes in the catalyst core.

Gas flow through the catalyst core (in yellow, fig.2) is described by an empirical algebraic correlation to take account of flow resistance in the following form:

$$\dot{m} = f(\Delta p, \rho, \mu, geometry) \tag{5}$$

giving mass flow rate from p and T in the two adjacent volumes (in green). In eq.(5) fluid properties are calculated from the averaged values of p and T taking into account the composition of the flowing gas. Catalyst geometry includes dimensions and morphological characteristics (i.e, honeycomb/foam, porosity, etc.) of the substrate.

Gas temperature at the core exit is estimated integrating the energy conservation equation in 1D in the following form:

$$\dot{m} \cdot c_p \cdot \frac{dT}{dx} = h \cdot P \cdot (T_{MON} - T) + q_{gen} \tag{6}$$

where heat flux in gases in axial direction [24] and kinetic energy are neglected. Convective heat transfer between gas and substrate is considered as usual through a convection coefficient h obtained from Nu estimated from Re and Pr using empirical correlations [19,20]. To limit calculation time wall temperature T_{MON} of the monolith is assumed constant with length.

Since molecular diffusion of different species in the gas mixture and in the washcoat and chemical reactions have not been considered, their overall effects have been quantified in terms of heat from oxidation processes of unburnt species (CO and HC) estimated from pollutants mass flow rates through the following expression:

$$q_{gen} = f(x, L, N, \dot{Q}_{gen}), \text{ in [W/m]} \tag{7}$$

where

$$\dot{Q}_{gen} = \sum_{i=1}^{n} \dot{m} \cdot X_{mi} \cdot LHV_i \cdot \eta_i \tag{8}$$

Mass concentration X_{mi} of each component in the exhaust gas mixture is determined from experimental look-up tables with reference to equivalence ratio φ and rotational speed n [25]. Conversion efficiencies η_i are estimated from experimental look-up tables with reference to substrate temperature T_{MON} and flow velocity [25].

The correlation for gas temperature changes in the axial direction obtained integrating eq.(6) can be used in the model to calculate gas temperature at the catalyst outlet T_{out} which allows to estimate heat flux between exhaust gas and monolith in the time step. Since properties of gas mixture are determined with reference to the average value of temperature within the substrate, a do-while procedure is used with a 0.1K threshold.

Changes in mean temperature of the monolithic substrate T_{MON} can be evaluated from energy conservation knowing substrate thermal inertia $m_s \cdot c_s$, heat flux exchanged by convection with gases and heat flux to external ambient air. The latter can be estimated taking account of conductive heat transfer through insulating layer and external metallic walls of the catalyst and of convective (natural or forced depending on vehicle speed) and radiation heat transfer to ambient air (fig.3). Evaluation of the overall thermal resistance $R_{t\ tot}$ through related heat transfer mechanisms [17] allows to calculate the heat flux to external ambient air. Coefficients for forced and natural convection can be obtained by considering fluid properties at the average temperature of the film. Also in this case a do-while loop on the wall temperature T_p is required with a specific threshold (0.1 K). The described procedure has been used for the simulation of different catalyst substrates (honeycomb or foam) by using suitable correlations to link mass flow rate and pressure change (eq.(5)) and to determine Nu.

A proper correlation should also be defined to describe catalytic oxidation of unburnt pollutants, i.e., a distribution of heat from oxidation processes in the substrate as $q_{gen}(x)$ [W/m] in eq.(7) to calculate exit temperature. Reaction rate in the catalyst is influenced by many processes: assuming temperatures high enough, chemical species can be supposed to react instantly as soon as they reach substrate walls. Starting from the analogy with heat exchange processes, changes in the concentration of chemical species can be assumed to decrease exponentially along the catalyst. This leads to an exponential distribution of heat generated from unburnt compounds, i.e., $q_{gen} = a \cdot e^{bx}$. Coefficients a and b can be determined imposing that the integral of q_{gen} along the substrate length is equal to the heat generated in the whole catalyst and that the ratio of heat generation between inlet ($q_{gen}(0)$) and exit ($q_{gen}(L)$) is equal to 100. Therefore the following expressions can be obtained:

$$a = \frac{N \cdot \ln(N) \cdot \dot{Q}_{gen}}{L \cdot (N-1)} \quad \text{and} \quad b = -\frac{\ln(N)}{L} \tag{9}$$

From eq.(9), eq.(6) can be written as follows:

$$\dot{m} \cdot c_p \cdot \frac{d\Delta T}{dx} + h \cdot P \cdot \Delta T = -q_{gen}(x), \text{ where}$$

$$\Delta T = T_{MON} - T \tag{10}$$

which can be solved giving gas temperature at substrate exit. Since thermodynamic properties of exhaust gas mixture are calculated with reference to averaged values of p and T inside the substrate, an iterative procedure is used.

745

2.3 Physical Characterization of the Catalyst Model

The presented model of the after-treatment system has been then calibrated with reference to a defined device. Flow resistance (eq.(5)) and heat transfer processes (eq.(6)) has been identified from correlations available in literature [12,22] and standard physical and geometrical properties have been used. The total volume of the catalytic reactor is assumed to be 1.5l with a reactor length of 15cm. A standard honeycomb structure, identified in the following as "h_Giani", is used as reference case and is characterized by a porosity of ε=63%, a characteristic channel diameter of D_c=1mm and a specific surface area of SSA=2700 m^2/m^3. The honeycomb structure is compared to two open cell foam like structures, one real foam [12] identified as "f_Giani" and one synthetic Kelvin cell structure [22] identified as "f_Lucci". Both cell structures have porosity of ε=73%, higher than the honeycomb, a lower surface area of SSA=1000 m^2/m^3 and characteristic pore dimension of d_h=2mm.

In Table 1 the different correlations for the flow resistance and the transfer properties for all the structures are reported. Further details on them can be found in the referenced literature [12,22].

Table 1 - Flow resistance and transport correlations used.

Structure	Pressure drop ($\Delta p / L$)	Transfer coefficient (Nu)
Honeycombs (h_Giani) [12]	$\dfrac{28.46}{Re} \rho u^2 \dfrac{1}{D_c}$	$2.977\left(1 + 0.095 Re \cdot Pr \dfrac{D_c}{L}\right)$
Foam (f_Giani) [12]	$\dfrac{2}{d_s}\left(0.87 + \dfrac{13.56}{Re}\right)\left(\dfrac{1}{1-G(\varepsilon)}\right)^4 \dfrac{G(\varepsilon)}{4} \rho u^2$	$1.2 \cdot Re^{0.43} Pr^{\frac{1}{3}}$
Kelvin cell (f_Lucci) [22]	$SSA \cdot \dfrac{\rho u^2}{2} \dfrac{\chi^2}{\varepsilon^3}\left(0.4 + \dfrac{30}{Re^{0.8}}\right)$	$1.28 \cdot Hg^{0.32} Pr^{\frac{1}{3}} \varepsilon^{2.34}$

2.4 Development and validation of the engine model

The exhaust system and the catalyst models have been coupled with a 0D "crank-angle" model of a turbocharged Diesel engine (layout is sketched in fig.1 and main technical data are reported in tab.2). The structure of the model (alternating volume and non-volume blocks) avoided numerical problems and algebraic loops [7].

Table 2 – Technical specifications of the considered engine.

Displaced volume	1598.4 cm^3
Stroke	80.5 mm
Bore	79.5 mm
Cylinders	4
Compression ratio	16.5

The model has been identified on the basis of steady-state experimental data from the OEM, which were used to define look-up tables and coefficients of interpolating functions through a least-square method (i.e., flow coefficients of intake/exhaust valves, pressure loss coefficients of air filter and exhaust system, etc.). Compressor and turbine models were identified on the basis of their characteristics from the Manufacturer [7]. The algorithm developed for the integration of model equations uses a constant principal time step of 2ms and a variable time step for in-cylinder processes to keep an angular step of approx. 1deg CA independently of engine speed n. In this application, on a 2GHz PC with 2Gb RAM, the ratio of simulation time to physical time was always remarkably lower than 0,65.

Input parameters are engine rotational speed, fuel mass flow rate, driving signals for VGT and EGR, ambient temperature and pressure. Outputs can be every single one of parameters estimated by the engine model, e.g., torque, *bmep*, effective power output, state parameters in the intake and exhaust manifolds (i.e., p, T, X_{mi}), etc. Once identified, the engine model has been tested comparing calculated results with experimental data measured on a test bench in steady-state operating conditions by the OEM (other than those used for the identification), giving a good agreement [7].

2.5   Operating conditions from the driving cycle

In order to highlight the influence of substrate characteristics on engine behaviour the Extra Urban Driving Cycle (EUDC) section of the New European Driving Cycle (NEDC) was chosen. To this extent input parameters (rotational speed, fuel mass flow rate, VGT and EGR driving signals) were defined through an inverse model of the vehicle (developed in [26]). Vehicle data were identified with reference to the Alfa Romeo Giulietta 1.6JTD. From time histories of speed and gear prescribed for the 400s EUDC, instantaneous requested values for rotational speed and torque were calculated and used as model input. Difference between target and actual engine torque was used to estimate through a closed-loop PID control algorithm the injected fuel mass flow rate.

3 Simulation results for the EUDC

3.1 Intake and exhaust systems behaviour

Thermodynamic parameters in the intake and exhaust systems obtained with different substrates were compared. As an example, in the following several results are plotted with reference to the EUDC assuming the honeycomb substrate as a baseline ("h_Giani", in solid red) and calculating differences between the two open cell foam like structures (the real foam "f_Giani", in dashed green, and the Kelvin cell structure "f_Lucci", in dash-dot blue).

As expected, foam substrates lead to higher pressure losses. In fig.4 static pressure difference through the catalyst Δp_{DOC} is reported, showing a max increase of about 10kPa for both considered foams. However, significant non-linearities due to typical processes in the intake and exhaust system give rise to an overall non-trivial behaviour. As a matter of fact, pressure drop through the turbine Δp_{tur} is slightly lower (fig.5) and therefore changes in the exhaust manifold pressure p_{exh_man} (fig.6) are lower than expected (i.e., lower than the increase in pressure drop Δp_{DOC}, fig.4). This lead to the conclusion that higher pressure losses induced by foams may be partly counterbalanced by the turbocharger effects, at least at high engine loads.

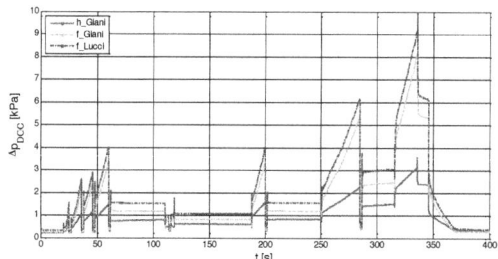

Fig.4 – Calculated pressure losses through different catalysts substrates.

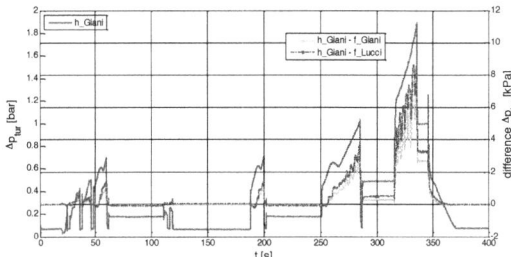

Fig.5 – Calculated pressure changes through turbine.

Temperature profiles inside the catalytic reactor block are presented in fig.7. Due to the higher porosity, open cell structures have a lower thermal inertia and present shorter thermal transients. Fig.7 shows that both foams and Kelvin Cell structures are able to reach a light off temperature of 550K approximately in half the time of the honeycombs. However, for the same reasons they are characterized by a faster cool down phase.

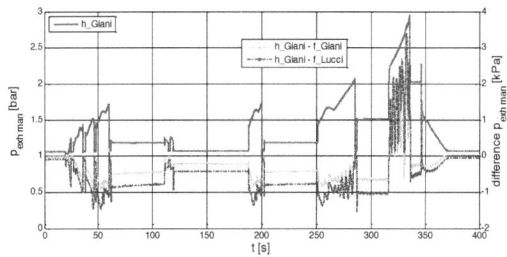

Fig.6 – Calculated pressure in the exhaust manifold.

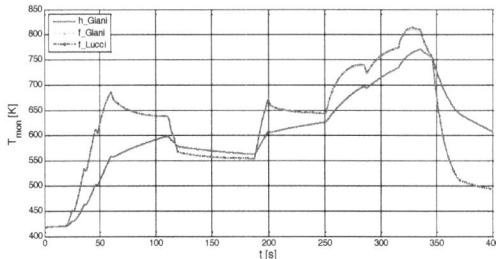

Fig.7 – Calculated temperature of the substrates.

3.2 Prediction of Fuel Economy

The model allowed to estimate instantaneous and cumulative fuel consumption on the considered EUDC: results are plotted in fig.8. The red solid line represents the cumulative fuel consumption for the engine with the honeycomb substrate ("h_Giani"), which is assumed as reference to highlight the effects of open cell substrates. Therefore in fig.8 blue and green lines show the percentage deviation when the foam (green dashed line, "f_Giani") and Kelvin Cells substrates (blue dash-dot line, "f_Lucci") are used with reference to the honeycomb one ("h_Giani").

The analysis of the instantaneous fuel consumption $\dot{m}_f$ shows that, within the assumed conditions, it is lower for the honeycomb than for both open cell structures. However, differences in cumulative fuel consumption between the cases is lower than 0.20%. Furthermore, among the open cell substrates, fuel consumption with real foams ("f_Giani") is slightly lower than that with Kelvin Cells structures ("f_Lucci").

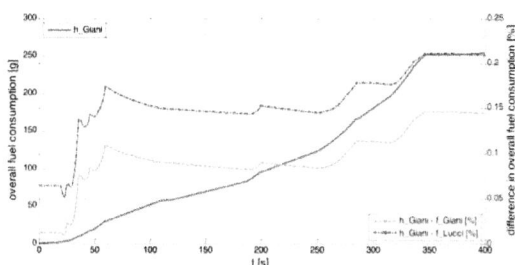

Fig. 8 – Calculated cumulative fuel consumption during the EUDC.

As previously shown, pressure drop through the catalytic converter is higher for open cell structures (fig.4) for all the analysed cases, confirming that open cell structures are characterized by higher flow resistance. This is the main reason of higher fuel consumption for the considered open cell structures especially during accelerations, when higher torque is requested and pressure drop in the exhaust manifold increases. It should be noted, however, that in the present study increase in fuel consumption is caused by the replacement of a honeycomb substrate with open cell foams assuming equal shape and volume. But higher mass transfer properties of open cell structures allow for more compact reactors compared to honeycomb ones, and this lead to a decrease in their flow resistance thus rebalancing the drawback in fuel consumption.

The maximum deviation in instantaneous fuel consumption between all the cases is 0.35%, appearing only during accelerations when higher torque is requested. During steady driving condition at constant velocity the increased instantaneous fuel consumption due to the open cell structure substrate is lower (0.10% approximately). These variations result in an increase of only 0.20% in total injected fuel over the overall 400 s of the cycle.

4 Conclusions

Mathematical models represent an interesting (and often unavoidable) way to get a proper understanding of the behaviour of complex systems. As a matter of fact, development of theoretical tools require a good compromise between physical and empirical approaches to limit the CPU time.

In the paper a fast model of a catalytic converter for automotive application has been built up and integrated in a 0D "crank-angle" model of a turbocharged Diesel engine. After improving the heat exchange model for the exhaust manifold (to take account of thermal dynamics during transients), a 0D model of the catalyst has been developed to simulate related mass flows and thermal processes. Then the catalyst model has been linked to the engine model to investigate the behaviour of the overall system and the effects of catalyst substrate characteristics.

To this extent an actual 1.6l turbocharged Diesel with EGR has been simulated within an EUDC driving cycle comparing engine performance with different catalyst substrates. The behaviour of three different catalytic structures was analysed: honeycomb, open cell foams and open Kelvin cell structures. It has been show that, using reactors of the same volume, the increased pressure drop caused by open cell structures results in a total fuel consumption increase of only 0.20%. On the other side open cell structures show faster thermal transients due to their lower thermal inertial and thus are able to reach quickly light-off temperatures.

Finally it should be noted that higher mass transfer properties of open cell structures may allow for more compact reactors compared to honeycombs. This may help to reduce the overall flow resistance of foams opening new possibilities to improve the efficiency of the after-treatment system lowering at the same time specific fuel consumption. The presented mathematical tool proved to be very effective to simulate the behaviour of the comprehensive system (engine+after-treatment system) and will be used in the next future to explore exhaustively these topics.

References

1. C.Bach, P.Dimopoulos Eggenschwiler, "Ceramic Foam Catalyst Substrates for Diesel Oxidation Catalysts: Pollutant conversion and operational Issues", SAE paper no.2011-24-079, 2011.

2. F.Lucci, A.Della Torre, G.Montenegro, P.Dimopoulos Eggenschwiler, "On the catalytic performance of open cell structures versus honeycombs", Chemical Engineering Journal, vol.264, pp.514–521, 2015.

3. J.von Rickenbach, F.Lucci, C.Narayanan, P.Dimopoulos Eggenschwiler and D.Poulikakos, "Effect of washcoat diffusion resistance in honeycomb and foam based catalytic reactors", Chemical Engineering Journal, DOI:10.1016/j.cej.2015.03.132.

4. A.Gambarotta, G.Lucchetti, "Control-oriented "crank-angle" based modeling of automotive engines", 10th International Conference on Engines for Automobiles, SAE paper no.ICE2011-24-0144, Capri, 9/2011.

5. A.Gambarotta, G.Lucchetti, I.Vaja, "Real-time Modelling of Transient Operation of Turbocharged Diesel Engines", Proc. I.Mech.E., Part D: J.Automobile Engineering, Vol. 225, 2011, ISSN 0954-4070, DOI:10.1177/0954407011408943.

6. L.Guzzella, C.H.Onder, "Introduction to Modeling and Control of Internal Combustion Engine Systems.", Springer-Verlag, Berlin, 2010.

7. A.Gambarotta, G.Lucchetti, "A Crank-angle model for the "real-time" simulation of Diesel engines in HiL/SiL applications.", 13th Stuttgart International Symposium on Automotive and Engine Technologies, Stuttgart, 2013.

8. J.B.Heywood, "Internal Combustion Engines Fundamentals", McGraw-Hill, New York, 1988.

9. G.Gaiser, J.Oesterle, J.Braun, P.Zacke, "The progressive spininlet – Homogeneous flow distributions under stringent conditions", SAE Technical Paper no.2003-01-0840.

10. A.P.Martin, N.S.Will, A.Bordet, P.Cornet, C.Gondoin, X.Mouton, "Effect of flow distribution on emissions performance of catalytic converters", SAE Technical Paper no.2000-05-0175.

11. K.Zygourakis, "Transient operation of monolith catalytic converters: a two-dimensional reactor model and the effects of radially nonuniform flow distributions", Chemical Engineering Science, vol.44 (9), pp. 2075–2086, 1989.

12. L.Giani, G.Groppi, E.Tronconi, "Mass-transfer characterization of metallic foams as supports for structured catalysts", Industrial and Engineering Chemistry Research, vol.44 (14), pp.4993–5002, 2005.

13. M.Twigg, J.Richardson, "Fundamentals and applications of structured ceramic foam catalysts", Industrial and Engineering Chemistry Research, vol.46 (12), pp.4166–4177, 2007.

14. P.Dimopoulos Eggenschwiler, D.Tsinoglou, J.Seyfert, C.Bach, U.Vogt, M.Gorbar, "Ceramic foam substrates for automotive catalyst applications: Fluid mechanic analysis", Experiments in Fluids, vol.47 (2), pp.209–222, 2009.

15. G.C.Koltsakis, D.K.Katsaounis, Z.C.Samaras, D.Naumann, S.Saberi, A.Bohm, I.Markomanolakis, "Development of metal foam based aftertreatment system on a diesel passenger car", SAE Technical Paper no.2008-01-0619.

16. A.Gambarotta, A.Ruggiero, M.Sciolla, G.Lucchetti, "HiL/SiL System for the Simulation of Turbocharged Diesel Engines", MTZ Worldwide, vol.73, 02/2012, ISSN 0024-8525.

17. F.P.Incropera, D.P.Dewitt, T.L.Bergman, A.S.Lavine, "Principles of heat and mass transfer", 7th edition, John Wiley & Sons, New York, 2013.

18. C.Depcik, D.Assanis, "A Universal Heat Transfer Correlation for Intake and Exhaust Flows in an Spark-Ignition Internal Combustion Engine", SAE Paper 2002-01-0372, 2001.

19. P.A.Konstantinidis, G.C.Koltsakis, A.M.Stamatelos, "Transient heat transfer modelling in automotive exhaust systems", Proc.Inst.Mech.Engrs., vol.211, part C, 1997.

20. I.P.Kandylas, A.M.Stamatelos, "Engine exhaust system design based on heat transfer computation", Energy Conversion & Management, n.40, pp.1057-1072, 1999.

21. J.von Rickenbach, F.Lucci, C.Narayanan, P.Dimopoulos Eggenschwiler, D.Poulikakos, "Multi-scale modelling of mass transfer limited heterogeneous reactions in open cell foams", Int.J.of Heat and Mass Transfer, n.75, pp.337-346, 2014.

22. F.Lucci, A.Della Torre, J.von Rickenbach, G.Montenegro, D.Poulikakos, P.Dimopoulos Eggenschwiler, "Performance of randomized Kelvin cell structures as catalytic substrates: Mass-transfer based analysis", Chemical Engineering Science, vol.112, pp.143–151, 2014.

23. T.Shamim, H.Shen, S.Sengupta, S.Son, A.Adamczyk, "A comprehensive model to predict three-way catalytic converter performance", Journal of Engineering for Gas Turbines and Power, vol.124, pp.421-428, 2002.

24. G.N.Pontikakis, G.S.Konstantas, A.M.Stamatelos, "Three-way catalytic convereter modeling as a modern engineering desing tool", Journal of Engineering for Gas Turbines and Power, vol.126, pp.906-923, 2004.

25. P.Fiorani, A.Gambarotta, G.Lucchetti, F.P.Ausiello, M.De Cesare, G.Serra, "A detailed Mean Value Model of the exhaust system of an automotive Diesel engine", SAE Technical Paper , n.2008-28-0027, 2008.

26. L.Guzzella, A.Sciarretta, "Vehicle Propulsion Systems", Springer Verlag, 2005.

A PID and state space approach for the position control of an electric power steering

Vivan Govender*, Grigoriy Khazaridi*,Thomas Weiskircher*, Daniel Keppler*, Steffen Müller**

*Vehicle Automation and Chassis Systems, Daimler AG
Hanns-Klemm-Str. 45, D-71034 Böblingen, Germany;
E-mail: Vivan.govender@Daimler.com

**Department of Automotive Engineering, Technical University of Berlin
13355 Berlin, Germany

1 Introduction

The steering system is one of the primary controls for a vehicle. With the increase in driver assistance the demands of this subsystem have drastically increased. This has brought along much greater control intelligence. As with some assistance functions, autonomous driving requests a front steer angle for the vehicle to conduct a manoeuvre. The front steer angle is realized through the position control of the electric power steering (EPS). Thus far, the control of EPS systems in literature has focused heavily on generating a desired driver hand torque as is shown in works by Mehrabi et al. (2011), Fankem et al. (2014) and Dannöhl et al. (2011). The absence of a driver in the loop yields new challenges for control of an EPS. For instance, autonomous driving means that the steering wheel is free moving without the hands of the driver to control it. The resonance frequency caused by the free motion dynamics has an adverse effect on the position control. Moreover, both internal and external disturbances which are normally compensated by the driver have to be regulated by the EPS controller. Work by von Groll et al. (2006) shows that the most of the relevant frequencies for driver's inputs are below 4Hz. This is the frequency range where the controller should perform well in order to achieve all the relevant manoeuvres. Additionally, the steering wheel should move in a smooth and non-erratic way.

One of the controller types proposed in this work is an "advanced" PID. Proportional-Integral-Derivative (PID) controllers are the most used type of controllers for industrial purposes due to the low costs and easy implementation. Despite of their simple structure, they may provide good control performance for broad range of applications. The main disadvantage of a classical PID controller is that it cannot provide optimal set-point following (tracking) and disturbance rejection (regulation) simultaneously as presented in Araki & Taguchi (2003). In this regard advanced PID structures such as two-degree-of-freedom (2DOF) PID or cascade PID control may offer a better alternative. Since both requirements, tracking as well as regulation performance are crucial for control of an EPS, classical PID controllers are not considered in this work.

Another control design method for an EPS is the state space approach. This requires a complete understanding of the dynamics of the system states and not simply an input/output transfer function used in the PID control approach. This method however has the advantage of allowing multiple states to be easily controlled which would be useful in our problem. Moreover, through the computation efficiency of matrices this approach allows well developed optimal control theory to be applied. This in turn provides a quick and efficient way of placing the control poles to provide a multiple output control solution.

This paper presents a both a PID as well as a state space approach to the problem of front steering angle control. A detailed modelling and analysis of the steering system is presented as well as a discussion of the non-linearities it contains. This system is then linearised and a suitable linear system is described. The linear system is then reduced to show the minimum viable plant description for this control problem. Thereafter various PID strategies are applied and their effectiveness in controlling the high-fidelity non-linear system is presented. Next two state space methods are used, namely pole placement and linear quadratic (LQ) control. Pole placement was chosen to demonstrate the effectiveness of achieving desired performance on the front steer angle when the system is well described. The optimal control method is well known to maintain a good level of performance as well as robustness when the system shows unmodelled behaviour as shown in Anderson & Moore (1990). An implementation of this control method for electric drives is shown in Carrière et al. (2008). Various results are compared against each another and their advantages and disadvantages are discussed. The aim of this study is to demonstrate the effectiveness of well-established PID control against state space methodology for the purpose of position control within a steering system.

2 System Modelling and Analysis

The steering system description is presented in Figure 1. This description divides the system into three parts, namely the primary side, the secondary side and the front axle. The primary side consists of all components from the steering wheel until the pinion driving the steering rack. The secondary side consists of the steering gear which is the steering rack coupled with the EPS motor. The last part is the front axle which includes the front wheel.

At the outset, the described steering system with the accompanying vehicle model were modelled and identified. Here, specific measurement data was used to identify and validate system parameters. This process is presented by Diebold et al. (2006). This high fidelity non-linear model may be referred to as Model-based Testing (MbT).

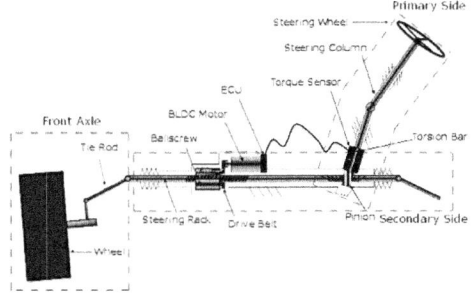

Figure 1: Overall description of the main steering system components. Modified from Pfeffer & Harrer (2013)

2.1 Non-Linearities

The steering system contains several non-linear effects. These arise mainly from elasticities, friction, variable gear ratios and the self-aligning torque. The elasticities for our analysis are concentrated at the torsion bar, coupling between steering rack and at the front axle. It was found that within our analysis these could be considered as pure linear springs. This assumption simplifies our modelling. Three friction models, one for every part of the steering system, were developed. These non-linear models were identified by combining all friction effects within each part. This was a pragmatic simplification and the result correlated to the measured behaviour relatively well. The variable gear ratios are present between the steering rack and pinion, steering rack and front wheel as well as between the pinion and steering wheel. These variable ratios were identified using appropriate measurements. Lastly the self-aligning torque was derived from the non-linear bicycle model. A full description of the steering system can be found in Mastinu & Ploechl (2014).

2.2 Linearised Model for Controller Design

For a linear control approaches proposed this system must be linearised. To this effect, friction is modelled as viscous damping, constant gear ratios and a linear vehicle model are used. Lagrangian dynamics provides an elegant way to formally derive this system as described by Parmar & Hung (2004). The mechanical model of the system is shown in figure 2.

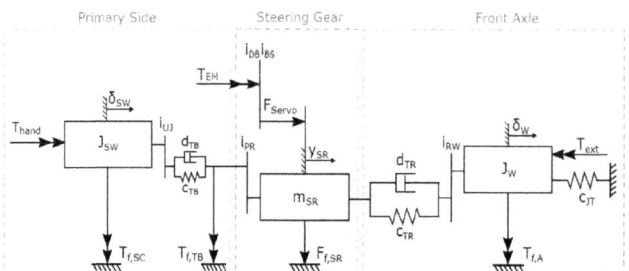

Figure 2: Mechanical model description of the steering system

The state vector is $q = [\delta_{SW} \quad y_{SR} \quad \delta_W]'$, where the elements represent the steering wheel angle, the steering rack position and the front wheel angle respectively. The electric motor is modelled as a non-elastic coupling onto the steering rack. This is justified by the high stiffness parameters at this coupling point. These parameters were provided by the steering gear suppliers. Furthermore, the electric motor was modelled as a 2nd order low pass filter. This adds two states to the system. A full description of all the model parameters is provided in Table 1 in the Appendix.

The system differential equations are presented as follows,

$$J_{SW}\ddot{\delta}_{SW} + \frac{1}{d_{i_{UJ}}}c_{TB}\left(\frac{\delta_{SW}}{i_{UJ}} - i_{PR}y_{SR}\right) + d_{SC}\dot{\delta}_{SW} + \frac{1}{d_{i_{UJ}}}d_{TB}\left(\frac{\dot{\delta}_{SW}}{d_{i_{UJ}}} - d_{i_{PR}}\dot{y}_{SR}\right) = 0 \quad (1)$$

(The right-hand side is equal to zero if the torque at the steering wheel $T_{hand} = 0$)

$$m\ddot{y}_{SR} - d_{i_{PR}}c_{TB}\left(\frac{\delta_{SW}}{i_{UJ}} - i_{PR}y_{SR}\right) + c_{TR}(y_{SR} - i_{RW}\delta_W) - d_{TB}d_{i_{PR}}\left(\frac{\delta_{SW}}{d_{i_{UJ}}} - \right.$$
$$\left. d_{i_{PR}}\dot{y}_{SR}\right) + d_{SR}\dot{y}_{SR} + d_{TR}\left(\dot{y}_{SR} - d_{i_{RW}}\dot{\delta}_W\right) = i_{DB}i_{BS}T_{EM} \quad (2)$$

$$J_W\ddot{\delta}_W - c_{TR}d_{i_{RW}}(y_{SR} - i_{RW}\delta_W) + c_{JT}\delta_W - d_{TR}d_{i_{RW}}\left(\dot{y}_{SR} - d_{i_{RW}}\dot{\delta}_W\right) + d_A\dot{\delta}_W = -T_{ext} \quad (3)$$

Finally the system equations can be written in the form

$\dot{x} = Ax + Bu,$

$$\begin{bmatrix} \dot{q} \\ \ddot{q} \\ \dot{T}_{EM,out} \\ \ddot{T}_{EM,out} \end{bmatrix} = \begin{bmatrix} 0_{3\times3} & I_{3\times3} & 0_{3\times2} \\ -M^{-1}K & -M^{-1}D & \alpha \\ 0_{2\times3} & 0_{2\times3} & A_{EM} \end{bmatrix} \begin{bmatrix} q \\ \dot{q} \\ T_{EM,out} \\ \dot{T}_{EM,out} \end{bmatrix} + \begin{bmatrix} 0_{7\times1} \\ \frac{1}{\tau^2} \end{bmatrix} T_{EM,in}$$

and $y = cx$, $y_{sr} = \begin{bmatrix} 0_{1\times2} & 1 & 0_{1\times5} \end{bmatrix} \begin{bmatrix} q \\ \dot{q} \\ T_{EM,out} \\ \dot{T}_{EM,out} \end{bmatrix}$ \quad (4)

Where $\alpha = \begin{bmatrix} 0 & 0 \\ \frac{i_{DB}i_{BS}}{m_{SR}} & 0 \\ 0 & 0 \end{bmatrix}$ and $A_{EM} = \begin{bmatrix} 0 & 1 \\ -\frac{2}{\tau^2} & -\frac{2}{\tau} \end{bmatrix}$, M, K, D are the motor dynamics, mass, stiffness and damping matrices respectively.

The model represented by (1-4) is an 8th order linear system. The number of states presents challenges for retrieving the measurements for each state as well as adding processing load when generating the controller response. For this reason, effort has been made to reduce the model to the simplest possible model for controller synthesis.

2.3 Model Reduction

The first reduction is one of practical considerations. The front wheel angle given by δ_w is defined as the effective steer angle combining the angles from the left and right wheel. This angle is currently not measured on our series vehicles. However, the kinematic ratio between steering rack position and front wheel angle is approximated. Using this approximation the states δ_w and $\dot{\delta}_w$ can be reduced from the model and the mass of the front axle is then transferred onto the steering rack. With this reduction a 2 mass system is obtained which neglects the elasticity between the steering rack and front axle. As a further simplification it is possible to reduce the steering wheel angle δ_{SW} and its velocity $\dot{\delta}_{SW}$. This is also achieved by transferring the inertia of the steering wheel as an additional mass onto the steering rack which results in 1 mass system. As with the reduction of the front axle the elasticity between the steering wheel and steering rack is neglected. With these reductions we are able to analyse the behaviours of the 3, 2 and 1 mass system model. An additional simplification would be to model the resistive load from the vehicle as a simple spring which produces a linear self-aligning torque based on rack position. The implication of these simplifications is presented in the next section.

2.4 System Analysis

Figure 3 shows the Bode plot of the reduced system models described previously. The Bode plot describes the response of steering rack position to motor Torque input. The influence of the modelled vehicle on the steering system can be seen between 1~2 Hz. The reduction without the vehicle model shows attenuation in the gain. However this difference is minimal and therefore this reduction seems justified.

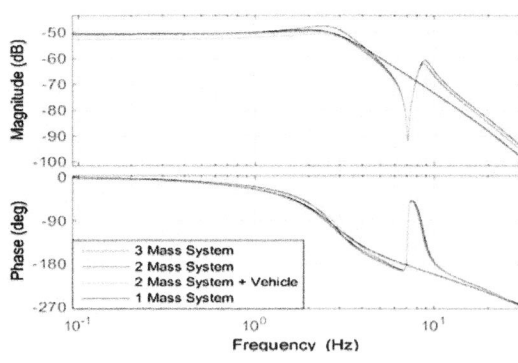

Next, the system was reduced from a 3 mass to a 2 mass and lastly to a 1 mass system. The reduction of system order seems to show minimal difference until 4 Hz. Thereafter the 1 mass system exhibits normal second order low pass characteristics and continues to fall away with reduced gain. The 2 and 3 mass systems clearly

Figure 3: Bode Plot of the various steering system models

show the resonance caused by the steering wheel. Until about 20 Hz the 2 and 3 mass systems respond very similar to each other with the resonance frequency of the steering wheel at 8.5 Hz well modelled in both cases. The resonant peak in the 7~8 Hz range from the steering wheel dynamics is lost in the 1 mass reduction.

Similarly the resonance frequency of the front axle modelling is lost in the reduction from 2 to 3 mass systems. Due to the elasticity of the front axle an offset in the amplitudes after 8.5 Hz is clearly seen. The 3 mass system does not have any additional resonance peaks compared to the 2 mass system. Therefore it can be claimed that the 2 mass system models all the relevant dynamics for the controller design. The validation of this claim is proved in the control design section.

3 Control Design Approaches

3.1 Performance Criteria

In order to evaluate whether a controller design is successful or not objective performance criteria are designed. These are normally based on very specific controller behaviour such as step response, rise time and overshoot. Here however we decided to evaluate the behaviour of the autonomous driving front angle controller based on overall required system behaviour. This system behaviour uses the vehicle's trajectory-following ability to evaluate what the performance of the front steering angle controller needs to be to ensure good path following during autonomous driving.

It was found that the steering system and front angle controller could be represented as a 3rd order system whose fastest and slowest response can be shown by two transfer functions below.

$$\left(T_{PT2_{Slow}}^2 s^2 + 2T_{PT2_{Slow}}D_{slow}s + 1\right)\left(T_{PT1_{Slow}}^2 s + 1\right)y_{SR} = y_{ref\_lower}$$

$$\left(T_{PT2_{Fast}}^2 s^2 + 2T_{PT2_{Fast}}D_{Fast}s + 1\right)\left(T_{PT1_{Fast}}^2 s + 1\right)y_{SR} = y_{ref\_upper} \tag{5}$$

3.2 2DOF PID Controllers

The main advantage of a two-degree-of-freedom PID controller over the classical PID structure is that it enables to adjust the complementary sensitivity as well as sensitivity transfer function independently. This allows the optimisation of the controller performance regarding set-point tracking as well as disturbance rejection thus providing additional flexibility to the controller design as presented in Araki & Taguchi (2003). Both aspects are important for control of an EPS in an autonomous driving vehicle since it has to follow the driving trajectory as precise as possible and should not be influenced by unexpected disturbances e.g. road or weather conditions.

In this work a 2DOF PID controller shown in Figure 4 is tested. The advantage of using this structure may be explained as follows.

Abrupt changes in reference signal, for instance a step signal, are passed on by the proportional controller and amplified by the derivative action, which causes large amplitude peaks in the control signal. This results in large overshoot of the system response and may produce a control signal which reaches the saturation limits of the actuator. The use of a 2DOF PID controller circumvents these effects. Instead of implementing a PID controller in $C_1(s)$, it is possible to move the proportional and derivative parts to $C_2(s)$ and implement only an integral or PI controller in $C_1(s)$. This measure allows suppressing the overshoot in the system.

A set-up where $C_1(s)$ is an integral and $C_2(s)$ a proportional-derivative controller is called an I-PD controller. Similarly, a PI-D controller is obtained if $C_1(s)$ is a proportional-integral and $C_2(s)$ is a derivative controller.

The best acceptable system reduction is the 2 mass system which is represented by the transfer function of 6th order. Since analytical derivation of controller gains for a system of such a high order is very tedious, the tuning was done using the SISO

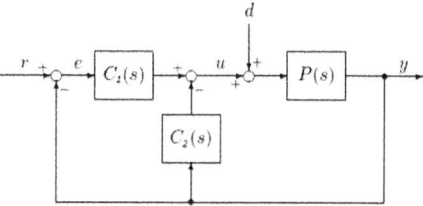

Figure 4: Feedback type (FB-type) 2DOF PID

Design Tool in MATLAB©. The gains were adjusted manually in order to achieve best possible rise time and low overshoot to overcome the friction effects in the nonlinear model.

3.3 Cascaded PID Control

Cascaded control loops offer an alternative control strategy when dealing with systems with single input and multiple outputs as described in Aström and Hägglund (1995). It is possible to control the velocity of the steering rack additionally to the position. In this way a multi-loop structure is obtained as seen in Figure 5. The inner loop controls the velocity of the steering rack and outer loop controls the position. Since the dynamics of the inner loop are faster than that of the outer loop, the velocity sig-

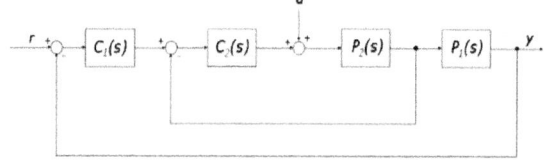

Figure 5: Cascade control loop

nal will respond much faster to disturbances. Therefore the idea is to design the inner loop with main emphasis on minimizing the negative effect of disturbances before they influence the outer loop. This simplifies the task of the outer controller which aims to achieve best control accuracy as possible.

The strategy to design a cascaded control loop is as follows. In a first step the inner (faster) controller is designed without considering the outer loop. The objective is to tune the loop for a high sensitivity in order to achieve good disturbance attenuation and reduce the effect of nonlinearities as well as model uncertainties. The position loop is tuned for best accuracy.

There are several possibilities to choose the appropriate controllers for both loops. In this case the combinations of PI-controller for position loop and P-controller for velocity loop (PI-P controller) as well as P-controller in $C_1(s)$ and PID-controller in $C_2(s)$ (P-PID-controller) achieved the best performance. Regarding the design considerations P-PID controller should guarantee better robustness. Since the inner loop responds faster to disturbances than the outer loop, the PID controller in $C_2(s)$ will achieve better robustness to sensitivity than a proportional controller.

SISO Design Tool in MATLAB was used for the tuning procedure. After the tuning for the linear 2 mass system the controller gains were adjusted manually for the nonlinear plant in order to overcome the friction effects.

3.4 State Space Controller Synthesis

Since the linear system has been well described previously, pole placement can be easily used to achieve a desired linear system performance. This allows a transparent understanding of the system's closed loop performance and provides a benchmark for the optimal control approach. Using the fact that the system is stable, a zero-pole cancellation can be performed by moving the dominant poles at 8.5 Hz to the zeros at 7.1 Hz. In this way, the negative effects caused by the steering wheel resonance can be theoretically eliminated on the closed loop rack position performance. In essence the LQ approach follows the standard full state feedback control method where the poles of the system described by equation (8), are moved using state feedback, u = −KX. The LQ approach however places the poles optimally to solve the weighted criteria set out by (6).

$$J = \int_0^\infty (X^T Q X + U^T R U)dt \tag{6}$$

where Q and R are state and control weighting matrices and are always square and symmetric. J is a scalar quantity, see Anderson and Moore (1990). The Hamilton-Jacobi Equation can be used to solve and rearrange (6) to give the optimal control law for the system,

$$U_{opt} = -R^{-1}B^T Px \qquad (7)$$

where P is the positive solution to the Riccati equation,

$$PA + A^T P + Q - PBR^{-1}B^T P = 0 \qquad (8)$$

The merit of this method lies in the tuning of Matrices Q and R in order to fulfil the performance criteria for multiple states of interest. For our purposes matrix R is an identity matrix since there is no limiting of the input signal. Weighting of the states was placed purely on the position and velocity of the rack, y_{SR}. This ensures the best tracking performance.

Furthermore due to the low velocity friction effects which are not well modelled with viscous damping a feedforward pre-compensation is implemented as an inverse transfer function of the steady state system,

$$\frac{u}{y_{ref}} = C^{-1}(A - BK)B^{-1} \qquad (9)$$

Lastly, an integration effect is included to minimise the offset in tracking the desired value as well as providing an amount of robustness against unmodelled effects and disturbances. This makes the controller a linear quadratic integrator (LQI). The integration effect is also weighted within the Q matrix as an augmented state x_i with the controller output

$$U = -K[x; x_i] + u_{pre-comp} \qquad (10)$$

This controller strategy for the state space controllers can be seen in Figure 6.

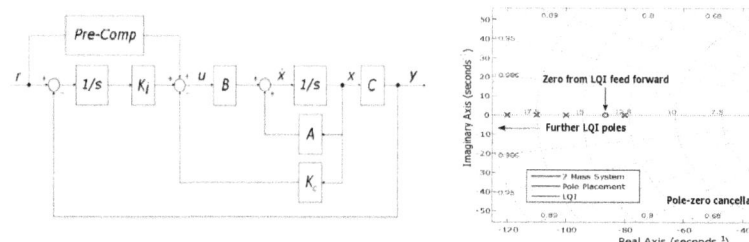

Figure 6: Control diagram layout for the full state feedback controller design

Figure 7: Close Loop Pole Zero map

Figure 7 shows the final pole-zero plot of the closed loop system. Here it can be seen that the optimal control solution also placed the pole very closed to the zero at 7 Hz. This gives support to the choice of poles used during pole placement.

3.5 Controller Performance with Non-Linear System

Figure 8 shows the performance of the 4 modified PID controllers introduced in section 3.2 and 3.3 applied to the nonlinear steering system. All the controllers were tuned in order to achieve the best rise time and to keep the overshoot low. However none of the controllers could match the desired performance regarding the design criteria. The rise times in Figure 8 show that the I-PD controller has the worst performance. The reason for this is that the proportional and the derivative part act on the output and not on the error signal. Therefore the system reacts slower to a change in reference. When the proportional part acts on the error signal as in PI-D controller the rise time improves significantly. However, the fast response comes at the cost of oscillations in the transient phase. Cascaded PID controllers perform better that 2DOF PID controllers regarding the rise-time and oscillatory behaviour. Consequently 2DOF PID controllers were not considered for the further analysis in this work.

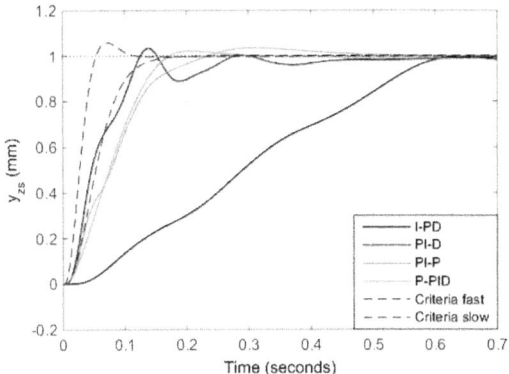

Figure 8: Step response of the various PID controllers

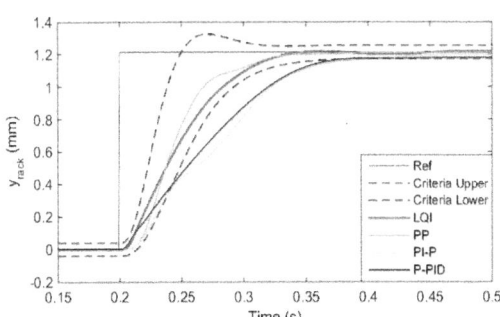

Figure 9: Step response of PID vs. State Space controllers

Figure 9 shows the step response of the cascaded PID controllers against the state-space controllers (pole placement method and LQI). The state space controllers show a much better performance than both cascaded PID controllers and both do not violate the performance criteria.

The controller tracking performance was also simulated for two different driving manoeuvres which are shown in Figure 10.

The first scenario should test the controllers during fast steering manoeuvre. The front steer angle is variated by 2° with the frequency of 1 Hz at the vehicle speed of 40 m/s.

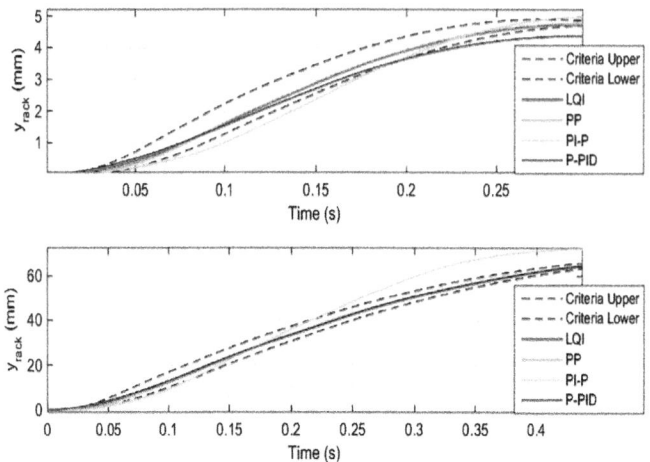

The second scenario reproduces the controller behaviour at a low speed and slow change in the reference. This may simulate a parking manoeuvre. Here the requested front steer angle is 30° with the frequency of 0.5 Hz at the vehicle speed of 5 m/s.

Figure 10: Controller response for two driving manoeuvres

For the fast steering manoeuvre similar to the step response both cascaded controllers do not achieve the desired performance according to design criteria. However during the simulated parking the P-PID control performance was within the performance criteria. From the results we see that during high dynamic manoeuvre the PID control structure does not perform as well as the state space controllers. This may be down to the fact that the PID controllers were manually tuned rather than analytically solved as in the case of the State Space controllers. P-PID controller shows the best performance of all modified PID structures and can be considered for the validation during real driving conditions along with the state space control solutions which showed good performance throughout

3.6 Robustness Tests

Robustness is also of vital importance when evaluating the performance of the controller. Unmodelled aspects such as changes in internal friction and external forces, for example side wind, provide additional challenges for the controller. Robustness, in essence, is a measure of how well the controller will perform in real world scenarios. With this in mind a variation analysis with the non-linear model was conducted. The friction at the primary, secondary and axle side was doubled and halved respectively. This provides an upper and lower bound for the friction variation. Next, a side wind force of 2000N over a frequency range of 0.1 to 2Hz was applied to the vehicle model. These variations were applied to three front steer angle test signals, namely a step input at 0.5 deg, sine input 2deg at 1 Hz and sine input at 30deg at 0.5Hz. The analysis shown in Table 1 presents the maximum deviation in steering rack position from the nominally parametrised signals without an addition side wind force.

Table 1: Parameter variation for the various controllers

	Requested front steer angle signal	Step 0.5 (in deg)	$2Sin(2\pi)$ (in deg)	$30Sin(0.5(2\pi))$ (in deg)
PI-P	Low Friction	$1.435*10^{-1}$	$2.968*10^{-1}$	$3.199v10^{-1}$
P-PID	(Rack position error	$6.676*10^{-2}$	$1.314*10^{-1}$	$1.493*10^{-1}$
LQI	in mm)	$2.177.10^{-2}$	$4.040*10^{-2}$	$8.315*10^{-2}$
PI-P	High Friction	$2.216*10^{-1}$	$7.342*10^{-1}$	$7.954*10^{-1}$
P-PID	(Rack position error	$1.217*10^{-1}$	$3.448*10^{-1}$	$3.836*10^{-1}$
LQI	in mm)	$3.147*10^{-2}$	$8.385*10^{-2}$	$1.265*10^{-1}$
PI-P	External Force	$7.823*10^{-2}$	$2.389*10^{-1}$	$4.340*10^{-1}$
P-PID	Disturbance	$3.997*10^{-2}$	$1.069*10^{-1}$	$2.043*10^{-1}$
LQI	(Rack position error in mm)	$5.612*10^{-3}$	$1.541*10^{-2}$	$3.404*10^{-2}$

The results above show some clear trends. The P-PID cascade controller shows roughly half the sensitivity of the PI-P controller throughout the variation analysis. The LQI controller shows at least a further 4 times smaller error to variation than the P-PID controller. As mentioned earlier P-PID controller design provides better robustness performance since the PID controller in $C_2(s)$ will achieve better sensitivity than a proportional controller for the inner loop. Overall the LQI achieved the best robustness performance which can be credited to solving the optimal function which guarantees a phase margin of ±60 deg.

4 Conclusion

This paper presents the comparison of the state-space methods and modified PID structures for front steer angle control in an autonomous driving vehicle. The modified PID controllers were realized as 2DOF PID and cascaded PID loop. It was shown that the performance of 2DOF PID is not sufficient for the presented control challenge. The cascaded PID controllers do not perform as well as state-space method but offer an acceptable alternative. Regarding the advantages of PID controllers as cheap cost and easy implementation the latter may be considered for testing in a real vehicle.

The state-space approaches are realized with pole-placement method and LQI controller. Both controllers show very good performance when applied to a nonlinear steering model. Further work on the modified PID control as friction compensation techniques may improve the controller performance and should be considered in future.

Bibliography

1. B. Anderson & J. Moore (1989). *Optimal control. – Linear quadratic methods.* Prentice Hall Intl.

2. M. Araki & H. Taguchi (2003). Two-Degree-of-Freedom PID Controllers, *International Journal of Control, Automation, and Systems,* Vol. 1, No. 4, 401-411

3. K. Aström & T.Hägglund (1995). *PID Controllers: Theory, Design and Tuning,* Instrum. Soc. Amer.

4. S. Carrière, S. Caux & M. Fadel (2008). Optimal LQI Synthesis for Speed Control of Synchronous Actuator under Load Inertia Variations, *Proceedings of the 17th World Congress, The International Federation of Automatic Control Seoul,* Korea, July 6-11, 5832–5836.

5. L. Diebold, W. Schindler, J. Haug, C. Daesch and M. Lahti (2006). *Application of a Single-Track Model for Simulation and Analysis of Vehicle Dynamics,* ATZ 11.

6. C. Dannöhl , S. Müller & H. Ulbrich (2012). H∞-control of a rack-assisted electric power steering system, *Vehicle System Dynamics: International Journal of Vehicle Mechanics and Mobility,* 50:4, 527-544.

7. S. Fankem & S. Müller (2014). A new model to compute the desired steering torque for steer-by-wire vehicles and driving simulators, *Vehicle System Dynamics: International Journal of Vehicle Mechanics and Mobility,* 52:sup1, 251–271.

8. G. Mastinu & M. Ploechl (2014), *Road and Off-Road Vehicle system dynamics Handbook,* CRC Press, 919-938.

9. N. Mehrabi, N. L. Azad, and J. McPhee (2011). Optimal Disturbance Rejection Control Design for Electric Power Steering Systems, *2011 50th IEEE Conference on Decision and Control and European Control Conference (CDC-ECC) Orlando, FL, USA,* December 12-15, 6584–6589.

10. M. Parmar, J. Y. Hung (2004). A Sensorless Optimal Control System for an Automotive Electric Power Assist Steering System, *IEEE Transactions on Industrial Electronics,* vol. 51, no. 2, April, 290–298.

11. M. Von Groll , S. Müller , T. Meister & R. Tracht (2006). Disturbance compensation with a torque controllable steering system, *Vehicle System Dynamics: International Journal of Vehicle Mechanics and Mobility*, 44:4, 327–338.

5 Appendix

Table 1: Linear model parameters

Parameter Name	Symbol	Unit
Driver hand torque	T_{hand}	Nm
Steering wheel & steering column inertia	J_{SW}	Kg.m^2
Steering column friction torque	$T_{f,SC}$	Nm
Steering wheel angle	δ_{SW}	rad
Universal joint ratio	i_{UJ}	-
Torsion bar angle	δ_{TB}	rad
Torsion bar stiffness	c_{TB}	Nm/rad
Torsion bar friction torque	$T_{f,TB}$	Nm
Pinion-to-rack ratio	i_{PR}	rad/m
Electric motor torque	T_{EM}	Nm
Drive belt-to-ballscrew ratio	i_{DB}	-
Ballscrew-to-steering rack ratio	i_{BS}	rad/m
Force on steering rack	F_{Servo}	N
Steering rack displacement	y_{SR}	m
Steering rack mass	m_{SR}	kg
Steering rack friction force	$F_{f,SR}$	N
Tie rod stiffness	c_{TR}	N/m
Tie rod damping	d_{TR}	Ns/m
Steering rack-to-front wheel ratio	i_{RW}	rad/m
Front wheel inertia	J_W	Kg.m^2
Front wheel angle	δ_W	rad
Front axle friction torque	$T_{f,A}$	Nm
External torque acting on wheel	T_{ext}	Nm
Spring stiffness for jacking torque	c_{JT}	Nm/rad
Motor Torque Request	$T_{EM,in}$	Nm
Actual Motor Torque	$T_{EM,out}$	Nm
Motor Period	τ	sec

Identification and evaluation of the real temperature loading of steering electronics

Dipl.-Ing. Ulrike Weinrich[a], Stefan Walz[b], Dr. Gerd Baumann[a],
Prof. Hans-Christian Reuss[a]

[a]   Research Institute of Automotive Engineering and Vehicle Engines
Stuttgart (FKFS)

[b]   Robert Bosch Automotive Steering GmbH

Introduction

Within a few years, electric power steering (EPS) almost completely replaced hydraulic steering assistance in all vehicle segments. This is largely due to a more compact design – because the hydraulic unit is no longer needed – and a significant reduction in fuel consumption for the end customer. [1]

The EPS is becoming increasingly important as a central actuator in an electronic ECU network for vehicle guidance. This means the whole EPS system must be extremely reliable. Errors with a high potential for danger must be avoided at all costs: An important parameter is the probability of failure (Failures in Time, FIT rate) of the electronics module in the steering system. This is the main focus of the cooperation between Robert Bosch Automotive Steering GmbH and the FKFS. The aim of the project is a reliable prediction of the probability of failure under realistic operating conditions for the EPS, when used by the end customer ("Real world driving").

The FIT rate of electronics modules is influenced by numerous parameters. The thermal loading is of central importance, since the probability of failure of electronic components increases exponentially with increasing temperature (Arrhenius equation). Various vehicle categories with different technical steering designs, their geometrical arrangement in the engine compartment, and different engine types are to be taken into account here. Also of particular significance is the global application of the EPS in different regions/climatic zones.

The investigation method used will be an extended form of the representative volunteer study in vehicle testing on public roads. The fundamental approach was jointly developed in 2001 by Bosch and the FKFS [2]. Since then, the procedure has been improved several times and applied to different problems. For example, proving reduction in fuel consumption due to introducing EPS [1], the achievable CO_2 saving by using ACC [3] and the increase in the range of electric vehicles due to automated longitudinal control [4].

In this article, this procedure will be applied to the problem of thermal loading of the electronics during operation by the end customer. The method will be extended, enabling a quantitative transfer of the results (gained from vehicle trials in central European climatic conditions) to global markets in other climatic zones.

Simulation of End Customer Operation with Volunteer Studies

Motivation

There are numerous standardized component tests to validate the function of safety-related systems. In the automotive sector, these can be complemented by driving cycles – synthetic cycles and/or cycles derived from reality.

Reliable results can be calculated using numerical simulations based on the specified speed, starting temperature, switching points, vehicle preparation, payload etc. However, it is problematic that many factors which occur in real-life operation cannot be incorporated, or can only be incorporated to a certain extent. A further disadvantage of driving cycles compared with driving on public roads is the lack of longitudinal incline of the road, which has been proven to influence driver behaviour, driving performance and consumption [5].

But the individual behaviour of the driver of a vehicle in real traffic cannot be documented using the simulations listed above. Here, realistic loading is of central importance for the design of the components.

Methodological Approach for Investigating End Customer Operation

A strictly scientific approach is essential to acquire meaningful results for the temperature loading at defined points in the engine compartment (in this case around the steering control unit) during operation by the end customer. The approach used by the FKFS can be divided into the steps shown in Figure 1.

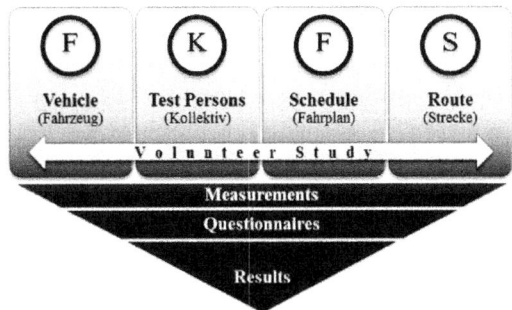

Figure 1: Methodological Approach for a Volunteer Study

The selection of the **vehicle** is, above all, based on the statistics for the vehicles to be used and/or the distribution of the component being studied in the appropriate country. The vehicle(s) being studied is/are equipped with identical high-performance measuring equipment, to prevent measurement errors and to ensure the results are comparable.

The aim of a measurement run study is to achieve the greatest possible statistical meaningfulness. Here the number of **test persons** is very important. Before carrying out the test, the distribution of characteristics in the population must already be known from earlier trials, or specified by guidelines [6]. This will be used to determine the minimum system effect expected, to calculate the size of the effect and to define the level of significance. The optimum sample size is determined in accordance with [6]. This comprises a representative group of normal drivers. The distribution of age, gender and vehicle performance must be characteristic for the vehicle type/component being studied.

The **test plan** of the study has a significant influence on the result of the investigation. This means several important points must be taken into account during its creation. This includes the weather and conditions, the amount of traffic (peak/off-peak, traffic during holiday periods), the number of trips in one day and the length of the breaks between these trips.

The **route** on the public roads is to be selected with regard to a representative distribution of motorway, highway and city driving, as well as a representative distribution of vehicle speeds. To represent realistic end customer driving, parking and turning manoeuvres are also of significance. However, if they can be proved to have no influence on the aims of the study, these can be left out.

An interview of the test persons in the form of one or several **questionnaires** on demographic data, the behaviour of the vehicle/component, and to investigate acceptance, is absolutely essential for later conclusions.

After completing all the preparations, the **measurement runs** are carried out. Here the vehicles are driven by the acquired drivers on the representative route, for the times defined in the test plan. To ensure the driving style of the test persons does not change, in previous studies it has been found useful for the driver to be accompanied by someone with local and vehicle knowledge [7]. Furthermore, this passenger documents driving times, traffic conditions and problems or faults in the recording of the measurement data, in a trip log.

The documented measurement data will be analysed with regard to the problem being studied. In addition, to ensure statistical reliability, the assumptions made in the selection of the sample size must be verified. In the last step, the **results** for the subsequent investigation will be put together in a presentable format.

Test Planning

The goal of the volunteer study in summer 2015 is representative documentation of the temperature distribution around the steering control unit, for all German drivers. Therefore, the selection of the described parameters is primarily oriented towards values from statistical surveys.

Alongside measurement-related constraints, the selection of the vehicle is, above all, based on statistics regarding the stock of vehicles in Germany and the distribution of the Servolectric® electric power steering. [8] It is intended to use a mid-size car with a diesel engine (Vehicle 1) for electric power steering with an axially parallel servo unit – EPSapa. A compact car with a gasoline engine (Vehicle 2) was selected as the test vehicle for electric power steering on a second pinion – EPSdp. The conscious selection of different drive configurations should enable variance of the results. An investigation of the variant with a servo unit on the steering column – EPSc – was not included in this study, due to its significantly different installation location.

To be able to make valid statements about the German population with the study, the sample size (test persons) must be representative. Its composition must be as similar as possible to that of the population [6]. The criteria age, gender and years of driving experience were used to preselect the test persons. Thanks to a uniform distribution of experience in traffic, it should be ensured that the measurement runs are driven with different driving styles. Taking [6] into account, the sample size for the population-describing investigation was specified as 50 test persons. Taking the age profile of the population into account, the distribution of age groups for the volunteer study was selected as shown in Table 1.

Since, due to the higher external temperatures, more heat generation is to be expected during summer, the trips for the volunteer study were performed in August and September. Furthermore, the influence of varying traffic conditions is to be determined by performing testing during and after the school summer holidays in Baden-Württemberg. To achieve comparable traffic conditions, all the measurement runs will be performed during defined periods in the daytime. Furthermore, for evaluation later on, a difference will also be made between cold trips (vehicle at rest > 6 hours) and hot trips (vehicle at rest < 1 hour). The test plan also captures the anonymized details of the test person and passenger, the vehicle (F1: Vehicle 1, F2: Vehicle 2), the direction of travel (M: motorway, C: city), the vehicle condition (C: cold trip, H: hot trip) and the vehicle manoeuvre (RP: reverse parking, PP: parallel parking, TPT: three-point turn).

Table 1: Distribution of Age Groups

	Part of the Population in %		Number of Test Persons	
Age Group	Male	Female	Male	Female
20-29	9,31 %	9,00 %	5	5
30-39	9,34 %	9,09 %	5	5
40-49	12,54 %	12,56 %	6	6
50-59	10,59%	10,61%	5	5
60-69	8,24 %	8,73%	4	4

The test route is selected to correspond to the driving profile of the average driver in Germany, in terms of route types and their proportion of the overall journey. The basis for the proportions of the individual route types are statistical data on the use of roads and annual mileage in Germany [7] [8] [9] [10]. Based on this data, the approx. 60 km long "Stuttgart circuit" was selected. This was developed by the FKFS in 2001 [11]. Figure 2 shows the route and the percentage distribution of road types. Each trip ended with a parking manoeuvre (parallel parking/reversing) or turning manoeuvre (three-point turn).

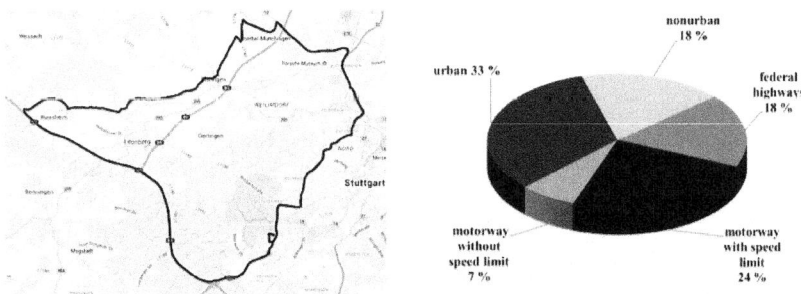

Figure 2: Stuttgart Circuit (left) and Distribution of Route Types (right)

To determine the influence of the route on the development of the temperature in the engine compartment, the circuit is driven in an anticlockwise direction by half the test persons (direction motorway) and clockwise by the other half (direction city).

The measurements include multiple sensor signals, e.g. the ambient air temperature of the steering control unit, the torque sensor and in the wheel arch, as well as tie rod forces, GPS signals and fieldbus communication.

Results

In Figure 3, the top graph shows the chronological sequence of the coolant and oil temperature as well as the ambient air temperature of the steering control unit (EPS ECU) for Test Person No. 33. The lower graph shows the vehicle speed profile and the engine torque for the same test. The "→M" in the left corner indicates that the test person drove around the circuit in a clockwise direction (towards the motorway). Based on the temperature profiles, the graph can be divided into four sections:

1. Minute 0 to 17: Warming up the engine to operating temperature

2. Minute 17 to 50: Temperatures are almost constant, peaks in the ambient air temperature of the Servolectric®

3. Minute 50 to 70: Increase in the temperature level on the observed ECU by approximately 20 °C, occasional peaks; coolant and oil temperature constant in the range between 90 and 100 °C

4. Minute 70 until end: EPS ECU ambient air temperature constant at 40 °C

Based on the chronological sequences, it can be shown that the peaks in the ambient air temperature in the engine compartment always occur at the same time as a drop in engine temperature. An explanation for the occurrence of this temperature fluctuation lies with the electric cooling fan. At these points in time, the steering system is briefly surrounded by warmer air.

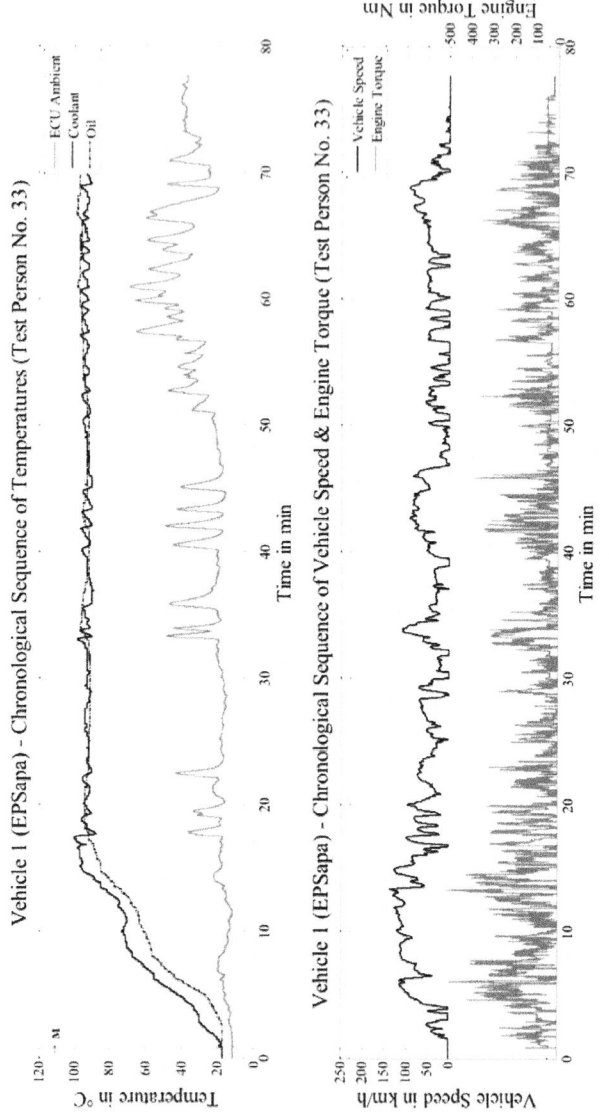

Figure 3: Chronological Sequences of Vehicle 1 (EPSapa), Test Person No. 33,
top: Temperatures, bottom: Vehicle Speed and Engine Torque

Figure 4 shows the temperature distribution of the measurement points (ambient air EPS ECU, coolant, oil) for all test persons driving Vehicle 1. The weighted averages are shown in addition.

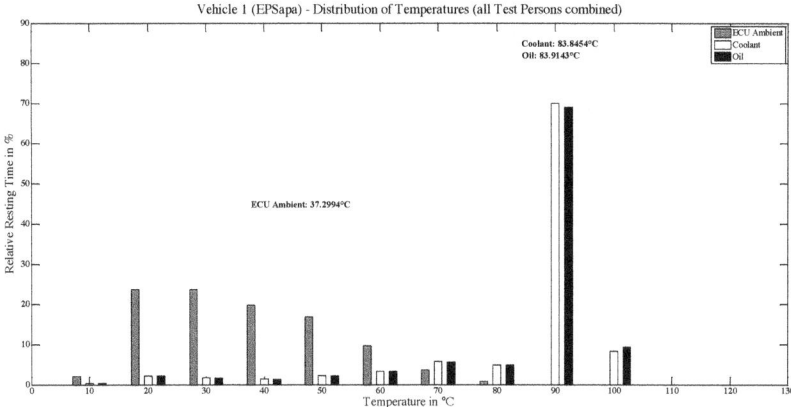

Figure 4: Distribution of Temperatures of Vehicle 1, all Test Persons combined

Applying the Results to Other Climate Zones

Motivation

The design of the volunteer study described above, with regard to test persons, vehicle and route parameters, means that the collection of temperature results is only representative for Germany. However, the driving cycle determined could potentially be used on the test rig, for example, for warmer climatic regions.

Figure 5 shows the FKFS thermal wind tunnel. Here the engine load and the speed can be realistically simulated according to the representative driving profile. The temperature level can be adjusted in the range between 20 and 50 °C. The influence of the external and ambient air temperature directly around the steering control unit on the heating of the components is measured – not simulated. The total occurring temperature collective is determined.

The validation of the measured temperature distributions from the thermal wind tunnel compared with the measurement results from the study can be performed using the test vehicles from the study.

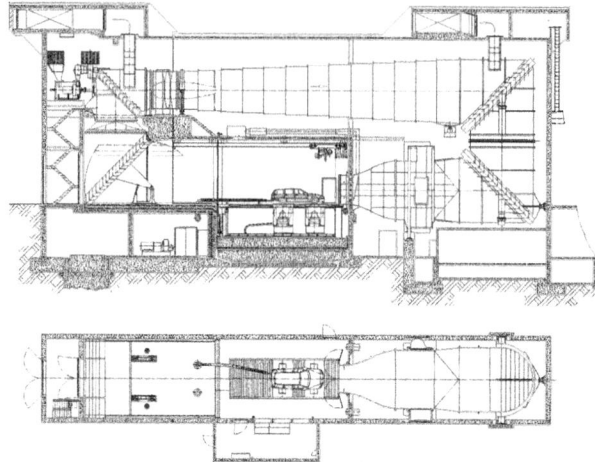

Figure 5: Thermal Wind Tunnel of the FKFS

Test Planning

Driving on the road is simulated in the wind tunnel. Therefore, there may be deviations compared with reality. Qualitatively, the resulting errors can be detected through analysis of the original procedure [12].

A driving cycle and a road profile are specified for the tests to be performed on the test rig, taking the data from the volunteer study into account. The driving cycle consists of time profiles of vehicle dynamics parameters, such as vehicle speed and longitudinal acceleration. To validate the measurement data from the thermal wind tunnel, the actual driving profile of a test person is used which best represents the average of the measured speed profiles for all the drivers. The road profile results from the altitude profile of the Stuttgart circuit. Combined with the driving resistances, this results in a necessary wheel and engine torque. This leads to the corresponding engine load and the associated heating effects on the steering electronics.

The measurement data from the same test person is also used for the ambient air temperature. An air temperature of 20 °C is specified for the validation. An air temperature of 40 °C is used for the tests to apply the results to other climatic zones.

The vehicles and the measurement equipment used in the volunteer study are also used for these tests.

Results

For ease of presentation, the following chronological profiles are each limited to the section with the highest ambient temperature of the steering control unit.

In the top graph, Figure 6 shows the chronological sequences of the coolant temperature, oil temperature and the EPS ECU ambient air temperature, for the measurement in the thermal wind tunnel with an external temperature of 20 °C. The graph below shows the vehicle speed and the engine torque. Just like the measurement in Figure 3, in the thermal wind tunnel there are also temperature peaks near the control unit at the same time as a drop in coolant temperature. Overall, in the test rig measurement with an identical external temperature, the temperature level in the engine compartment is somewhat higher than the measurements taken on the road.

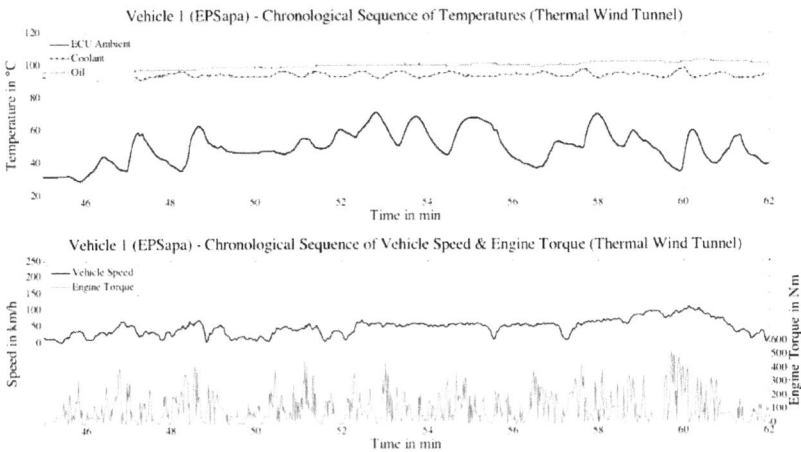

Figure 6: Chronological Sequences of Vehicle 1 (EPSapa), Thermal Wind Tunnel at 20 °C, top: Temperatures, bottom: Vehicle Speed and Engine Torque

Responsible for this is, firstly, that the test rig driving cycle contains shorter breaks/stops. The reduction in ambient temperature due to a longer rest period is demonstrated in Figure 3, in the region from 55 to 57 minutes. Secondly, due to boundary layer formation in the thermal wind tunnel, the airstream does not flow completely around the vehicle, particularly in the underbody region. Therefore, the exhaust air flow from the engine compartment back into the surrounding airstream is interrupted.

Figure 7 shows the distribution and weighted averages of the EPS ECU ambient air, coolant and oil temperatures, for the test in the thermal wind tunnel with an external temperature of 20 °C.

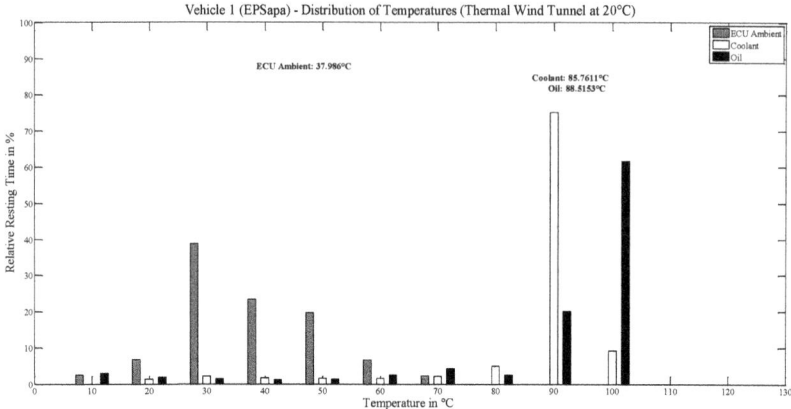

Figure 7: Distribution of Temperatures of Vehicle 1, Thermal Wind Tunnel at 20 °C

Figure 8 shows the bar chart for the test in the thermal wind tunnel with an external temperature of 40 °C as well as the weighted averages.

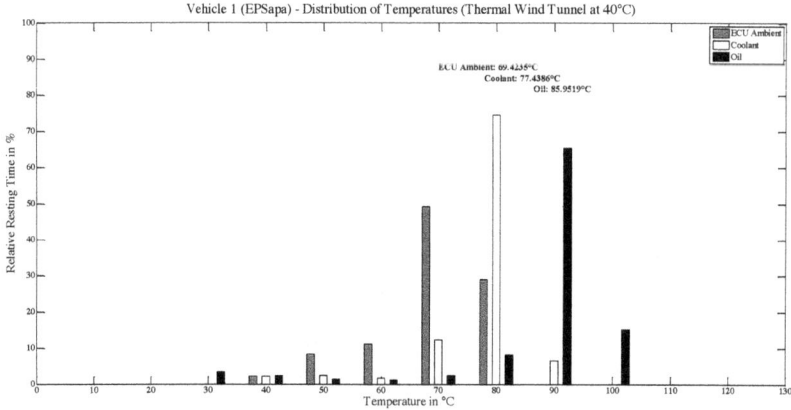

Figure 8: Distribution of Temperatures of Vehicle 1, Thermal Wind Tunnel at 40 °C

Comparing Figures 7 and 8, it becomes clear that the temperature collective of the EPS ECU ambient air for Vehicle 1 moves to the right with increased external temperature and tends to be left skewed.

Table 2 summarizes the weighted averages of both vehicles for the measurements of the Test Person No. 33 as a part of the volunteer study with an external temperature of 20 °C ("Study"), in the thermal wind tunnel at 20 °C (cold trip: "TWT20C", hot trip: "TWT20H") as well as the measurements carried out at 40 °C (cold trip: "TWT40C", hot trip 1: "TWT40H1", hot trip 2: "TWT40H2").

Table 2: Weighted Averages of ECU Ambient Temperatures of Vehicles 1 (EPSapa) and 2 (EPSdp)

Vehicle	Study	TWK20C	TWK20H	TWK40C	TWK40H1	TWK40H2
1	27,3 °C	38 °C	47,9 °C	69,4 °C	75,1 °C	78 °C
2	33,6 °C	46 °C	49,9 °C	69,3 °C	76,5 °C	76,3 °C

The measurements in the thermal wind tunnel show that the EPS ECU ambient air temperature is, given the same external temperature, slightly higher on the test rig than on the road. Thus, the temperature distribution of the thermal wind tunnel is more pessimistic compared to end customer driving. Altogether this is an advantage for the transfer of the results to global markets in other climatic zones and confirms the procedure.

Summary and Outlook

This publication presents a strictly scientific approach to investigating the temperature loading at defined points in the engine compartment during operation by the end customer. It was shown how a representative volunteer study should be designed with regard to the vehicle, the group of volunteers, the test plan and the route.

In the thermal wind tunnel at the FKFS, road trips were simulated according to the representative driving profile at external temperatures of 20 and 40 °C. The ambient air temperature directly around the steering control unit was determined. When comparing the measurement results for both vehicles from the road tests and on the test rig, it becomes clear that the weighted averages at the same external temperature have a difference of 10-12 °C. These deviations in the thermal wind tunnel compared with reality were expected, but the errors can be identified qualitatively.

Based on the method presented here, the driving cycle can also be used for other vehicles. Due to the individual packaging situation and air flow conditions, it is necessary to create a representative temperature collective in the engine compartment for each vehicle. Here, the more vehicles measured, the more precisely a characteristic curve for the measured temperature collective can be created for different components, such as the Servolectric®, for example, or even for vehicle classes. This can be used to read off ranges where the temperature collectives for new vehicles with similar constraints are to be expected.

Considering the transfer of the measurement results to the usage profiles in other global markets an analysis of the mobility behaviour and an evaluation of the influence on the temperature distribution have to be carried out. The characterisation of parallels respectively differences of the original parameters (test persons, vehicles, and route) presents a further development of the presented method to maintain a representative thermal loading of the electronics during operation by the end customer.

Bibliography

1. Vähning, A.; Heger,, M.; Gaedke, A.; Runge, W.; Reuss, H.-C.: Ganzheitliche Wirkungsgradoptimierung von elektromechanischen Lenksystemen. VDI-Berichte Nr. 2075, 14. Internationaler Kongress Elektronik im Kraftfahrzeug, 2009

2. Fried, O.; Bargende, M.; Hötzer, D.: Kraftstoff-Einsparpotenziale für elektromechanische Antriebsstränge im realen Fahrbetrieb (4. Internationales Stuttgarter Symposium Kraftfahrwesen und Verbrennungsmotoren). Bargende, M.; Wiedemann, J. (Hrsg.), Renningen-Malmsheim: expert-Verlag, 2001

3. Wagner, C.; Salfeld, M.; Knoll, S.; Reuss, H.-C.: Quantifizierung des Einflusses von ACC auf die CO2-Emissionen im kundenrelevanten Fahrbetrieb, 10. Internationales Stuttgarter Symposium Automobil- und Motorentechnik, 2010

4. Becker, G.; Reuss, H.-C.: „Efficient Cruise Control – A Method for Increasing the Range of Electric Vehicles". 10th Symposium Automotive Powertrain Control Systems, 11.-12.09.2014, Berlin

5. Fried, O.: Betriebsstrategie für einen Minimalhybrid-Antriebsstrang. Universität Stuttgart, Dissertation, 2003. Shaker-Verlag. ISBN 3-8322-2496-3

6. Bortz, J.; Döring, N.: Forschungsmethoden und Evaluation für Human- und Sozialwissenschaftler. 2006. Springer Medizin Verlag Heidelberg. ISBN-13 978-3-540-33305-0

7. Baumann, G.; Rumbolz, P.; Piegsa, A.; Grimm, M.; Reuss, H.-C.: Analyse des Fahrereinflusses auf den Energieverbrauch von konventionellen und Hybridfahrzeugen mittels Fahrversuch und interaktiver Simulation. VDI-Berichte Nr. 2107, Tagung: SIMVEC - Berechnung und Simulation im Fahrzeugbau, 2010

8. Kraftfahrt-Bundesamt: Bestand an Personenkraftwagen am 1. Januar 2015 nach Herstellern, Handelsnamen und ausgewählten Merkmalen. Juli 2015

9. Deutsches Institut für Wirtschaftsforschung: Mobilität in Deutschland 2008, Tabellenband, Juli 2008

10. Geschwindigkeitsbegrenzungen auf Autobahnen. http://www.autobahnatlas-online.de/Limitkarte.pdf

11. Janisch, V.: Untersuchung des Fahrverhaltens. Studienarbeit, Universität Stuttgart. Stuttgart, 2002

12. W.-H. Hucho (ed.): Aerodynamik des Automobils, Springer Fachmedien Wiesbaden 2008

Fully automatic and haptic test of electric power steering in a virtual environment

Jörg Paschedag, Marc Scherer

ITK Engineering AG

1 Why Testing in a Virtual Environment?

In recent years, the amount and level of requirements for component development in automotive industry has severely increased. This also applies to electric power steering. On the one hand, steering dynamics and driving feeling are to be improved continuously. On the other hand, the components are getting more and more complex, especially because of enclosed actuators and electronic control units (ECUs). Advanced driver assistant functions like lane keeping, stability programs and parking assistants have to be implemented. Furthermore, the communication and interaction of the different control units in a vehicle get more and more complex. Besides all this, the number of variants is increasing, while the development cycles are getting shorter. In order to assure the fulfillment of the requirements, also the effort of component testing has to be significantly increased.

However, suitable test vehicles are often limited in numbers. Especially in early development stages, there may not be any available at all. As a result, more and more tests have to be shifted from the road to test benches (see e.g. figure 1). Those have to be able to realistically reproduce the conditions that occur at real driving. Conventional performance and stress tests can be carried out by applying prefabricated test sequences to the component under test. For a more realistic analysis, which also takes into account the interaction of component and vehicle, also the environment of the component must be included in the test. The environment consists mainly of vehicle, driver, road and environmental conditions. Since the real environment is not available at the test bench, it has to be replaced by a virtual model. A suitable procedure for this purpose as well as relevant variants and their properties are proposed in this contribution.

In the following chapter 2, the principle of classic component testing is described. Electric power steering is considered as an example. In chapter 3, the test system is extended by the virtual environment. The approach of testing with manual steering is explained in chapter 4. This special method allows a real person to get a haptic impression of the steering feeling. In chapter 5, an outlook is provided on the potential that lies in a further increase of the degree of virtualization. A summary of the contribution is finally given in chapter 6.

Figure 1: Steering test bench[1]

2 Classical Component Test

An electric power steering is a mechatronic system with mechanical properties on the one hand and an electronic control unit on the other hand. For a comprehensive validation, both aspects have to be tested. The respective techniques as well as their integration are explained in the following.

2.1 Testing the Mechanical Component Behavior

To test the mechanical behavior of a vehicle component at a test bench, physical values like angle and force have to be set at the component by applying feedback control. A respective system is shown in figure 2. Besides relatively slow stress and behavior tests, the dynamic behavior can be tested by using synthetic test sequences with quickly changing set point values as well as with set point data from measurements in a real car.

---

[1] ©Klaus Junk, blickfang

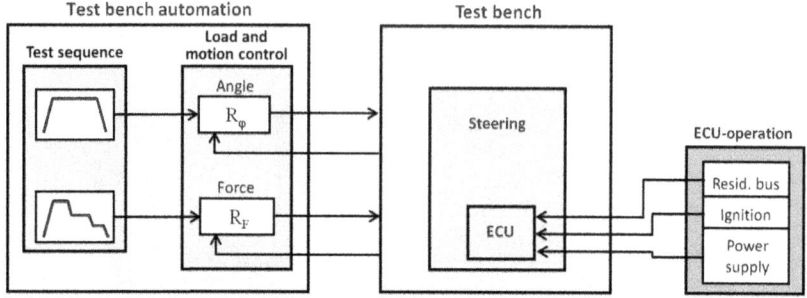

Figure 2: System for the test of mechanical component behavior

Figure 3 shows the result of a typical test sequence for validation of the mechanical component behavior. In the upper part of the diagram, the set point and measured signal of steering wheel angle control can be seen. The lower part shows the respective signals of the load force control at the tie rod.

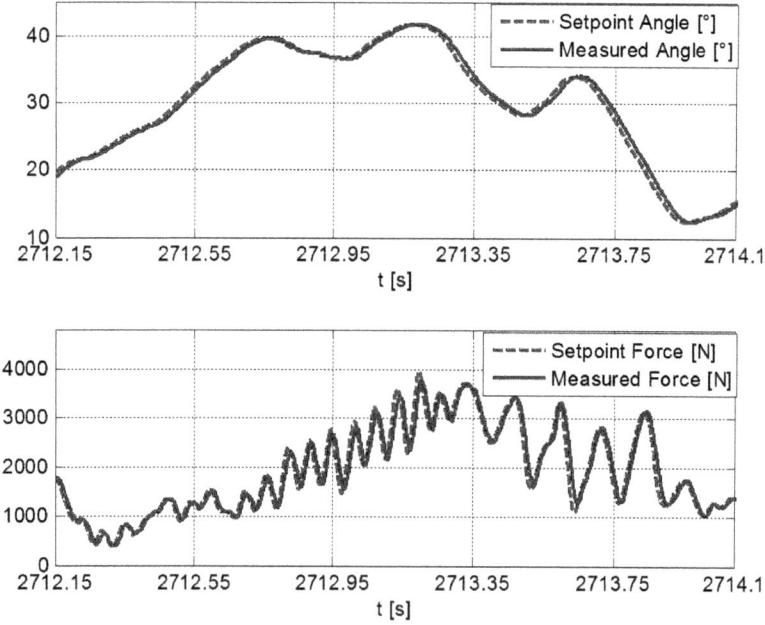

Figure 3: Test sequence result

2.2 ECU Test

The assist force of the electric power steering is controlled by the included electronic control unit (ECU). The testing of the ECU is possible even before integration in the car by so called Hardware-in-the-Loop (HIL) tests. The signals and messages that the ECU normally gets from the vehicle bus are in that case provided by a residual bus model. The residual bus model already represents a virtual environment for the ECU. However, at least in conventional test benches for steering systems, its implementation is rather rudimentary. The fully virtual environment is explained in Chapter 3.

In classical test benches for validation of mechanical system behavior, the system that provides the residual bus for the ECU operation is typically separated from the main test bench automation system (see figure 2). But, to validate the behavior of the complete component thoroughly, integrated testing together with the test of the mechanical component behavior is necessary, as described in the following section.

2.3 Integrated Testing

A comprehensive validation of the complete steering system is possible by combining and parallelizing the test of the mechanical component behavior with the ECU test. A suitable test system setup for this is shown in figure 4. The functionalities for operation and test of the ECU are integral part of the test system in this case. This increases the complexity of the test bench automation system. Know how in the area of „conventional" test bench systems is necessary as well as in modern ECU technology. Since this knowledge is often divided among different development departments, ITK Engineering AG developed an automation software that allows comprehensive component testing in one system. The software TACware®, which is especially designed for the automation of component test benches, includes all functionality that is necessary for highly dynamic tests of mechanical component behavior as well as tests of ECU functionality. Besides that, standard interfaces for the integration of simulation models are available. Therefore, a comfortable and safe connection to a virtual environment is possible.

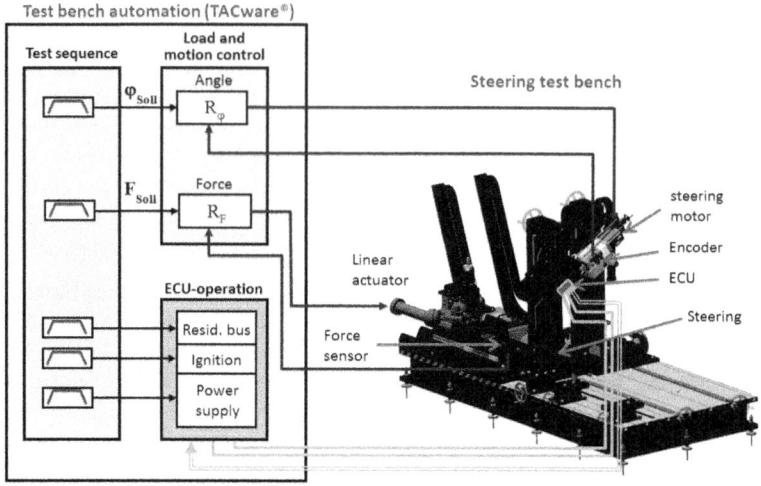

Figure 4: Test bench for fully integrated testing

3   Providing the Complete Virtual Environment

As described in chapter 1, the goal is to establish a test system that allows testing as in the real vehicle. Therefore, the component, i.e. the steering, is set in a loop with a vehicle model. In order for the virtual car to be able to actually drive, some additional model parts are necessary: A model of the driving lane in form of road data has to be available. Furthermore, a driver model is required that keeps the vehicle on the street with the desired velocity, by accelerating, braking and applying steering motion. Finally, an environmental model is necessary that provides values for e.g. wind and temperature. With this full environment, maneuver based testing becomes possible.

In figure 5, the resulting test system is shown. The complete component is set in a loop with a model of its environment. The load and motion control sets the set point values from the model at the steering system. With the given setup, the mechanical interaction between component and environment model can be analyzed as well as the integration of the ECU in the overall vehicle bus. The test gets the more valid the more realistic the environment model is. In the ideal case, the test bench could replace the actual vehicle completely. Specific advantages of the considered system are that the performed maneuvers are completely reproducible regarding driver behavior and environmental conditions. Furthermore, objective evaluation criteria can be thoroughly applied.

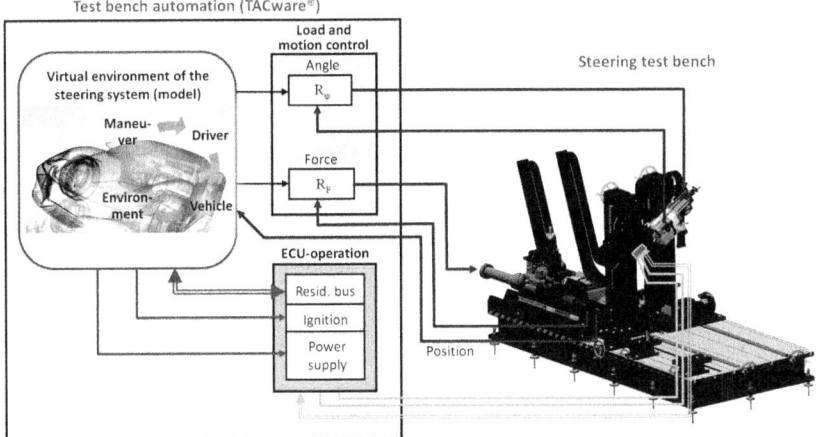

Figure 5: Test bench system with complete virtual environment[2]

For successful model integration, it has to be assured that the set point values provided by the model (e.g. angle and force) are precisely adjusted by the load and motion control at the physical component. On the one hand, this requires highly effective control algorithms. On the other hand, the mechatronic test bench set up has to meet high demands on dynamics and precision. This is achieved e.g. by using cutting edge technology as high quality hydraulic double-rod cylinders for force generation and mechanical structures that are optimized regarding structural behavior.

4   Manual Steering Test with a Real Driver

A further step towards testing under realistic circumstances is the "Driver-in-the-Loop" test. This special kind of test allows an analysis of haptic component behavior and subjective driving feeling. It is carried out at the test bench with the real steering system in the virtual environment. The test setup is almost the same as in figure 5, but an actual steering wheel is connected to the steering column, and a real person performs the steering motion at the wheel. At the same time, the vehicle model calculates the load force from the street that would occur at the steering rack via the tires in real-time. The load force value is sent to the test bench load and motion control as a set point and is in that way applied to the steering system. In this case, the demands on

---

[2] ©ag visuell – fotolia.de

the load and motion control are especially high since the driver should only experience the behavior of the steering system and not effects from insufficient control.

With the manual steering test, the haptic behavior of the active steering can be analyzed as well as the effects of driver assist functions. The results can also be used for an objective evaluation. This can be done by identifying mathematical descriptions of the steering dynamics. Basis for the system identification can e. g. be simple step responses of the force signal at the steering. By comparing the parameters of the identified descriptions with the subjective results of the "Driver-in-the-Loop" test, correlations between both can be determined. In this way, criteria can be obtained, that can mathematically be tested and at the same time indicate subjective properties. A simple subjective property can be "the steering feels good". The results of the comparison can e.g. be used to include subjective criteria in End-of-Line tests for high numbers of components.

Essentially, the test with manual steering represents a reduction of the virtualization degree compared to the system in figure 5, since the driver model is replaced by a real person. In the following chapter, cases are described where the degree of virtualization is actually increased.

5 Increasing the Degree of Virtualization (HIL and SIL)

So far, this text has considered the testing of complete steering systems in a virtual environment (see figure 5). One advantage of this approach is that realistic tests are possible even when no test vehicle is available. If the applied virtual environment is sufficiently realistic, valuable and meaningful test results can be obtained.

However, the virtualization of the test system can even be extended. This makes sense e. g. in early development stages, when the ECU of a steering system already exists, but the remaining hardware part of the steering or the corresponding mechatronic test bench setup is not yet available. The non-existing, or for lack of test bench not testable physical part of the steering can in that case be replaced by a model. As shown in figure 6, the input values of the respective steering model are the set point values that were originally fed to the test bench load and motion control (see figure 5). With this setup, an ECU test can be carried out. Furthermore, also the modeled part of the component can be tested regarding behavior and qualification as part of the complete steering system. The significance of the results depends again strongly on the accuracy of the implemented model.

Please note that in practice it may not always be feasible or sufficient to test the ECU completely separated. In that case, e.g. the whole module consisting of ECU and assist motor can be tested. The steering model and signal interfaces have to be modified accordingly.

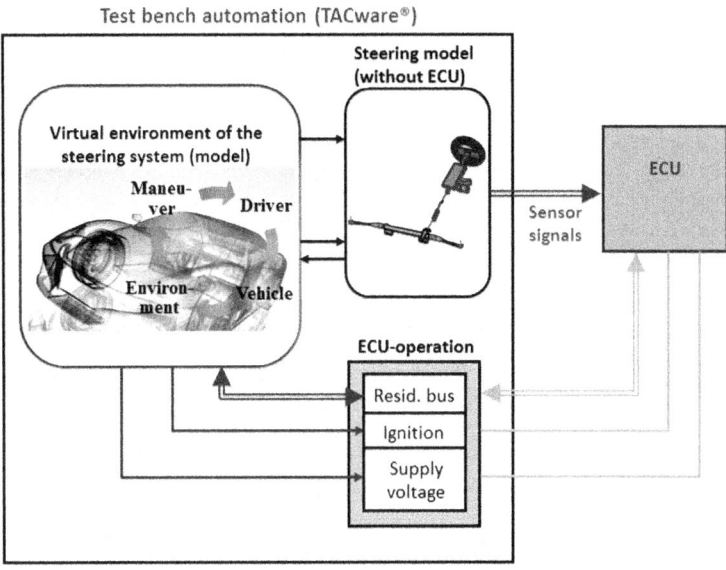

Figure 6: Test system with steering model[3]

In principle, figure 6 shows a HIL test bench (see section 2.2), but with a complete virtual environment and the same full extent of test functionalities as the mechatronic test system described in chapter 3. For the practical application, this means that for both cases the same test bench automation system can be used. I. e. the same vehicle models, test sequences, etc. can be applied consistently for different development stages. Such a unified testing is desirable for reasons of quality assurance.

The same automation system can furthermore be used for cases when also the hardware of the ECU is not yet available. In that case, the ECU software can be executed on the same computer as the test bench automation. By doing so, also SIL (Software-in-the-Loop) tests can be carried out in a consistent way.

---

[3] ©ag visuell – fotolia.de

6 Summary

This contribution points out the chances and possibilities that arise from including virtual environments in component test bench systems. First, the "classic" testing system is explained. Such a system is already quite sophisticated for contemporary active components, but neglects the interaction of component and surrounding vehicle. Then, the approach of setting the component in a loop with a virtual environment via the test bench is presented. With the given method, testing in the same way as in an actual vehicle becomes possible.

After that, a method is presented, where the steering wheel is actuated by a real person. By doing so, information on the haptic feeling of the steering is received. With system identification methods, mathematical descriptions of the steering behavior can be generated. By comparing these descriptions with the results of the subjective driver tests, criteria for an automated test of subjective properties can be generated.

Finally, test systems are discussed where not only the environment of the steering system is virtualized, but also parts of the steering itself are replaced by a model. The results are HIL and SIL test systems with a complete virtual environment and the same set of test functionalities as the full mechatronic steering test bench. Sufficiently high modeling accuracy assumed these systems can not only be used for ECU and software test, but also for the evaluation of the modeled steering parts. A test of the whole integrated steering system in early development stages becomes possible.

Ergonomical sequencing with state of the art tube lifters

Tilmann Hilbert, J. Schmalz GmbH

1 Motivation

The development of lifting assistance devices that moves the parts from one position to another position independent of the geometry of the part is an important task within the industry. Solving this task supports companies of every size. One of the most important indicators for the use of handling devices is the acceptance of the operators. For example, no operator will use a manipulator, if the lifting assistance device is slowing the operator down or the use of the tool is too complex. J. Schmalz GmbH was looking into developing a standard product that has the maximum freedom and flexibility in gripping the product and has at the other hand all the advantages of a standardized product.

Quick change adapter for maximum standardization and flexibility (1)

Due to the fact, that the requirements of a handling system and an assembling system diverge, Schmalz was looking into the tasks of a handling system for logistical applications (Pick & Place) only. The requirements of an assembling system that could not be handled with a tube lifter (or at least that could not be handled optimal) are requirements like handling with high demands to the positioning (tolerances) or high demands to momentum forces e.g. if you have to reach into a car. Considering ergonomic aspects getting more and more important in the daily work business, it is surely short sighted only to focus on assembling areas. Logistical work processes need to be considered as well.

2 Methods

The used methods can be clustered into three different groups: Design, Assessment and Validation. For the design of the lifting device, the Ramsis system was used. Ramsis is based on a large body dimension database and supports the design engineer by determining sizes for human interfaces. The Assessment was done with EAWS Method and with summarizing observations. Last but not least, there was done an evaluation with a 3D measuring handle.

3   Fundamentals of manual handling devices

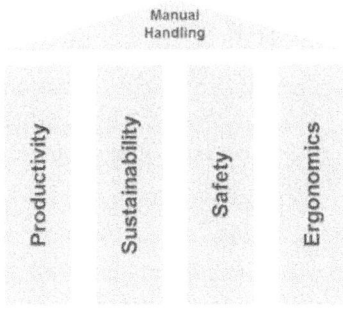

Manual Handling fundamentals (2)

In order to have a manual handling device, that is successful, Schmalz believes in four fundamentals that have to be fulfilled. Following sections will explain this strategy.

3.1   Productivity

Productivity of the overall process and the influence of the lifting device are driven by throughput, uptime and yield. Especially when the handled good is not too heavy and could be lifted by the operator himself without any support, it is mandatory that the lifting aid is not influencing the throughput in a negative matter. This means, that the handling device should be as fast as the operator without the handling device. The contemplation period is one shift.

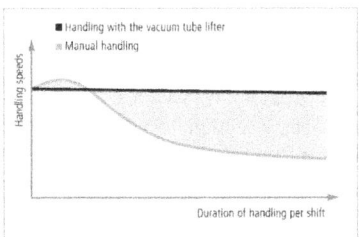

Comparison of handling speeds between tube lifter and manual handling (3)

Looking to the uptime of a system in an abstract way, there is a correlation of number of mechanical parts to the uptime. By having "only" an empty hose, a handle with a defined leakage and a vacuum generation, there is not much that can break. Concerning the Yield, there was no further evaluation, because there is no clamping or scratching, that could damage the part.

3.2 Sustainability

Schmalz views sustainability as an integral system comprising economic success, ecological responsibility and social engagement.

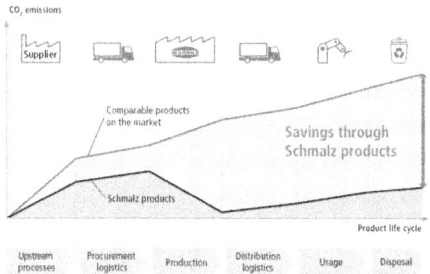

Development of CO_2 emissions during the product life cycle (Product Carbon Footprint) (4)

The CO_2 emissions that occur during the creation of a product are in large part subject to the design features of a product. For example, the materials and production techniques that are used and the energy that is later required are already determined during the development and design of a product. The main factors that influence the product carbon footprint have already been determined long before production of the first part begins.

3.3 Safety

The safety concept of a tube lifter follows DIN EN 14238:2010-02 "Manually controlled load manipulating devices". Section 3.9 is defining the tube lifters as a separate category of manually controlled load manipulating devices. Especially chapter 5.3 is taking care of vacuum tube lifters. Holding force, power failure, pressure failure and mechanical requirements to the tube and the fastening of the system are the main concerns and the safety concept respects those issues, by using a reservoir with a safety valve. The forces are correlating with the actual pressure difference and the area that is used for sucking. Schmalz is calculating the dimensions of all suction cups at least with a safety factor of 2.

3.4 Ergonomics

In order to have an ergonomic Work system, you have to take care of the tool, the workspace, the surrounding, the task and the organization of work. In 3.4.1 the efforts in improving the immediate interface with the human are explained. But having an ergonomic tool does not mean, that the work is ergonomically, too. For example having a screw driver with a perfectly designed handle, but working overhead or with a bended back is still not an ergonomically work process. How this could be done is explained in 3.4.2 JumboFlex ergonomics in lifting.

3.4.1 Design of human-machine-interface

The design of the human-machine-interface of the tube lifter series at Schmalz was basically a result of an evolution. This evolution started at the end of the last century in 1999, when Schmalz had developed the first tube lifters called Handy.

Operating unit in 1999 (5)

The design was driven by functionality and known components by the time. Therefore it was a product, that was improving the ergonomics at the work process by taking over the work load, but for the product itself, there was a potential to improve under ergonomic aspects.

In 2009 Schmalz had realized, that the product ergonomics and the operating concept could be upgraded. Therefore Schmalz had worked together with well recognized institutes in the field of Ergonomics. With the Ramsis model, the optimum of the diameter of the handle itself was determined. The angle for the wrist was turned and the materials were picked under the consideration of haptic influences. The wrist should be in a natural position. The immediate access to the buttons in every position (pressed button and released button) without changing one's grip was one of the tasks for the design engineer. It was important to design a user interface, that could be used by people that are left handed and by people that are right handed at the same time.

Definition of the handle and push button (6)   Angle of the wrist (7)

Workload assessment is considered a crucial criterion, especially from an occupational safety and health perspective. All those influences are "mechanical" influences, but the influences of the operating concept are at least as important as the mechanical influences as well. The handling device will definitely not be used, if the concept is to complex. Therefore one of the tasks to the engineers was to design something that could be operated by one push button. The intelligent Joystick for the control panel was the result.

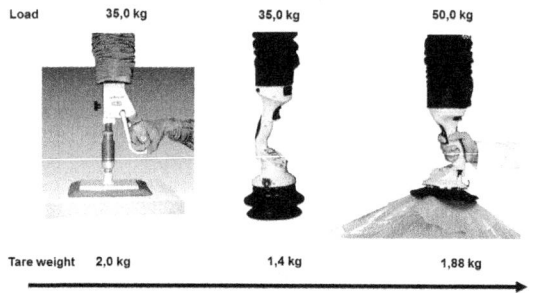

Evolution of the one handed handle design (8)

Human factors and ergonomic design strategies (e.g. task orientation) and principles (e.g. compatibility) to be considered during the design of work systems. This also applies to the design of effective, efficient and safe human-system interaction.

3.4.2 JumboFlex for an ergonomic handling work space

On the contrary to the product approach is the system approach not only looking on parameters of the product itself, but looking in how the product is interacting in the surroundings. In this case the specific workspace for handling goods.

To have the tool in a proper height, Schmalz integrated a height adjustment wheel to the tube lifters. With this wheel you can adjust the height of the workspace, where you pick up the tube lifter. But it is still important, to organize the height of the picking and placing location. With the tube lifter you are able to reach a wide range; you have the freedom to bend your back or to reach above shoulder level. It is still mandatory to organize the complete workplace in an ergonomic way. The work process could be optimized by combining the tube lifter with e.g. a scissor table.

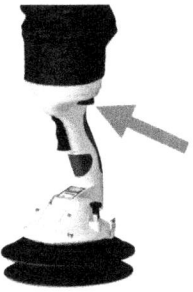

Adjustment of the height (9)

The main requirements for ergonomic tools and supplies for work compared to the new PINA studies are the noise emissions, the push back forces, vibrations and the load. The load of the JumboFlex itself is below 2kg for the operating unit and below 5kg for the complete lifting device. Compared to traditional manipulators, this is by far more light weighted. This has not only an influence to the movement in vertical direction. The load has an influence to the horizontal movement as well. You could reduce the forces for pushing and pulling by optimizing the crane systems. With the EAWS method, a workspace with a Jumbo Tube lifter was evaluated, and you can see very clearly the improvement in ergonomics. The workload is reduced for the worker.

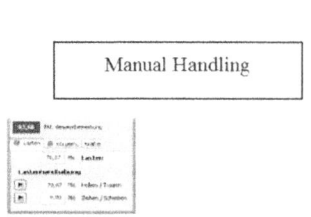

Evaluation of a Jumbo Tube lifter by EAWS method (10)

3.4.3 Rail Systems as a complementary product

Having a lifting device, that could be easily used in the lifting direction, but is mounted to a manual crane system with heavy rails, could be worthless, because moving the system is too heavy. Therefore it is necessary to have an ergonomically optimized rail system as well. The task is, to reduce the weight, increase stability, avoid vibrations and look for a perfect fit in terms of material pairing because of the forces of the starting torque.

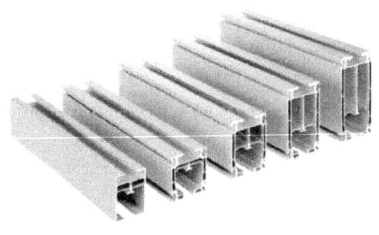
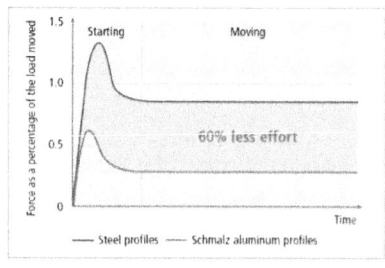

Aluminum light weighted Rail System (11)   Comparison of handling speed (12)

3.4.4 Overall examination of the workspace

In order to judge the work system, measurements with a 3D measuring handle were done.

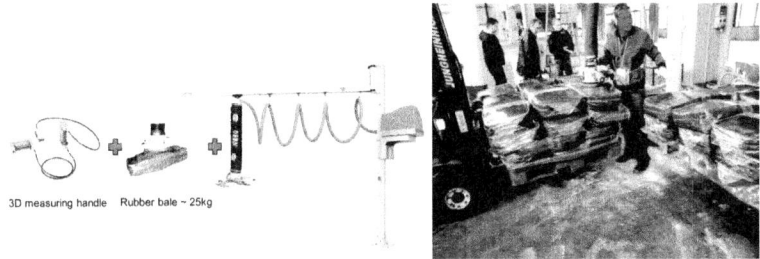

Setup of trials (13)   Measuring the handling forces (14)

The handling of rubber bales (~25kg) was chosen for the examination. The 3D measuring handle was mounted immediately to the operating unit.

4 Results and Discussion

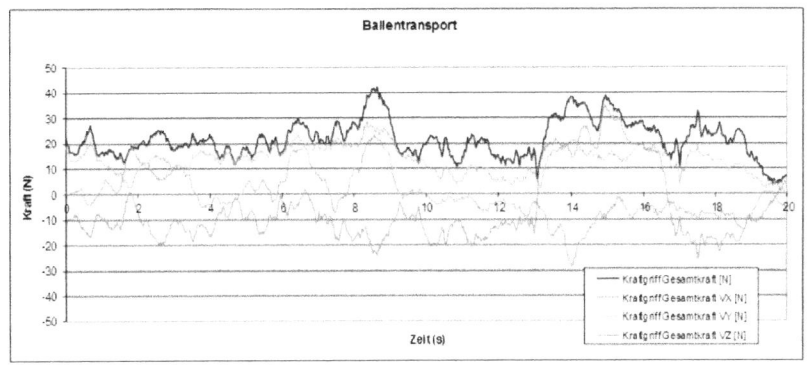

Measurements of a 3D handle (15)

The measurements show, that the forces are not exceeding 40N. The design of the JumboFlex and the aluminum rail system were both tweaked concerning the weight. The evaluation of workspaces that have this setup compared to conventional workspaces shows, that the ergonomics could be improved from red to green (83,56points compared to 12 points). By using Ramsis Software, it was possible to improve interface of the operating unit. Because of the functionality of the quick change adaptor the actual gripping device could be changed. This leads to a universal use of the lifting unit independent from the geometry of the part. The JumboFlex is a universal lifting device that improves the ergonomics of many workspaces.

Future ergonomics tools – From the prototype to the serial product by comprehensive product optimization

Tanja Schembera-Kneifel, Mathias Keil

Industrial Engineering Methoden, Audi AG

Abstract

Due to the demographic change, the requirements of generation Y as well as the rising proportion of women, the automobile industry is facing new challenges. The reduction of physical stress and the improvement of the workplace attractiveness are of special importance. In order to face these challenges in an active way, the AUDI AG initiated the ergonomics strategy "Wir für uns. Aktiv in die Zukunft" in 2013. The goal of the field of action „The workplace of the future" within the ergonomics strategy is the identification, pilot testing and development up to serial production readiness of new ergonomics tools. The active involvement of affected employees is highly important for the acceptance of ergonomics tools. In this context Audi started diverse projects in 2014 which will be discussed in detail below.

1 Starting point

In order to face new challenges caused by the demographic change, the requirements of generation Y as well as the rising proportion of women strategically and comprehensively, the AUDI AG initiated the ergonomics strategy "Wir für uns. Aktiv in die Zukunft" (figure 1) in 2013. „Wir für uns" means that conditions are created jointly by optimal framework conditions to meet the corporate requirements and to ensure health, commitment and motivation of the employees (Unger et al. 2014). The goals of the ergonomics strategy are pursued by diverse projects within six fields of action.

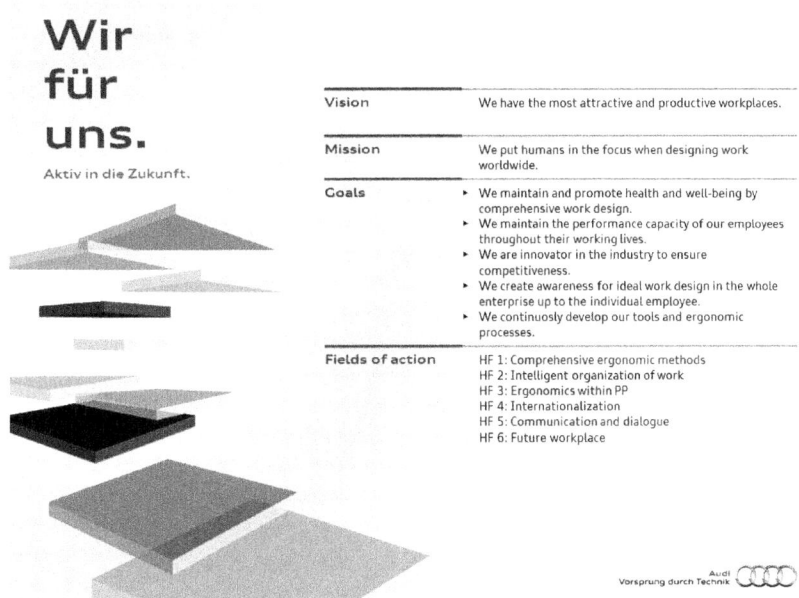

Wir für uns. Aktiv in die Zukunft.		
Vision		We have the most attractive and productive workplaces.
Mission		We put humans in the focus when designing work worldwide.
Goals		▸ We maintain and promote health and well-being by comprehensive work design. ▸ We maintain the performance capacity of our employees throughout their working lives. ▸ We are innovator in the industry to ensure competitiveness. ▸ We create awareness for ideal work design in the whole enterprise up to the individual employee. ▸ We continuosly develop our tools and ergonomic processes.
Fields of action		HF 1: Comprehensive ergonomic methods HF 2: Intelligent organization of work HF 3: Ergonomics within PP HF 4: Internationalization HF 5: Communication and dialogue HF 6: Future workplace

Figure 1: „Wir für uns. Aktiv in die Zukunft." – the ergonomics strategy of the Audi AG

The content of the first field of action is the identification of the framework conditions for attractive workplaces and well-being at Audi. Moreover, it includes making relevant criteria measurable and controllable within the company by using a practical tool. The second field of action focusses on the optimization of existing levers in terms of work organization to ensure attractive working conditions. The objective of the third field of action is the further improvement of the working conditions by target-oriented product modifications within the scope of the product development process. Within the fourth field of action the AUDI AG pursues the objective of globally uniform ergonomics standards. The fifth field of action is targeted on strengthening the presence of excellent ergonomics within the company and beyond corporate limits. For this purpose the ergonomics strategy has been exported to the Audi locations in the context of a kick-off event. By means of an interactive market place the employees got the opportunity to test examples of different trades and production areas of all Audi production sites and to get insights into the goals and the content of the fields of action of the ergonomics strategy.

The youngest field of action is represented by the sixth field of action. It follows the dynamics connected with the term industry 4.0 and the ongoing digitalization of the production which Audi is facing by the "smart factory" approach. Smart factory is the intelligent interaction of machines (artificial intelligence) and humans using the respective strengths in an optimal way. Robots assist humans taking over monotonous, physically exhausting and stressful tasks. At the same time humans can perform value-adding activities through creativity and innovative skills (Waltl 2015). The smart factory will fundamentally change the workplaces in the production sector. To accompany these changes with regard to ergonomic influences is the task of the field of action „the workplace of the future" of the ergonomics strategy.

2 Objectives for the future workplace

Within this field of action ergonomic tools are developed by active employee involvement from the prototype up to the serial product in the course of different projects. In this context the following objectives are pursued:

– Identification, pilot testing and extension of ergonomics tools for the production of the future.

– Reduction of physically and psychologically caused down time.

– Increase of the well-being of the employees.

– Integration of employees with constraints relevant to the production as well as performance-related constraints.

– Monitoring of ergonomically relevant projects in the context of the smart factory.

A standardized process for the analysis, the pilot testing and the implementation has been established for the development of innovative ergonomics tools (figure 2). The process is accompanied by an interdisciplinary team consisting of experts from the affected departments and if necessary external partners.

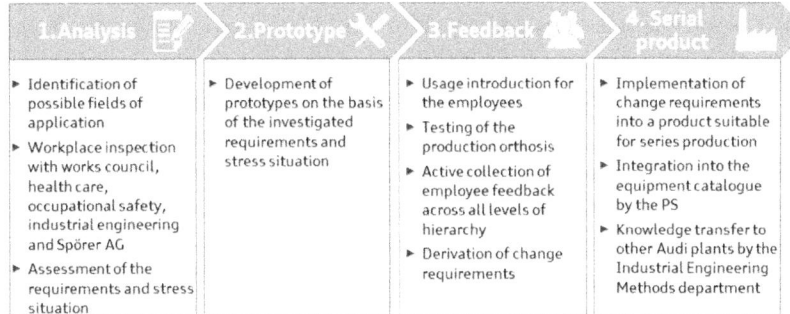

Figure 2: Ergonomics tools process from the prototype up to the serial product

In the phase of analysis innovative ideas are initially identified and possible fields of implementation tested. In the next step the requirement and stress situation is analyzed in order to develop a prototype based on this information. Afterwards, the prototype is pilot tested in the production and feedback of the employees actively collected. The feedback directly influences the further development of the prototype. Due to the early involvement of the employees the practical requirements and demands of the employees are considered in an optimal way resulting in a high acceptance of the ergonomics tool. After the successful implementation the serial product is standardized and exported to all Audi locations.

3 Examples for the implementation for the future workplace

Since the end of 2014 the first ergonomics tools – Chairless Chair and production orthosis - which will be described in more detail in the following, pass through the process from the prototype up to the serial product. Moreover, an innovative ergonomic tool for manual handling of loads worn on the human body is to be developed within a new project in collaboration with the Fraunhofer IPA.

3.1 Chairless Chair

The Chairless Chair is a high-tech construct made out of titanium which allows employees to sit down temporarily without using a chair. At the same time it ensures an improved body posture for assembly processes and reduces the stress on the legs.

In collaboration with the Swiss start-up noonee AG Audi further developed the Chair-less Chair. As visible in figure 3 the Chairless Chair is a support structure worn on the human body which is worn on the backside of the legs and attached by belts to the hip, the knees and the ankles. A plastic seat enabling the air circulation supports the buttock and the thighs. The exoskeleton is equipped with joints in the height of the knees and consequently dynamically adjustable to the body height and the desired seating position. The body weight is transferred into the ground through these adjust-able elements. The Chairless Chair itself only weighs approximately three kilograms.

Figure 3: Chairless Chair in the assembly

The employee wears the Chairless Chair during the work process like a second pair of legs. For many activities in the production, it enables to sit down in an ergonomically correct position instead of standing – even for short assembly intervals. At the same time the high-tech support structure improves the body posture and reduces the stress on the legs. With the use of the Chairless Chair Audi expects a continuous improve-ment of the ergonomics for assembly activities, new possibilities for placing employ-ees with impaired capacity and an increase of the well-being of the employees.

In February 2015 the Chairless Chair was pilot tested in three divisions in Neckar-sulm. The employees, who could only perform their tasks while standing on the ground, experienced considerably less physical stress and fatigue thanks to the possibility to alternate between sitting and standing while installing the components. The feedback of the employees was recorded and included into the input for the further development. In a second pilot testing phase, employees in Ingolstadt could test the enhanced prototype in September 2015. Due to the positive feedback an eight to ten week long testing is planned.

3.2 Production orthosis

The production orthosis is an orthotic glove for the production which reduces the pressure stress for many assembly activities by more than 50 percent. The decisive reason for the development of the production orthosis is the partly increased strain on wrists and palms in certain assembly sectors. One example is the assembly of the trim strips when pre-assembling doors where about 300 trims are to be installed in one workplace per shift.

In order to reduce the stress for the employees, Audi developed in collaboration with the orthopedics technology specialist Spörer AG from Ingolstadt a joint-protective orthosis. As a result the hand is opposed to less strain but keeps its tactile perception and maximum flexibility.

The production orthosis was developed in close collaboration with the direct production employees. An interdisciplinary team consisting of company doctors, occupational safety, planning and production regularly collected feedback of the employees and consequently adapted the gloves to their requirements. As a result a breathable production glove made out of Teflon was created. In the area of the hitting surface it is equipped with an absorbing foam which reduces the forces resulting out of the assembly on the hand and joints. A highly elastic closure (Bandage) stabilizes the wrist joint and has an orthotic effect.

Figure 4: Production orthosis in the pre-assembly of the doors

In order to identify further fields of application and to collect feedback from the long-term testing, 1,000 production orthoses have been handed over to all Audi production sites. The feedback from the Audi production sites will be collected in the beginning of 2016. As soon as all information from the feedbacks is evaluated the production orthosis will be transferred to mass production. The procurement and distribution will be executed by the in-plant equipment warehouse.

3.3 Ergonomics tool for the handling of loads

Thanks to large investments by the Audi AG into the production, most workplaces could be designed ergonomically and body-protective for the worker. In some areas there are ergonomic challenges which could not be meet sufficiently so far. For that reason a workshop was conducted in collaboration with the Fraunhofer Institut für Produktionstechnik und Automatisierung (IPA), the orthopedics technology specialist Spörer AG from Ingolstadt and an interdisciplinary team of the AUDI AG. Content of this workshop was the analysis of the ergonomic stress in three workplaces which were not directly in the focus of ergonomic serial optimizations:

1   Commissioning in the distribution center (manual tire handling into the vehicle of the customer)

2   Mounting and demounting of the wheels during test drives (manual mounting of the wheels)

3   Loading of crash vehicles with weights in the context of crash tests

All three workplace have similar stress features – high loading weights partly have to be manipulated in non-ergonomic body postures. Space restrictions on the spot as well as limitations due to complex product geometries make the application of common tools for the manipulation of loads hard. Goal of the workplace analysis was the description of requirements and possibilities for ergonomic tools for the handling of loads and the development of ideas for solutions. For this purpose individual workflows were examined and analyzed on spot or by video. Additionally, employees working in these workplaces described their experiences. Out of all this information the requirements on ergonomic tools were weighed and summarized. The stress was visualized by the participants of the workshop. As visible in figure 5, it became clear that the stress occurs in similar body areas.

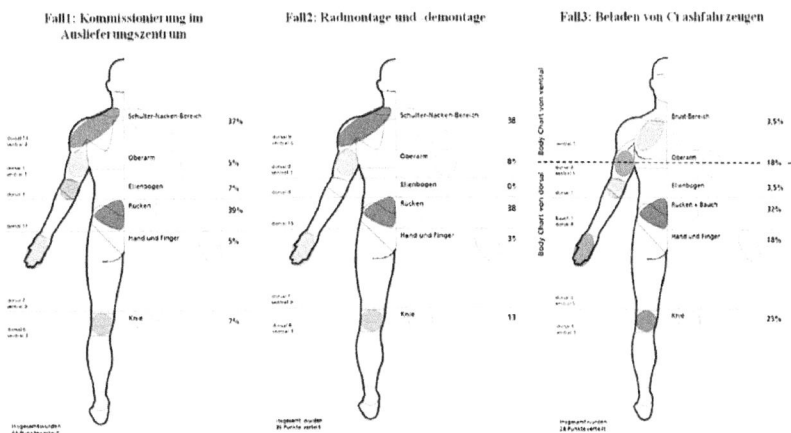

Figure 5: Valuation of the stress incurred in the three cases considered

All three of the cases described are ergonomic challenges that are not suited to standard solutions. Possible approaches range from a simple revision of existing workflows to the development of completely new processing systems that can be mounted on existing equipment, or are free to move or, ultimately, can be worn on the human body.

As the next step, solutions have to be specified in greater detail and pilot testing carried out on prototypes according to the process shown in diagram 2.

4 Summary and outlook

Health and well-being of the employees is one of the most important goals for Audi in the production. In the context of the smart-factory ergonomic tools for the workplace of the future are currently developed, pilot tested and transferred to mass production. The processes implemented for this purpose from the prototype to the serial product has proven to be ideally suited with the example of the presented examples. Due to this, Audi plans the pilot testing and handing over to series of further ergonomic tools in the next year.

5 Literature

1. Unger K, Becker M, Keil M (2014) „Wir für uns. Aktiv in die Zukunft" – Herausforderungen an die Ergonomie-Strategie der AUDI AG. In: Gesellschaft für Arbeitswissenschaft (Hrsg) Gestaltung der Arbeitswelt der Zukunft. Dortmund: GfA-Press.

2. Waltl H (2015) Dialoge Magazin Smart Factory: Audi denkt Produktion neu. Die Zukunft der Fabrik hat begonnen. Intelligente Systeme, innovative Technologien, effiziente Strukturen. Audi AG.

Loading of the musculoskeletal system in automotive assembly – The potential of an applied biomechanics analysis concept

Christian Lersch
Velamed GmbH – Science in Motion

Lukas Hausmanninger
Audi AG

Prof. Dr. Wolfgang Potthast
Deutsche Sporthochschule Köln

1 Introduction

The physical work in automotive assembly plants can often be characterized by repetitive movement patterns. In combination with a continuous mechanical loading of the biological tissues that are involved in the movement patterns, overuse injuries may occur as a result. Therefore, it is the aim of modern ergonomic approaches to a) identify the tasks that are involved in overload accumulation, b) to quantify the (induvial) biomechanical loading of the subjects in the assembly situation and c) to develop strategies to counteract overuse injury accumulation. This presentation reviews an applied biomechanical analysis concept behind the above mentioned background. Furthermore, it summarizes the results of a study investigating the biomechanics of overuse accumulation in different clip assembly techniques.

2 Applied Biomechanics Analysis Concept

Biomechanics is defined as the science that analyzes forces acting upon and within a biological structure and that determines the effects produced by such forces. Traditionally, a biomechanical analysis consists of a kinematical, a kinetical and a neuromuscular (Electromyography) component. The combination of these components allow for a detailed insight into the loading situation of the musculoskeletal system. Due to the complexity of these measurement techniques, the described analyses usually involve a laboratory site rather than the investigation in the field (e.g. at the assembly plant). The following concept is designed to allow for investigations in the field and therefore in real life working situations. Wireless data transmission of the applied sensors is used to even reduce the effect of constraints of a measurement situation. The biomechanics of overuse in clip assembly techniques was investigated using a kinematical representation of the upper body and the elbow and wrist joints in particular, electromyography of the lower arm muscles and reaction forces of the clip application point as well as the ground reaction forces acting upon the subjects.

2.1 Kinematics

The kinematics component is represented by a wireless IMU system in the current experiment. It consisted of the segments hand, lower arm, upper arm, cervical spine, thorax and lower bag. Furthermore, a synchronized video documentation was applied.

2.2 Kinetics

The ground reaction force at the clip application point was measured with a three-dimensional load cell while the reaction forces under the feet of the subjects were measured with a three-dimensional force plate.

2.3 Neuromuscular Component

Electromyography was applied to the M. flexor carpi radialis and the M. extensor carpi radialis longus.

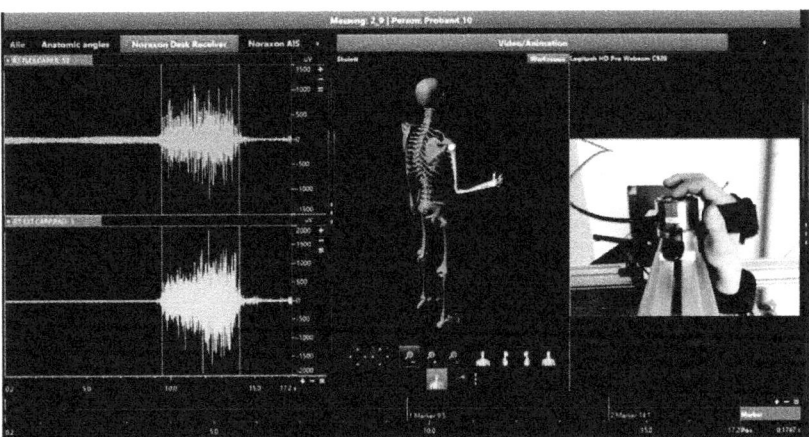

Figure 1. Synchronized data collection in real life working situation (from left to right: EMG, IMU kinematics, video documentation of the clip application point).

2.4 Outlook

The Applied Biomechanics Analysis concept allowed for the investigation of different clip assembly techniques in real life working situations. The ongoing data analysis revealed first insights into the musculoskeletal loading of the subjects and may therefore contribute to a better understanding of overuse injury etiology in automotive assembly processes.

New approaches for analysis in ergonomics: From paper and pencil methods to biomechanical simulation

Florian Blab, Okan Avci, Urban Daub, Urs Schneider

Fraunhofer Institute for Manufacturing Engineering and Automation IPA

1 Introduction

Ergonomics and prevention of work-related diseases are an increasing issue in many companies. The demographic change, "War for Talent", political demand, financial benefits and upward healthcare expenditures are the assumed reasons, why an increasing of this issue is being expected by German companies (Booz & Company 2011). In Germany musculoskeletal diseases (MSDs) are the most frequent reason for incapacitation for work (DAK Forschung 2014).

To lower injury rates and MSD incidences, the goal of ergonomics is to design work tasks fitting to the worker, respecting the capabilities and limitations of the human body.

To get the most accurate understanding of the loads on the human body, modern approaches make use of technological measuring instruments, in order to derive specific interventions with maximum performance.

2 Analysis in ergonomics: „Integrative process chain"

2.1 Definition of ergonomic challenge

Any intervention will only be accepted by the company and the workers, if it responds to the needs and to the local conditions. Also it may not hinder the work process to be a sustainable improvement. Thus, a close and open collaboration with the workers in the department, the foreman, works committee, administration and company doctor is strongly recommended. The definition of ergonomic challenges is hence preferably dealt with in the form of a workshop with members of all disciplines to gather all requirements into the developing process.

The success of an ergonomic intervention can only be evaluated on the basis of quantitative parameters. Measured variables allow the comparison with reference values and provide economic justification for investments in ergonomic issues.

2.2 Analysis of working movements

2.2.1 Standard ergonomic evaluation scores

To figure out, which work places are most relevant for ergonomic adaptations, a vast variety of ergonomic assessment tools is available. Common assessments aim on the analysis of postures during the working process to generate a comparable score, e.g. Ovaka Working Posture Analyzing System – OWAS (Salvendy 2001). Most of these assessments combine postures with physical loads on the worker, distances to walk or lifting of weights, e.g. Ergonomic Assessment Worksheet EAWS, Key Indicator Method Manual Handling Operations (Leitmerkmalmethode Manuelle Arbeitsprozesse), WISHA Lifting Calculator, NIOSH Lifting Equation and many more.

These tools are very common, because they don't need much equipment besides the assessment sheet and a pencil. On the other hand this fact makes these methods also vulnerable to errors (Denis et al. 2015). A general source of inaccurate data collection lies in the method of observation (Spielholz et al. 2001; Jones und Kumar 2007). The biggest discrepancies arise in the estimation of the applied forces and the posture (Steinberg et al. 2007; Steinberg 2012). Thus, these methods are very dependent on the actual rater and their interrater reliability might vary a lot.

Paper and pencil assessments can also be supported by biomechanical recorded data. Specified post-processing scripts in Matlab can for instance export and display EAWS-relevant data and parameters in a useful manner. This combination of a traditional ergonomic tool with biomechanical analysis data is a practical application that allows implementing the results into an existing ergonomic workflow. Furthermore, this kind of computed analysis speeds up the overall rating process and raises its reliability.

2.2.2 Biomechanical measurements

Quantitative analysis of human movements is frequently used in different fields and applications, for instance in sports sciences, medicine, physiotherapy or animation film industries. There is a wide range of measuring systems to determine kinematics, kinetics and possibilities to calculate internal loads and forces to the human body in a very accurate way.

With regards to applications in ergonomics, the challenges are:

– to simplify complex working tasks and simulate it within a laboratory area,

– to be able to adapt and use high-tech measuring systems for tests in the authentic working environment, and

– to create an economic and task-specific test setup that allows the recording of highly accurate motion and load data.

Within the last two years, Fraunhofer IPA equipped their laboratories and biomechanical systems tremendously to meet these challenges.

Two laboratory sites are available for movement analysis at Fraunhofer IPA. The Motion Lab with an area of 90 m² and a 12 m walkway is used for complex biomechanical measurements. Within the Training Factory Hall, a laboratory field is equipped for movement analysis measurements with larger test setups (see figure 2).

A modular floor system (patent pending by RK Rose+Krieger) allows a highly flexible integration of force plates to determine ground reaction forces and moments. The modular floor can be adapted to various test settings within a few minutes. Furthermore, measurements in working environments are possible with any changes to the original floor (Pat.no.: WO2015139734A1).

Kinematics in working environments is often difficult to determine. Therefore, a combination optical motion capture systems, inertial sensors and specified electromechanical sensors systems were used at Fraunhofer IPA Motion Lab to measure 3-dimentional motions:

– Qualisys Motion Capture System with 19 IR-cameras with passive reflective markers (Qualisys AB, Goteborg, Sweden)

– Xsens MVN Awinda wireless inertial sensor full-body system with 17 sensors (Xsens Technologies, Enschede, Netherlands)

– Delsys Trigno biaxial goniometers to determine knee, shoulder, elbow and wrist joint flexion angles (Delsys Inc., Natick/MA, USA)

Case-specific real time data streams and post-processing algorithms were used to analyze and to report these data in an adequate way.

Most working tasks were performed with a body-environment interaction by contacts of one or both hands with the environment (e.g. tools, work bench, assembling parts, …) that results hand contact forces. Reliable and valid calculations of the loads within the human body (joint torques and forces) are only reasonable if three-dimensional forces of any contact point can be determined by using specified and custom-made 6-DoF force sensors. Fraunhofer IPA uses different sorts of force sensors and designed adapted testing rigs, for instance a force-sensor box for lifting tasks. Pressure sensor systems provide further information about the force distribution within the contact area (e.g. Novel PLIANCE, Novel GmbH, München, Germany).

Physiological data such as muscle activity, heart rate and transpiration can be recorded with wireless systems. Due to their compact design and wireless data recording setups, these systems can be used for research topics in a laboratory field as well as for long-term evaluation studies during real working shifts:

- Delsys Trigno Eletromyography system with 16 sensors (Delsys Inc., Natick/MA, USA)
- Equivital Senor Belt with ECG, temperature and skin conductance sensors (Hidalgo Ltd., Cambridge, UK)

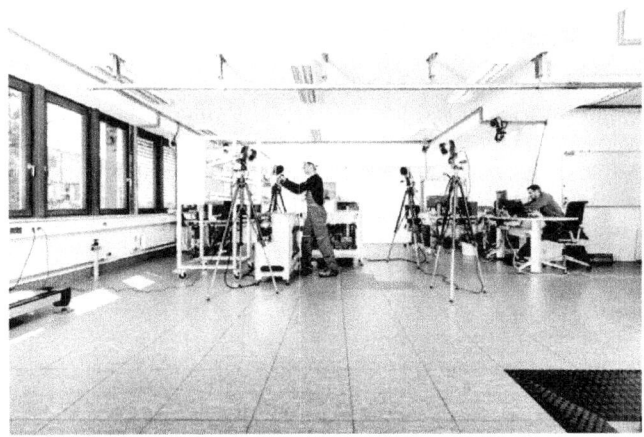

Figure 1: Fraunhofer IPA motion lab

Figure 2: Fraunhofer IPA motion lab test setup in the training factory hall

2.2.3 Human loads by inverse dynamic calculations

By using accurate data of body segment motion and contact forces, human joint forces and torques can be determined.

There are different calculation methods that can be used for determination of joint forces. Besides various mathematic approaches, main difference is the kind of human body model that is applied for the calculations.

Simple human models, like the HANAVAN model (cp. Hanavan, 1964) create rough geometric objects (e.g. cones, cylinders, ellipsoids) with a defined length, a distal and a proximal radius with inertial properties for each body segment. The whole body is subdivided in 15 segments. The length and mass of the segments is scaled based on individual dimensions of each subject. Analysis tools like Visual3D (C-Motion, Inc., Germantown, USA) use such models for clinical biomechanics analysis as well as for performance analysis in sports. At Fraunhofer IPA, Visual3D is commonly used for gait analysis and kinematic movement analysis.

A challenging aspect with regards to kinetic analysis of human working tasks is the appearance of external contact forces that need to be taken into consideration for proper results. Interaction of the worker with tool or materials results mainly in hand contact forces or contact forces with external devices (e.g. exoskeletons). These contact forces have to be added to the used human model model within the calculations of the human body kinetics. Simple body models can only use ground reactions forces (e.g. force plate data). Furthermore, these models cannot be manipulated and adapted to specific environments.

Fraunhofer IPA uses the AnyBody Modeling System to handle complex inverse kinematics calculations (AnyBody Technology, Aalborg, Denmark). AnyBody uses a musculoskeletal modeling approach with the possibility to use complex human body models. Modification of the human body models is also possible. The tool "Solidworks-To-AnyBody" can be used for importing CAD designs and inertial data of objects (e.g. tools), that need to be used for calculations in client-specific measurement tasks.

2.3 Developing process

After the load peaks and the timing when they occur have been identified, specific ergonomic interventions can be developed, according the requirements of the company. They may vary from training advices, adaptions of working tools to wearable motor driven assistive devices. In the end they must be accepted by the workers, pass the practical test in the working process and be able to show better results on a retest to be a successful intervention.

2.3.1 Training advices

With the help of biomechanical data collected, special physiotherapeutic exercises for compensation of one-sided load and activity can be developed. These exercises must be conceived to be highly included into the daily workflow. This is the prerequisite for a high level of acceptance among workers.

If specific MSDs are known to occur more frequently at the considered workplace, causative strains may be specifically sought, with the help of biomechanical parameters. For this purpose, a basic orthopedic understanding is essential.

2.3.2 Adaption of working tools and environment

High repetition rates of hand arm movements, working in non-physiological postures and high action forces of hands and arms increase the probability of the occurrence of MSDs significantly (Bernard 1997; Dragano 2007). Non-ergonomic tools may even increase that problem and amplify the load on the upper extremities. For that reason kinematics and force data of the hands can be used to develop special tools or adaptions on common tools, which realize more ergonomic processes. On the other hand redesigns of tools can reduce inconvenient postures during the working process.

As a basic principle, the working posture has to be considered in the early planning phase of the workflow or even in the design of the work piece. This concept is increasingly being considered in modern manufacturing engineering.

2.3.3 Assistive devices

Next to visual assistive devices as guidance for ergonomic behavior or to show intuitively special exercises, new concepts of wearable assistive devices can be a helpful intervention to lower strains on the workers' body. For example, this can be smart adaptions of standard working gloves to prevent lateral epicondylitis on tasks with high repetitions and big forces on hand and elbow muscles (Schönherr et al. 2014), but also complex motor driven exoskeletal systems, that actively support the worker in his physical work. Depending on the ergonomic challenge and the local conditions in the work, the proper effort has to be defined.

In the development of exoskeletal systems, many things have to be respected. They may not be heavy, uncomfortable or restrict the worker's range of motion. Safety has to be the highest priority. Wearable devices may not raise hazards to the body. Therefore the system itself has to be safe and it must be ensured that there are no shear forces in the joints caused by the device. In addition to the challenge of a reliable and intuitive control, recent developments have to consider these potential sources of danger.

Figure 3: Example of a shoulder exoskeleton kinematic prototype

2.4 Integration into working environment

After the development, the intervention has to pass the practical test in the company. This part is very delicate. It has to be well-prepared and communicated with the workers and all further involved persons in the company. In the very first phase an expert as guide and contact person has to be available for instructions and questions.

This very close guidance in the beginning can be reduced gradually as soon as a certain degree of assurance with the new system ensues. The more all involved persons (workers, foremen, works committee, administration, company doctor) were considered at an early stage of development process, the easier integration into working environment will proceed.

2.5 Evaluation

There are many parameters to measure the success of a solution. In addition to the reduction of sick days, the subjective rating of the workers is an important indicator. For this reason an employee survey about the intervention should take place after a certain time. Also a further biomechanical analysis can be used.

Depending on the complexity of the ergonomic challenge the solution may be a continuous and iterative process.

3 Examples of current projects

3.1 Biomechanical analysis of height adaptive work benches

Height adaptive work benches are commonly used in a couple of workstations. The worker and task specific adjustment of the work bench is a critical aspect that is often unclear to manufacturers as well as users of these stations. Aim of this project was the evaluation of a packing task with various setups (3 box sizes, max. 5 kg handling).

Based on ergonomic and biomechanical methods, a quantitative rating of an optimal adjusted work station versus a standard table in two different subjects was performed (m/1.76 m/65 kg, m/1.96 m/89kg). Motion data was captured by an optical Qualisys QTM system (9 cameras, 240 Hz) with passive-reflective markers (diameter 10 mm). Ground reaction forces of each foot were record with 2 AMTI AccuGait force plates (2400 Hz).

Kinematics of the whole body were analyzed and relevant joint angles (e.g. back sagittal/lateral flexion angle) were calculated. Inverse dynamics of the lower extremities were calculated with a full-body model by the Visual3D (V4.96.11). Each working task was rated based on the EAWS with the software EAWSdigital+ (V2.01.00).

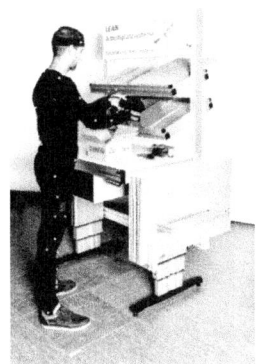

Figure 4: Height adaptive work bench

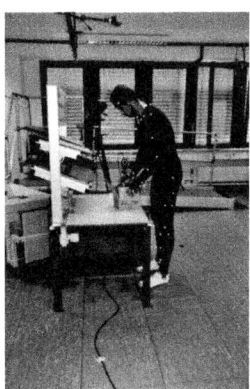

Figure 5: Static work bench

These are the most expressive results of this study:

- An optimal adapted height of the work bench reduces critical back joint angle (sagittal flexion and lateral flexion). Especially for tall workers and varying box sizes (small boxes), height adaptive work benches could avoid critical extreme positions effectively

- A static work bench provokes load peaks in joint torques of the lower extremities as a consequence of inappropriate position, often with a unilateral stress, e.g. up to 100 Nm hip joint torque at static station compared to 50 Nm at optimal adapted height (see figure below).

- An optimal adapted height of the work bench results in a load reduction up to 32.8% (21 EAWS points)

- Besides the height of the work bench, gripping distance is an important factor for ergonomic adjustments of a work station that should be focused on

- A systematic variation of movements in periodic working task can be used for an effective reduction of load peaks

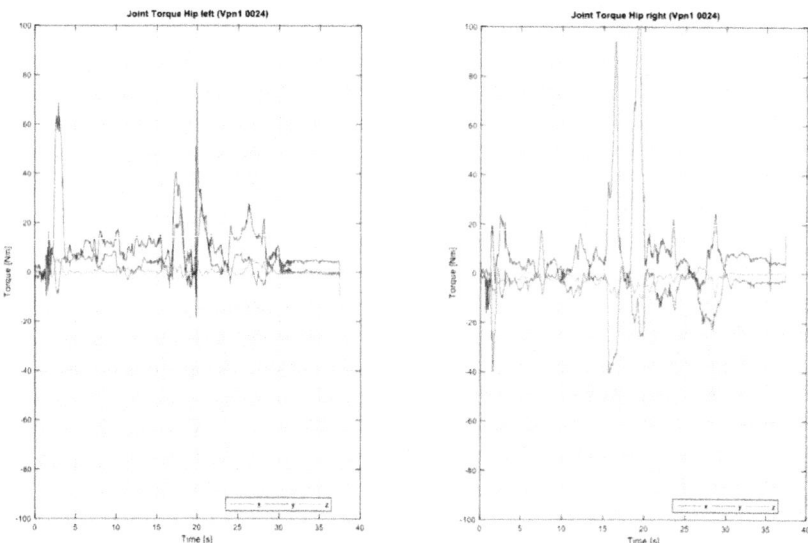

Figure 6: Peaks in hip joint torque at a static work bench

3.2 Evaluation of human body loads by handling heavy weights at a welding machine

Unloading of a welding machine is a stressful task were high weights need to be lifted manually. Due to some restrictions in the current manufacturing (machine design, environment factors, etc.), assistance by handling systems or effective tools is not easy handled without the knowledge about the exact loads to the human body.

In a current industrial project, Fraunhofer IPA established an experimental test setup and a biomechanical analysis based on AnyBody Modelling System (AnyBody Technology) to determine loads with focus on the upper extremities of worker.

The experimental setup at Fraunhofer IPA Motion Lab were built up with all original parts (pallet cages for transportation, shoring bars with 3.0 m maximum length) and the exact distances of the original work station. Motion data was captured by an optical Qualisys QTM system (9 cameras, 240 Hz) with passive-reflective markers (diameter 10 mm). Ground reaction forces of each foot were record with 4 AMTI AccuGait force plates (2400 Hz). For biomechanical calculations, a human body model was modified and a corresponding geometry and inertial data shoring bar were imported using the tool "Solidworks-to-Anybody".

Result of the study will compare differences in handling movements and will focus on the collection of possible interventions.

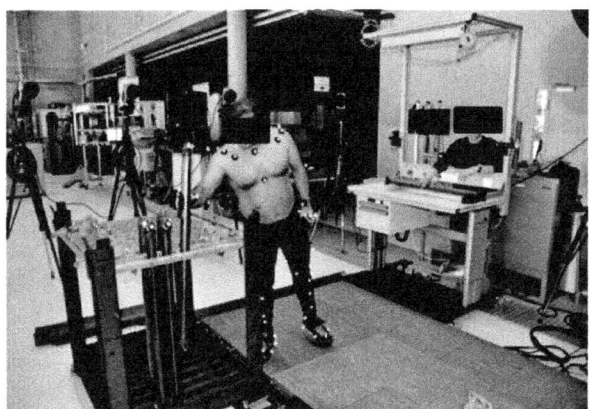

Figure 7: Experimental setup with a worker

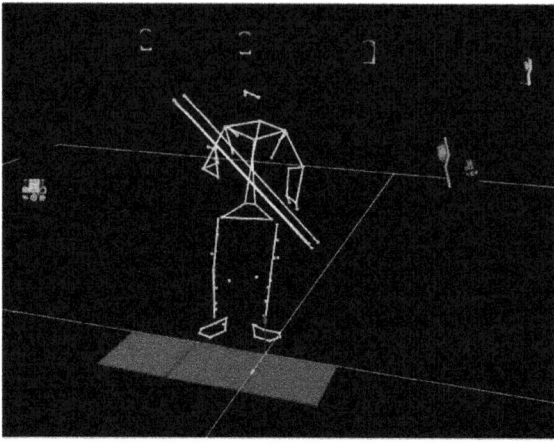

Figure 8: Three-dimensional position data of the passive markers in Qualisys QTM

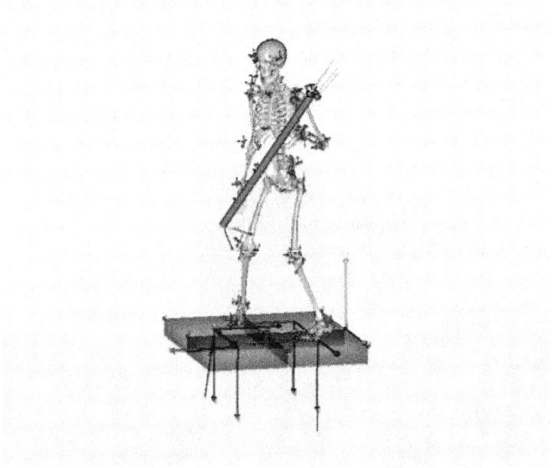

Figure 9: Modified human model in AnyBody with a linked shoring bar

3.3 EMMA-CC: Ergo-dynamic moving manikin with cognitive control

The modelling and efficient simulation of human motions for applications in the fields of ergonomics, medicine and computer graphics is a big challenge. The objective of this Fraunhofer research project EMMA-CC is to develop a digital human model for ergonomic assessment of dynamic motions by validated simulation to support the design of healthier and safer work places in future product development and product planning processes.

The numerical modelling and simulation of multi-skeleton systems with multi-body and finite-element tools is a sophisticated task and an ongoing process of research. However, in the EMMA-CC project, we will additionally combine ergonomic considerations and cognitive capability with numerical simulation. Therewith, the work process can be improved by a better adaptation of prospective ergonomic demands on the workstation. Flexible and employee-specific work processes will be more and more important due to the steady increase in the average age of the employees. The intention of the EMMAC-CC project is to develop healthier work places and to integrate of ergo-dynamic aspects which will help improve the quality of life.

Fig. 10 for example, shows the installation process of a car cockpit. The installation with difficult working environment has different demands on the employees concerning the motion. Using ergonomic assessment with numerical analysis of motion and forces for example on the elbow joint or wrist, it will be possible to define a better motion sequence in order to avoid high loads on joints.

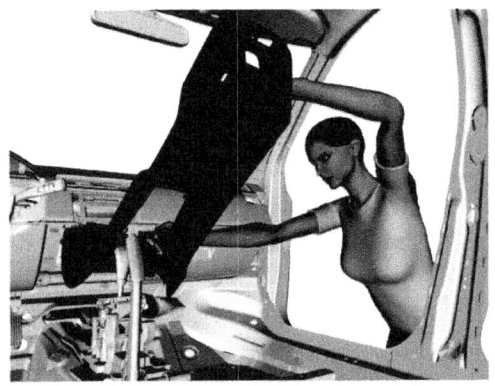

Figure 10: Ergonomic simulation in car assembling

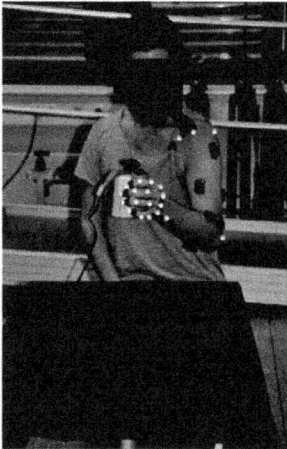

Figure 11: Basic grasping test

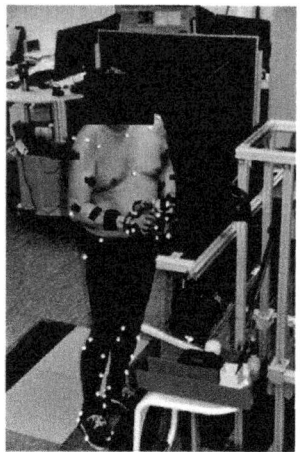

Figure 12: Applied test setup in car assembling

4 Outlook

For a realistic description of limb motion, the respective muscle forces and the result-ing force directions have to precisely predict, which are developed by muscle contrac-tions. Due to the complexity of the 3D anatomy (muscle fiber distribution, interac-tions between muscles, bone, skin and other tissues), the phenomenological 1D muscle models are only able to compute the force direction and value in a simplistic way. Using detailed 3D models, on the other hand, more realistic force directions and values can be computed. Additionally, we are exploring the use of forward dynamic simulations of 3D muscles which are computationally very challenging, but provide an accurate estimate of joint forces and moments induced during motion. Another challenge is the simulation and modeling of complex and larger muscle systems, where motion results from the mutual interaction of several muscles. To overcome this challenge, 1D simulations were replaced step-by-step with more realistic 3D sim-ulation, for example, the hand and forearm.

5 Sources

Bernard, Bruce P (1997): Musculoskeletal Disorderss and Workplace Factors. A Critical Review of Epidemiologic Evidence for Work-Related Musculoskeletal Disorders of the Neck, Upper Extremity, and Low Back. In: DHHS (NIOSH).

Booz & Company, Aussagen von deutschen Unternehmen im Rahmen der Befragung zum Thema betriebliche Prävention, 2011.

DAK Forschung (2014): DAK Gesundheitsreport 2014. Die Rushhour des Lebens. Gesundheit im Spannungsfeld von Job, Karriere und Familie

Denis, Denys; Lortie, Monique; Rossignol, Michel (2015): Observation Procedures Characterizing Occupational Physical Activities. Critical Review. In: International Journal of Occupational Safety and Ergonomics 6 (4), S. 463–491. DOI: 10.1080/10803548.2000.11076467.

Dragano, Nico (2007): Arbeit, Stress und krankheitsbedingte Frührenten. Zusammenhänge aus theoretischer und empirischer Sicht. 1. Aufl. Wiesbaden: VS, Verl. für Sozialwiss.

Jones, Troy; Kumar, Troy (2007): Assessment of physical exposures and comparison of exposure definitions in a repetitive sawmill occupation: trim-saw operator. In: Work (Reading, Mass.) 28 (2), S. 183–196.

Salvendy, G. (2001): Handbook of Industrial Engineering: Technology and Operations Management: Wiley.

Schönherr, Ricardo; Wächter, Michael; Bullinger, Angelika C.; Schneider, Urs (Hg.) (2014): Entwicklung eines Arbeitsschutzhandschuhs zur Epicondylitis-Prävention. Gestaltung der Arbeitswelt der Zukunft, 12. - 14.03.2014. TU und Hochschule München. Gesellschaft für Arbeitswissenschaft e.V

Spielholz, P.; Silverstein, B.; Morgan, M.; Checkoway, H.; Kaufman, J. (2001): Comparison of self-report, video observation and direct measurement methods for upper extremity musculoskeletal disorder physical risk factors. In: Ergonomics 44 (6), S. 588–613. DOI: 10.1080/00140130118050.

Steinberg, Ulf (2012): Leitmerkmalmethode Manuelle Arbeitsprozesse 2011. Bericht über die Erprobung, Validierung und Revision; Forschung Projekt F2195. Dortmund, Berlin, Dresden: Bundesanstalt für Arbeitsschutz und Arbeitsmedizin.

Steinberg, Ulf; Dr. sc. med. Caffier, Gustav; Dipl.-Ing. Schultz, Karin; Dr.-Ing. Jakob, Martina; Behrendt, Sylvia (2007): Leitmerkmalmethode Manuelle Arbeitsprozesse. Forschung Projekt F 1994. Dortmund, Berlin, Dresden: BAuA